Plasmonic Catalysis

Plasmonic Catalysis

From Fundamentals to Applications

Edited by

Pedro H.C. Camargo
Emiliano Cortés

WILEY-VCH

Editors

Prof. Pedro H.C. Camargo
University of Helsinki
Department of Chemistry
A.I. Virtasen aukio 1
PO Box 55
00014 Helsinki
Finland

Prof. Emiliano Cortés
University of Munich (LMU)
Faculty of Physics
Nanoinstitute Munich
Königinstr. 10
80539 Munich
Germany

Cover

Kindly provided by Few Good Geeks Oy

Library of Congress Card No.:
applied for

British Library Cataloguing-in-Publication Data
A catalogue record for this book is available from the British Library.

Bibliographic information published by the Deutsche Nationalbibliothek
The Deutsche Nationalbibliothek lists this publication in the Deutsche Nationalbibliografie; detailed bibliographic data are available on the Internet at <http://dnb.d-nb.de>.

Print ISBN: 978-3-527-34750-6
ePDF ISBN: 978-3-527-82695-7
ePub ISBN: 978-3-527-82696-4
oBook ISBN: 978-3-527-82697-1

Typesetting SPi Global, Chennai, India
Printing and binding CPI Group (UK) Ltd, Croydon, CR0 4YY

Printed on acid-free paper

C103602_250421

Contents

Prologue

Naomi J. Halas[1] *and Peter Nordlander*[2]

[1] *Department of Chemistry, Laboratory for Nanophotonics, Department of Electrical and Computer Engineering, and Department of Physics and Astronomy, Rice University, Houston, Texas 77005, United States*
[2] *Laboratory for Nanophotonics, Department of Electrical and Computer Engineering, and Department of Physics and Astronomy, Rice University, Houston, Texas 77005, United States*

Humans have long been fascinated by the chemical reactions of nature: from plant growth to the putrefaction of waste, we have marveled, explored and hoped to mimic nature's processes in the hopes of achieving similar chemical transformations. Many of the chemical mysteries that fascinated ancient humans were frequently reactions facilitated by enzymes, nature's catalysts, which had been highly optimized, some for many millions of years, accelerating reaction rates with tremendous specificity. But the energy sources of nature's chemical reactions were not initially obvious. So when first alchemists, then chemists, began to attempt chemical transformations, the only energy sources they could rely upon were those they could control: fire, for raising temperatures, and eventually, pressure, with the advent of strong materials in which to contain and confine chemical reactions.

Chemistry became a vast and mature science in the last two centuries, opening entire new fields ranging from organic chemistry, allowing us to synthesize the molecules of nature in our own ways, to physical chemistry, advancing our understanding of the phases of matter, the elements, and eventually, the atom and quantum mechanics. But the energy toolkit of the chemist – the way energy was applied to scale reaction barriers – remained largely unchanged. With the dawn of our understanding of electricity came electrochemistry. While electrical current could now be harnessed to drive chemical reactions, our understanding of electromagnetism had not progressed far enough at that point for us to understand how to use light to directly deliver energy into a chemical reaction. Einstein's insight, followed by the advent of modern, controllable light sources, brought us new ways to deposit energy into chemical reactions. Still, the ultralow efficiencies that haunted the first chemists largely remained. The catalysts developed for conventional energy sources brought advances, but still proved to be so limited that the chemical industry became the single largest consumer of energy on the entire planet.

Our modern understanding of how to deliver light energy efficiently to chemical reactions, to lower reaction barriers and direct chemical outcomes, is still in its infancy. But advances made possible by the nanoscience revolution of the past twenty years, in particular in nanophotonics, brought us the concept of nanoscale, optical frequency "antennas," capable of capturing light from the far field and localizing it in confined volumes whose dimensions are compatible with chemical processes. Ironically, it was Faraday, one of the fathers of electrochemistry, who also advanced the first nanoscale antennas, in the form of gold colloid that he synthesized. Noble metal nanoparticles, each with their characteristic plasmon resonances, form the foundation of localized light delivery that is the central theme of this book. Shape modification to control the photon energy that can be coupled into nanoparticles provides an important "tuning knob" to control localized energy delivery even further, in new ways.

Our modern knowledge of condensed matter physics opened the door to understanding the specific processes that illuminated noble metal nanoparticles could deliver to chemical reactions. The collective electronic, or plasmon, resonance, a coherent oscillation of its delocalized electrons, is responsible for an extremely strong coupling between incident light waves and the nanoparticles. The quanta of light energy deposited into a metal nanoparticle can be dissipated in several important, and ultimately useful, ways. The coherent oscillation can be damped by coupling to phonons, resulting in highly efficient photothermal heating at the nanoparticle surface and in its direct surroundings. The plasmon can also decay in two ways by emitting a photon: direct radiative decay, i.e., scattering, an efficient process particularly for large nanoparticles; and the radiative recombination of hot carriers, also known as plasmon-induced photoluminescence, an indirect and relatively rare process, but one that can provide important information regarding the actual electron temperature within the nanoparticle. But perhaps most importantly, plasmons can decay by the excitation of an energetic "hot" electron-hole pair within the nanoparticle. The nonequilibrium carriers excited by this process can transfer from the surface of the metal nanoparticle, to, or from, the molecular orbitals of an adsorbate molecule on the metal nanoparticle's surface. The transfer of an electron to an otherwise unoccupied orbital of an adsorbate molecule, creating a transient negative ion state, can substantially lower the barrier to molecular dissociation. Hot carriers can also transfer their energy to adsorbed molecules through Auger-like shake-up processes, leaving the molecule in an excited state. Thus we see how optically excited "plasmonic" nanoparticles can, in relatively simple processes, photocatalyze a chemical reaction that would normally require high temperatures to achieve the same molecular rearrangement. What is required, however, is the energy of the plasmon excitation exceeds that of the energy of the unoccupied molecular orbital with respect to the Fermi energy of the metal. It is a substantial challenge for theoretical chemists to calculate these energy offsets with high accuracy: it is the key to using this process for many more plasmon-mediated chemical reactions, and a current critical challenge for this field. This effect is also further complicated by the electronic structure of the metal itself: for example, metals with prominent occupied d-bands, such as gold can produce both hot and warm electrons, and cold and warm holes, for a substantial range of excitation energies. The energy required for a straightforward charge transfer process to an adsorbate orbital may dictate a preference for certain noble metal nanoparticles over others.

While noble metal nanoparticle "antennas" possess several properties that can efficiently drive chemical reactions with light excitation, they also have a fundamental limitation: low chemical affinities for most adsorbate molecules of interest for chemical reactions. So judiciously incorporating more reactive materials, in the form of individual atomic reactive sites, multi-atomic reactive regions, or entire layers of new materials with greater chemical affinities, plasmon-induced chemistry can be substantially extended to more types of chemical reactions and processes. In these more complex, "antenna-reactor" constructs, one can also observe other important electronic processes: specifically, the desorption of adsorbates from surfaces due to the hot electrons excited by plasmon decay. Electronic desorption processes were discovered and studied decades ago on bulk metal surfaces, but with metal nanoparticles, where plasmons can readily be excited by direct illumination, these processes become almost universal. The plasmon-induced desorption of adsorbates from binding sites can eliminate the irreversible binding, known as "poisoning" of reactive sites, an extremely common problem for conventional thermocatalysts. This process can also be used to modify reactive chemical outcomes, removing reactive adsorbates abruptly before a reaction can proceed further offers a new way to control the product of a chemical reaction. Thermocatalysts have no similar ability for active chemical control.

This book embodies an entirely new, twenty-first century path forward for the control of chemical processes. The low reaction temperatures and record efficiencies that have been observed thus far in plasmonic photochemistry truly tantalize us. The incredible boost in the efficiencies of visible light sources, such as LEDs, have made photons extraordinarily inexpensive. Is this the new path forward for chemistry? Will this direction transition into a new, light-based chemical industry where low-temperature, low-pressure light-induced photocatalytic reactions ultimately make the massive chemical plants of the twentieth century obsolete? If that is the case, what does the next generation of chemists and chemical engineers need to learn about light-particle interactions to master this new type of light-based chemical reactivity? Will sunlight, a free source of photons energetic enough to drive these reactions, be used to directly drive these types of chemistries instead of using photovoltaics as intermediaries? Perhaps photosynthesis, the object of our millennia-old fascination, will finally be realistically copied – along with the low temperatures and high efficiencies characteristic of nature's processes – with precisely designed and engineered nanoantennas and nanoreactors. We eagerly look forward to many exciting advances in this field, for years to come.

Houston, TX, USA | *Naomi J. Halas*
14 October 2020 | *Peter Nordlander*

Introduction

Pedro H.C. Camargo[1] *and Emiliano Cortés*[2]

[1]*Department of Chemistry, University of Helsinki, Helsinki, Finland*
[2]*Faculty of Physics, Nanoinstitute Munich, University of Munich (LMU), Munich, Germany*

Catalysis is central to move toward a more sustainable future and enable our society to transition to a circular economy. For this reason, the possibility of harvesting sunlight to drive, accelerate, and control chemical reactions via photocatalysis has fascinated scientists for years. The unique optical properties of metal nanoparticles in the visible and near-infrared ranges turn them into ideal candidates for sunlight-activated catalysts. In fact, is has been recently established that the excitation of the localized surface plasmon resonance (LSPR) in these systems can be employed to drive and accelerate a variety of chemical reactions. This has led to the rise of plasmonic catalysis as a new frontier in catalysis, photocatalysis, and photoelectrocatalysis.

Plasmonic catalysis is the acceleration of a chemical reaction due to a plasmon excitation. To understand this simple definition is necessary to incorporate concepts from various research fields such as heterogeneous catalysis, nano-optics, physical chemistry, and material science. This book emerged as a necessity for unifying concepts, ideas, techniques, and advances in the rapidly growing field of plasmonic catalysis. To the best of our knowledge, this is the first book dedicated to this emerging area of research, covering its most important concepts and recent developments. To do so, a big, diverse, and heterogeneous group of world leaders in the field prepared exciting contributions for this book. The book comprises 10 chapters encompassing topics such as theoretical considerations of using plasmons for catalysis, optical and catalytic properties in plasmonic nanoparticles and hybrid systems, their synthesis, the fundamentals and mechanisms by which plasmonic excitation leads to the acceleration of reaction rates, examples and discussion of plasmonic catalysis applied to important chemical transformations, plasmonic catalysts based on earth-abundant materials, plasmonic electrocatalysis, and plasmonic metal–semiconductor heterostructures. Here is a quick overview of the main aspects covered in each chapter.

The book starts by describing in **Chapter 1** the theoretical framework of plasmon excitation and decay in the context of plasmonic catalysis. Energy conversion from photons to molecules and transfer from the plasmonic catalyst to the environment are the fundamental processes that take place in a plasmon-catalyzed chemical reaction. The chapter provides a detailed theoretical analysis of plasmon excitation, decay mechanisms, energy transfer, and carrier injection across different interfaces, near-field and scattering enhancements, and photoheating. Toward the end of the chapter and in order to exemplify these theoretical concepts, the chapter overviews a series of applications and experiments where these phenomena can be spotted. As such, this chapter sets the ground to understand the physics behind the uses of plasmons for chemistry.

We move next to the characterization and properties of plasmonic catalytic systems in **Chapter 2**. From conventional heterogeneous catalysis methods to plasmonic techniques, this chapter tackles the integration of conventional methods as well as new methods able to unravel the optical, electronic, and chemical properties of these systems. Different approaches can be followed in order to study chemical reactions mediated by plasmons either at the ensemble level or at the nanoscale, as well as to disentangle the role of light, heat, and carriers in the underlying mechanism. This chapter groups techniques with different temporal, spatial, and chemical resolution in order to gain deeper insight of the behavior of plasmonic catalysts under light illumination.

It has been recognized that the optical properties arising from the LSPR excitation are strongly dependent on several physical and chemical parameters that define the plasmonic nanoparticles. These include size, shape, composition, and structure (solid or hollow interiors) of the nanoparticles. Because these properties are related to the performances in plasmonic catalysis, the synthesis of plasmonic nanoparticles where these parameters can be tightly controlled has gained increased attention. In fact, this is important not only to optimize performances, but also to unravel–structure performance relationships that may aid on the rational design of plasmonic catalysts with desired performances for a reaction of interest. In this context, **Chapter 3** discusses the fundamentals and important examples on the controlled synthesis of metal nanoparticles that are relevant for plasmonic catalysis. The chapter begins by focusing on several methods for the controllable synthesis of Ag, Au, Cu, and Al nanoparticles. The chapter pays particular attention on shape control, in which morphologies such as quasispheres, nanocubes, nanowires, among others, are described. Then, different assemblies having these nanoparticles are presented. These colloidal assemblies are important as they often outperform their individual counterparts due to the formation of electromagnetic hot spots, which can enhance plasmonic catalytic activities. The chapter then moves to bimetallic nanoparticles. This is attractive because nanoparticles having a plasmonic and a catalytic metal enables one to marry optical and catalytic properties in a single nanoparticle. This way, the plasmonic metal can harvest energy from light through the LSPR excitation, which can then be used to accelerate and control reactions at the sites containing the catalytic metal (which may not display LPPR in the visible or near-infrared ranges). This enables one to extend application of plasmonic catalysis to metals that do not display LSPR resonance in the visible or near-infrared ranges. The chapter ends by discussing applications of controlled nanoparticles in LSPR mediated oxidation, reduction, and dehalogenation reactions as well as electrocatalytic applications (such as alcohol oxidation, H_2 evolution, and O_2 reduction reactions).

Chapter 4 covers plasmonic catalysis toward hydrogenation reactions. Hydrogenations are important transformations in industry. Thus, the possibility to enhance performances with visible or near-infrared light – enabling milder or greener reaction conditions or controlled selectivity – are highly relevant and expect to lead to high economic and environmental impacts. The chapter begins by discussing the main mechanisms by which the LSPR excitation can lead to higher reaction rates in the context of catalytic hydrogenations. Then, several important examples involving the hydrogenation of alkanes and alkynes are explained. These examples include the use of Ag and Au NPs as well as bimetallic plasmonic-catalytic NPs having Pd and Pt as the catalytic part in plasmonic catalysis. The chapter also discuss examples including the hydrogenation of aldehydes and ketones, which included the use of Cu NPs as plasmonic catalysts. Next, the reduction of nitro-compounds is covered, and examples of bimetallic plasmonic-catalytic NPs are highlighted.

Chapter 5 discusses the harvesting of LSPR-excited charge carriers for driving multielectron redox reactions. The chapter focuses on the fundamentals and applications of this effect toward important energy relevant transformations: the water splitting reaction and the CO_2 conversion to hydrocarbon fuels. While water splitting can supply green H_2, the CO_2 conversion to fuels can help to alleviate its alarming high levels in the atmosphere while producing important molecules. The chapter starts by discussing the fundamentals of the plasmon-induced generation of charge carriers and its subsequent extraction for redox processes. In this case, the mechanisms for charge transfer are explained in detail. Then, the chapter discusses the energetics and kinetics of the carrier harvesting, which is followed by a description of the chemical potential of the plasmonic excitations. Then, several study cases are covered toward the plasmon-excitation-assisted charge transfer reactions. These include the photodriven growth of Ag and Au nanoparticles and the switching of redox states in metal complexes. After establishing these strong foundations, the chapter focuses on the plasmon-excitation-driven processes relevant for fuel generation. In this context, both the H_2O splitting reaction and the CO_2 reduction are discussed, which includes the possibility of using the LSPR to control selectivity and thus enable the generation of C_{2+} products from CO_2. Interestingly, the chapter includes a discussing on thermodynamic insights into plasmon-excitation-driven CO_2 reduction before the future perspectives regarding photoredox reactions and multi-carrier processes that can be addressed by plasmonic catalysis.

Chapter 6 discusses plasmonic catalysis for N_2 fixation reactions. In addition to improvements in performances due to LSPR, the chapter focuses on the underlying reaction mechanisms. In this case, the role of near-field enhancement and transfer of LSPR-excited charge carriers (both hot electrons and holes) are presented. The chapter begins by showing the importance of N_2 fixation for synthesis of ammonia (NH_3) from dinitrogen (N_2), and the attractive features of being able to drive this reaction via photocatalysis using sunlight as the energy input. Then, the chapter moves to the discussion of the catalytic enhancements and mechanisms in the plasmonically enhanced NH_3 photosynthesis. Three general reaction pathways are presented: N_2 fixation by near-field enhancement, transfer of LSPR-excited charger carriers directly from the plasmonic NPs to N_2, and transfer of LSPR-excited charger carriers N_2 and molecules via a semiconductor, in which the generation of charge carriers is enhanced by the plasmonic nanoparticles. Next, associative and dissociative pathways are discussed for N_2 fixation, followed by the analysis and quantification of plasmon-induced NH_3 evolution. The chapter continues and present representative examples and a deeper discussion on N_2 fixation by the proposed mechanisms, highlighting the use of sacrificial reagents and water as electron donors for reactions mediated by the transfer of LSPR-excited hot charge carriers.

Chapter 7 is devoted to disentangle the critical role of photothermal heating in plasmonic catalysis. Some recent experimental approaches are described in order to compute for the thermal and nonthermal contributions in plasmonic catalysis. In the same way, the already demonstrated chemical selectivity in some plasmonic catalytic systems is shown and discussed in this chapter. These results, concepts, and methods are central to understand the benefits and advantages of exciting plasmon modes for catalysis, beyond photothermal heating.

One important emerging direction in plasmonic catalysis is to replace the commonly employed plasmonic metals, such as Au and Au, by earth-abundant alternatives. Interestingly, in addition to metals such as Cu and Al, it has been shown that some metal nitrides and doped semiconductors also display plasmonic response. This topic is covered in detail in **Chapter 8**, focusing on the use of earth-abundant doped semiconductors plasmonic catalysts. The chapter begins by discussing the importance of earth-abundant plasmonic materials and a comparison between several nanostructures having plasmonic properties. The fundamentals for the plasmonic type of properties in semiconductors are also highlighted. The chapter then brings representative examples on the application, characterization, origins of plasmonic responses, and model plasmonically enhanced reactions for different classes of earth abundant materials in plasmonic catalysis. This includes MoO_{3-x} and WO_{3-x}, H_xMoO_3 ($0< x \leq 2$) and H_xWO_3 ($0< x <0.6$) (Mo and W based bronzes), and $Cu_{2-x}E$ (E = S, Se, Te). In these systems, a variety of hybrid plasmonic-catalytic materials (containing the semiconductors decorated with Pd, for example) are presented as a strategy to marry plasmonic and catalytic properties, enabling the use of the LSPR excitation from the doped semiconductors to enhance the rates of reactions promoted by the catalytic metal.

The incipient advances in plasmonic electrocatalysis are discussed in **Chapter 9**. As a natural extension of traditional photoelectrocatalysis (PEC), the role of plasmon excitation in these systems has recently attract increasing attention. Electrically driven reactions are usually highly efficient in comparison to photo-driven ones. In this sense, the combination of plasmons into this system may open new opportunities toward sustainable, highly efficient, and selective control of chemical reactivity. Some examples of important electroreduction and electro-oxidation reactions under plasmon excitation (H_2O, CO_2, N_2, and fuel cells) are discussed in this chapter.

The last chapter of the book, **Chapter 10**, covers the plasmonic metal–semiconductor assembly and properties. These hybrid systems emerged as a natural way to increase the lifetime of plasmonic carriers by engineering the Schottky barrier at the interface between the metal and the semiconductor. The chapter describes the important role of the interface in these heterostructures and their constituting materials. From the fabrication to the design and successful applications of these systems are detailed in this chapter.

It is clear that plasmonic catalysis has a bright future and will become an essential part of chemistry, physics, materials science, and engineering. We expect that this book serves as a guideline to students, professors, researchers, and anyone working or interested in the field. Our aim was also to provide a comprehensive and complete resource in plasmonic catalysis encompassing fundamental understanding, case studies, and potential, while inspiring scientists to further develop and move the field forward.

1

Theory of Plasmonic Excitations

Fundamentals and Applications in Photocatalysis

Lucas V. Besteiro[1,2], *Xiang-Tian Kong*[1], *Zhiming M. Wang*[1] *and Alexander O. Govorov*[1,3]

[1] *Institute of Fundamental and Frontier Sciences, University of Electronic Science and Technology of China, Chengdu, China*
[2] *Institut National de la Recherche Scientifique, Centre Énergie, Matériaux et Télécommunications, Varennes, QC, Canada*
[3] *Department of Physics and Astronomy, Ohio University, Athens, OH, USA*

1.1 Introduction

Plasmonic systems have become a common tool in a variety of subfields of nanotechnology, be it to manipulate the propagation of light or to harvest its energy. The fundamental properties that make them attractive and versatile are their large interaction cross-sections and a great degree of spectral tunability, both arising from the resonant nature of their interaction with light [1]. We can therefore use them as nanoantennas capable of effectively controlling the flow of light at frequencies ranging from the ultraviolet (UV) to the infrared (IR). Although, historically, a good amount of interest in plasmonics was directed toward creating optical metamaterials [2, 3], the fundamentally lossy nature of plasmonic materials, such as noble metals, introduced practical limits on their use in optical devices [4]. This invited many researchers to find ways for taking advantage of the nonradiative losses of plasmonic systems [5, 6]. Some examples include their usage as photoheaters [7–10], photodetectors [6, 11–13], or, most relevantly to the contents of this book, as photosensitizing elements in photocatalysis and photoelectrochemistry [14–22]. In order to understand why plasmonic nanoparticles (NPs) are of interest in the context of chemical catalysis, we should explore the fundamental properties of metallic materials, and how these impact their interaction with light and with their environment, so that they can drive chemical reactions. In this chapter, we will discuss the above points, providing a general theoretical perspective of the dynamics of plasmonic excitation in NPs, and the mechanisms by which energy can be transferred from a coherent plasmonic oscillation to the environment.

Before commencing a detailed description of the internal dynamics in plasmonic systems, it would be convenient to contrast the fundamental properties of these materials with those of semiconductors, more conventional heterogeneous photocatalysts [24]. In doing so, we will highlight the relevance of electronic structure

Plasmonic Catalysis: From Fundamentals to Applications, First Edition.
Edited by Pedro H.C. Camargo and Emiliano Cortés.

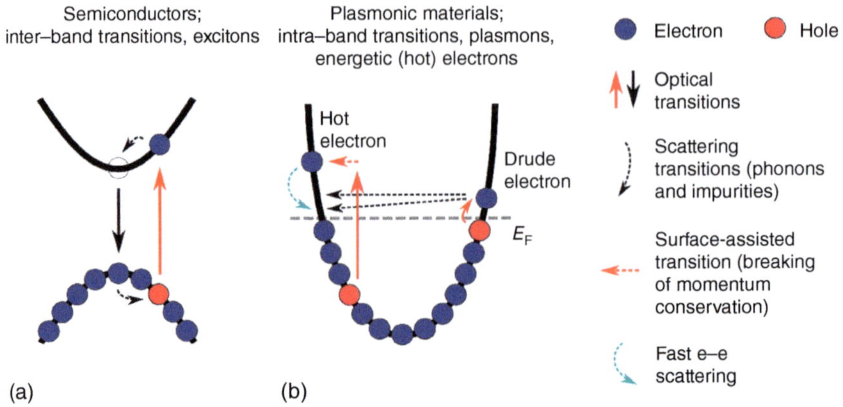

Figure 1.1 Schematic representation of the typical electronic states in two relevant types of materials used in heterogeneous photocatalysis: (a) semiconductors and (b) plasmonic crystals. The diagrams include the main optical excitation channels connecting the electronic states in these materials, as well as characteristic relaxation mechanisms. Plasmonic crystals are conductive and can be metals, degenerately doped semiconductors, or conducting oxides. *Source*: Adapted with permission from Ref. [23] Copyright 2019 Elsevier.

in determining the optical properties of a given material. Semiconductors are crystalline solids, and the periodicity of their structure gives rise to bands with a continuum of electronic states. These bands are key quantum properties of the periodic lattice of a crystal, and contrast with the discrete electronic states in atoms and molecules. A schematic representation of this can be seen in Figure 1.1a, which depicts the electronic structure of a typical bulk semiconductor, showcasing also its characteristic bandgap separating occupied and unoccupied states. Optical transitions can excite electrons across the bandgap, generating relatively long-lived electron–hole pairs that can drive surface chemistry in photocatalytic systems, targeting different redox reactions depending on their relative energy alignment with the valence or conduction band of the semiconductor. They are widely used in photoelectrochemical cells [25, 26] and other photocatalytic setups [27, 28], employing both bulk and nanostructured semiconductor materials as a repository of optically excited charge carriers for driving the target reaction. These long-lived carriers can be extracted via contacts since semiconductors, when doped, possess high conductivities and are suitable for building electrochemical circuits [25, 29]. At the same time, the bandgap introduces a threshold on the energy of the photons that can be absorbed, which limits the fraction of the solar spectrum available for conversion. This shortcoming of semiconductors as photocatalysts under solar irradiation has been one of the motivations for combining them with plasmonic systems [15].

Plasmonic materials are also crystalline solids, but their defining characteristic is their large number of mobile carriers. The paradigmatic case for this is a metal, whose electronic bands overlap in energy across the reciprocal space and therefore lacks a bandgap, although materials like conductive oxides and highly doped semiconductors can also support plasmonic resonances [30–32]. Direct optical transitions such as those occurring in semiconductors are not possible within a metal conduction band, due to conservation of momentum, unless it is allowed through quantum confinement in the system or the interaction of the electron with crystal defects or phonons [33, 34]. This is sketched in Figure 1.1b, but we will discuss it in more detail when describing the excitation of hot electrons (HEs) and hot holes (HHs). On the contrary, the distinctive mode of interaction of plasmonic systems with light occurs through the collective excitation of the electrons in the system. The electrons in the conduction band oscillate coherently, with small changes in energy and momentum for each of them, collectively constituting the excitation of the plasmonic particle.

It is illustrative to consider such a particle in a fully classical way, where the cloud of free electrons moves in a continuous manner following an external electric field. If the frequency of the driving field is slow enough, the mobile charges in the conductive particle remain in phase with the external field and equilibrium between field and internal polarization is maintained, resembling a typical electrostatic picture (Figure 1.2a). As we consider higher frequencies, however, we will approach the natural or intrinsic resonance of the particle. This resonant frequency will depend on the number and mobility of the carriers in the conduction band, as well as the geometry of the particle. When the external field excites the system at its resonant frequency, the charge oscillation in the particle will lag behind the external field by a phase of $\pi/2$, resulting in an internal polarization that does not cancel out the driving field, but is instead amplified by it (Figure 1.2b). This is of course analogous to the behavior of classical mechanical oscillators when driven resonantly at their natural frequency [36]. Likewise, a plasmonic NP excited at its plasmonic resonance will increase the amplitude of its oscillation, i.e. will show a larger charge polarization, giving rise to the well-known plasmonic enhancement of the particle's near field. It will also continue oscillating if the driving force is stopped, to then reduce its amplitude as the oscillation is dampened or, in a complementary perspective, the plasmon decays. Understanding how different mechanisms contribute to dissipating the energy in the plasmon will be crucial to discuss how plasmonic systems can contribute to photocatalysis, so we will describe them separately in different sections of this chapter.

This resonant excitation of a large number of carriers in the conduction band is responsible for the large interaction cross-sections of plasmonic NPs, much larger than those of semiconductor QDs. This difference can reach several orders of magnitude, as exemplified by the data in Figure 1.2c. Furthermore, and because the movement of electrons will be constrained by the size and shape of the particle, the collective excitation of the mobile carriers also provides metal NPs with a strong spectral tunability of their plasmonic resonances. Using a given conductive material, we can determine the resonant frequency of a nanocrystal (NC) by considering different geometries. The shape and size of the NC sets the physical extension of the effective electromagnetic resonator composed by the electrons moving freely within the metal. Figure 1.2c,d illustrate this by showing the large shift between the plasmonic resonance of an Au spherical NP and an Au nanorod (NR). Importantly, it is not just the aspect ratio of the structure that determines its resonant frequency, but its curvature is critical as well. In Figure 1.2d, we see how the spectrum of an Au cube differs significantly than that of a sphere with the same volume. The curvature

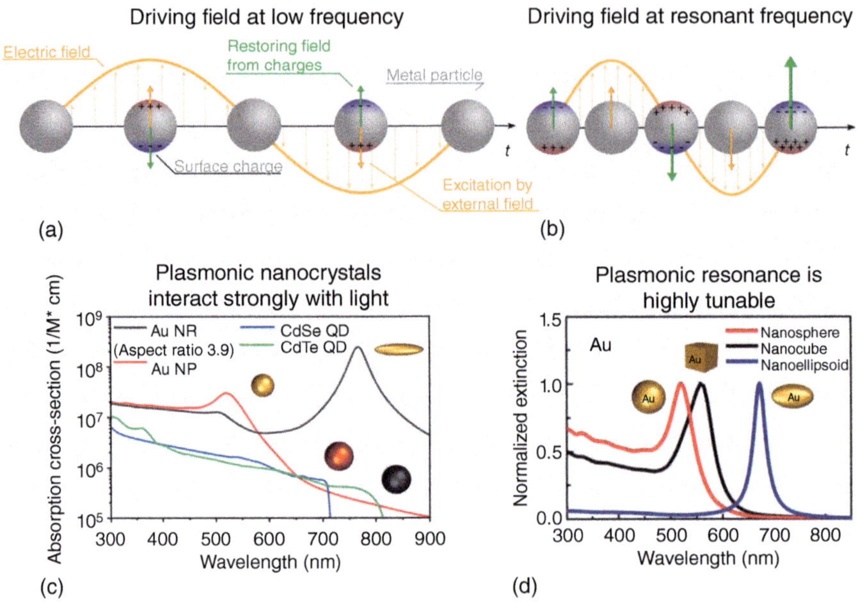

Figure 1.2 (a) Schematic diagram of metal nanoparticles driven at a low frequency. The electrons in the conduction band can follow the fields, with the resulting polarization field being in opposition to the driving field. (b) Same system under a higher frequency external field, exciting the plasmonic resonance of the particles. In this case the polarization field lags behind the driving field, is amplified by the driving field, and creates an enhanced near field around the particle. (c) Calculated absorption cross-sections of NCs made of different materials, but preserving the same volume. Plasmonic NCs absolutely dominate in terms of light–matter interaction strength. (d) Extinction cross-sections for gold particles with different shapes. By using different NC geometries, we can tailor the response of a given material. *Source*: Adapted with permission from Refs. (c) [33] Copyright 2019 American Chemical Society, (d) [35] Copyright 2014 Elsevier.

of the surfaces will also impact the distribution of the enhanced electric field around the particle. Combining their large cross-sections and their tunability, we can design plasmonic systems that can strongly interact with light across the full spectrum of solar radiation. In this way we can capture a large proportion of the incoming light flux with relatively low particle concentrations.

It should also be noted, however, that real metals also exhibit single-electron optical excitations. Electrons in fully occupied bands below the Fermi energy can be promoted to higher-energy empty states through direct optical transitions if the photon energy is sufficiently large. Such a promotion is referred to as an interband transition, and contrasts with the intraband transitions that occur within the semi-occupied conduction band. We include a schematic illustration of interband transitions in Figure 1.3a, taking Au as an example material. Optical transitions of this type are commonplace in real materials and are responsible for their response in the short wavelength part of the spectrum. Interband transitions are the reason why the spectra of Au NCs in Figure 1.2c,d are not a simple plasmonic resonance and show instead a significant absorption profile at wavelengths below ~540 nm.

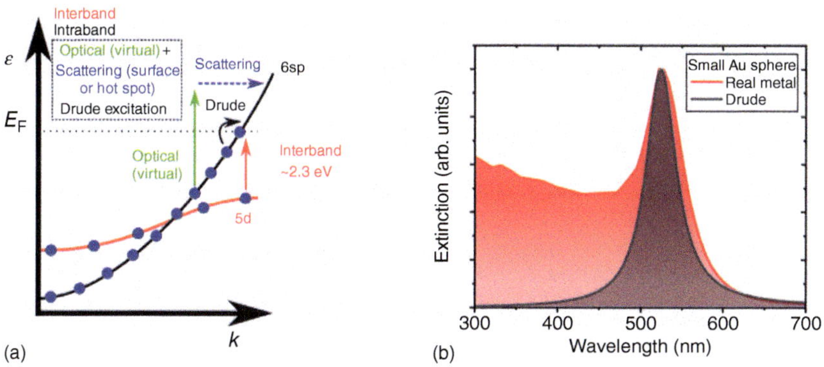

Figure 1.3 (a) Schematic diagram of the band structure of gold, depicting the conducting band and the fully occupied *d* band. We also note relevant optically-induced electronic transitions. (b) Extinction of a small Au sphere in water, calculated under the quasistatic approximation. We contrast results obtained with experimental bulk permittivity data from gold (red) and from the Drude model (black). The Drude data only models the carriers in the conduction band, and does not include interband transitions. *Source*: Panel (a) was adapted with permission from Ref. [23] Copyright 2019 Elsevier.

These interband processes are single-electron transitions, they scale linearly with the volume of the crystal, and start at photon energies that depend on the band structure of the material. However, they can also be used for our advantage in a photocatalytic context, as we will comment when discussing carrier injection. Then, metal NPs have two main modes of interaction with light. A collective response of their free carriers, which can support plasmonic resonances driven by an external field, and a single-electron response where carriers from low-lying, fully occupied bands are optically promoted to vacant states. This distinction is illustrated in Figure 1.3b, which shows two extinction spectra for a small gold NP, one obtained by using the experimental permittivity of bulk gold [37] and a second one with a permittivity obtained by modeling only its free carriers in the conduction band, using the Drude model. We describe this model in more detail in the next section, as we discuss the internal dynamics of the collective oscillation modes and how the electromagnetic energy of the driving field is moved into different degrees of freedom of the system.

1.2 Dynamics of Plasmon Excitation and Decay

1.2.1 Collective Charge Dynamics

In this section, we will describe the characteristics of carrier dynamics in a plasmonic oscillation, which will give us the opportunity to examine the different mechanisms by which the coherent oscillation can dephase. As discussed in the previous section, the materials that support a plasmonic oscillation are conductors. Therefore, they have a band that is only partially full, where the quasifree electrons can move in reaction to the external field. Within this context, interband transitions that overlap with the plasmonic resonance will be presented as another dephasing mechanism. In modeling the behavior of the carriers in the conduction band, we start by considering an infinite, boundless gas of free electrons. Of course, this model can be connected with a realistic metal when we consider electron quasiparticles instead, which behave as free electrons with an effective mass $m \neq m_0$ that depends on the curvature of the band, with m_0 being the mass of a free electron in vacuum. These carriers, being fermions, occupy states of growing energy, E, following Fermi–Dirac statistics characterized by the distribution:

$$f_{\mathrm{F}} = \frac{1}{1 + e^{(E-E_{\mathrm{F}})/k_{\mathrm{B}}T}}. \tag{1.1}$$

Here T and k_{B} are the temperature of the system and the Boltzmann constant, respectively. In such a system, the Fermi energy, E_{F}, is a critical magnitude, indicating the energy of the highest occupied electronic state at $T = 0$ K. At nonzero temperatures, the electrons spread across the states with energies around this value, following Eq. (1.1). An equivalent picture can be presented in terms of their momentum, which will be useful in discussing the excitation of the gas by external fields. For free electrons, their energy is a direct function of their linear momentum p, expressed in terms of the electron wavenumber k as $E_k = \hbar^2k^2/2m$, where $\hbar k = p$. We can thus derive a Fermi wavenumber from the Fermi energy, $k_{\mathrm{F}} = \sqrt{2mE_{\mathrm{F}}}/\hbar$. So, let us now consider a three-dimensional gas of free electrons in this representation, where the concept of Fermi surface corresponds to a spherical surface of constant wavenumber containing the occupied electronic states at equilibrium (assuming $T = 0$ K to simplify its description). A two-dimensional representation of this idea is presented in Figure 1.4a. In this figure, the black circle centered at $\mathbf{k} = 0$ is the Fermi surface at equilibrium. When an external electric field $\mathbf{E}$ interacts with the electrons, these will gain momentum in the axis containing the polarization of the field. This

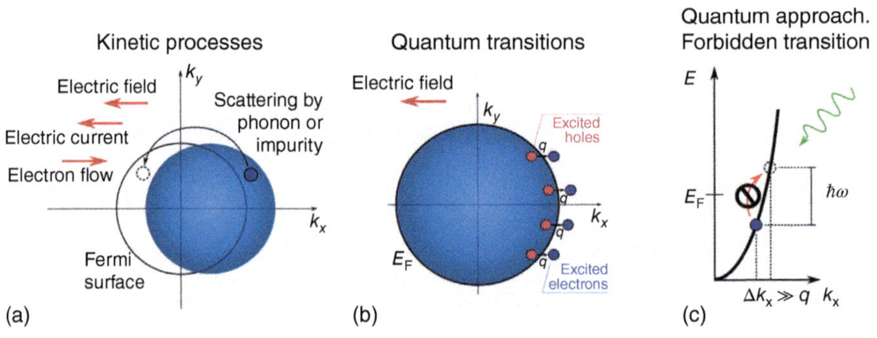

Figure 1.4 Free electron model in a system without boundaries. (a) Classical picture of the electron gas. Fermi sea in momentum space. The Fermi sea at equilibrium (empty black circle) is displaced by an external electric field, and the scattering of electrons with the crystal impose a friction-like force on their movement. (b) Quantum picture of the electron gas. Excitation of a Fermi sea by an external classical field. Each excited electron acquires a momentum $\hbar q$, with q being the wave vector of the external field, and a small amount of energy. (c) Diagram illustrating the impossibility of a mobile electron in the conduction band absorbing a photon in the quantum picture. The required change in its linear momentum is much larger than that carried by the photon. *Source*: Panels (a)–(c) were adapted with permission from Ref. [33] Copyright 2019 American Chemical Society.

is depicted in Figure 1.4a, where the large blue circle represents the Fermi sea and its displacement occurs in response to a field $\mathbf{E} \parallel -\hat{\mathbf{x}}$. The resulting displacement in the momentum space generates an electron flow and, consequently, a current. It is worth mentioning that this current will be finite, i.e. the acceleration of the charges by the field will be compensated by scattering events, like the one depicted in the diagram, until reaching a steady state.

In this picture, the particles in the electron gas are excited collectively. One can easily extend it to a situation where the external field oscillates harmonically, as $\mathbf{E}(t) = \mathbf{E}_0 e^{-i\omega t}$, such that the sphere of occupied electronic states will also oscillate in the same axis that the driving field does. This picture is a useful representation of the charge dynamics of a 3D degenerate gas of electrons under weak optical excitation. This is also a good framework to discuss the Drude model under a semiclassical perspective. Let us start by considering the average electronic velocity in the Fermi sea. By symmetry considerations, it is evident that this will be zero for the undisturbed electron gas, but it will increase as an external electric field displaces the sphere of occupied states in the reciprocal space. We express the change of the average velocity of the electron gas in terms of the balance between the driving force of the oscillating field and the opposing friction-like effect of electronic scattering, as

$$m\frac{d\overline{\mathbf{v}}}{dt} = e\mathbf{E} - m\ \gamma_{\mathrm{D}}\ \overline{\mathbf{v}},$$

where $\gamma_{\mathrm{D}} = 1/\tau_{\mathrm{D}}$ is the effective scattering rate of electrons with the crystalline matrix and other electrons. From this framework we can recover the expression for the Drude permittivity of a metal [1, 34, 38]. For this we have to, again, consider that the average velocity of the electrons follow the same harmonic relationship as the field, to obtain its expression in the frequency domain

$$\overline{\mathbf{v}}_{\omega} = \frac{e\mathbf{E}_0}{m\left(\gamma_{\mathrm{D}} - i\omega\right)}.$$

At this point, we can calculate the electrical conductivity predicted by the Drude model, σ_{D}, through Ohm's law and then use it in the expression for the complex-valued permittivity, $\varepsilon_{\mathrm{D}} = \varepsilon_{\mathrm{c}} + i\sigma_{\mathrm{D}}/\omega\varepsilon_0$, to obtain Drude's expression for the permittivity

$$\varepsilon_{\mathrm{D}} = \varepsilon_{\mathrm{c}} - \frac{\omega_{\mathrm{p}}^2}{\omega\left(\omega + i\gamma_{\mathrm{D}}\right)}, \qquad \omega_{\mathrm{p}}^2 = \frac{e^2 n_0}{\varepsilon_0 m}.$$

This expression includes the term ε_{c} to account for the optical response of lower-energy bands in the material, n_0 is the charge density in the conduction band, and ε_0 is the permittivity of free space.

In order to progress toward a more realistic description of the charge dynamics in a metal, especially if we are concerned with the description in microscopic systems, we should adopt explicitly quantum descriptions. In the first place, we can discretize the electronic transitions in the Fermi gas. Each electronic state is characterized by a wave vector $\mathbf{k}$, and an external electric field with wave vector $\mathbf{q}$ will be responsible for triggering the electronic transitions between states. Figure 1.4b depicts schematically the excited Fermi sea under this perspective, where the quantized electron gas is excited by a classical external field. Using the random phase approximation (RPA) to simplify the many-body interaction in this framework, one arrives at Lindhard dielectric function [39, 40]

$$\varepsilon_L(\omega, q) = \varepsilon_c - \frac{e^2}{\varepsilon_0 q^2 V} \sum_k \frac{f_F(E_{\mathbf{k}}) - f_F(E_{\mathbf{k+q}})}{\hbar\omega - E_{\mathbf{k+q}} + E_{\mathbf{k}} + i\hbar\gamma_{rel}},$$

where V is the crystal's volume, $f_F(\boldsymbol{E}_{\mathbf{k}})$ and $E_{\mathbf{k}}$ are the Fermi distribution function and the electron energy corresponding to the wave vector $\mathbf{k}$, respectively, and γ_{rel} is the phenomenological electronic relaxation rate. Here each electron changes its linear momentum by $\hbar q$ and gains a small amount of energy, i.e. the fastest electrons remain close to the Fermi surface (see Figure 1.4b). The currents generated by the collective excitation of the single carriers give rise to the plasmonic oscillation, similar to the previously described case.

Now, in a fully quantum-mechanical description of the system, we also have to consider the quantization of the external field. Under such perspective, the electrons are illuminated by discrete photons with energy $\hbar\omega$ and wave vector $\mathbf{q}$. Importantly, in this description of light interacting with the free electron gas (or equivalently, and more pertinently to our interests, the quasifree conductive electrons in a metal), we encounter that the absorption of a photon is impossible. Figure 1.4c illustrates this idea. Fundamentally, the small energy dispersion of the conduction band does not offer any available end state satisfying $E_{\mathbf{k}+\mathbf{q}} - E_{\mathbf{k}} = \hbar\omega$ [41], and for that transition to be possible the electron would either require to acquire additional momentum (through, e.g. inelastic scattering with a defect or a phonon), or occur in a confined system with boundaries. This latter situation is of particular relevance in the context of studying plasmonic resonances in nanostructures.

1.2.2 Confined Systems

After this progressive introduction to the dynamics of charge carriers in a metal through the exploration of a boundless electron gas, we can now discuss specific details of what happens when we consider finite systems, including the effects of confinement in small particles. The first immediate consequence of this transition is that the current created by the driving field will accumulate electric charge at the surface of the particle (inset in Figure 1.5a). In an ideal classical perspective, these charges would accumulate solely at the surface of the particle, but in realistic systems the charges at the interface extend into a non-zero volume inside the metal, screened and in interaction with other mobile charges. As we approach small particle sizes and electrons are not well approximated as point-like particles anymore, we can see Friedel oscillations caused by electron interference in the charge distribution close to the surface (see Figure 1.5b), which acts as a disturbing impurity [40, 44]. It is therefore not surprising that, as we decrease the size of the particle, the quantum effects at the surfaces become more relevant over the plasmon dynamics. In particular, they contribute to its dephasing, with the coherent collective oscillation of the carriers in the metal decaying into incoherent electronic excitations. This is observable through the broadening of the plasmonic peak in small NPs, with an increased plasmon decay rate scaling as the inverse of their size [45, 46],

$$\gamma_{\mathrm{surf}} = A \cdot \frac{v_{\mathrm{F}}}{D},$$

where A is a numerical constant, D stands for the particle's diameter, and v_{F} is the Fermi velocity of the metal. Figure 1.5c illustrates such a trend for silver NPs in different substrates [43], but this is a general property that broadens the resonant peak in small particles of different materials [43, 46–50]. One can interpret this phenomenon either classically, arising from the collisions of ballistic electrons with the boundaries of the NC [51, 52], or in terms of quantum mechanics, as the boundary discretizes the electronic states inside the metal [45, 46, 52, 53]. Then, the surfaces allow the breaking of the momentum conservation described above (Figure 1.4c) by discretizing the electronic states in the particle, in what is known as surface-assisted plasmon decay or Landau damping. An important consequence of this is that particles' surfaces allow the excitation of high-energy HEs (Figure 1.5a,d), with energies up to the total photon quantum $\hbar\omega$ [53, 54], even in crystals with no defects and assuming an absence of electron–phonon interaction. Then, in the context of plasmonic photocatalysis, the surface effect in the plasmon dephasing is of particular importance because it is responsible for the internal generation of high-energy hot charge carriers in a manner that we can affect by controlling the shape and size of the metallic particle (Figure 1.5a,d). We should also bear in mind that another factor allowing the promotion of electrons to high-energy states is the existence of strong field gradients in the electromagnetic "hot spots" of the material [23, 55]. These are also related to the surface of the particles, more specifically to its shape and local curvature [54, 56], but hot spots can also arise by the near-field interaction of two or more plasmonic particles [55, 57–59], as we will see in more detail when discussing near-field enhancement.

1.2.3 Plasmonic Decay Channels

In this section, we have discussed the fundamental carrier dynamics in a plasmonic resonances, and we have described how these can decay through electron scattering and surface damping. In later sections of this chapter, we will discuss additional damping mechanisms for the plasmon, both internal and also depending on its environment. Before doing so, here we present a summary of the processes limiting the lifetime of the plasmon. Jointly, all of them contribute to broadening the plasmon resonance, and can be counted toward the total decay rate of the plasmon:

$$\gamma_{\mathrm{plasmon}} = 1/\tau_{\mathrm{plasmon}} = \gamma_{\mathrm{NC}} + \gamma_{\mathrm{rad}} + \gamma_{\mathrm{env}}.$$

In this equation, we have separated the decay processes between those which are internal to the particle (γ_{NC}), those where the plasmon decays while exciting its immediate environment (γ_{env}), and the radiative losses (γ_{rad}). In Figure 1.6a, we provide a summary diagram of the rates involved in the plasmon and single-particle dynamics inside the particle. For the internal losses, we should distinguish between bulk mechanisms and those arising from the geometry of the nanostructure [23, 54].

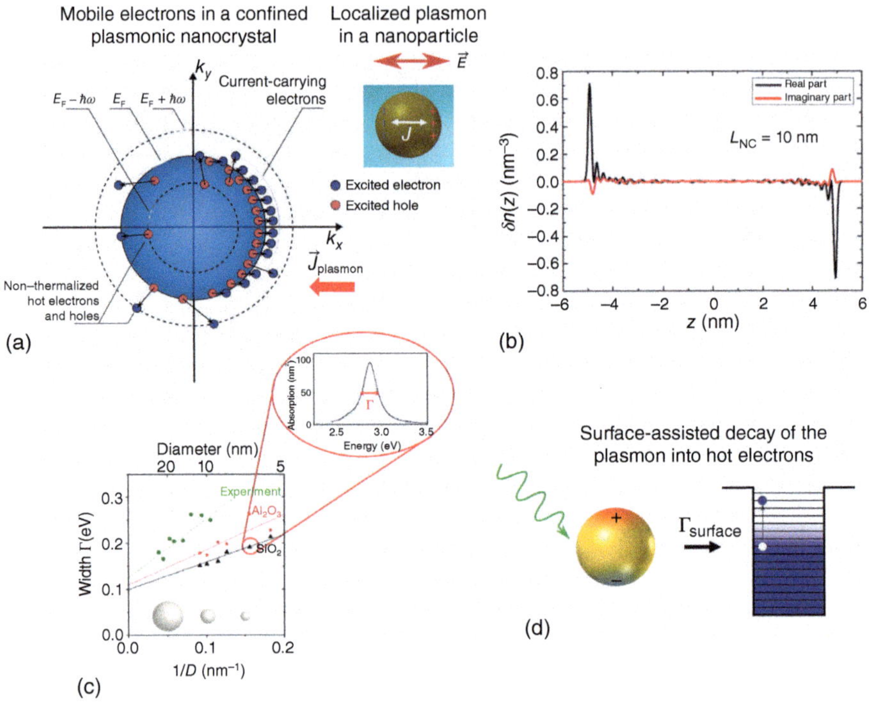

Figure 1.5 Effects of boundaries in the plasmon dynamics. (a) Excitation of the Fermi sea in a metal NP. Carriers can now be excited to energies up to the total energy of the photon thanks to their interaction with the particle's surface. (b) Real and imaginary part of the nonequilibrium complex electron density in an optically excited gold slab. We can see Friedel oscillations next to its surfaces. (c) The surface broadening of the optical features of an NC increase with decreasing particle size. (d) Diagram of the surface-assisted promotion of high-energy electrons. The presence of the surface allows the absorption of a photon by a single electron, breaking the limitation depicted in Figure 1.4c. *Source*: Adapted with permission from Refs. (a) [33] Copyright 2019 American Chemical Society, (b) [42] Copyright 2015 American Chemical Society, (c) [43] Copyright 2011 American Chemical Society, (d) [23] Copyright 2019 Elsevier.

The latter allow the internal generation of HEs, either because of the change of linear momentum allowed by the surfaces, γ_{surf} (Figure 1.6b), or because of the strong internal field gradients in hot spots, $\gamma_{\text{hot spot}}$:

$$\gamma_{\text{NC}} = \gamma_{\text{bulk}} + \gamma_{\text{surf}} + \gamma_{\text{hot spot}},$$

and where the decay channels relating to the bulk material are further subdivided into two terms:

$$\gamma_{\text{bulk}} = \gamma_{\text{Drude}} + \gamma_{\text{interband}},$$

one for the rate of electron–electron and electron–phonon scattering, γ_{Drude}, which enters the description of the system under the Drude model to give a friction-like force relaxing the electronic momentum (Figure 1.6b); and another quantifying the rate of optical interband transitions, $\gamma_{\text{interband}}$(Figure 1.6b) [1]. Lastly, we also include the effects of the environment into the decay of the plasmon,

$$\gamma_{\text{env}} = \gamma_{\text{charge transfer}} + \gamma_{\text{near field}},$$

where the $\gamma_{\text{charge transfer}}$ term refers to processes where an electron is promoted from the particle into a excited state in the environment, or vice versa, be it in an adsorbed molecule or an interfacing semiconductor. Lastly, the $\gamma_{\text{near field}}$ term would quantify the decay rates arising from the near-field coupling of the plasmonic NC with another system.

The collective electronic oscillation that is sustained in a metallic particle when exciting a plasmonic resonance stores energy in a relatively ordered many-body state. As this phenomenon, involving the coherent motion of carriers, interacts with the material supporting it or its environment, this energy will degrade until it diffuses as heat (Figure 1.6c). However, most of these decay mechanisms provide ways for the plasmonic NP to share its energy or charge with other material elements in the system. By using plasmonic NPs in a photocatalytic context, we aim to create a reactor in which other elements can efficiently utilize this energy in subsequent chemical transformations, so that we can direct its flow as it proceeds downstream in terms of its coherence. In the rest of the chapter we will explore different mechanisms that can make this possible.

1.3 Hot Electrons: Energy Distribution and Mechanisms of Generation

As the coherent plasmonic oscillation evolves in time and loses energy to single-electron states, several types of excited electrons coexist inside the particle, characterized by their energies and their origin. The first kind, and most common, are the electrons that carry the current of the plasmonic oscillation and have small energies above the Fermi energy (see Figure 1.5a). These are found in the large peaks with energies just above E_F in Figure 1.7a, and we refer to them as Drude electrons, with the equivalent populations with energies below E_F being Drude holes. This figure shows a diagram with the general features appearing in the change in electronic occupation, with respect to the equilibrium population at room temperature, in the steady state of a small Au NP excited at its plasmonic resonance in the linear regime, i.e. with relatively low light fluxes [33, 54]. The other main feature in this plot is the quasiflat population that extends to carrier energies up to the total energy of the incoming photons. These are HEs and HHs that can be excited due to the interaction of the plasmon with the surfaces of the NP, and to which we refer as nonthermal hot carriers, because their populations cannot be well described by an effective temperature. These arise in systems with boundaries that break the translational symmetry of the material, allowing for the electron scattering that permits the full absorption of the photon. Another important category of excited carriers are the thermalized HEs, which also contribute to the main low-energy peaks in Figure 1.7a. As the nonthermal HEs lose energy via rapid electron–electron scattering, they become thermalized HEs, contributing with the

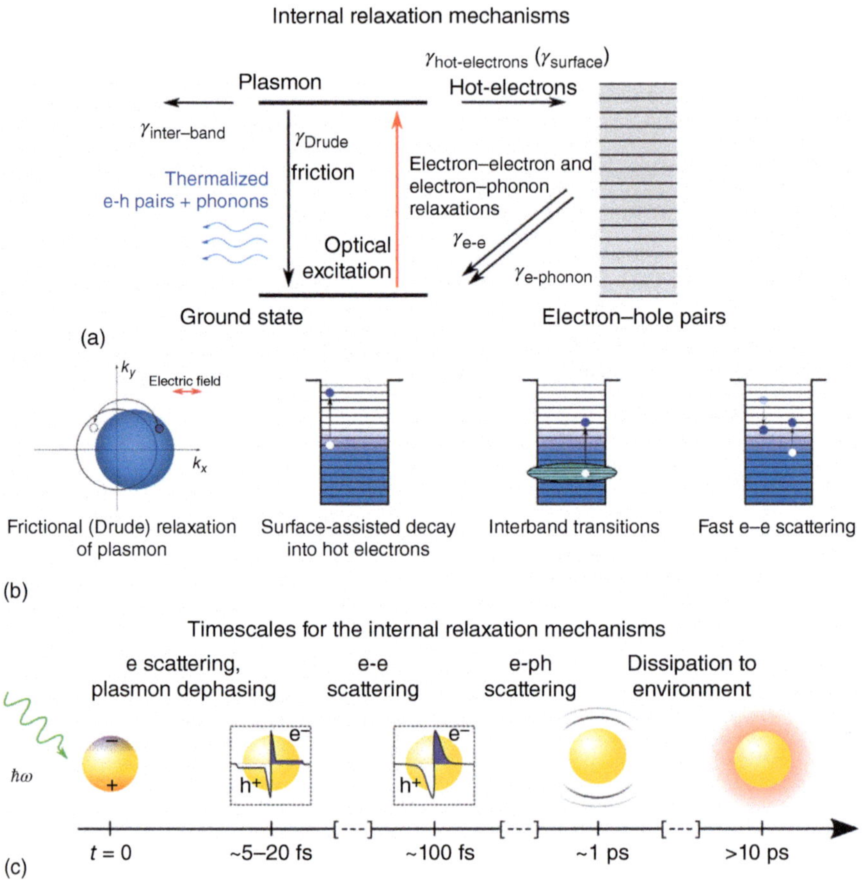

Figure 1.6 (a) Internal channels for the decay of the plasmon in an NC, as well as single-particle relaxation mechanisms. (b) Schematic diagram of the main mechanisms depicted in panel a. The first three are channels for plasmon decay, which transfer energy to single-particle states, while the fourth is the main relaxation mechanism for high-energy HEs and HHs. (c) Diagram with the sequence of dominant mechanisms involved in plasmon decay and electronic relaxation, indicating at which approximate timescales after excitation they become relevant. Of course, this picture presumes a pulse excitation, and under continuous illumination all of these processes happen concurrently, achieving a nonequilibrium steady state that balances Eq. (1.2). *Source*: Adapted with permission from Refs. (a) [54] Copyright 2017 American Chemical Society, (b) [33] Copyright 2019 American Chemical Society, (c) [23] Copyright 2019 Elsevier.

Drude electrons to an excited population that can be described by an effective electronic temperature T_e, higher than that of the atomic lattice [23].

Developing the notion of these electronic subsystems within the plasmonic NP, several multitemperature models have been proposed to characterize its internal time dynamics. In Figure 1.7b, we illustrate the main elements of our proposed quantum two-temperature (Q2T) model [23], which expands on existing extended 2 T models [60–64]. This panel illustrates the distinct subsystems, the couplings between them, and the energy inputs to the system. The Q2T model codifies this information

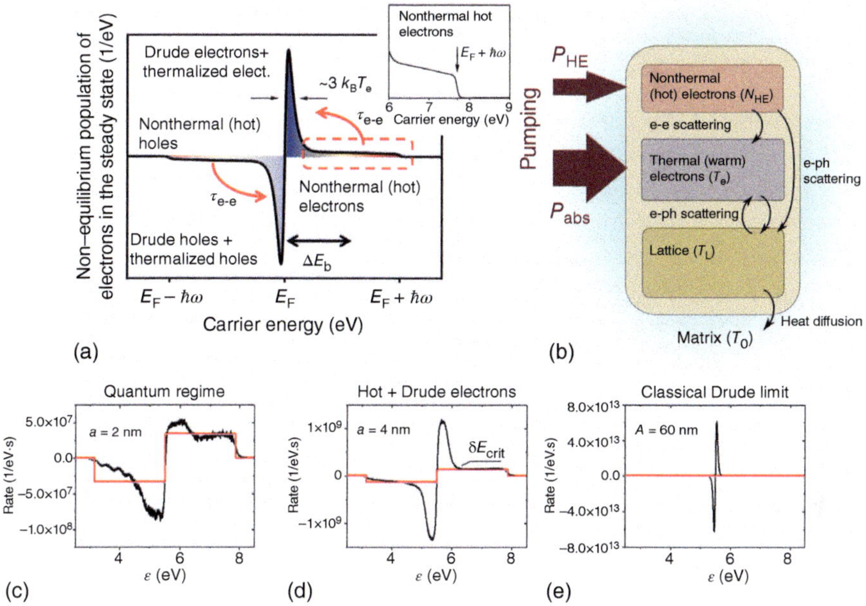

Figure 1.7 (a) Typical distribution of excited electrons in an optically driven plasmonic NC under continuous illumination. There are two types of carriers: low-energy electrons and holes (Drude-like and thermalized) and nonthermal high-energy HEs and HHs. The latter can have up to the total energy of the photon (see inset). Here, T_e is the effective temperature of the optically driven Fermi gas, and τ_{e-e} is the average time between electron–electron scattering events that distribute and homogenizes their energies. (b) Diagram with the different subsystems exchanging energy in a plasmonic nanostructure, as well as the relevant energy inputs and output. This diagram corresponds to the quantum 2 T model [23], a revision to previous extended 2 T models [60–64]. (c–e) Profiles for the rates of excitation of electrons in gold NPs of different diameters. Each particle is excited at its plasmonic resonance by light with a flux of $I_0 = 3.6 \times 10^3$ W/cm^2. The surface-assisted excitation of HEs and HHs is prominent for small NPs, but its relative relevance decreases with the reduction of the particle's surface-to-volume ratio. For larger sizes we recover the behavior of a classical plasmon, with negligible rate of HE and HH excitation. *Source*: Adapted with permission from Refs. (a) [33] Copyright 2019 American Chemical Society, (b) [23] Copyright 2019 Elsevier, (c) [54] Copyright 2017 American Chemical Society.

in the following set of coupled equations, accompanied by full description of the notation used:

$$
\begin{aligned}
\frac{dE_{HE}}{dt} &= -a_{e-e}E_{HE} + P_{HE} \\
\eta_e T_e \frac{dT_e}{dt} &= -G\left(T_e - T_L\right) \\
&\quad + a_{e-e}E_{HE} + P_{abs} \\
C_L \frac{dT_L}{dt} &= G\left(T_e - T_L\right) \\
&\quad - C_L \frac{T_L - T_0}{\tau_{heat\ \ transfer}}
\end{aligned}
\quad
\left\{
\begin{array}{l}
E_{HE} \equiv \text{total energy of nonthermalized HEs in NC} \\
T_e \equiv \text{electronic temperature} \\
T_L \equiv \text{lattice temperature} \\
T_0 \equiv \text{ambient temperature} \\
a_{e-e} \equiv \text{e} - \text{e relaxation rate of HEs} \\
P_{abs} \equiv \text{classical optical absorption rate of NC} \\
P_{HE} \equiv \text{power exciting HEs} \\
G \equiv \text{e-ph coupling constant} \\
C_e = \eta_e T_e \equiv \text{electronic heat capacity of NC} \\
C_L \equiv \text{lattice heat capacity of NC;}\ \ C_L >> C_e \\
\tau_{heat\ \ transfer} \equiv \text{time of heat transfer NC} \rightarrow \text{matrix}
\end{array}
\right.
\tag{1.2}
$$

We should note that the above equation is written only for the intraband excitations and, therefore, it is applicable for Au NCs with plasmonic resonances in the red region such as nanorods, nanocubes, and nanostars. It can also be applied to Ag NCs and other plasmonic systems whose plasmonic resonance does not overlap with the interband transitions of the material. For spherical Au NCs, one needs to modify the above equations to account for the interband transitions. The physical significance of most of the variables involved should be clear by contrasting the equations with Figure 1.7b, but it is worth making explicit the expressions for the two terms that model the energy input to the system. The classical absorption rate contributes directly to the effective electronic temperature, and it is calculated as the time-averaged losses arising from the electron scattering described by the Drude model:

$$P_{\text{abs}} = \left\langle \int_{\text{NC}} \mathrm{d}V \ \mathbf{j}\cdot\mathbf{E} \right\rangle_{\text{time}} = \mathrm{Im}\left(\varepsilon_{\text{Drude}}\right) \frac{\omega}{8\pi} \int_{\text{NC}} \mathrm{d}V \ \mathbf{E}_{\omega}\cdot\mathbf{E}_{\omega}^{*},$$

while the generation of nonthermal HEs should be calculated following a quantum formalism as [54, 65]

$$P_{\text{HE}} \approx \frac{1}{4}\frac{2}{\pi^2} \times \frac{e^2 {E_{\text{F}}}^2}{\hbar} \frac{1}{(\hbar\omega)^2} \int_{S_{\text{NC}}} \left| E_{\omega,\text{normal}}\left(\mathbf{r}\right)\right|^2 \mathrm{d}s, \tag{1.3}$$

where $E_{\omega,\,\text{normal}}$ is the electric field inside the particle and normal to its surface, and the physical field is defined here as $\mathbf{E} = \mathrm{Re}\,\mathbf{E}_{\omega}\mathrm{e}^{-i\omega t}$. Again, the generation of nonthermal HEs, with energies up to the total energy of the photon, is allowed by the surface effect discussed above. Without it this transition would be forbidden, as it is the case for a free electron (Figure 1.4c). As a consequence of this, small particles with large surface-to-volume ratios will use a larger proportion of the total absorbed power, $P_{\text{abs}} + P_{\text{HE}}$, to excite nonthermal HEs, in comparison to Drude electrons. The three last panels in Figure 1.7 show precisely this, by contrasting side to side the theoretical rates of intraband carrier excitation for three different sizes of spherical gold NPs as a function of carrier energy. While for the large particle the volume effect dominates and almost all carriers are excited with low energies around E_{F}, the balance changes as we decrease the particle size, until we reach a system where the nonthermal HEs are excited at rates comparable with the Drude-like carriers. From Eq. (1.3) we also learn that the amplitude of the electric field at the surfaces of the particle is key to increase the rate of generation of HEs, so plasmonic hot spots will increase the number of hot carriers in the particle. We will discuss electromagnetic hot spots in more detail in the section dedicated to coherent energy transfer.

It is important to note, in the context of the transfer of energy between subsystems in the plasmonic particle (Figure 1.7b), the relevant timescales for the different transitions. Figure 1.6c illustrates the evolution of a plasmonic NP excited by a pulse excitation [23]. In particular, after the rapid dephasing of the plasmon into a population of single-electron states, their relaxation through electron–electron scattering will proceed at rates that depend on their energy

$$a_{\text{e-e}} \approx \frac{1}{\tau_{\varepsilon,\text{e-e}}(\varepsilon)}, \qquad \tau_{\varepsilon,\text{e-e}}(\varepsilon) = \tau_{0,\text{e-e}} \frac{E_{\text{F}}^2}{\left(\varepsilon - E_{\text{F}}\right)^2}, \tag{1.4}$$

where $\tau_{0,\,\text{e-e}}$ is a constant depending on the metal. The expression for $\tau_{\varepsilon,\,\text{e-e}}$ in Eq. (1.4) holds for Fermi liquid theory [66], and entails that high-energy electrons will lose energy very rapidly, after a few collisions [67]. At this point, the relaxation of the thermalized excited carriers, sharing their energy with the crystalline matrix, will be controlled by electron–phonon scattering events, which occur at a significantly slower rate than electron–electron collisions, $\tau_{\varepsilon,\,\text{phonons}} \gg \tau_{\varepsilon,\,\text{e-e}}$. Of course, the picture described in Figure 1.6c corresponds to a pulse excitation. This is useful to understand the carrier dynamics inside the particle, but in general we will be interested in the steady state under continuous illumination when studying photocatalysis. In the steady state, the nonthermal hot carriers will coexist with thermalized hot carriers, as more are continuously excited and the system continues to dissipate energy into the crystalline matrix and its environment. It is important, however, to keep in mind the short lifetimes of the nonthermal hot carriers, as this will limit the probability of them leaving the plasmonic particle before thermalizing.

1.4 Charge Transfer Mechanisms Associated with Plasmons

We have seen how, as the plasmon dephases inside a plasmonic NP, its energy is distributed among single-electron degrees of freedom. Some of these charge carriers can be transferred to the environment, either reaching molecular adsorbates directly [17] or using semiconductor catalysts in an intermediate charge separation step [15]. The consequent ionization of the molecules can contribute to drive redox processes [68, 69], or excite vibrational states through an intermediate charged excited state in what is often referred as desorption induced by electronic transition (DIET) [70–72]. In contrast with the thermal activation of the reaction, on which we will comment in more detail later in the chapter, charge transfer mechanisms can offer access to new reaction products, as well as product selectivity [73–76]. Although the transfer processes are concurrent with the heating by the plasmonic particle, and they can act synergistically [73, 77–79], it is in principle possible to estimate their relative contributions to the reaction rates [78, 80]. For instance, two properties have been used to recognize charge transfer processes in experimental photocatalytic systems with metal NPs: the transition from linear to superlinear dependence of the reaction rates with light's intensity as multitransfer events become more probable [17, 81], and the higher rates induced in reactions with lighter isotopes [17, 22]. Beyond these, additional procedures have been proposed for teasing apart the relative importance of hot carrier injection and photoheating in driving chemical reactions. [82]

Although charge transfer mechanisms play an important role in the effective design of plasmonic photocatalytic systems, the efficiencies achieved to date with systems exploiting them remain low [83]. There are different paths that can lead to larger efficiencies, such as tailoring the plasmonic system to increase its generation rates of high-energy electrons. This can be done by considering shapes and sizes that maximize the surface-mediated promotion of these carriers [23], or creating electromagnetic hot spots through particle–particle interaction [55, 57] or the particle's curvature [65]. But in addition to studying the efficiency of these systems in generating such excited carriers capable of migrating out of the plasmonic particle, we should pay close attention to the injection process itself. Now, in order to discuss the mechanisms of charge transfer between plasmonic particle and environment, it is useful to consider the distinction between indirect and direct carrier transfer. In the former the electronic transition is internal to the metal and a secondary injection step is required, while in the latter the dephasing of the plasmon triggers an electronic transition between the metal particle and the environment.

1.4.1 Indirect Hot Carrier Injection

We can start by briefly discussing the indirect injection of excited carriers, as it follows straightforwardly from the discussion in the previous section. In it, we discussed how the decay of the plasmon populates a collection of excited electronic states, including both low- and high-energy carriers. Now, for any of these to contribute to the photocatalytic process, they need to leave the particle. A fraction of them will be injected into the environment, a process labeled as "indirect" because it represents an independent step from the excitation of the carrier. The energy of the internally excited carrier is very relevant in processes of indirect injection, because in general there will be a potential barrier that the carriers need to surpass to be ejected from the particle (Figure 1.8a) [84–86]. For metal–semiconductor interfaces, the relevant concept is that of a Schottky barrier [87]. This barrier resulting from the bending of the semiconductor bands when it contacts the metal has two effects: first, it prevents low-energy excited carriers to leave the metal; second, it limits the back-transfer of charge carriers injected from the metal, making semiconductors useful for limiting the rapid relaxation of high-energy electrons in the metal. One should note, however, that different combinations of materials can result in ohmic contacts or small Schottky barriers at their interfaces, opening new possibilities in designing efficient photocatalytic reactors [88]. Furthermore, by creating electrically gated devices, it is possible to actively control the barrier's height and thus filter carrier injection. [89] The simplest approach to quantifying the amount of excited carriers that will traverse the interface is to just consider whether they have enough kinetic energy to surpass the barrier. This approach sets a threshold value on the carrier's energy for its injection, but ignores consideration of the electron's momentum at the interface and can therefore only give us a ceiling on the real injection rates. The diagram in Figure 1.8a shows the energy barrier height ΔE_b, setting this threshold. Under this perspective, any estimation of the injection rate should include some function of the overbarrier energy, $\hbar\omega - \Delta E_\mathrm{b}$ [65, 67, 84, 86].

More detailed models for electronic injection through a Schottky barrier take into account other factors limiting electron transmission at the interface. For instance, energy profile and width of the potential barrier should be considered in order

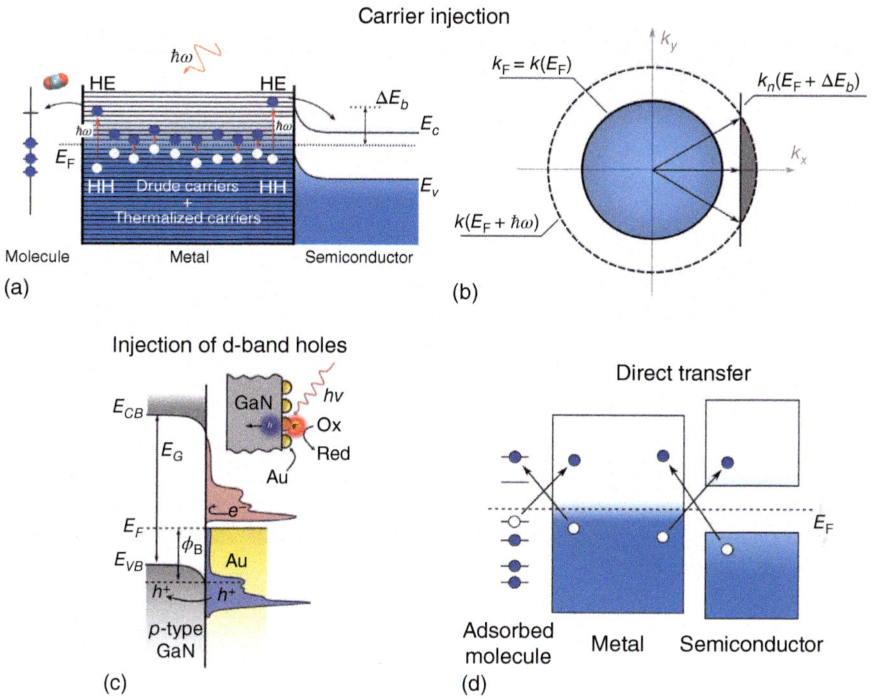

Figure 1.8 (a) Diagram showing the different types of excited carriers inside a NP under optical excitation. The carriers with enough energy to traverse the potential barriers at the boundaries of the NP are susceptible of being injected into its environment. (b) Diagram of electron injection in momentum space. The inner circle represents the ground state of the Fermi sea, and the dashed circle marks the maximum momenta of optically driven electrons. The grayed-out section contains the excited electrons satisfying, within the Fowler theory, the condition for injection at a boundary extending along the *yz* plane. The function used in the panel computes the wave number as $k(E) = \sqrt{2mE/\hbar^2}$, and $k_n(E)$ denotes the component of the wave vector normal to the surface. (c) Diagram illustrating the injection of holes into a *p*-type semiconductor. Although HEs and HHs are also generated here through plasmonic excitation, by illuminating at a wavelength that permits interband transitions of electrons from the *d*-band, one can obtain large numbers of high-energy HHs [19]. (d) A direct transfer to the environment is also possible, contrasting with the indirect, two-step injection shown in panels (a)–(c). If the plasmonic NP is in contact with a material providing an external density of states, be it an adsorbed molecule or an adjacent semiconductor, the plasmon can decay by exciting a single carrier across the two systems. An electron excited in such a way can go from the metal to the neighbor, or from the external system into the metal, resulting in the latter case a hole injection to the adsorbate or semiconductor. *Source*: Adapted with permission from Refs. (a) [23] Copyright 2019 Elsevier, (c) [19] Copyright 2018 American Chemical Society.

to account for the possibility of tunneling by electrons with lower energies than ΔE_b, although this is only be relevant for very thin interfacial regions [87]. More commonly, the study of photoinjection at Schottky barriers takes a ballistic model of the electron and examines the limitations that the barrier imposes on an injection event so that it respects both energy and momentum conservation. This is the framework under which the Fowler theory models the photoinjection in thick films [90]. Figure 1.8b shows a geometrical interpretation of the momentum conservation constraining the injection, which defines a region of allowed momenta by imposing the condition $(\hbar k_n)^2/2m > E_F + \Delta E_b$ for injection, where k_n is the component of the carrier's wave vector normal to the barrier [91]. The probability of carrier injection according to Fowler theory is of the form

$$\eta_{\text{Fowler}} \propto \frac{\left(\hbar\omega - \Delta E_b\right)^2}{\hbar\omega E_F}.$$

Extensions of this model use semiclassical methods to include additional considerations, like multiple electron reflection at surfaces, to explain the enhanced injection observed in Schottky photodetectors with thin metal films [92, 93]. It should be noted that these models of injection typically assume the generation of high-energy electrons across the volume of the metal, not only at its surfaces [94]. A complete account of the photoinduced injection at a Schottky barrier should also consider these carriers and their drift inside the particle. However, in the previous section of the chapter we have focused on the generation of HEs through a surface-assisted mechanism. This is so because this is the fundamental quantum mechanism that can provide high-energy carriers in a plasmonic resonance, and is dominant for small NPs [94]. Nonetheless, we direct the interested reader to a recent detailed discussion of these and other factors in Ref. [67].

A similar discussion can be presented for the case of molecules adsorbing to the plasmonic material. In this context, the threshold to the injection of excited carriers results from the alignment between the electronic states of metal and the hybridized electronic states of the adsorbed molecule. If the lowest unoccupied molecular orbital (LUMO) has a higher energy than the metal Fermi level, the electrons in the metal will require additional energy to reach it. Therefore, one can also consider an energy-threshold approach to find upper bounds for the rates of injection to adsorbates. Of course, in real systems there are also more factors playing a role in regulating the electronic injection from plasmonic NCs to molecules, such as the affinity of the molecule to the different crystal facets of the particle [95], and the details of the hybridization of states between metal and molecule [74].

At this point, it is useful to refer again to the interband transitions occurring in metals (Figure 1.3c). As mentioned in the introduction, the part of the absorption spectrum due to interband transitions cannot be tuned through changes in the geometry of the NP. Nonetheless, given the large density of states in the *d*-bands of noble metals, the absorption from interband transitions can also be large, and they can be useful in a photocatalytic context. Given that the *d*-bands in plasmonic materials are deep below the Fermi energy, optically excited electrons originating in them will have low kinetic energies, while the corresponding holes will be very energetic. Therefore, one can devise photocatalytic schemes that take advantage of this situation. As an example, Figure 1.8c illustrates a photoelectrochemical system using a *p*-type semiconductor that can extract the energetic holes, while the excited electrons drive CO_2 reduction [19]. High-energy HHs, either in the conduction band or in the *d* band, can also target the direct molecular oxidation of adsorbates while exciting the plasmonic NP at low wavelengths [20, 96].

1.4.2 Direct HE Injection

In addition to the internal decay of the plasmon into single-electron states, we should take into account that the dephasing of the plasmon is also influenced by the environment of the plasmonic structure. In particular, when the NP is in contact with the electronic density of states of an adjacent semiconductor or adsorbed molecule, the energy of the plasmon can trigger electronic transitions that cross the interface of these materials, in what is often referred to as direct charge transfer events [97–99]. From the perspective of an optical analysis of the plasmonic NCs, these mechanisms can become apparent as the additional broadening of the plasmon resonances in the spectrum of the sample, and was therefore included in our general discussion on the plasmon decay rate, γ_{plasmon}, as $\gamma_{\text{charge transfer}}$. This is a general term that aims to group what in the literature is typically referred to as chemical interface damping (CID) [100, 101] and plasmon-induced interface charge transfer [97], depending on whether the transfer occurs with adsorbed molecules or adjacent semiconductors, respectively. Its magnitude will depend on the materials composing the plasmonic particle and its neighbor, as well as on their surface and interfacial electronic states.

In contrast with the previously discussed indirect charge transfer, this is a process that excites a charge carrier into the particle's environment in a single step, instead of a two-step excitation–injection process. Therefore, this charge transfer connects a donor (acceptor) state in the plasmonic particle with an acceptor (donor) state in its environment, be it in an adsorbed molecule [99, 101, 102] or in an adjacent semiconductor [97–99] (Figure 1.8d). An important implication of this distinction is that the excited electron or hole does not stay for any length of time inside of the metal, where it can lose its energy before being able to leave the particle, due to the rapid scattering events with other electrons (see Figure 1.6c). In other words, this type of hot carrier excitation has an intrinsic 100% injection efficiency, and designing plasmonic catalysts with interfaces that promote this effect would be very desirable. This can prove to be a challenging goal, given the difficulty to experimentally distinguish between mechanisms of carrier injection, although it has been suggested that polarization-dependent injection efficiency in metal–QD junctions could be evidence of direct HE injection [97]. Therefore, most current efforts for understanding this mechanisms are theoretical, including the use of atomistic ab initio methods to probe the charge dynamics of coupled metal–semiconductor [98] or metal–molecule [99] systems. Interestingly, recent theoretical results suggest that upward of 40% of total plasmonic decay in small Ag NPs coupled to CdSe NPs, under pulsed excitation, is due to direct transfer [98]. This is a very promising result, inviting more research toward the possibility of exploiting this mechanism in practical setups to enhance the overall efficiency of plasmonic photocatalysis.

1.5 Plasmonic Near-Field Enhancement

As discussed in the introduction, the resonant coherent charge oscillation of a plasmon results in the generation of a strongly enhanced electric field inside and near the particle [1]. This stronger field can boost optical and photocatalytic processes in molecules or QDs nearby the plasmonic particle (Figure 1.9). A commonly mentioned example of this phenomenon is surface-enhanced Raman spectroscopy (SERS), an early application of plasmonics in modern scientific practice. It exploits the field enhancement of a surface plasmon supported by a metallic surface to increase the Raman signal of its molecular adsorbates, leading to large enhancements due to the nonlinear dependency of Raman scattering on the intensity ($\propto|\mathbf{E}|^4$) [106, 107]. Therefore, SERS benefits from the design of plasmonic systems with strong near fields (see Figure 1.9a) [103]. Other processes that can be enhanced by a strong plasmonic near field are fluorophore and semiconductor emission [108–111], photon upconversion [112–114], molecular chiral signal [115, 116] or, importantly, photocatalysis [18, 104, 105, 114], where the energy receptor can be molecules (Figure 1.9b) or semiconducting crystals (Figure 1.9c).

The local enhancement factor is a useful magnitude for quantifying the local strengthening of the field with respect to the background illumination, and is defined as

$$P_{\text{local}}(\mathbf{r}) = \frac{|\mathbf{E}_\omega(\mathbf{r})|^2}{E_0^2},$$

where $\mathbf{E}_\omega(\mathbf{r})$ is the total complex-valued electric field and E_0 is the amplitude of the impinging radiation. Using this magnitude, it is easy to characterize the near-field enhancement of different plasmonic NPs, which is relevant to understand the impact of the particle's geometry to the spatial distribution and strength of its near field. The image in Figure 1.10a maps the values of $\sqrt{P_{\text{local}}(\mathbf{r})}$ for a cross-section of a spherical Au NP of 10 nm in diameter, thus showing the enhancement on the electric field's amplitude. It is strongest along the direction of polarization of light, as it excites the particle as a dipole antenna, and decays rapidly as we get away from the surface of the particle [1, 117]. It is also worth noting that the enhancement factor is greater than one inside the particle, which is a signature of the resonant excitation that drives the internal energy dissipation and excitation of hot carriers. The geometry of a plasmonic NC influences its near field in terms of form and intensity. Most notably, it can give rise to electromagnetic hot spots, regions with a particularly large electric field amplitude. Hot spots appear at sharp features in a particle, because the small local radius of curvature does increase the local surface charge density. Figure 1.10b presents such an example, where we can see the hot spots that result at the sharp edges of an Au nanocube. Although real nanocubes cannot have perfectly sharp edges, they nonetheless show strong hot spots at edges and corners [118–120]. Metallic nanostars are another good example of this type of hot spots [121], and one that has also been shown to be useful in a photocatalytic context by increasing the internal generation rates of HEs [122]. Hot spots also appear where the near field of two or more plasmonic NPs interact. Figure 1.10c exemplifies this with a dimer of 10 nm Au spheres. These are the same NPs as the one shown in Figure 1.10a, but the increased field strength at the dimer gap is apparent, with an approximate five-fold increase between the strongest points in each map. Of course, one can create plasmonic systems combining these two types of hot spots, such as in the so-called

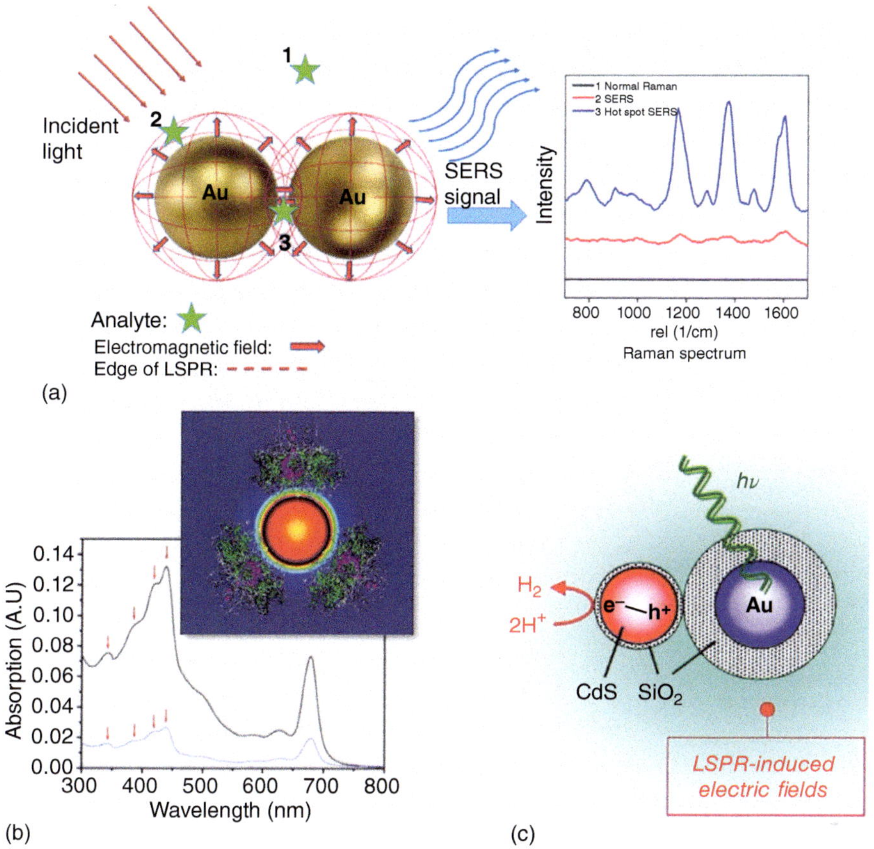

Figure 1.9 (a) Plasmonic systems increase the molecular signal in Raman spectroscopy through near-field enhancement, so that plasmonic hot spots are especially effective. Plasmonic field enhancement can also drive photocatalytic processes, by exciting neighboring molecules and semiconducting QDs. (b) Comparison between the absorption of suspensions of photosynthetic protein Photosystem I (PS I), unbounded (blue curve) and bounded to Ag NPs (black curve). The top inset depicts PS I molecules attached to a plasmonic NP. The background color gradient represents the enhancement factor of the NP, which is strongest close to the NP. (c) Schematic diagram of a system designed for enhancing the photocatalytic properties of CdS QDs with the near field of Au NPs. Both particles are separated by a SiO_2 layer preventing charge transfer between them. *Source*: Adapted with permission from Refs. (a) [103] published by the Royal Society of Chemistry, (b) [104] Copyright 2010 American Chemical Society, (c) [105] Copyright 2011 American Chemical Society.

bowtie antennas, a commonly used system designed to study the effect of plasmonic hot spots [57, 59, 123].

Given the strong decay of the enhanced near field with the distance from the particle's surface ($\propto 1/r^3$ for the dipolar plasmonic modes), it is clear that we will be interested in positioning the acceptor system as closely as possible to the plasmonic particle. But doing this also come at a disadvantage. Balancing enhancement and quenching is crucial to efficiently promote photoemission in fluorophores with plasmonic NCs, as their nonradiative losses increase when they come too close

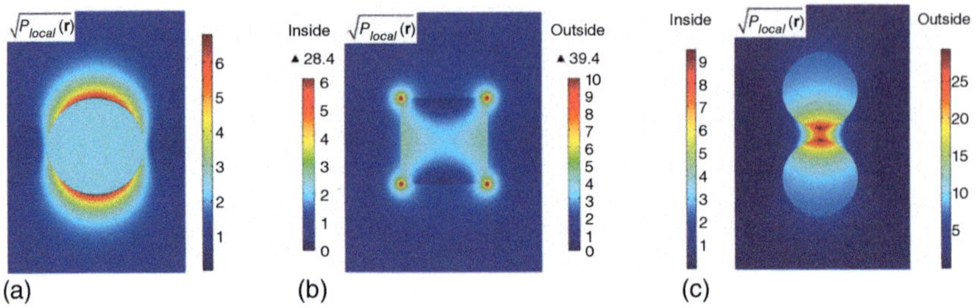

Figure 1.10 Enhancement factor maps of Au NCs with different geometries, immersed in water and excited at their resonant frequencies. (a) Au sphere with a diameter of 10 nm. (b) Au cube with an edge length of 7 nm, with strong hot spots at its vertices. This cube has sharp edges, in order to demonstrate the limits of near-field enhancement through increased surface curvature. (c) Dimer composed by 2 Au NPs of 10 nm in diameter, separated by a gap of 1.7 nm. Although the constant curvature of the spheres does not create hot spots, as in panel (a), the interaction between both generates a very strong hot spot at the gap. *Source*: Panel (b) was adapted with permission from Ref. [54] Copyright 2017 American Chemical Society.

to the NC and dominate over the radiative decay rates [124–126]. Consequently, it is common practice to use dielectric spacers to reduce nonradiative losses in systems of plasmonically enhanced photoemission [127, 128]. Similarly, while the enhanced near field can boost the absorption of a nearby molecule, if it comes in contact with the metal the latter can also act as an electron drain and decrease the final reaction yield. Then, photocatalytic strategies that rely on the coherent near-field excitation by the plasmonic particle can also use spacers to prevent the possibility of charge exchange with the plasmonic material [18, 105]. As we have seen in previous sections, charge transfer is an important mechanism in plasmonic catalysis, but whether is advantageous or disadvantageous will depend on different factors, including the relative energy alignment between the electronic states of the catalyst and the reactants.

Lastly, processes that are enhanced by the increased amplitude of the plasmonic near field will have to be not only spatially but also spectrally close to the resonance. The situation is slightly different, however, in the case of the near-field enhancement of the processes in other conductive particles [129, 130] or in multimetallic particles [131]. The strong, oscillating near field of the plasmonic NP will create currents in the other conductive material, or effectively resonate together in a hybrid coupled mode, and this excitation will drive catalytic activity at the surface of these nonplasmonic metal particles. In this case, the plasmonic resonance does not need to be close to, for instance, the absorption lines of the molecule, because the conducting co-catalyst mediates in the transfer. Taken as a single catalytic unit, the hybrid multimetallic system would work in a way more closely resembling monometallic plasmonic hot carrier injection, while providing additional control over reaction selectivity [132, 133].

1.6 Plasmonic Scattering

Let us now consider a mechanism of plasmonic decay connecting the particle with the far field, instead of with its immediate environment. In the same way that the plasmon can be excited by an incoming plane wave, it can re-radiate electromagnetic energy in the reciprocal process. This process, to which we typically refer to as radiative decay or plasmonic scattering, is perhaps the one for which the analogy of a plasmonic NP as an antenna is most clarifying. In the same way that macroscopic antennas show reciprocity between reception and transmission, the dipolar modes excited in a plasmonic system can radiate back into space. For plasmonic NPs, an important property to note is that their scattering cross-sections depend strongly on their geometry and size, so that small particles can radiate back a negligible amount of energy, even though they show a significant coupling with incoming light because of their plasmonic resonance. Looking at the expression of classical absorption and scattering cross-sections for small particles will help us discuss this more precisely.

Absorption and scattering cross-sections quantify the proportion of the incident light flux that is dissipated inside the particle and reradiated, respectively. Therefore, $\sigma_{\text{abs}} = P_{\text{abs}}/I_0$ and $\sigma_{\text{scat}} = P_{\text{rad}}/I_0$, where P_i stands for power, absorbed and radiated, and I_0 is the incident light's intensity. If we assume that the particle is much smaller that the wavelength of light, we can approximate its response as that of a point dipole, in what is known as the quasistatic limit or approximation. In such a case, it is useful to characterize a NC simply through its polarizability, $\alpha = \mu/E$, which tells us how strong an electric dipole μ is induced by the external field. We can then express both absorption and scattering cross-sections of an arbitrary small particle as a function of its polarizability [134, 135]

$$\sigma_{\text{abs}} = \frac{4\pi k}{\varepsilon_{\text{env}}}\text{Im}\,[\alpha], \qquad \sigma_{\text{scat}} = \frac{8\pi\, k^4}{3}|\alpha|^2,$$

where ε_{env} is the permittivity of the environment where the particle is embedded. The polarizability of a particle will depend on its shape, but for spherical systems it simple scales linearly with its volume. Therefore, for spherical particle of small sizes (e.g. low tens of nanometers for Au), the linear term on the volume, absorption, will dominate over the quadratic, scattering. At larger sizes, this trend will reverse. Of course, satisfying the quasistatic approximation means that we are also within the Rayleigh scattering limit, with its familiar $1/\lambda^4$ dependency, so that scattering will be more relevant at high frequencies. So, materials with large plasma frequencies, such as Al, will scatter more strongly that materials such as Au, which resonate in the visible to IR region. We should also remember that even small particles can show significant scattering cross-sections if they have large aspect ratios, as is the case of NRs, and therefore a larger polarizability than spheres of equivalent volume. Although far from being complete, this short discussion should give some intuition for the relevant factors impacting scattering on small plasmonic NPs, and the interested reader can learn more about the radiative properties of plasmonic systems in other resources on the topic [1, 135, 136].

Now, to consider how plasmonic scattering can play a role in photocatalytic strategies, we have to consider that, although the reemission will be biased toward the direction of the incoming light, the scattered photons will in general travel along a different path than the incoming radiation [136, 137]. Therefore, the presence of scatterers in a medium will increase the effective optical thickness of a system, as sketched in Figure 1.11a. As such, plasmonic NPs can be of use in photocatalytic [16] and photovoltaic [138, 139] systems, with the goal of achieving significant light

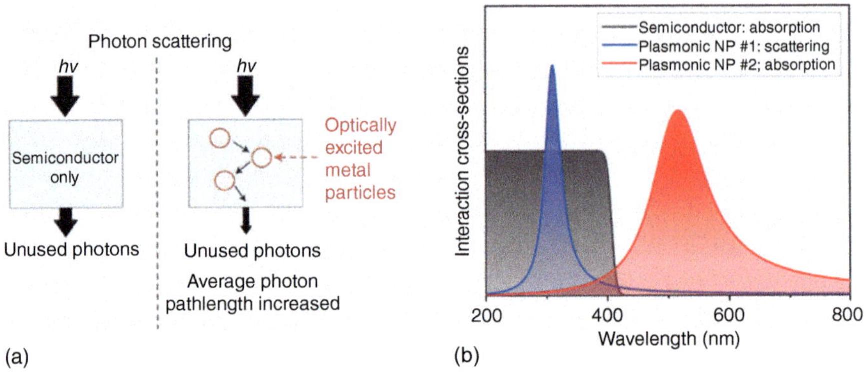

Figure 1.11 (a) Scattering plasmonic nanoparticles can be used to extend the effective optical thickness of an optically active material or device. However, for this mechanism to be useful, these particles need to scatter light at frequencies that the surrounding medium can absorb. In panel (b) we contrast this situation, in the context of a semiconducting absorbing material (modeled with the bandgap of TiO_2), with that of using other mechanisms like charge transfer or photoheating, where the plasmonic resonance does not need to overlap with the semiconductor's absorption. Consequently, using plasmonic nanoparticles purely as scatterers can increase the optical thickness of a system, but not its operative spectrum. *Source*: Panel (a) was adapted with permission from Ref. [16] Copyright 2011 Springer Nature.

absorption with thinner layers of material. It is important to keep in mind, however, that scattering is an elastic process, so the reemitted photons will have the same energy than those exciting the plasmonic resonance. Therefore, using plasmonic NPs effectively as scatterers depends on the presence of other mechanisms converting light's energy into a form that is useful for the chemical process. This can be either accomplished at the plasmonic NP itself – absorption and scattering can coexist in a particle resonating at a given wavelength, i.e. a fraction of the interacting photons will be absorbed – or by optical absorption events at the molecules or co-catalysts. An enhancing strategy that uses plasmonic NPs only as scatterers will then typically require that they operate at high frequencies, where other elements in the system can absorb the scattered photons. Figure 1.11b illustrates this situation in the context of a semiconductor medium, and contrasts it with an approach where the plasmonic NP uses other mechanisms to transfer energy – such as HEs or photoheating – in a manner that takes advantage of regions of the spectrum unavailable for direct absorption in the semiconductor.

1.7 Photoheating

Heating always occurs when we illuminate plasmonic NPs, as the energy of the plasmonic resonance dissipates internally due to electrons scattering with the atomic lattice of the metal. The previously discussed mechanisms can extract part of this energy and transfer it to other elements in the environment before this happens, contributing to the reaction [17, 69, 78]. But, of course, photoheating is also a useful process and can contribute to drive the chemical process through an increase in local temperature, and the nonthermal mechanisms discussed above can also be understood as acting in reducing the activation energy of the reactions, so that they can proceed at lower temperatures [77–79, 140]. Moreover, even plasmonic NPs used exclusively as nanoheaters can provide significant efficiencies in setups designed around this strategy, e.g. exploiting photogenerated temperature gradients where the resulting thermophoretic forces separate the reaction products from the active reaction sites [141].

When considering plasmonic heating as a photocatalytic mechanism, we have two main ways of describing the local effects of the plasmonic NC, depending on whether we focus on the phononic or electronic transfer of energy. In the first case, we can simply consider the plasmonic NP as a dissipative system, behaving like a localized photoheater [142]. As the metal particle is illuminated, it will heat up and raise its temperature and that of other materials with large thermal conductivity in contact with it. The environment is subsequently heated up due to the photoinduced temperature gradient. Figure 1.12a shows the steady state temperature obtained by illuminating an isolated gold NP immersed in water. The increase in temperature outside a small isolated particle follows the expression [142]

$$\Delta T(r) = \frac{V_{\text{NP}}Q}{4\pi k_0 r},$$

where V_{NP} is the volume of the particle, Q is the total heat it dissipates, k_0 is the thermal conductivity of the environment, and r is the distance to the center of the particle. With each particle serving as a photoheater, the macroscopic local temperature will increase at a rate controlled by the optical and thermal properties of the ensemble and matrix, as well as the light's intensity [7, 142, 144–146]. In Figure 1.12b, we can see theoretical values comparing the temperature increase of one particle and a small ordered ensemble of 16 NPs. Of course, one can also exploit interaction effects between plasmonic NPs, effectively creating a more efficient photothermal metamaterial [7]. Figure 1.12c exemplifies this with a model system created to study the effect of interparticle interaction in ensemble photoheating [143], but one can also create photoheaters by considering ordered planar structures functioning as perfect absorbers [5, 147].

On the second perspective, we can evaluate the thermal effects of plasmonic NPs insofar they impact charge exchanges with their environment, under a perspective that connects directly with the one used to describe the mechanisms of HE injection [33]. For this, we can consider the energy distribution of the carriers inside the plasmonic NP, and how the temperature of the lattice affects their injection possibilities in two different contexts. In the first, the electrons in the particle are separated from their environment by a relatively high potential barrier of $\Delta E = 0.8$ eV, a value compatible with typical values of Au and Ag Schottky barriers [84]. In such a situation, the total energy of a plasmon excited in the visible or near-IR spectral range would be sufficient to drive charge injection into its environment, be it through direct promotion or through the two-step process illustrated in Figure 1.13a. On the contrary, if in the context of a large barrier we consider electronic occupation of states through Fermi–Dirac statistics at the temperature of the lattice, the probability of injection will be very small [68]. Again, for a barrier $\Delta E = 0.8$ eV, the probability of injection would be in the order of 10^{-14} at low light fluencies, a situation in which the

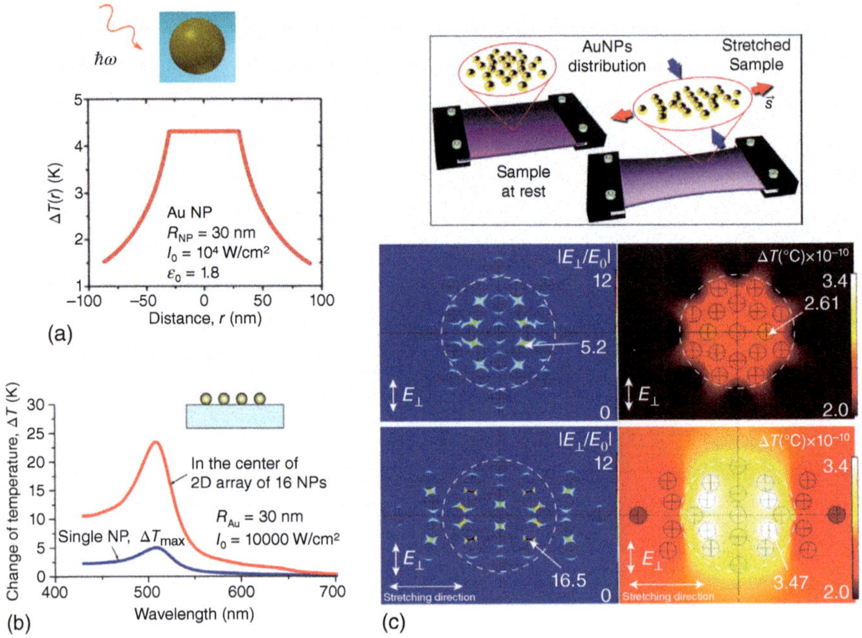

Figure 1.12 Plasmonic nanoparticles as photoheaters. (a) Single particle. Diagram of an optically excited gold nanoparticle immersed in water, and theoretical data of the temperature increase inside and around it as a function of the distance to its center. (b) Collective heating. Increasing the number of photoheaters will increase the heat deposited in the region, leading to higher phototemperatures. Comparison between the temperature increase of a single NP and a 4×4 NP planar array. (c) Electromagnetic coupling. Two or more plasmonic nanoparticles in close interaction will create hot spots that enhance the total absorption in the spheres, with the consequent increase in dissipated heat. The top schematic diagram shows an experimental setup used to quantify the effect of electromagnetic coupling between AuNPs on photoheating. The four lower panels show theoretical data for such a system, with field enhancement maps on the left and temperature maps on the right. The data in the top row shows the results under no tension, while the bottom row correspond to a stretched system. The dashed line outlines the laser beam spot. *Source*: Adapted with permission from Refs. (a) [142] Copyright 2007 Elsevier, (b) [7] Copyright 2006 Springer, (c) [143] Copyright 2018 Royal Society of Chemistry.

particles remain relatively close to room temperature. And although the probability of injection will of course increase when considering a hotter system, for realistic temperatures in colloidal systems, the energy of these thermalized electrons will remain significantly lower than the energy barrier, $T_e \ll \Delta E_b/k_B$. For this barrier height, one could estimate that the thermionic electrons would compare with the number of over-barrier plasmonic HEs only at $T_e \sim 500$ K, by using the expressions in Figure 1.13a,b [33]. On the other hand, other material combinations offer small potential barriers between metal and environment, such as the TiN/TiO_2 ohmic junction reported in Naldoni et al. [88]. In such a case, illustrated in Figure 1.13c, the picture is notably different, and realistic increases in the effective electronic temperature for nonconcentrated solar radiation can still notably increase the number of carriers that can leave the plasmonic NC.

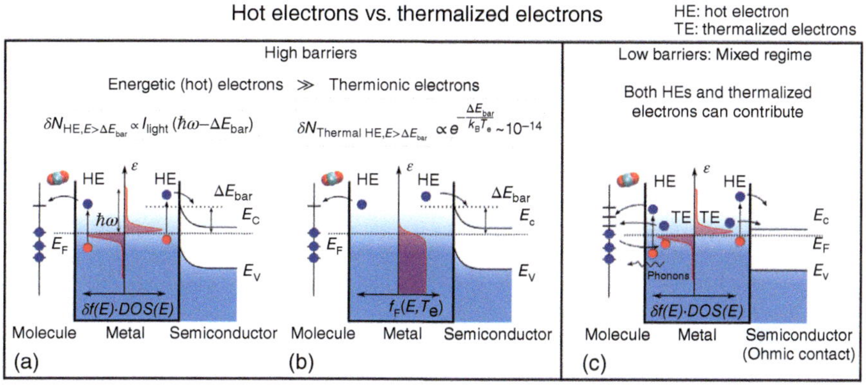

Figure 1.13 Comparison of thermal effects and indirect HE injection. These illustrations sketch the fundamental differences between the impact of thermalized and nonthermalized carriers toward charge injection, in two distinct situations at the metal–environment interfaces: having a high (panels a and b) or low (panel c) potential energy barrier. (a) Diagram focusing on the internal generation of hot carriers. The injection barrier is significantly larger than the effective electron temperature T_e, i.e. $\Delta E_{bar} > > k_B T_e$. The function inside the metal shows the distribution of nonequilibrium electrons in a metal under optical excitation. The number of HEs with energies $E > \Delta E_{bar}$ is linearly proportional to the intensity of light. (b) Still assuming a high barrier, we show the distribution of carriers inside a photoheated metal, under low-intensity illumination. We assume that its temperature is higher than room temperature by 20 K. The probability of injection over a barrier scales exponentially, and it is very low for realistic values of Schottky barrier (we take $\Delta\varepsilon_{bar} = 0.8$ eV, as in an Au–TiO_2 interface). (c) In the case of low interfacial barrier, such as that of TiN–TiO_2 junctions [88], both high-energy HEs and thermalized electrons are susceptible of being injected to the environment and contribute to the reaction. Regardless of barrier height, the crystal heats the environment through the exchange of phonons. We illustrate this in panel c only, to simplify the overall presentation. *Source:* Adapted with permission from Ref. [33], Copyright 2019 American Chemical Society.

1.8 Example Applications

The mechanisms discussed here provide different opportunities for using plasmonic NCs to our advantage in photocatalytic and photoelectrocatalytic systems, enabling different strategies for the pursuit of greater efficiencies in the light-assisted catalysis of chemical reactions of technological relevance. In this section, we will briefly describe different applications, including some on which other chapters will expand upon, that highlight the relevance of specific ideas discussed in this chapter. Perhaps most notably, plasmonic NPs have been proposed as useful tools driving solar water splitting for the clean production of hydrogen fuel [21, 152], CO_2 reduction [153, 154], and nitrogen fixation [141, 155], but the list of potential applications is much broader [22, 156, 157]. Across these, plasmonic NPs can be exploited in different ways. A well-known strategy is the combination of different materials to create hybrid reactors, with a prime example being the loading of Au NPs in TiO_2 to enhance the charge separation capabilities of the semiconductor and target specific reactions [15]. Other material combinations have also shown great promise, including the use of plasmonic NPs with good catalytic metals, such as Pt and Pd [129, 131, 158–161], with semiconductor nanostructures [16, 157, 162–164], or with

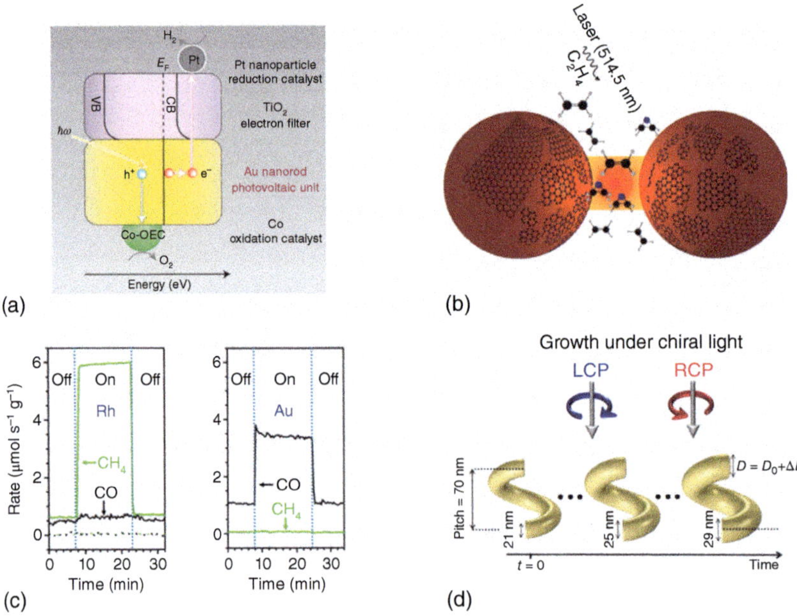

Figure 1.14 (a) Autonomous water-splitting photocatalytic unit, where charge carriers are generated in a Au NR and different materials separate and direct them to distinct reaction sites [148]. (b) Schematic diagram of the photocatalytic formation of graphene on Ag nanoparticles, as an intermediate process promoting ethylene epoxidation [149]. (c) Results from photocatalytic systems for CO_2 hydrogenation using Rh or Au NPs, showing a strong product selectivity. (Left) Rates of production of CO and CH_4 in the dark and under UV light, using Rh. (Right) Rates of production of CO and CH_4 in the dark and under white light, using Au [150]. (d) Diagram depicting hot-electron induced growth of metallic crystals, which can be controlled through the circular polarization of light when using chiral plasmonic nanostructures [151]. *Source*: Adapted with permission from Refs. (a) [148] Copyright 2013 Springer Nature, (b) [149] Creative Commons Attribution 4.0 International License, (c) [150] Creative Commons Attribution 4.0 International License, (d) [151] Copyright 2020 American Chemical Society.

materials offering rapid carrier extraction and transport, such as CNTs [165, 166]. An interesting example of this kind of strategies is the work presented by Mubeen et al. [148], where a complex of four materials is used to complete both oxidation and reduction steps in water splitting (Figure 1.14a). A gold NR is used to collect radiant energy within a broad range of wavelengths, up to the near-IR, and the excited electron and holes are routed to two distinct reaction sites functionalized with two different catalytic materials, each targeting specific semireactions of water splitting.

This serves as a reminder of the crucial importance of the detailed surface characteristics at the active reaction elements, which are critical for targeting a given chemical reaction. Among the many ways of exemplifying the nontrivial dependency on the catalyst's surface, let us briefly look at studies of metal catalysts in ethylene epoxidation, another chemical reaction of societal significance [167]. Silver is a common catalyst used in temperature-driven ethylene epoxidation [167, 168], and the shape of nanostructures of this metal – determining the crystal facets contributing to the reaction – was shown to be relevant for controlling the selectivity of this catalytic process [95]. Silver was subsequently shown to also be effective as a photocatalyst in enhancing this reaction [156]. Interestingly, Zhang et al. have recently presented results showing that, in the photocatalytic epoxidation of ethylene using Ag NPs, the reaction proceeds through an intermediate step synthesizing graphene at the NPs' surfaces (Figure 1.14b), which mediates the injection of HEs photoexcited at the NP and whose defects become the reaction sites for the epoxidation [149]. Such results invite the question of whether we could design different complexes with advanced surface functionalization that improves the efficiency of different plasmonic photoreactors.

Plasmonic particles also facilitate the design of photocatalytic systems with product reaction selectivity. This is a property that we have not explored in depth in this chapter, but it is nonetheless of great importance in practical applications. The spectral tunability of plasmonic structures, in combination with material-dependent properties such as its work function and crystalline surface facets, make them a versatile tool for targeting specific reaction paths [17, 22, 169]. As an example, we can highlight the work by Zhang et al., where they contrast the evolution of systems with either Rh or Au NPs loaded on Al_2O_3 supports for the hydrogenation of CO_2 [150]. Besides the overall increased efficiency of the photoexcited systems for both materials with respect to the thermal catalysis in dark conditions, the comparison between both photocatalysts showed that using Rh NPs offered a greater degree of control over the reaction products. As shown in Figure 1.14c, the CH_4 reaction path was substantially promoted over the CO path when the Rh-loaded system was illuminated. This kind of product selectivity is presumably allowed by the excitation of plasmonic hot carriers to promote the otherwise kinetically dispreferred reaction path [69, 75, 150, 170], and underscores the relevance of plasmonics in catalytic applications.

We can also gain additional degrees of control over a photocatalytic system by designing asymmetric plasmonic nanostructures. For instance, plasmonic structures with chiral geometries interact differentially to opposite polarizations of circularly polarized light (CPL) [171, 172]. The optical asymmetries that can be achieved with chiral plasmonic structures, measured through circular dichroism or through the dissymmetry or anisotropy factor g, greatly exceed those of chiral molecules. This is true for planar metamaterials [173–175], structured particle complexes [174, 176–179], and colloidal chiral particles [151, 172]. Therefore, using nanostructured metals we can also tailor the optical response of a photocatalytic system and engineer its response to circularly polarized light. It bears clarifying that these processes are fundamentally different from traditional chiral photocatalysis, which pertains to the differential synthesis of molecular enantiomers. On the contrary, these mechanisms are not intended to offer control over the chiral asymmetry of the molecular reaction, but to grant us another variable to control photocatalytic processes through the properties of incoming light. Recently, we have proposed this as a mechanism in surface photochemistry induced by CPL, to either control the evolution of chemical reactions downstream of the plasmonic excitation [175, 178] or trigger differential crystalline growth of chiral particles through their surface chemistry [151, 180]. Figure 1.14d schematically represents this idea, and stands as another example of a distinctive research direction associated with plasmonic catalysis.

1.9 Outlook

Plasmonic NPs offer a variety of ways to enhance photocatalytic processes. Their great spectral tunability and strong coupling with light make them excellent tools to either absorb light across the whole solar spectrum or to create custom spectral profiles to offer reaction selectivity. The stages in the life of a plasmon, from its initial excitation to the dissipation of its energy in the form of heat, offer different and distinct mechanisms for the transfer of its energy to the environment of the particle. This chapter has described the fundamental properties of plasmon dynamics in a metal NP and the associated energy-transfer mechanisms. Designing successful strategies for exploiting plasmonic materials in a photocatalytic context will depend on the careful design of hybrid systems that manage the flow of energy and charge from the resonant plasmonic modes to the molecular species. Having a detailed description of plasmonic dynamics and transfer mechanisms is crucial for such a design process, and we expect to see new exciting approaches and opportunities as we improve our theoretical understanding of these processes and this active field continues to develop.

Acknowledgements

We gratefully acknowledge the support provided by the following funding institutions. L.V.B, was supported by the Institute of Fundamental and Frontier Sciences, University of Electronic Science and Technology of China and the National Natural Science Foundation of China (12050410252). Z.M.W. was funded by the National Key Research and Development Program (No. 2019YFB2203400), the "111 Project" (B20030), and the UESTC Shared Research Facilities of Electromagnetic Wave and Matter Interaction (Y0301901290100201). A.O.G. was supported by the U.S.–Israel Binational Science Foundation (BSF).

References

1 Maier, S.A. (2007). *Plasmonics: Fundamentals and Applications*. New York: Springer.
2 Cai, W., Chettiar, U.K., Kildishev, A.V., and Shalaev, V.M. (2007). *Nat. Photonics* 1: 224–227.
3 Yu, N., Genevet, P., Kats, M.A. et al. (2011). *Science* 334: 333–337.
4 Khurgin, J.B. (2015). *Nat. Nanotechnol.* 10: 2–6.
5 Landy, N.I., Sajuyigbe, S., Mock, J.J. et al. (2008). *Phys. Rev. Lett.* 100: 207402.
6 Li, W. and Valentine, J.G. (2016). *Nanophotonics* 6: 177.
7 Govorov, A.O., Zhang, W., Skeini, T. et al. (2006). *Nanoscale Res. Lett.* 1: 84–90.
8 Hao, J., Zhou, L., and Qiu, M. (2011). *Phys. Rev. B* 83: 165107.
9 Wang, X., Li, G., Ding, Y., and Sun, S. (2014). *RSC Adv.* 4: 30375–30383.

10 Hogan, N.J., Urban, A.S., Ayala-Orozco, C. et al. (2014). *Nano Lett.* 14: 4640–4645.

11 Akbari, A., Tait, R.N., and Berini, P. (2010). *Opt. Express* 18: 8505.

12 Li, W. and Valentine, J. (2014). *Nano Lett.* 14: 3510–3514.

13 Kong, X.-T., Khosravi Khorashad, L., Wang, Z., and Govorov, A.O. (2018). *Nano Lett.* 18: 2001–2008.

14 Hung, W.H., Aykol, M., Valley, D. et al. (2010). *Nano Lett.* 10: 1314–1318.

15 Primo, A., Corma, A., and García, H. (2011). *Phys. Chem. Chem. Phys.* 13: 886–910.

16 Linic, S., Christopher, P., and Ingram, D.B. (2011). *Nat. Mater.* 10: 911–921.

17 Linic, S., Aslam, U., Boerigter, C., and Morabito, M. (2015). *Nat. Mater.* 14: 567–576.

18 Shaik, F., Peer, I., Jain, P.K., and Amirav, L. (2018). *Nano Lett.* 18: 4370–4376.

19 DuChene, J.S., Tagliabue, G., Welch, A.J. et al. (2018). *Nano Lett.* 18: 2545–2550.

20 Schlather, A.E., Manjavacas, A., Lauchner, A. et al. (2017). *J. Phys. Chem. Lett.* 8: 2060–2067.

21 Mascaretti, L., Dutta, A., Kment, Š. et al. (2019). *Adv. Mater.* 31: 1805513.

22 Gellé, A., Jin, T., de la Garza, L. et al. (2020). *Chem. Rev.* 120: 986–1041.

23 Besteiro, L.V., Yu, P., Wang, Z. et al. (2019). *Nano Today* 27: 120–145.

24 Fox, M.A. and Dulay, M.T. (1993). *Chem. Rev.* 93: 341–357.

25 Jiang, C., Moniz, S.J.A., Wang, A. et al. (2017). *Chem. Soc. Rev.* 46: 4645–4660.

26 Li, J. and Wu, N. (2015). *Cat. Sci. Technol.* 5: 1360–1384.

27 Kisch, H. (2017). *Acc. Chem. Res.* 50: 1002–1010.

28 Belver, C., Bedia, J., Gómez-Avilés, A. et al. (2019). *Nanoscale Materials in Water Purification*, 581–651. Elsevier.

29 Grätzel, M. (2001). *Nature* 414: 338–344.

30 Naik, G.V. and Boltasseva, A. (2010). *Phys. Status Solidi RRL - Rapid Res. Lett.* 4: 295–297.

31 Faucheaux, J.A., Stanton, A.L.D., and Jain, P.K. (2014). *J. Phys. Chem. Lett.* 5: 976–985.

32 Taliercio, T. and Biagioni, P. (2019). *Nanophotonics* 8: 949–990.

33 Chang, L., Besteiro, L.V., Sun, J. et al. (2019). *ACS Energy Lett.* 4: 2552–2568.

34 Tanner, D.B. (2019). *Optical Effects in Solids*. Cambridge: Cambridge University Press.

35 Govorov, A.O., Zhang, H., Demir, H.V., and Gun'ko, Y.K. (2014). *Nano Today* 9: 85–101.

36 Landau, L.D. and Lifshitz, E.M. (1976). *Mechanics*, 3e. Elsevier, Butterworth Heinemann: Burlington, MA.

37 Johnson, P.B. and Christy, R.W. (1972). *Phys. Rev. B* 6: 4370.

38 Wooten, F. (1972). *Optical Properties of Solids*. New York: Academic press.

39 Lindhard, J. (1954). *Dan. Mat.-Fys. Meddeleiser* 28: 1–57.

40 Mahan, G.D. (2000). *Many-Particle Physics*, 3e. Boston, MA: Springer US.

41 Ashcroft, N.W. and Mermin, N.D. (1976). *Solid State Physics*. New York: Holt, Rinehart and Winston.

42 Govorov, A.O. and Zhang, H. (2015). *J. Phys. Chem. C* 119: 6181–6194.

43 Lermé, J. (2011). *J. Phys. Chem. C* 115: 14098–14110.
44 Friedel, J. (1952). *Lond. Edinb. Dublin Philos. Mag. J. Sci.* 43: 153–189.
45 Kawabata, A. and Kubo, R. (1966). *J. Physical Soc. Japan* 21: 1765–1772.
46 Genzel, L., Martin, T.P., and Kreibig, U. (1975). *Z. Für Phys. B Condens. Matter Quanta* 21: 339–346.
47 Kreibig, U. and Genzel, L. (1985). *Surf. Sci.* 156: 678–700.
48 Link, S. and El-Sayed, M.A. (1999). *J. Phys. Chem. B* 103: 4212–4217.
49 Derkachova, A., Kolwas, K., and Demchenko, I. (2016). *Plasmonics* 11: 941–951.
50 Baida, H., Billaud, P., Marhaba, S. et al. (2009). *Nano Lett.* 9: 3463–3469.
51 Doremus, R.H. (1965). *J. Chem. Phys.* 42: 414–417.
52 Shahbazyan, T.V. (2016). *Phys. Rev. B* 94: 235431.
53 Manjavacas, A., Liu, J.G., Kulkarni, V., and Nordlander, P. (2014). *ACS Nano* 8: 7630–7638.
54 Besteiro, L.V., Kong, X.-T., Wang, Z. et al. (2017). *ACS Photonics* 4: 2759–2781.
55 Besteiro, L.V. and Govorov, A.O. (2016). *J. Phys. Chem. C* 120: 19329–19339.
56 Schuller, J.A., Barnard, E.S., Cai, W. et al. (2010). *Nat. Mater.* 9: 193–204.
57 Cortés, E., Xie, W., Cambiasso, J. et al. (2017). *Nat. Commun.* 8: 14880.
58 Negrín-Montecelo, Y., Comesaña-Hermo, M., Kong, X.-T. et al. (2018). *ChemCatChem* 10: 1561–1565.
59 Ding, W., Bachelot, R., Kostcheev, S. et al. (2010). *J. Appl. Phys.* 108: 124314.
60 Sun, C.-K., Vallée, F., Acioli, L.H. et al. (1994). *Phys. Rev. B* 50: 15337–15348.
61 Carpene, E. (2006). *Phys. Rev. B* 74: 024301.
62 Avanesian, T. and Christopher, P. (2014). *J. Phys. Chem. C* 118: 28017–28031.
63 Della Valle, G., Conforti, M., Longhi, S. et al. (2012). *Phys. Rev. B* 86: 155139.
64 Zavelani-Rossi, M., Polli, D., Kochtcheev, S. et al. (2015). *ACS Photonics* 2: 521–529.
65 Kong, X.-T., Wang, Z., and Govorov, A.O. (2017). *Adv. Opt. Mater.* 5: 1600594.
66 Pines, D. and Nozières, P. (1989). *Theory of Quantum Liquids*. Reading, MA: Addison-Wesley Pub. Co., Advanced Book Program.
67 Khurgin, J.B. (2020). *Nanophotonics* 9: 453–471.
68 Yu, S. and Jain, P.K. (2020). *Angew. Chem.* 132: 2101–2104.
69 Yu, S., Mohan, V., and Jain, P.K. (2020). *MRS Bull.* 45: 43–48.
70 Avouris, P. and Walkup, R.E. (1989). *Annu. Rev. Phys. Chem.* 40: 173–206.
71 Gadzuk, J.W. (1983). *J. Chem. Phys.* 79: 6341–6348.
72 Gadzuk, J.W. (2000). *Chem. Phys.* 251: 87–97.
73 Bonn, M. (1999). *Science* 285: 1042–1045.
74 Kale, M.J., Avanesian, T., Xin, H. et al. (2014). *Nano Lett.* 14: 5405–5412.
75 Landry, M.J., Gellé, A., Meng, B.Y. et al. (2017). *ACS Catal.* 7: 6128–6133.
76 Kumar, P.V. and Norris, D.J. (2017). *ACS Catal.* 7: 8343–8350.
77 Kim, Y., Dumett Torres, D., and Jain, P.K. (2016). *Nano Lett.* 16: 3399–3407.
78 Zhou, L., Swearer, D.F., Zhang, C. et al. (2018). *Science* 362: 69–72.
79 Gargiulo, J., Berté, R., Li, Y. et al. (2019). *Acc. Chem. Res.* 52: 2525–2535.
80 Kamarudheen, R., Castellanos, G.W., Kamp, L.P.J. et al. (2018). *ACS Nano* 12: 8447–8455.
81 Olsen, T. and Schiøtz, J. (2009). *Phys. Rev. Lett.* 103: 238301.

82 Baffou, G., Bordacchini, I., Baldi, A., and Quidant, R. (2020). *Light Sci. Appl.* 9: 108.

83 Ratchford, D.C. (2019). *ACS Nano* 13: 13610–13614.

84 Sze, S.M. and Ng, K.K. (2007). *Physics of Semiconductor Devices*, 3e. Hoboken, NJ: Wiley-Interscience.

85 Govorov, A.O., Zhang, H., and Gun'ko, Y.K. (2013). *J. Phys. Chem. C* 117: 16616–16631.

86 Knight, M.W., Sobhani, H., Nordlander, P., and Halas, N.J. (2011). *Science* 332: 702–704.

87 Tung, R.T. (2014). *Appl. Phys. Rev.* 1: 011304.

88 Naldoni, A., Guler, U., Wang, Z. et al. (2017). *Adv. Opt. Mater.* 5: 1601031.

89 Sistani, M., Bartmann, M.G., Güsken, N.A. et al. (2020). *ACS Photonics* 7: 1642–1648.

90 Fowler, R.H. (1931). *Phys. Ther. Rev.* 38: 45–56.

91 Mooney, J.M. and Silverman, J. (1985). *IEEE Trans. Electron Devices* 32: 33–39.

92 Vickers, V.E. (1971). *Appl. Optics* 10: 2190.

93 Dalal, V.L. (1971). *J. Appl. Phys.* 42: 2274–2279.

94 Brown, A.M., Sundararaman, R., Narang, P. et al. (2016). *ACS Nano* 10: 957–966.

95 Christopher, P. and Linic, S. (2010). *ChemCatChem* 2: 78–83.

96 Peng, T., Miao, J., Gao, Z. et al. (2018). *Small* 14: 1703510.

97 Wu, K., Chen, J., McBride, J.R., and Lian, T. (2015). *Science* 349: 632–635.

98 Kumar, P.V., Rossi, T.P., Marti-Dafcik, D. et al. (2019). *ACS Nano* 13: 3188–3195.

99 Kumar, P.V., Rossi, T.P., Kuisma, M. et al. (2019). *Faraday Discuss.* 214: 189–197.

100 Hövel, H., Fritz, S., Hilger, A. et al. (1993). *Phys. Rev. B* 48: 18178.

101 Foerster, B., Joplin, A., Kaefer, K. et al. (2017). *ACS Nano* 11: 2886–2893.

102 Therrien, A.J., Kale, M.J., Yuan, L. et al. (2019). *Faraday Discuss.* 214: 59–72.

103 Wei, H., Hossein Abtahi, S.M., and Vikesland, P.J. (2015). *Environ. Sci. Nano* 2: 120–135.

104 Carmeli, I., Lieberman, I., Kraversky, L. et al. (2010). *Nano Lett.* 10: 2069–2074.

105 Torimoto, T., Horibe, H., Kameyama, T. et al. (2011). *J. Phys. Chem. Lett.* 2: 2057–2062.

106 Gersten, J. and Nitzan, A. (1980). *J. Chem. Phys.* 73: 3023–3037.

107 Schatz, G.C. (1984). *Acc. Chem. Res.* 17: 370–376.

108 Munechika, K., Chen, Y., Tillack, A.F. et al. (2010). *Nano Lett.* 10: 2598–2603.

109 Okamoto, K., Niki, I., Shvartser, A. et al. (2004). *Nat. Mater.* 3: 601–605.

110 Akselrod, G.M., Argyropoulos, C., Hoang, T.B. et al. (2014). *Nat. Photonics* 8: 835–840.

111 Srinivasan, V. and Ramamurthy, S.S. (2016). *J. Phys. Chem. C* 120: 2908–2913.

112 Wu, D.M., García-Etxarri, A., Salleo, A., and Dionne, J.A. (2014). *J. Phys. Chem. Lett.* 5: 4020–4031.

113 Park, W., Lu, D., and Ahn, S. (2015). *Chem. Soc. Rev.* 44: 2940–2962.

114 Xu, Z., Quintanilla, M., Vetrone, F. et al. (2015). *Adv. Funct. Mater.* 25: 2950–2960.

115 Lieberman, I., Shemer, G., Fried, T. et al. (2008). *Angew. Chem. Int. Ed.* 47: 4855–4857.

116 Govorov, A.O. and Fan, Z. (2012). *ChemPhysChem* 13: 2551–2560.

117 Jain, P.K. and El-Sayed, M.A. (2010). *Chem. Phys. Lett.* 487: 153–164.

118 Sherry, L.J., Chang, S.-H., Schatz, G.C. et al. (2005). *Nano Lett.* 5: 2034–2038.

119 Agrawal, A., Kriegel, I., and Milliron, D.J. (2015). *J. Phys. Chem. C* 119: 6227–6238.

120 Goris, B., Guzzinati, G., Fernández-López, C. et al. (2014). *J. Phys. Chem. C* 118: 15356–15362.

121 Tsoulos, T.V., Atta, S., Lagos, M.J. et al. (2019). *Nanoscale* 11: 18662–18671.

122 Sousa-Castillo, A., Comesaña-Hermo, M., Rodríguez-González, B. et al. (2016). *J. Phys. Chem. C* 120: 11690–11699.

123 Nien, L.-W., Lin, S.-C., Chao, B.-K. et al. (2013). *J. Phys. Chem. C* 117: 25004–25011.

124 Anger, P., Bharadwaj, P., and Novotny, L. (2006). *Phys. Rev. Lett.* 96: 113002.

125 Carminati, R., Greffet, J.-J., Henkel, C., and Vigoureux, J.M. (2006). *Opt. Commun.* 261: 368–375.

126 Govorov, A.O. and Carmeli, I. (2007). *Nano Lett.* 7: 620–625.

127 Khanal, B.P., Pandey, A., Li, L. et al. (2012). *ACS Nano* 6: 3832–3840.

128 Naiki, H., Masuhara, A., Masuo, S. et al. (2013). *J. Phys. Chem. C* 117: 2455–2459.

129 Swearer, D.F., Zhao, H., Zhou, L. et al. (2016). *Proc. Natl. Acad. Sci.* 113: 8916–8920.

130 Spata, V.A. and Carter, E.A. (2018). *ACS Nano* 12: 3512–3522.

131 Li, H., Wu, H., Zhai, Y. et al. (2013). *ACS Catal.* 3: 2045–2051.

132 Li, Z. and Kurouski, D. (2020). *J. Phys. Chem. C* 124: 12850–12854.

133 Li, Z., Wang, R., and Kurouski, D. (2020). *J. Phys. Chem. Lett.* 11: 5531–5537.

134 Jackson, J.D. (2007). *Classical Electrodynamics*. Wiley.

135 Pelton, M. and Bryant, G.W. (2013). *Introduction to Metal-Nanoparticle Plasmonics*. Hoboken, NJ: Wiley; Science Wise Publishing.

136 Bohren, C.F. and Huffman, D.R. (1998). *Absorption and Scattering of Light by Small Particles*. Wiley-VCH Verlag GmbH: Weinheim, Germany.

137 Wang, H., Fu, K., Drezek, R.A., and Halas, N.J. (2006). *Appl. Phys. B* 84: 191–195.

138 Atwater, H.A. and Polman, A. (2010). *Nat. Mater.* 9: 205–213.

139 Catchpole, K.R. and Polman, A. (2008). *Appl. Phys. Lett.* 93: 191113.

140 Zhang, X., Li, X., Reish, M.E. et al. (2018). *Nano Lett.* 18: 1714–1723.

141 Li, X., Zhang, X., Everitt, H.O., and Liu, J. (2019). *Nano Lett.* 19: 1706–1711.

142 Govorov, A.O. and Richardson, H.H. (2007). *Nano Today* 2: 30–38.

143 Palermo, G., Cataldi, U., Condello, A. et al. (2018). *Nanoscale* 10: 16556–16561.

144 Keblinski, P., Cahill, D.G., Bodapati, A. et al. (2006). *J. Appl. Phys.* 100: 054305.

145 Mazzanti, A., Yang, Z., Silva, M.G. et al. (2019). *Proc. Natl. Acad. Sci.* 116: 8161–8166.

146 Baffou, G. and Quidant, R. (2013). *Laser Photonics Rev.* 7: 171–187.

147 Yu, P., Besteiro, L.V., Huang, Y. et al. (2019). *Adv. Opt. Mater.* 7: 1800995.

148 Mubeen, S., Lee, J., Singh, N. et al. (2013). *Nat. Nanotechnol.* 8: 247–251.
149 Zhang, X., Kumari, G., Heo, J., and Jain, P.K. (2018). *Nat. Commun.* 9: 3056.
150 Zhang, X., Li, X., Zhang, D. et al. (2017). *Nat. Commun.* 8: 14542.
151 Khorashad, L.K., Besteiro, L.V., Correa-Duarte, M.A. et al. (2020). *J. Am. Chem. Soc.* 142: 4193–4205.
152 Zhang, Q., Thrithamarassery Gangadharan, D., Liu, Y. et al. (2016). *J. Materiomics* 3: 33–50.
153 Yu, S., Wilson, A.J., Kumari, G. et al. (2017). *ACS Energy Lett.* 2: 2058–2070.
154 Kumaravel, V., Bartlett, J., and Pillai, S.C. (2020). *ACS Energy Lett.* 5: 486–519.
155 Thangamuthu, M., Santschi, C., and Martin, O.J.F. (2019). *Faraday Discuss.* 214: 399–415.
156 Christopher, P., Xin, H., and Linic, S. (2011). *Nat. Chem.* 3: 467–472.
157 Kochuveedu, S.T., Jang, Y.H., and Kim, D.H. (2013). *Chem. Soc. Rev.* 42: 8467.
158 Guo, S., Li, J., Dong, S., and Wang, E. (2010). *J. Phys. Chem. C* 114: 15337–15342.
159 Bera, S., Lee, J.E., Rawal, S.B., and Lee, W.I. (2016). *Appl. Catal. Environ.* 199: 55–63.
160 Wen, M., Mori, K., Kuwahara, Y., and Yamashita, H. (2017). *ACS Energy Lett.* 2: 1–7.
161 Verma, P., Kuwahara, Y., Mori, K., and Yamashita, H. (2015). *J. Mater. Chem. A* 3: 18889–18897.
162 Qin, J. and Zeng, H. (2017). *Appl. Catal. Environ.* 209: 161–173.
163 Rajender, G., Choudhury, B., and Giri, P.K. (2017). *Nanotechnology* 28: 395703.
164 Cai, G., Yu, Z., Ren, R., and Tang, D. (2018). *ACS Sens.* 3: 632–639.
165 Xu, Y., Xu, H., Yan, J. et al. (2013). *Phys. Chem. Chem. Phys.* 15: 5821.
166 Hung, K.-C., Lai, Y.-H., and Lin, T.-W. (2016). *Cat. Sci. Technol.* 6: 4020–4026.
167 Özbek, M.O. and van Santen, R.A. (2013). *Catal. Lett.* 143: 131–141.
168 Park, D. (1987). *J. Catal.* 105: 81–94.
169 Zhang, X., Ke, X., and Yao, J. (2018). *J. Mater. Chem. A* 6: 1941–1966.
170 Yu, S., Wilson, A.J., Heo, J., and Jain, P.K. (2018). *Nano Lett.* 18: 2189–2194.
171 Hentschel, M., Schäferling, M., Weiss, T. et al. (2012). *Nano Lett.* 12: 2542–2547.
172 Fan, Z. and Govorov, A.O. (2012). *Nano Lett.* 12: 3283–3289.
173 Wang, Z., Cheng, F., Winsor, T., and Liu, Y. (2016). *Nanotechnology* 27: 412001.
174 Hentschel, M., Schäferling, M., Duan, X. et al. (2017). *Sci. Adv.* 3: e1602735.
175 Wang, W., Besteiro, L.V., Liu, T. et al. (2019). *ACS Photonics* 6: 3241–3252.
176 Fan, Z. and Govorov, A.O. (2010). *Nano Lett.* 10: 2580–2587.
177 Kuzyk, A., Schreiber, R., Fan, Z. et al. (2012). *Nature* 483: 311–314.
178 Liu, T., Besteiro, L.V., Liedl, T. et al. (2019). *Nano Lett.* 19: 1395–1407.
179 Shen, X., Zhan, P., Kuzyk, A. et al. (2014). *Nanoscale* 6: 2077.
180 Saito, K. and Tatsuma, T. (2018). *Nano Lett.* 18: 3209–3212.

2

Characterization and Properties of Plasmonic-Catalytic Nanostructures from the Atomic Scale to the Reactor Scale

Briley B. Bourgeois, Dayne F. Swearer and Jennifer A. Dionne

Department of Materials Science and Engineering, Stanford University, Stanford, CA, USA

2.1 Overview

This chapter introduces a variety of techniques that have been used to characterize plasmonic photocatalytic nanostructures. We highlight both traditional approaches to characterize heterogeneous catalysts and nanophotonic structures, in order to showcase historical best practices, as well as emerging technologies that promise new understanding of plasmonic photocatalytic behavior. The field of plasmonic photocatalysis is rapidly advancing, yet still relatively early in its development. Therefore, several characterization examples provided study plasmonic nanostructures in noncatalytic environments or nonplasmonic yet highly active heterogeneous catalysts; however, we also highlight more recent reports that are investigating both strongly plasmonic and strongly catalytic nanoparticles. Our goal is to provide a reference for what has been accomplished in the field so far, as well as a roadmap for future best practices in understanding and optimizing plasmonic photocatalysts.

Historically, nanophotonics and heterogeneous catalysis evolved independently, as two distinct disciplines. Their convergence to the nascent field of plasmonic photocatalysis promises to bring light-driven chemistry into the mainstream of chemical manufacturing. Yet, advances in the development of plasmonic catalysts will crucially rely on characterization techniques that can successfully bridge these disciplines. Standard characterization of plasmonic nanostructures, developed largely by the nanophotonics community, does not provide direct information about the catalytic activity of the nanostructure. Likewise, standard measurements from heterogeneous catalysis have historically not been amenable to optical inputs for characterizing the influence that light has on nanostructure reactivity. Another catch-22 is that a good plasmonic nanostructure does not necessarily make a good catalyst, and vice versa. Decreased nanoparticle size increases the surface-to-volume ratio, increasing the catalytic activity; generally, the most catalytically active nanoparticles have dimensions below 10 nm. However, decreased nanoparticle size also decreases the ratio between optical scattering and absorption. The reduced optical scattering

Plasmonic Catalysis: From Fundamentals to Applications, First Edition.
Edited by Pedro H.C. Camargo and Emiliano Cortés.

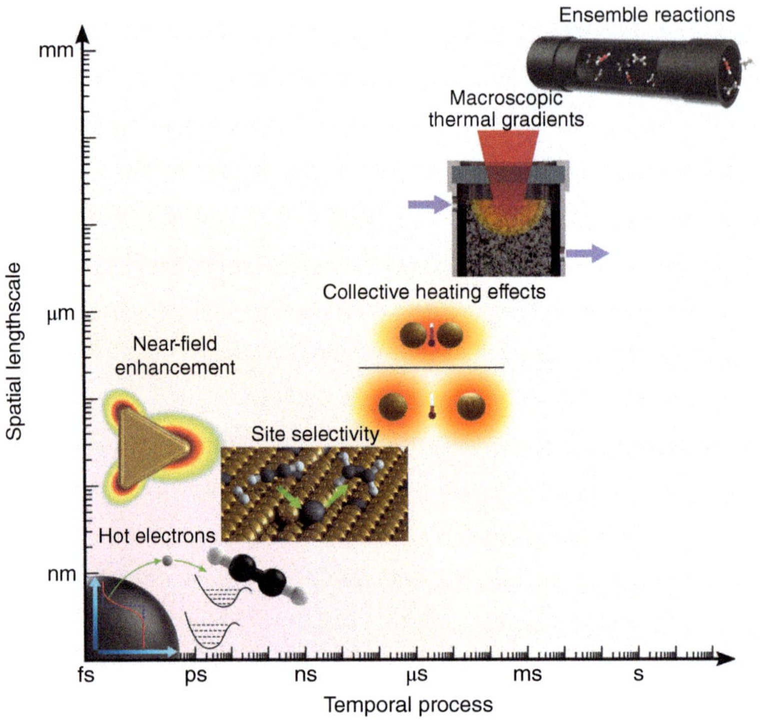

Figure 2.1 A spatiotemporal summary of plasmonic photocatalytic processes, highlighting the vast length and timescales involved, and hence the challenges in characterization.

of smaller, sub-10 nm particles challenges standard optical characterization of these plasmonic photocatalysts. Indeed, the competing effects of scattering vs. absorption have been experimentally tested but are not definitively understood [1].

Figure 2.1 summarizes some of the challenges in gaining a comprehensive grasp of plasmonic photocatalysis, showing the spatiotemporal processes that range from ultrafast femtosecond subparticle dynamics to ensemble chemical reactivity over the course of hours or days. At the single particle limit, individual plasmonic nanostructures provide ultrafast electronic, thermal, or photonic contributions to chemical activity on timescales of femtoseconds to nanoseconds [2]. Similarly, distinct atomic arrangements at nanoparticle surfaces and interfaces significantly influence reactant adsorption and reaction thermodynamics. Within nanoparticle ensembles, the interaction of nanoscale electromagnetic hot spots with regions of various chemical reactivity in plasmonic photocatalysis has been shown to yield significant enhancement to chemical reactions but is not fully understood. Studying plasmonic nanostructured systems also requires conscious thought to understand how nearest neighbor proximities influence local and collective heating [3]. When prepared in bulk, and measured within an ensemble reactor, characterizing plasmonic photocatalysts becomes increasingly complicated. In addition to standard challenges in chemical reactor design, effective integration of a light source and the three-dimensional thermal gradients induced by light absorption and scattering

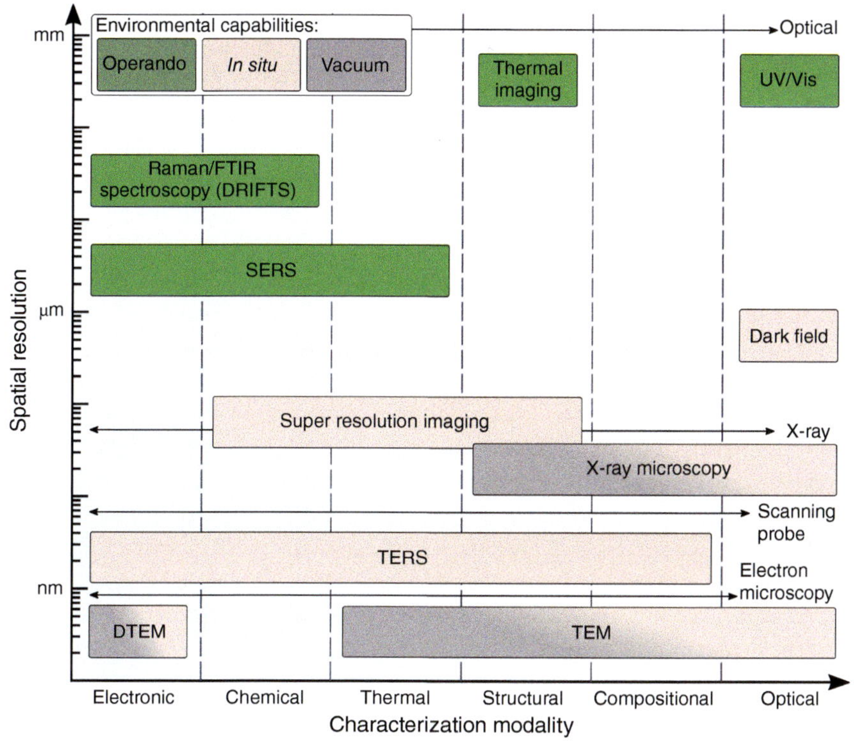

Figure 2.2 Key characterization techniques in plasmonic photocatalysis, spanning optical, x-ray, scanning probe, and electron microscopies.

dramatically influence reaction kinetics [4, 5]. Therefore, full characterization of plasmonic nanostructures requires both spatial and temporal understanding of competing factors across vastly different scales in order to engineer optimized systems.

Figure 2.2 presents an overview of the characterization techniques discussed in the remainder of this chapter along with their spatial resolution and observables. There is an immense number of characterization techniques used in material science and analytical chemistry, spanning optical, electron, x-ray, and scanning probe spectroscopies and microscopies. We highlight the techniques that have been particularly relevant to the plasmonic photocatalysis community to date and discuss emerging methodologies that could enable the field's translation from lab to industry.

2.2 Ensemble Studies and Mechanistic Mysteries

Plasmonic photocatalysis aims to enable chemical reactions by utilizing photons as an energy source rather than more traditional thermal heating. A critical characterization requirement is to observe and quantify the impact of optical illumination on a given chemical reaction. There are several reactor/chemical analysis setups that can be used to accomplish this goal. This section described the most pertinent analytical chemistry techniques utilized to study ensemble chemical reactivity by the plasmonic photocatalysis community to date and their basic operations. We will then explore the quintessential experiments performed in ensemble studies and the type of information gained from these methods. Finally, we highlight the shortcomings of ensemble measurements and discuss efforts to probe increasingly smaller length and timescales to uncover the mechanism of plasmonic photocatalysis.

As a brief introduction to analytical chemistry techniques, we begin by discussing UV/Vis monitoring of reactant decomposition as a simple and effective tool. We next introduce gas chromatography/mass spectrometry as the standard analytical tool in most recent ensemble scale plasmonic photocatalysis studies, providing substantially increased chemical and quantitative capability. After discussing the tools needed to observe a reaction, we will introduce the prominent experiments typically performed in plasmonic photocatalysis studies such as power dependence, wavelength dependence, and apparent external quantum efficiency measurement. We highlight the origins of recent controversy in these prototypical methodologies before discussing kinetic isotope effect measurements as a gold-standard technique. We conclude this section by discussing apparent activation barrier, reaction order measurements, and surface area analysis as proper steps toward incorporating more techniques from the heterogeneous catalysis community while once more highlighting the unique difficulties faced by the plasmonic photocatalysis community such as macroscopic photothermal temperature gradients.

2.2.1 Monitoring an Ensemble Reaction

In general, ensemble chemical reactions monitor the conversion of reactant chemicals to products under a given set of conditions. The first characterization task is to track this chemical transformation over the extent of the experiment. A particularly simple and effective approach to quantifying a chemical reaction is to quantify this conversion via spectroscopy. In semiconductor-based photocatalysis, where plasmonic particles have been used as sensitizers, these types of experiments are often performed to model pollutant degradation. A light absorbing dye, oftentimes methylene blue, is used as an easily observable reactant [6]. In accordance with Beer's law, the total absorbance in a solution will be proportional to the volumetric density of the absorbing molecules within the solution. As the solution volume is constant with time, the absorbance is directly related to the concentration of absorbing molecules. When the molecules are photocatalytically decomposed, their decomposed products no longer absorb light of the same wavelength. Therefore, the change in absorbed light intensity tracks the reaction dynamics and can be used as a means of observing photocatalysis. Zheng and coworkers demonstrates this behavior nicely in Figure 2.3a, showing a decrease in the 400 nm absorption peak of 4-nitrophenol over time in the presence of Pd-tipped Au nanorods [7]. They later demonstrate a faster decrease in this absorption peak when the plasmonic nanorods are irradiated with visible light. Awazu and coworkers utilized this technique to demonstrate the plasmon enhanced decomposition of methylene

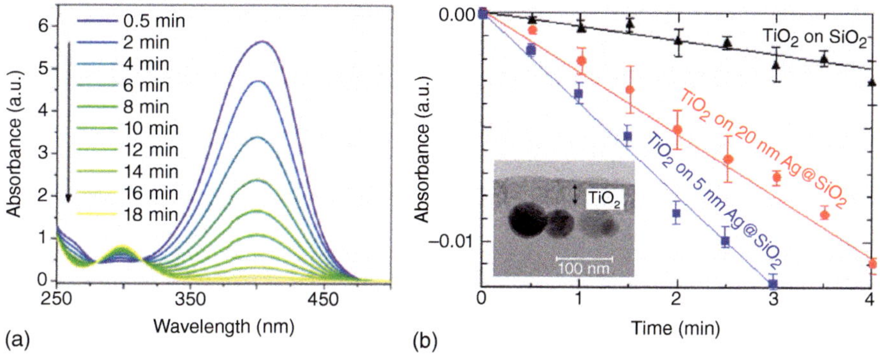

Figure 2.3 (a) A demonstration of decreasing absorption of molecules over time during a catalytic reaction. This measurement shows the absorption change during the reduction of 4-nitrophenol by $NaBH_4$ in the presence of Pd-tipped Au nanorods. The absorption peak decreases more quickly under white light illumination [7]. *Source*: Zheng et al. [7]. © 2015, American Chemical Society. (b) Example data from Awazu and coworkers showing the change in absorbance over time at a particular wavelength. As dye molecules break down, the absorbance of the reaction solution decreases, allowing more light to be collected and indicating the reaction rate of the decomposition [8]. *Source*: Reprinted with permission from Awazu et al. [8]. © 2008, American Chemical Society.

blue on a TiO_2 surface [8]. As seen in Figure 2.3b, the total absorbance of the reaction solution decreases more quickly upon the addition of plasmonic SiO_2 encapsulated Ag nanoparticles, then again as the SiO_2 shell thickness is decreased. While this technique is extremely simple and accessible, it is often limited to chemistries involving light absorbing dyes as most small molecules absorb in the UV. Historically, plasmonic photocatalysis has in part evolved from semiconductor photocatalysis [9, 10]. Early reports of plasmonic enhancement in photodecomposition utilized this type of measurement, but more advanced techniques are typically employed to probe the usefulness of plasmonic catalyst in other reactions.

A common technique for analyzing complex chemical streams is Gas Chromatography/Mass Spectrometry (GC/MS). Gas Chromatography (GC) and Mass Spectrometry (MS) are two separate but complimentary analytical tools, which can be implemented in conjunction with one another. In GC, a small portion of products from the reactor output stream are fed into a temperature-controlled column. In traditional GC, any compound that can be vaporized without decomposing can be analyzed [11]. An inert carrier gas (typically He) is used to aid flow through the column. The column is loaded with a stationary phase, which often consists of highly viscous or solid polymer possibly supported by a porous, inert solid. As the analyte travels through the column, each component chemical interacts with the stationary phase to a varying degree which can be adjusted by varying the column temperature. This molecule-specific interaction separates the analyte into physically and chemically similar components, each exiting the column at separate times. The now separate components enter a detector for quantification or identification. A wide range of GC analyzers exists, each with their own strengths and weaknesses. In MS, a small portion of the exit stream from the GC is collected, ionized, and

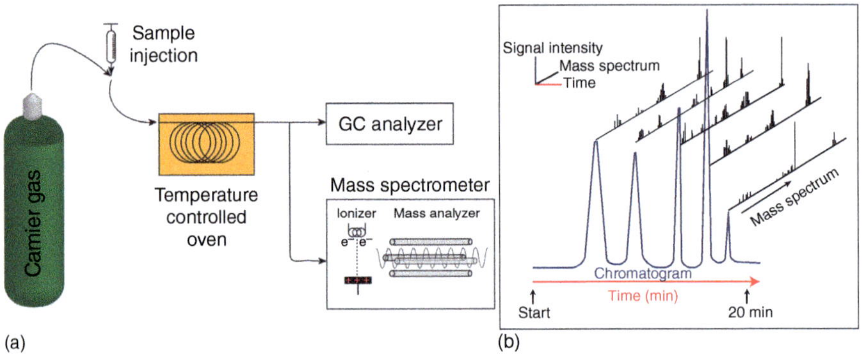

Figure 2.4 GC/MS graphical overview. (a) Schematic overview of the main components of a GC/MS system, (b) example of output data collected during GC/MS measurements.

analyzed. By ionizing the analyte (typically using an electron ionizer, though others exist), the interaction strength with a magnetic field can be determined, yielding the chemical's charge to mass ratio. This analysis is often performed by means of a quadrupole-based mass analyzer, though time-of-flight analyzers are becoming increasingly popular. This mass spectrum is then compared to a database for chemical identification. A basic schematic of a GC/MS system and an example of GC and MS data is depicted in Figure 2.4a,b, respectively. Further details of the theory and best practices of GC/MS in analytical chemistry can be found elsewhere [12, 13].

2.2.2 Ensemble Experiments

Once a system is in place to monitor the dynamics of a given reaction, there are a number of experiments that have become standard in the plasmonic photocatalysis community. In general, we are concerned with the systems response to incident light of various wavelengths and intensities. As discussed throughout the remainder of this book, researchers have applied plasmonic photocatalysis to many different reactions, demonstrating increases in selectivity and kinetics under milder reactor conditions than those found in traditional thermocatalysis [14–16]. The characterization challenge is to identify the magnitude of these reaction enhancements and the root cause which contributes most to improved performance. To this end, a great deal of controversy has surrounded the field as researchers attempt to parse apart the respective contributions of photothermal heating, electric field enhancement, and hot-electron effects [17]. Ensemble level measurements have helped to demonstrate the potential of the field while leaving much ambiguity as to the underlying mechanisms.

While previous work incorporated plasmonic nanoparticles as visible light sensitizers in semiconductor-based photocatalytic systems, it was around 2011 when photochemical reactions on the surface of purely plasmonic nanostructures began to gain prominence in the scientific literature [10, 18–20] The early work in plasmonic photocatalysis demonstrated that metal nanoparticles themselves could enhance chemical reactions, and scientists have sought to quantitatively explain these results since [18, 21, 22]. Figure 2.5 demonstrates early results from

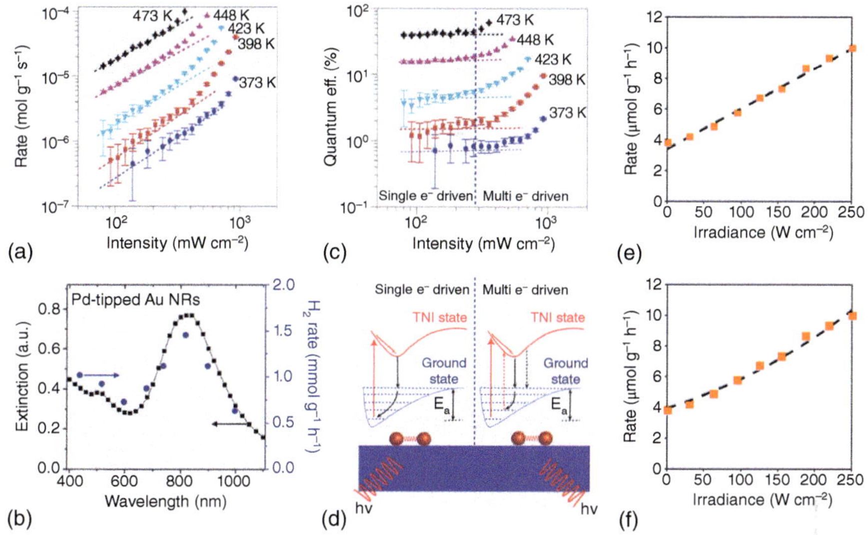

Figure 2.5 (a) Ensemble plasmonic photocatalytic rate enhancement as a function of white light intensity. The rate increases linearly with light intensity for modest intensities [21]. (b) Rate enhancement (blue dots) and extinction (black squares) as a function of wavelength showing a relationship between absorption and rate enhancement [23]. *Source*: Zheng et al. [23]. © 2015, American Chemical Society. (c) The apparent external quantum efficiency of the plasmon-enhanced epoxidation of ethylene showing constant EQE for low intensities with increased EQE beyond some threshold. (d) A schematic of the DIET and DIMET processes. Hot electrons interact with molecular adsorbates to form a transient negative ion (TNI), having an elongated average bond length. The decay of the TNI leaves the adsorbate in an elevated vibrational state which can either react more easily or be subsequently excited by a second hot electron [21]. *Source*: Christopher et al. [21]. © 2012, Springer Nature. (e) A linear fit applied to a plasmonic photocatalytic reaction and (f) a low-order exponential applied to the same data. Both show strong matching to the data [24]. *Source*: Data from Christopher et al. [18]; Baffou et al. [24].

these measurements and attempts to highlight the origins of recent controversy over the reaction enhancement mechanism. As seen in Figure 2.5a, plasmonic photocatalytic experiments often show a linear relationship (on a log-log scale) between reaction rate and increasing light intensity. These results are corroborated by the left most portion of Figure 2.5c, which shows a constant external quantum efficiency with increasing irradiance. Both of these examples are presented by Christopher et al. and demonstrate plasmonic photocatalytic ethylene epoxidation on Ag nanocubes [21]. Figure 2.5b demonstrates the wavelength dependence of plasmon-enhanced formic acid dehydrogenation on Pd-tipped Au nanorods by Zheng and coworkers [23]. By utilizing white light illumination with multiple band pass filters, the authors were able to show a relationship between increased extinction at the particles' plasmon resonance and increased catalytic activity, a relationship frequently found in reports of plasmonic photocatalysis. These results were originally interpreted as an indication of a photochemical process whereby every photon has some likelihood of assisting in a chemical reaction. Hence, the more photons (increased light intensity), the greater the rate. Curiously, deviation from linear behavior has been demonstrated in plasmonic photocatalytic reactions at higher light intensities. This behavior was originally explained with the concept of desorption induced by electronic transition (DIET), a schematic of which is shown in Figure 2.5d. The superlinear rate enhancement and increased quantum efficiencies observed at high light intensity were explained by desorption induced by multiple electronic transitions (DIMET) in which energetic adsorbates originally excited by hot electrons interact with a second hot electron while still in an elevated vibrational state. With a strong rooting in the surface science community, this mechanism has largely been accepted in the plasmonic photocatalysis community and used to explain similar ensemble scale behavior in other reports [25, 26]. However, the validity of the DIET and DIMET mechanisms as the primary contributor to plasmonic photocatalytic enhancement has received enhanced scrutiny recently [27–30].

It is well known that plasmonic excitation induces thermal heating as a result of the relaxation of excited charge carriers. This plasmonic heating should exponentially increase the reaction rate just as in traditional heterogeneous catalysis. Recently, Baffou and coworkers have pointed out that the seemingly linear reaction rate increases observed by many studies in the field could actually be a low-order exponential curve as demonstrated in Figure 2.5e using a sample of the data from Figure 2.5f [24]. For mild increases in temperature, it would not be surprising to see an apparently linear trend in reaction rate with light intensity. If we return to the EQE measurement presented in Figure 2.5c, it is unclear how one could differentiate between a multiphoton process and increasing thermal activation. In reality, thermal effects must be present in plasmonic photocatalysis, but quantifying the individual contributions of thermal- and non-thermal effects is exceedingly difficult. While there have been some experimental suggestions for differentiating thermal- and photoeffects using relatively simple, ensemble techniques [24], there is a great need for advanced characterization techniques to help distinguish the primary plasmonic photocatalytic mechanisms and guide further development of the field toward real-world applications.

In the realm of ensemble scale characterization techniques, kinetic isotope effect (KIE) measurements uniquely offer an ability for distinguishing thermal- and photoeffects in plasmonic photocatalysis. The DIET and DIMET photoelectric processes were first developed by the femtosecond surface science community as early as the 1970s [26]. According to DIET theory, a lighter isotope interacting with a hot electron will be accelerated along the reaction coordinate more than a heavier isotope. This effect is much more pronounced than any thermal differences among isotopes and has been used to identify electron-mediated chemical reactions in both the surface science and plasmonic photocatalysis communities [22, 25, 31, 32]. Figure 2.6a demonstrates the KIE experiment data from the previously discussed work from Christopher et al. [21]. The magnitude of increase in the KIE with increasing light intensity is indicative of a hot electron mechanism. For comparison, Figure 2.6b,c demonstrate KIE measurements from the surface science community. Using H_2 and

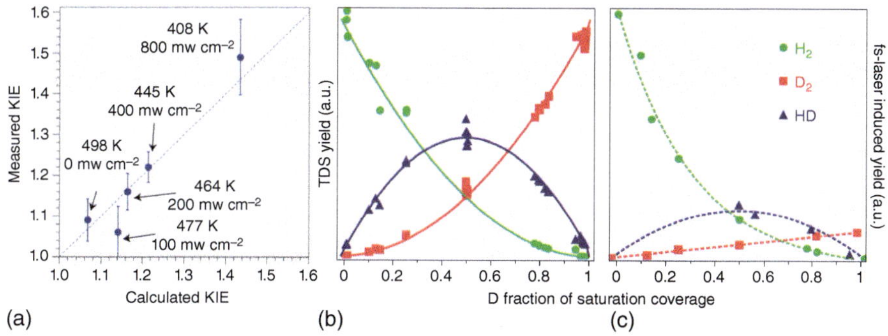

Figure 2.6 (a) KIE experiment from the previously discussed Linic's work demonstrated large enhancement with increasing light intensity [21]. *Source*: Christopher et al. [21]. © 2012, Springer Nature. (b) Thermal desorption of gases from a Ru(0001) surface as a function of D vs. H coverage. (c) The same experiment repeated using fs-laser instead of resistive heating. The results demonstrate a much larger KIE [25]. *Source*: Denzler et al. [25]. © 2003, American Physical Society.

D_2 desorption from a single crystal Ru(0001) surface, Ertl and coworkers demonstrate minimal KIE in thermal desorption (Figure 2.6b) and much higher KIE from femtosecond laser-induced desorption (Figure 2.6c) [25]. With its long-standing history in the field of surface science, KIE measurements strongly support the existence of hot electron effects in plasmonic photocatalysis, but the experiments and equipment can be expensive and inaccessible. Furthermore, KIE measurements do not necessarily quantify the relative contributions of photothermal and photoelectronic mechanisms.

2.2.3 Room for Growth in Ensemble Characterization Procedures

At this stage, we can see the overarching questions that persist in plasmonic photocatalysis and begin to understand their implications. Photothermal and excited charge carrier effects are both expected to exist within plasmonic photocatalysis, but it is critical to understand the extent to which each impacts a particular reaction and develop control over those effects. Attempts at quantifying the relative contributions of hot electrons and photothermal heating have been made using the increasingly popular apparent activation barrier measurement [33]. A chemical reaction occurring on the active site of a catalyst will have some activation barrier to transformation inherent to the system. This activation barrier determines the kinetics of the reaction in accordance with Arrhenius behavior. The activation barrier is commonly determined by fitting the reaction rate as a function of temperature in what is known as an Eyring plot. In plasmonic photocatalysis, the reaction rate is a function of temperature, light intensity, and incident wavelength. By maintaining a specific surface temperature in the reaction chamber under a given set of optical parameters, the apparent activation barrier of the reaction under those optical conditions is given. A very thorough example of this analysis is given by Zhou et al. in Figure 2.7 in which the apparent activation barrier is plotted as a function of both wavelength and intensity. This study also demonstrated that the reaction order differed under light

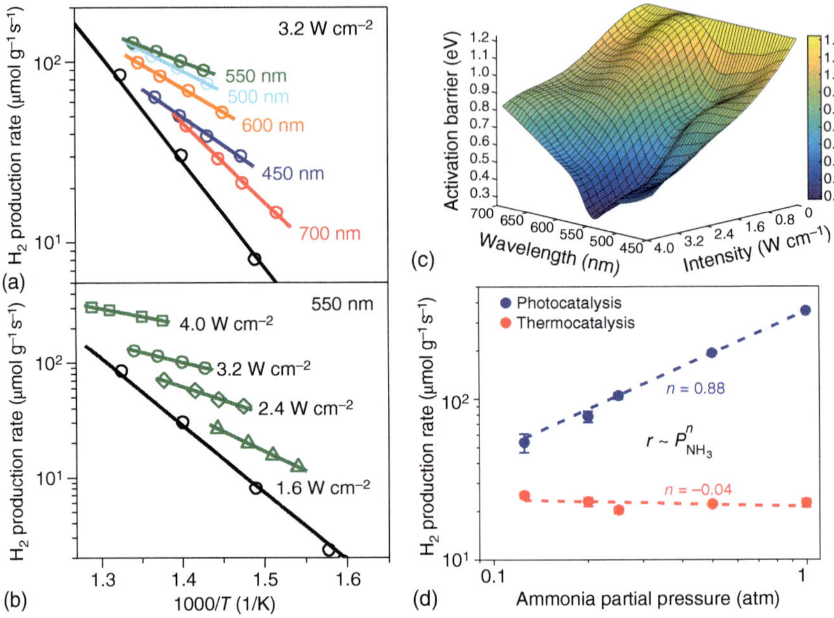

Figure 2.7 Eyring plots at various (a) wavelengths and (b) intensities. The slopes of these plots are used in the calculation of the apparent activation barrier and summarized in (c). (d) Reaction rate as a function of reactant pressure for both thermal- and photoenhanced reactions. The change in reaction order supports the hypothesis of a change in reaction mechanism between the two processes [16]. *Source*: Zhou et al. [16].

and dark conditions (Fig. 2.7d), definitive evidence of qualitative differences in the reaction mechanism depending on the type of external stimuli. In addition to attempting to develop a more mechanistic understanding, experiments such as this are a step in the right direction toward utilizing more of the methods developed in the thermocatalysis field, but it is important to consider the unique challenges faced in the photocatalysis community. In particular, this type of study relies on a clear understanding of the local temperature at the catalysts surface, which is often not well understood under a given set of illumination conditions.

Plasmonic nanoparticles are great nanoscale heat generators [3, 34]. As will be discussed shortly, the relaxation of excited charge carriers into heat can lead to inhomogeneous, nanoscale temperature distributions within nanoseconds of illumination. In addition, the photothermal heating capabilities of plasmonic nanostructures have a significant impact on the macroscopic temperature distributions within a catalyst pellet or reactor. Liu and coworkers demonstrated that the commercial Harrick *in situ* Raman reactor cell, commonly used in the community as ensemble photoreactors possess nonuniform temperature distributions under light illumination [4]. By placing thermocouples at both the top and bottom of their 3 mm catalyst pellet, they observed linear temperature gradients as high as 100 °C at high light intensities along the length of their packed bed reactor; radial temperature gradients are also possible under focused illumination. Their findings also highlighted disagreement between the set point of the device and temperatures modeled closer to the active component. Considering the effects of convective cooling, photothermal heating, and nonuniform thermal distributions is critical in understanding the true reaction conditions in an ensemble photoreaction.

Landmark studies of plasmonic photocatalysts at the macroscale have revealed a promising avenue for chemical processing using plasmonic nanoparticles, but also left behind major questions about the mechanisms of these processes. Heat generation, electric field enhancement, molecular adsorption, surface-mediated reactions, and electronic interactions are all processes which occur at small length scales and fast timescales. In the following section, we will explore advancements in the field which push toward higher spatial and temporal techniques for determining the underlying plasmonic photocatalytic mechanisms. Before transitioning, it is important to note the areas in which more ensemble testing would be beneficial. Ultimately the goal of this research is to develop a new method for synthesizing chemicals at a commercial scale with direct solar energy or renewable power, high-efficiency LEDs. As such, the activity, selectivity, and input requirements of these catalysts are of utmost concern, and this information can only be collected from ensemble scale experiments under realistic reaction conditions. However, heterogeneous catalysis and the investigation of thermocatalysts is a rich and well-developed field with a slew of standard characterization techniques [35]. The plasmonic photocatalysis community has largely underutilized these techniques, and a great deal of progress should come from incorporating traditional characterization techniques into this new field. For example, active surface area determinations using techniques such as the Brunauer–Emmett–Teller (BET) adsorption method are common place in thermocatalysis and needed for quantitatively comparing results from one experiment to another in which active site changes can influence rates in addition to optical, structural, and compositional changes in the individual catalysts themselves. In addition, thermal desorption techniques, like seen in Figure 2.6b, can give insight into adsorption behavior and determine inherent chemical interactions with catalysts. Figure 2.7d provides another example of an experiment standard to traditional thermocatalysis, the investigations of pressure dependence. Here, the authors report a change in the reaction rate order as a function of increased reactant pressure, strongly suggesting a change in the reaction mechanism rather than simply rate enhancement [16]. In particular, as the field pushes toward bimetallic catalysis, combining materials with strong optical properties and strong catalytic activity, it is necessary to quantitatively determine the chemical properties of our catalysts, such as binding energies and apparent activation barriers. As the field of plasmonic photocatalysis continues to develop, it is necessary to utilize proven and industry standard characterization techniques in order to effectively communicate our findings to the broader catalysis community.

2.3 Single/Subparticle Measurements – Toward Uncovering Mechanisms

Ensemble investigations of plasmonic photocatalysis have sparked interest in the field and demonstrated remarkable potential but have left mechanistic questions which characterize the fundamental nature of these materials unanswered. A great deal of recent research has focused on understanding the processes involved in plasmonic photochemistry. Electric field enhancement, nanoscale temperature distributions, electronic interactions, hot electron generation, preferential adsorption, and macroscale heating are all effects critical to plasmonic photochemistry but which occur across vast length and timescales. To fully understand the most important parameters for a promising plasmonic photocatalyst, it is necessary to dive into the smaller and faster characterization regimes.

In this section, we will explore prominent techniques used in characterizing both plasmonic nanostructures and plasmonic photocatalysis. We start off with optical characterization techniques such as UV/Vis spectroscopy and diffuse reflectance spectroscopies which can explore the optical properties of our catalysts and provide chemical and structural information at the diffraction limit. We next move on to dark-field techniques, which, with modern advancements, can simultaneously characterize hundreds of nanostructures at the single particle level in rapid timescales. Super resolution techniques are discussed for their ability to gather information at subwavelength spatial resolutions using purely optical techniques. Finally, we will discuss electron microscopy and recent advancements in *in situ* environmental electron microscopy for nanometer, multimodal characterization.

2.3.1 Diffraction-Limited Optical Characterization Techniques

The design of plasmonic photocatalysts has largely evolved from research in developing optically tunable plasmonic nanostructures [36]. As such, most plasmonic photocatalysts studied thus far consist of coinage metal nanomaterials of various shapes with dimensions between 10 and 150 nm. These small nanostructures interact strongly with light, allowing them to be easily seen with optical methods. However, plasmonic interactions are highly sensitive to material composition as well as the size, shape, and local dielectric environment of the nanostructures. For this reason, increasingly advanced techniques have been developed for examining plasmonic particles with ever higher spatial resolution. Thankfully, the field of plasmonic optics is well-developed and many of their optical characterization techniques can be utilized in the field of plasmonic photocatalysis [37].

In catalysis, we hope to take advantage of this strong optical interaction and use it to supply energy to chemical reactions. As such, we are primarily interested in asking how much light is absorbed or scattered by the catalyst. By determining the absorption behavior, we can quantify the hot electron generation and, ultimately, photothermal heat generation in the nanostructure. From light scattering, we hope to understand how the electric field is enhanced at the surface of the particle and how it may contribute to reactivity enhancements. The simplest method to characterize the optical behavior of plasmonic photocatalysts is to analyze the optical spectra of a bulk sample, either in a liquid dispersion or as a dried powder/film. This is commonly performed by UV/Vis spectroscopy in which a beam of white light impinges on a sample and the transmission or diffuse reflectance is subsequently collected via a photodetector. A grating is typically used to disperse the transmitted

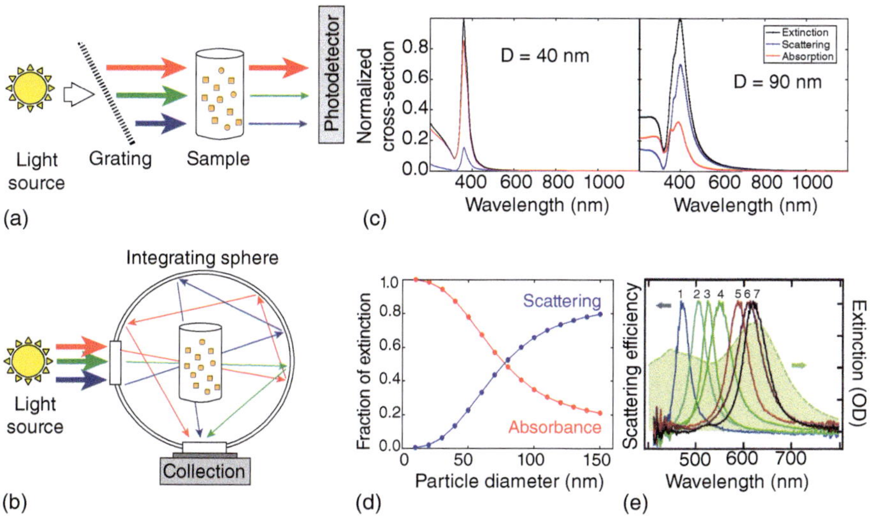

Figure 2.8 Schematic of UV/Vis spectroscopy in (a) extinction measurement mode and (b) absorption measurement mode. (c) Calculated extinction, absorption, and scattering for small and large particle plasmonic particles. (d) Calculated absorbance and scattering from Ag nanospheres as a function of edge length showing a transition from highly absorbing, small particles to highly scattering, large particles. (e) An ensemble extinction profile showing broad resonance plotted against single particle scattering spectra measured by dark-field spectroscopy. The annotations above each single particle spectra correspond to the particles depicted in the dark-field micrograph in Figure 2.10b [38]. *Source*: Based on Chen et al. [38] © 2007, American Chemical Society.

light into its component frequencies. The intensity of this transmitted light is compared to a reference beam and the difference in intensity corresponds to the wavelength-dependent extinction of the sample, demonstrating the combined absorption and scattering behavior of the sample. An example of this setup is shown in Figure 2.8a. The light removed from the beam path as it passes through the sample is proportional to the sample's extinction, quantifying the combined absorption and scattering behavior. Because the absorption, which results in hot-carrier generation and photothermal heating, and the scattering, which relates to electric field enhancement, will have different effects on the catalytic process, it is desirable to separate the relative contributions of each. By introducing an integrating sphere, as depicted in Figure 2.8b, the light scattered from the sample can be collected. Comparing the resulting spectra to that obtained from a blank sample produces the wavelength-dependent absorption properties of the material; this technique is also referred to as diffuse reflectance UV–Vis Spectroscopy (DRUVS). By acquiring a baseline DRUVS measurement on a highly reflective surface, the optical properties of a powdered sample can be determined [39]. Similar to the calculations shown in Figure 2.8c, the extinction and absorption information gained from UV/Vis can be used to determine the scattering properties of the sample.

Through UV/Vis spectroscopies, the average optical response of a sample can easily be characterized, but a wealth of information can be hidden in this measurement. For example, ensemble scale optical experiments can reveal information such as size-dependent absorption/scattering ratio demonstrated by the calculation in Figure 2.8d. In general, as the size of plasmonic nanostructures decreases, the plasmon resonance tends to blue-shift and a higher proportion of incident light is absorbed rather than scattered. While this information is useful for experimentalists, providing both an idea of hot charge carrier generation as well as ideal optical excitation spectral profiles, it does not tell the whole story. All nanomaterials syntheses have imperfections resulting in a distribution of shapes and sizes in the resulting nanostructures. Because the plasmon resonance is so closely related to the structure's geometry, this distribution of size creates a broadened optical response, as seen in Figure 2.8e. The ensemble optical response actually comprises a distribution of individual particles, each with their own, higher linewidth optical response. Furthermore, knowing the scattering behavior of an ensemble sample give no information about the nanoscale distribution of electric field enhancement, which is ultimately the information of interest for plasmonic photocatalysis [40].

Extending DRUVS to the near-infrared portion of the electromagnetic spectrum (Diffuse Reflectance Infrared Fourier Transform Spectroscopy; DRIFTS) allows for measurements of diffuse reflections from powder surfaces under *in operando* conditions. DRIFTS is an important tool for characterizing the vibrational variations of molecules at solid-state interfaces. This tool can be used to verify structural properties, such as atomic arrangements in bimetallic plasmonic photocatalysts [15] (Figure 2.9a), or understanding structural variations under dynamic reaction conditions on catalytic surfaces (Figure 2.9b) [41, 42].

2.3.2 Dark-Field Spectroscopy/Microscopy

Dark-field spectroscopy/microscopy is a technique with a long-standing history in the optics community and can be used to determine a nanostructure's optical

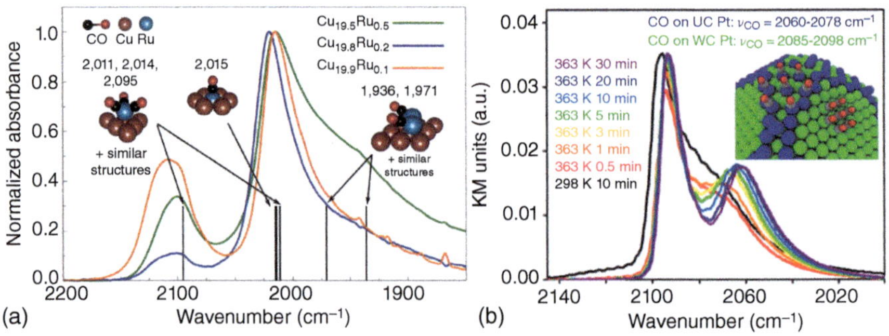

Figure 2.9 Diffuse reflectance infrared Fourier transform spectroscopy as a characterization tool for catalytic nanostructures. (a) DRIFTS of single-atom alloy Cu–Ru plasmonic photocatalysts showing the single atom dispersion as characterized by CO vibrations under different Ru loading concentrations [15]. *Source*: Zhou et al. [15]. (b) DRIFTS measurements of CO adsorption on various facets of Pt catalysts under *in operando* conditions. Structural changes induced by the gaseous atmosphere induce vibration changes of molecular adsorbates [41]. *Source*: Avanesian et al. [41] © 2017, American Chemical Society.

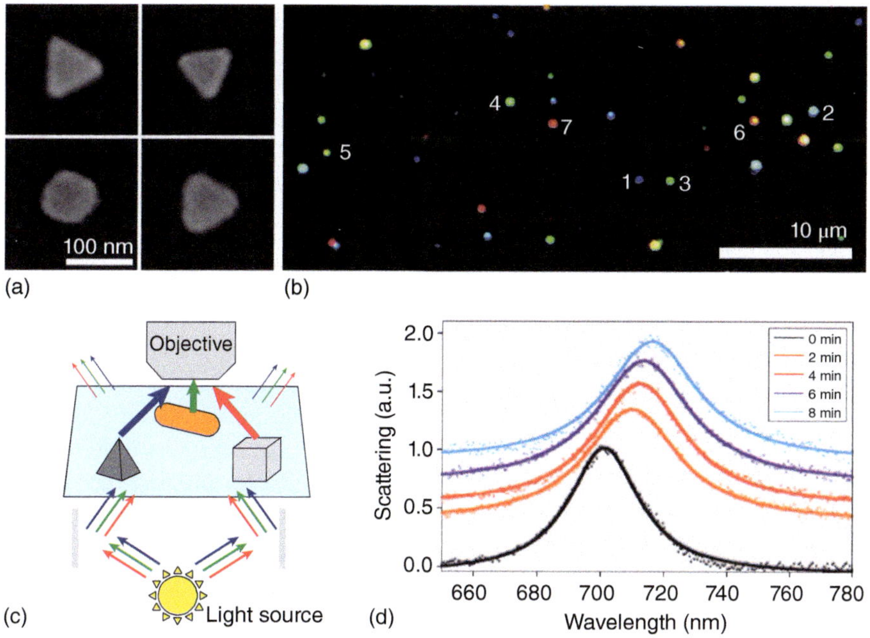

Figure 2.10 (a) Representative SEM micrographs of the sample. (b) Dark-field micrograph showing unique colors corresponding to the optical scattering efficiency of different shape and size particles. The annotated numbers correspond to the spectra located in Figure 2.8d [38]. *Source*: Reprinted with permissions from Chen et al. [38]. Copyright 2007 American Chemical Society. (c) Schematic of dark-field microscopy/spectroscopy setup. Scattered light from the nanostructures is captured by an objective lens while the transmitted light is either blocked or left uncollected. (d) Plasmon resonance of an individual Ag nanoparticle at various times under plasmon-enhanced growth. The red-shifting of the resonance feature is indicative of particle growth [20]. *Source*: Nova et al. [20].

properties at the single particle level [43]. Typically, particles are dispersed onto a planar, optically transparent substrate and irradiated with white light, and an image is captured from the scattered light, as depicted in Figure 2.10c. Chen and coworkers utilized the single particle resolution of dark-field microscopy/spectroscopy to demonstrate that fluorescence from molecules adsorbed to the particle's surface is impacted by the particle exact scattering properties [38]. Figure 2.10a,b show an image captured from the scattered field of plasmonic nanoparticles and representative SEM micrographs of the scattering particles. The various colors within the image correspond to the enhanced light scattering by different shaped nanoparticles. A spectrum can be captured from this dark-field scattering to determine the wavelength-dependent scattering. While the spatial resolution of dark-field microscopy is diffraction limited, the sample can be placed in either a transmission or scanning electron microscope to perform correlated electron microscopy. Typically, some sort of position marker is used to identify structures between instruments and both high spatial and spectral resolution data can be acquired from this combination of techniques.

In addition to characterizing plasmonic nanostructures, examples of dark-field techniques in observing plasmonic photocatalysis at the single particle level can be seen from the onset of the field. Mulvaney and coworkers were able to capture the time-dependent growth of nanoparticles in a photocatalytic Ag reduction reaction using dark-field microscopy/spectroscopy [20]. As seen in Figure 2.10d, the authors tracked the shifting, size-dependent plasmon resonance of the nanoparticles over time to demonstrate photocatalytic growth. Forrester et al. have used single particle dark-field microscopy to investigate how chemical interface damping (CID) influences the optical properties of single gold nanorods [44, 45]. Through careful analysis of full-width half-maximum (FWHM) of single particle spectra of individual gold nanorods with identical aspect ratios yet varying surface-to-volume ratios, the fraction components of plasmon decay, and most importantly the role of surface effects, were elucidated [44]. Similar experiments investigating the systematic exchange of surface ligands demonstrated how adsorbates induce metal electric dipoles which act as additional scattering centers for plasmon dephasing; no evidence for charge transfer was observed in this study [45]. Hyperspectral imaging combines the capture of spatial and spectral data using a number of acquisition schemes, such as line scanning or point collection, in which each pixel captured in space contains an associated optical spectra [46]. Recently, Landes and coworkers have applied a specific data acquisition methodology termed "Snapshot hyperspectral imaging (SHI)" to examine electrochemical oxidation of plasmonic Au nanorods [47]. Using this technique, they were able to record on the order of 100 individual particle spectra simultaneously with 20 ms time resolution, greatly improving the particle statistics which can be realistically acquired from this type of characterization. By tracking the shift in plasmon resonance in response to electrochemical oxidation, they observed the reaction dynamics of these particles under identical conditions. Their results found heterogeneity in reaction dynamics between otherwise similar nanostructures, demonstrating a need for understanding catalytic reactions with high spatial resolution.

2.3.3 Super-Resolution Microscopy – Beating the Diffraction Limit

The major disadvantage of dark-field microscopy/spectroscopy is the inability to capture spatial resolution beyond the diffraction limit. Super-resolution imaging is a technique also developed in the optics community which allows for subwavelength imaging of nanostructures and has produced exciting results in the plasmonic photocatalysis community [48–50]. The basic premise of super-resolution microscopy is to analyze light emission from fluorescent molecules adsorbed onto the surface of a nanostructure. These molecules are optically excited and stochastically relax and emit light. This light emission is distributed in time as the molecules do not relax simultaneously and the spatial distribution of the light emission from an individual molecule can be modeled using what is known as a point spread function. Fitting the light emission from individual fluorescent events to this point spread function allows one to approximate the emission origin to a region smaller

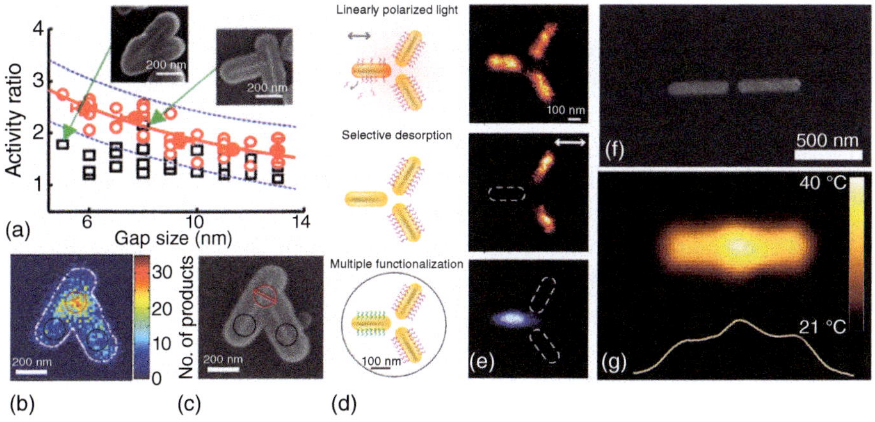

Figure 2.11 (a) Activity as a function of rod spacing. The black squares indicate the raw date. The red circles take into account a geometry factor that originates from the orientation of the two rods against each other. (b) Super-resolution image and (c) SEM micrograph of SiO_2 encapsulated gold nanobars [52]. *Source*: Reprinted with permissions from Zou et al. [52]. Copyright 2018 American Chemical Society. (d) Schematic of selective desorption and functionalization on plasmonic nanostructure. (e) Super-resolution map showing complete functionalization prior to illumination, removal of molecular adsorbate after illumination, and successful refunctionalization [53]. *Source*: Reprinted with permissions from Simoncelli et al. [53]. Copyright 2018 American Chemical Society. (f) SEM and (g) heat source density map of a gold dimer structure captured with via fluorescence polarization anisotropy [54]. *Source*: Reprinted with permissions from Baffou et al. [54]. © 2010, American Physical Society.

than the wavelength of light. The interested reader can find a detailed explanation of this process from a number of reviews [48, 51].

With super-resolution microscopy, plasmonic reaction enhancement can now be observed at subparticle resolutions. Rather than determining the optimal geometry for catalysts, super-resolution allows researchers to determine the most active regions of an individual plasmonic photocatalyst. An excellent example of this is demonstrated in Figure 2.11a–c, in which Chen and coworkers demonstrate electric field enhancement producing increased photocatalytic activity [52]. The authors use Au nanorods coated in a silica shell and record the catalytic activity by analyzing an increase of fluorescent events as resorufin is photocatalytically converted to highly fluorescent resazurin. By analyzing particles of different geometries, particularly those with different separation between the Au nanorods, the authors are able to demonstrate that the catalytic activity increases as a function of nanorod spacing, consequently changing the electric field enhancement within the gap between the particles. Another great example of the use of super-resolution microscopy in characterizing plasmonic chemical reactions is presented by Simoncelli and coworkers, as seen in Figure 2.11d,e [53]. Au nanostructures supporting polarization-dependent plasmonic resonances are coated with a fluorescent molecule. Polarized light is projected at the structure and the fluorescent molecules selectively desorb from the nanorod supporting a resonance corresponding to the incident light. The now bare nanorod is exposed to a second fluorescent molecule which adheres to the surface, effectively functionalizing the structure with nanoscale resolution. The major downside of super-resolution microscopy is the necessity for fluorescent molecules. Another interesting use of fluorescent molecules in subwavelength imaging is demonstrated by Baffou and coworkers in using molecular fluorescence polarization anisotropy for thermal microscopy [54]. The working principle of this technique is that the orientation-dependent fluorescent response of molecules in response to a polarized excitation will vary as a function of temperature. By analyzing the polarization of the fluorescent radiation, the temperature can be measured with spatial resolutions of approximately 300 nm (Figure 2.11f,g). The same authors have also written excellent reviews on nanoscale thermal mapping [3, 34, 54, 55]. While fluorescence-based, super-resolution techniques provide great potential for understanding plasmonic catalytic mechanisms at relevant length scales, directly translating these results to industrially relevant chemical reactions is not trivial.

2.3.4 Electron Microscopy

Electron microscopy offers subwavelength image resolution without the need for fluorescent tags, as well as a suite of complimentary spectroscopic tools. As discussed previously, correlated electron microscopy has often been used in tandem with dark-field spectroscopy/microscopy to obtain both high spatial resolution structural information and high spectral resolution optical behavior. Within the field of optics, it is usually sufficient to utilize scanning electron microscopy (SEM) for such correlated optical property studies. SEM uses a lower accelerating voltage (typically 0.5 kV –30 kV) relative to transmission electron microscopy (TEM), and images nanostructures via electron scattering. The key advantages of SEM are that nanoscale features, particularly sample topography, can be resolved with limited electron beam energy and without the need for electron transparent substrates. Conversely, TEM can be used to obtain higher resolution imaging. In TEM, higher accelerating voltages (typically 80–300 kV) are used and the transmitted electrons are collected. TEM is often the microscopy of choice in heterogeneous catalysis, where nanoparticles are typically in the sub-10 nm regime. Furthermore, TEM imaging can resolve atomic scale details of the catalysts, allowing correlation between optical properties and atomic structure, not only nanostructure. These atomic details are relevant in plasmonic optics but critical in catalysis where the most catalytically active sites are typically individual atoms, defects, or step edges on the surface of the particles [15, 56].

The advantages of electron microscopy in plasmonic photocatalysis extend far beyond high resolution imaging. Selected area electron diffraction (SAED) can be used to determine crystal structure within individual nanoparticles in conjunction with atomic resolution imaging. Inelastic scattering of the electron beam with atom-bound electrons in the sample produce x-ray radiation. These x-rays can be collected via energy dispersive spectroscopy (EDS) and used to determine the elemental composition of a sample. By focusing the electron beam to a small point, as is done in scanning TEM (STEM), these characteristic x-rays can be correlated to

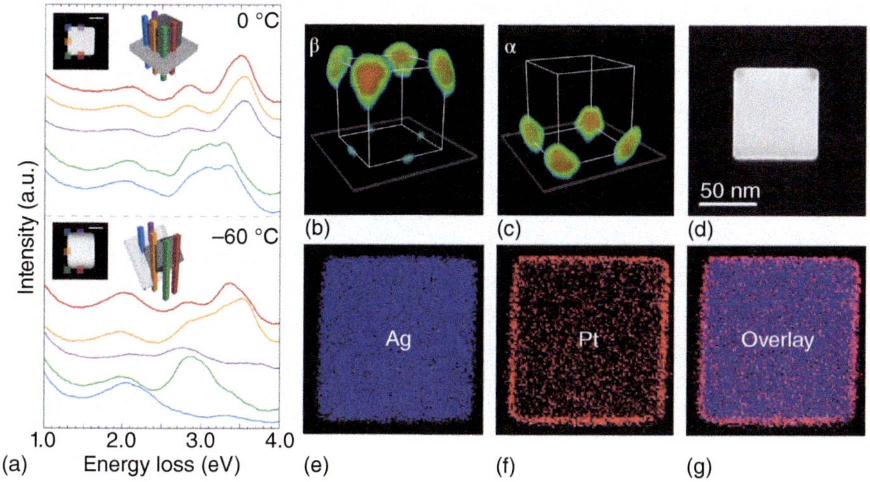

Figure 2.12 (a) Representative EELS point spectra for the electron beam located at the indicated position at 0 and 60° tilt. The inset provides a schematic of tomographic EELS spectra acquisition. (b) Proximal and (c) distal corner plasmon resonances of an Ag nanocube on a silicon nitride substrate mapped using tomographic STEM/EELS [57]. (c) STEM and (d) STEM/EDS maps of bimetallic AgPt nanocubes showing a clear core/shell structure [58]. *Source*: Reproduced with permissions from Nicoletti et al. [57] © 2013, Springer Nature.

highly localized regions in the sample, providing the ability to map out composition gradients within individual nanoparticles. An example of this is shown by Aslam et al. in Figure 2.12d–g in which the authors perform STEM-EDS to show the core–shell structure of a bimetallic plasmonic photocatalyst [58].

Electron energy loss spectroscopy (EELS) is a complimentary technique to EDS, which can be used not only for compositional analysis but has also recently been developed for understanding the nanoscale optical properties of plasmonic photocatalysts [59–61]. While EDS captures the emitted x-rays that result from an inelastic interaction with the electron beam, EELS directly analyzes the change in electron beam energy after passing through the sample. In addition to interacting with bound electrons, it is also possible for the electron beam to couple with the plasmon modes of a nanoparticle [62]. This interaction has been utilized for directly mapping the plasmon resonances of nanoparticles and is an excellent technique for drawing direct comparison between particle geometry and near-field optical properties. An excellent review of plasmon mapping with EELS spectroscopy was written by Camden and coworkers [59]. Figure 2.12b,c show examples of EELS plasmon maps taken by Nicoletti et al. In this specific example, the researchers also make use of a tomography holder, which has the capability of rotating the TEM sample to extreme angles with respect to the electron beam [57]. The authors were able to combine EELS plasmon mapping with electron tomography to capture a nanometer resolution, three-dimensional understanding of the plasmon modes of Ag nanocubes, the same system used in the ensemble scale reactions for ethylene epoxidation by the Linic group presented earlier in this chapter [22].

While electron microscopy is a great tool for characterizing the optical and structural properties of plasmonic catalysts, the technique is at a major disadvantage for characterizing reaction dynamics. Electron beam scattering from molecules is weak and adsorbates cannot be directly imaged with electron microscopy. Furthermore, electron beam effects are unavoidable, and the nature of the electron microscope requires high vacuum conditions. However, recent advances in *in situ* electron microscopy are continuously pushing the capabilities of the technique for characterizing complex systems. Figure 2.13a–e demonstrate two examples of the use of environmental TEM in the *in situ* characterization of plasmonic photocatalysis. Dionne and coworkers analyzed the hydrogenation of Pd nanocube reactors within the near field of Au nanoantennas [63]. The hydrogenation reaction chosen as the phase transformation of Pd to PdH provides an observable which can be imaged directly with the TEM. The authors found significant enhancements in the photocatalytic rate in response to plasmonic activation as well as evidence of spatial selectivity in response to the localized EM field. Sharma and coworkers similarly utilized *in situ* ETEM to characterize carbon monoxide decomposition on the surface of plasmonic Au nanoparticles [64]. The authors utilized deposition of amorphous carbon as a product of the CO decomposition as an observable for the surface reaction. By using EELS, the authors could quantify the reaction product on the surface of the particle after exciting the nanoparticle's plasmon resonance with the electron beam. This work also revealed spatial selectivity of the reaction again corresponding to the highest EM field enhancement on the particle, and importantly demonstrates the possibility of expanding *in situ* ETEM characterization to gas phase, industrially relevant chemical reactions. Figure 2.13f–h demonstrates plasmon energy expansion thermometry (PEET) as described by Mecklenburg and coworkers [65]. In PEET, the temperature-dependent bulk plasmon of a material can be analyzed using STEM EELS. By tracking the shift in the plasmon resonance, the temperature can be calculated with nanometer scale resolution.

2.3.5 A Note on Computational Tools

Using advanced optical techniques or electron microscopy, the optical properties of plasmonic photocatalysts can be experimentally determined with great accuracy. However, for predictive material design and theoretical understanding of the influences that environment has on nanostructure optics, it is common practice to compare experimental results with computational simulations. Mie theory, boundary element method (BEM), finite element method (FEM), and finite difference time domain (FDTD) methods are commonly deployed computational techniques for determining the optical behavior of plasmonic nanoparticles [66]. Similarly, calculations in quantum chemistry, namely density function theory (DFT) and its related variations, have been implemented widely to understand electronic effects in chemical reactions [67]. However, given the many-bodied physics and excited state dynamics that characterize plasmonic photocatalysis, current theories may not accurately describe the true phenomena, despite advances in theories such as embedded correlated wave function methods that more quantitatively describe

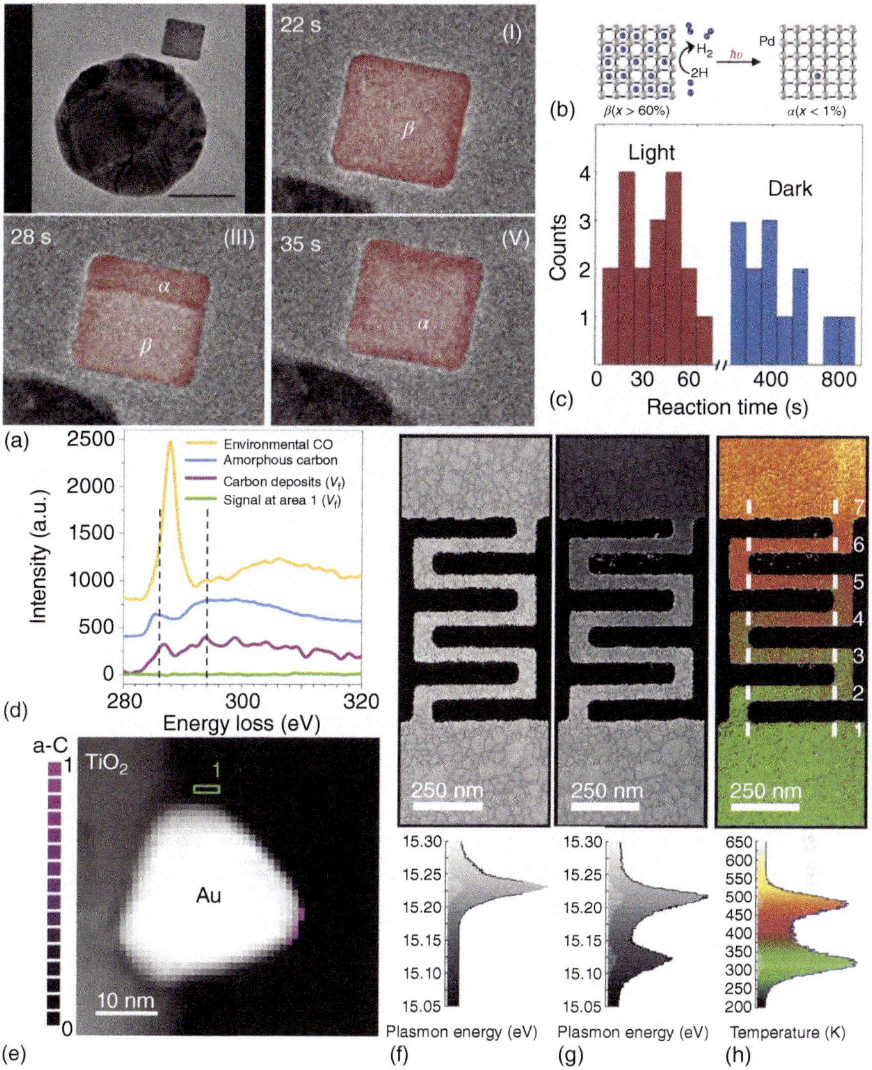

Figure 2.13 (a) TEM micrograph of Pd nanocube reactor adjacent to Au plasmonic antenna and frames from *in situ* movie of plasmon-assisted dehydrogenation of β phase PdH to α phase Pd. (b) Schematic of β to α phase transformation. (c) Reaction times for multiple single particles to transition from β to α phase [63]. *Source*: Reprinted with permissions from Vadai et al. [63]. Licensed under CC-BY-4.0. (d) EELS spectra for environmental CO signal in gaseous ETEM environment, am-C reference, C signature on particle surface, and background from vacuum. *Source*: Reprinted with permissions from Yang et al. [64]. © 2019, Springer Nature. (e) EELS mapping demonstrating spatial position of a-C deposition. Plasmon energy expansion spectroscopy (PEET), allows for nanometer resolution temperature mapping [64]. (f) The bulk plasmon map at room temperature. (g) The same region when 2 mW of heat is applied to the top surface of the wire by a heater outside the field of view. The EELS spectrum shows the maximum decrease in the bulk plasmon resonance with rising temperature. (h) False colored to demonstrate temperature calculations from these EELS measurements. *Source*: Reprinted with permissions from Mecklenburg et al. [65]. © 2015, American Association for the Advancement of Science.

excited state dynamics during molecule–metal interactions [68–70]. Many of the most well-developed theories have been reviewed thoroughly elsewhere, and we urge experimentalists interested in gaining full understanding of their plasmonic catalysts to integrate optical and catalytic simulation into future investigation.

2.4 Ultrafast Spectroscopy and Emerging Techniques – A Promising Future

The major challenge associated with characterizing plasmonic photocatalysts are the multiscale considerations of understanding these materials and their photochemical behavior. As discussed in the previous section, characterization techniques that can accurately provide insight with high spatial resolution are critical to understanding both the optical and chemical properties of plasmonic nanostructures. Unfortunately, these techniques rarely reveal the ultrafast temporal components of catalytic processes effected by the excitation and dephasing of plasmon resonances. To further understand the true influence of electronic and thermal processes in plasmonic photocatalysis, measurements must be made on the timescale of the plasmon-induced processes themselves.

The well-established ultrafast community is poised to make a significant impact in plasmonic photocatalysis research in the near future. We begin our discussion of advanced and emerging techniques by discussing ultrafast transient absorption spectroscopy for understanding electronic transitions on relevant timescales. Plasmons are particularly well equipped for this type of investigation due to the well-known surface enhanced Raman scattering effect. We discuss recent experiments that utilize this behavior to obtain high signal-to-noise ultrafast Raman spectroscopy data. Tip-enhanced Raman scattering represents a less utilized but extremely promising technique. By incorporating atomic force microscopy type tips with plasmon-induced near-field enhancements, TERS can provide chemical characterization at extremely precise spatial resolutions. Finally, advanced x-ray spectroscopies and uncommon ultrafast electron microscopes present incredible opportunities for achieving multimodal characterization at leading edge temporal and spatial resolutions.

2.4.1 Ultrafast Spectroscopy and Surface-Enhanced Raman Scattering

Observations of processes driven by hot carriers are most often studied using ultrafast laser-based pump/probe experiments. Of these, transient absorption spectroscopy is the most common form of pump/probe experiments utilized for looking at dynamic changes in reactivity with high temporal resolution. Here, the absorbance of a system is measured as a function of time and wavelength with respect to the initial laser pump excitation using an additional, time-delayed probe laser. Transient absorption measurements are highly sensitive to laser repetition rate, emission wavelength, pulse duration, polarization, intensity, sample concentration, and sample chemistry to name a few. For example, ultrafast transient absorption spectroscopy has been performed with a femtosecond visible pump and an infrared probe to measure electron transfer dynamics from 10 nm gold nanodots into TiO_2 as quickly as 50 fs [71, 72]. Similar ultrafast transient absorption measurements are critical characterization tools for understanding the temporal dynamics of chemical reactions driven by plasmonic photocatalysts.

Ultrafast surface-enhanced Raman scattering is another powerful tool that takes advantage of the high electromagnetic field enhancement of plasmonic nanostructures to directly probe molecule–metal interactions on plasmonic catalysts. Frontiera and coworkers have demonstrated several examples of how these techniques can be used to probe plasmon-induced chemistry. By studying the plasmon-induced dimerization of 4-nitrobenzenethiol to *p,p'*- dimercaptoazobenzene on electrostatically aggregated clusters of Ag nanoparticles with average diameters of 120 ± 20 nm, they revealed that plasmon-induced chemistry occurs with the highest rate at electromagnetic hot spots that show strong photoluminescence [73]. The observation of photoluminescence at regions of significant chemical conversion supports the hypothesis that a significant population of hot carriers exists on the Ag nanoparticle surface. These results suggest that reactivity is significantly influenced by hot electron transfer. Similarly, Frontiera and coworkers have used ultrafast nanoscale Raman thermometry to distinguish between electronic and thermal effects within electromagnetic hot spots [74] and the effect that oxide supports have on plasmonic heating [75]. By measuring the differences in Stokes and anti-Stokes vibrational energy deposition for coupled molecule–plasmonic systems (Figure 2.14a), Keller et al. were able to quantitatively measure an effective temperature increase less than 100 K with peak photon fluxes 10^8 times higher than solar fluxes, with energy dissipation to the local environment within 5 ps [74]. The relatively small temperature increases experienced under high photon fluxes in this study demonstrates that contributions to plasmonic photocatalysis are highly likely to be electronic in addition to thermal in nature.

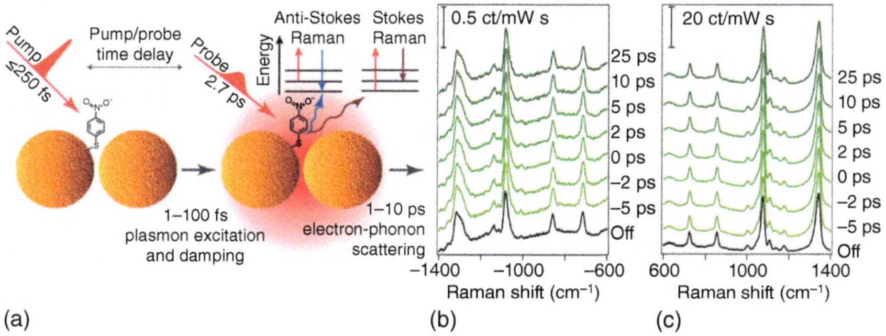

Figure 2.14 (a) Schematic of ultrafast spectroscopy vs. surface-enhanced Raman Scattering in the optical hot spot between two plasmonic particles. (b) Anti-Stokes and Stokes Raman scattering spectra allow for temperature probing with ps time resolution [74]. *Source*: Keller and Frontiera [74] © 2018, American Chemical Society.

Other multipulse ultrafast optical spectroscopy techniques are also possible and applicable to characterizing plasmonic photocatalysts. Halas and coworkers demonstrated a three-beam ultrafast process, surface-enhanced coherent anti-Stokes Raman scattering (SECARS), using an Au quadrumer nanostructure capable of detecting single molecules with Raman cross-sections as low as 10^{-30} cm^2 sr^{-1} using a bianalyte method [76]. Using lithographically fabricated substrates and exploiting Fano resonance effects, extremely high detection of molecular analytes within the hot spots were possible; however, these measurements were not made with high temporal resolution. Combining ultrafast spectroscopy with tailored plasmonic nanostructures, studying single-molecule reaction dynamics may be possible in the future.

2.4.2 Tip-Enhanced Raman Spectroscopy

Another approach that can yield both high spatial and temporal resolution on plasmonic catalysts is tip-enhanced Raman spectroscopy (TERS) [77–79]. TERS is similar to other scanning probe techniques such as atomic force microscopy (AFM) and scanning tunneling microscopy (STM) in the sense that a sharp tip (the "probe") is brought into close proximity with a substrate of interest; however, in TERS the tip and substrate are made of materials with strong electromagnetic interactions such that an electron magnetic hot spot is formed within the junction (Figure 2.15a). This technique has been demonstrated to reveal vibrational dynamics with a spatial resolution on the order of single molecules (Figure 2.15c) [81]. Molecules within the tip–substrate junction are then subject to the strong electromagnetic field enhancement and the Raman scattering signal is collected.

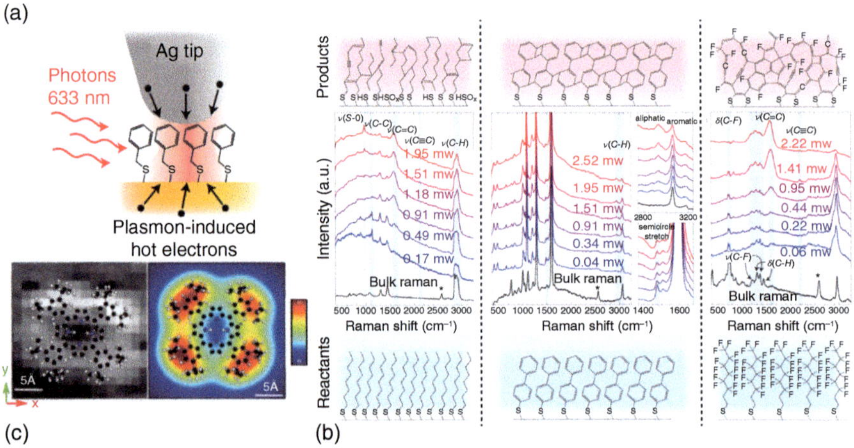

Figure 2.15 (a) Schematic of TERS experimental setup. (b) TERS signal from different SAMs after light illumination [80]. *Source*: Szczerbiński et al. [80] © 2018, American Chemical Society. (c) TERS mapping of vibrational signal from an individual molecule [81] *Source*: Reprinted with permissions from Chen et al. [81]. Licensed under CC BY 4.0.

Szczerbínski et al. have used TERS to study the mechanism of reactions within single hot spots [80]. In this report, a Ag tip was scanned over a substrate of aggregated Au nanoparticles with self-assembled monolayers (SAMs) of thiolates with different desorption temperatures and electron capture cross-sections. Changes to the Stokes scattering spectra of different SAMs primarily revealed electron-induced cross-linking included Raman signatures which have previously been reported in x-ray and electron-beam surface chemistry (Figure 2.15b). An important conclusion from this study is that plasmonic processes are likely a part of a larger family of electronic processes that uniquely have desorption induced by electron transitions (DIET) mechanisms in common (Figure 2.16). These processes have been studied for decades in the surface science literature, and therefore will be an important point of reference for future work trying to understand the unique chemistry that plasmonic photocatalysts may contribute to [82]. While TERS may be an important characterization tool for studying catalytic reactions at individual hot spots, this technique has not yet been combined with ultrafast temporal measurements. Several challenges may need to be overcome, such as minimizing the formation of amorphous carbon in the hot spots, but combining TERS with high temporal measurements could likely yield important information about dynamic catalytic processes and where they occur under illumination [78].

While femtosecond transient absorption spectroscopy and ultrafast Raman scattering can routinely be used to access temporal information for ensemble measurements, measuring electron dynamics in single nanostructures is complicated due to pulse broadening in microscope objectives. Link and coworkers have described a combined photoluminescence spectroscopy and dark-field microscopy approach to quantitatively measure electron transfer dynamics from gold between gold nanorods and graphene without an applied bias in the frequency domain [83]. Since this technique relies only on optical microscopies, the unpredictable influence of a

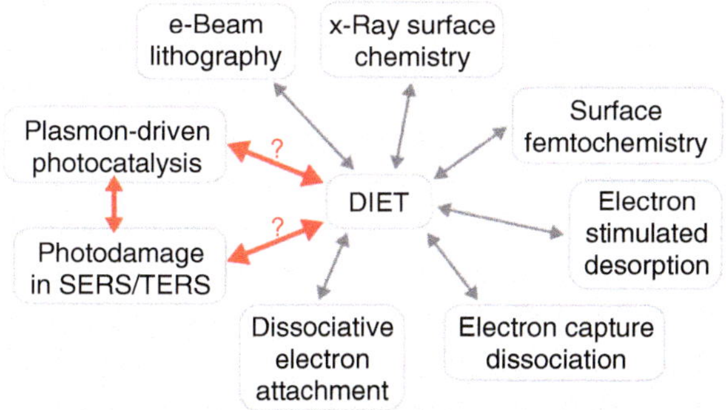

Figure 2.16 Summary of chemical processes known to under desorption by electronic transitions (DIET) mechanisms. Plasmon-driven catalysis and photodamage frequently found during SERS/TERS measurements are part of a larger family of metal surface electron phenomena [80]. *Source*: Szczerbiński et al. [80] © 2018, American Chemical Society.

nearby plasmonic tip and the formation of amorphous carbon is eliminated. By analyzing linewidth broadening of individual CTAB-coated Au nanorods on quartz and graphene, they were able to determine an average electron transfer time of 160 ± 30 fs from changes in the resonance peak width as small as 10 meV. The authors showed that this electron transfer process occurred with 10% efficiency, despite the CTAB monolayer encasing the nanorods. This ultrafast approach to quantitative single particle measurements of electron transfer dynamics may be utilized to optimize redox reactions occurring at the metal–molecule interface in plasmonic photocatalysts.

2.4.3 X-Ray and Ultrafast Electron Microscopy

While femtosecond laser pulses have significant utility in understanding electron transfer dynamics, advances in free electron laser sources and tabletop high harmonic generation open the doors to plasmon-induced chemistry with high temporal resolution and elemental and oxidation state specificity. To date, there have only been a few reports on extreme ultraviolet (XUV) reflection–absorption (RA) spectroscopy techniques as an interfacial probe [84]. However, despite the fact that there are currently no reports of XUV–RA spectroscopy probing plasmonic photocatalysis, it may be an important tool for future work that could reveal the ultrafast electron dynamics between molecules and plasmonic interfaces.

Recent advances in the development of dynamic transmission electron microscopy (DTEM) and ultrafast electron microscopy (UEM) hold promise for characterizing plasmonic photocatalysts with extremely high spatiotemporal resolution. Here, conventional electron microscopes are coupled to ultrafast UV laser sources which are focused onto the electron source. Short pulses of UV irradiation eject photoelectrons which are manipulated prior to interaction with the sample to form high resolution images and reveal important structural dynamics in solid state systems. In one of the first examples, UEM has been used for real-space imaging of nanoscale intra- and interparticle acoustic-phonon dynamics in single crystalline Au nanorods [85]. Flannigan and coworkers used a bright-field TEM imaging approach which is highly sensitive to slight changes in intraparticle strain-fields which can be visualized as diffraction contrast. Oscillations in diffraction contrast following the first 50 ps of photoexcitation revealed coherent acoustic phonons; a relevant phenomenon following plasmon excitation and decay through electron–electron and electron–lattice scattering that results in thermal effects in plasmonic photocatalysis. The authors subsequently demonstrated spatiotemporal mapping of intraparticle phonon modes and their vibrational dynamics (Figure 2.17). The utilization of UEM for real-space visualization of dynamic plasmonic processes in Au nanorods is an important tool to probe the fundamental properties of energy evolution and transfer in photocatalytic systems. While this is one of the first demonstrations of UEM being used as a tool for high spatiotemporal measurements for plasmonic nanostructures, there are numerous examples of fruitful research in this nascent field.

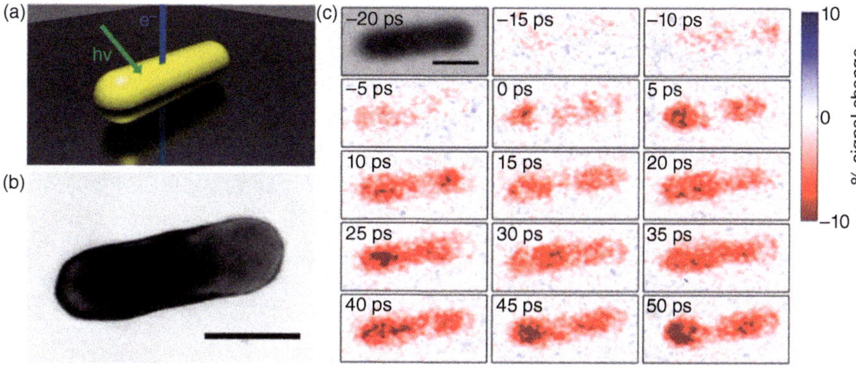

Figure 2.17 Ultrafast Electron Microscopy (UEM) can reveal acoustic-phonon dynamics in gold nanorods. (a) A simplified schematic of the experimental setup showing optical excitation (green) and electron probe (blue). (b) Bright-field TEM image of a gold nanorod; scale bar: 50 nm. (c) Intraparticle diffraction-contrast dynamics revealing changes to Au nanorod structure following pump excitation at 1030 [85]. *Source*: Reprinted with permissions from Valley et al. [85] © 2016, American Chemical Society.

2.5 Outlook

This chapter has summarized several key classical and emerging characterization techniques that are commonly used to understand plasmonic photocatalysis. The scientific challenges associated with this field arise from the numerous length and timescales that must be accounted for when studying plasmonic photocatalysts. The ultrafast timescales of plasmon excitation and decay lead to dynamic electronic, phononic, and chemical processes. The nanoscale size requirements of optically tunable catalysts test the limits of spatial resolution for most spectroscopic measurements. In addition, when plasmonic photocatalysts are utilized in photochemical reactors, competing factors such as scattering and absorption, electronic and thermal effects, local and collective heating, and temperature gradients all complicate the characterization and quantitative understanding of reactions driven by plasmonic photocatalysts.

We hope this overview serves to guide future researchers en route to mechanistic understanding and optimization of the multidimensional factors governing plasmon photocatalysts. A recent example from the literature highlights the promise and potential of the field: small, highly absorptive plasmonic photocatalysts based on oxide supported Cu—Ru surface alloys have been shown to be incredibly potent photocatalysts, challenging the status quo for plasmonic photocatalysis design [15, 16]. Their small size (~5–10 nm) and alloyed composition grown directly onto a mixed oxide support (0.05–3 μm) make many of the characterization tools discussed in this chapter incredibly challenging, if not physically impossible (though we use the word impossible with the greatest caution!). Their highly active catalytic surface area makes them effective catalysts but could limit their optical tunability. Still, they may have as-of-yet undiscovered electronic and optical properties that control their record efficiency. Therefore, they may represent a paradigm shift in the development of highly efficient, catalytically tunable plasmonic nanostructures for integration into the chemical industry at large. Future development of technologies to characterize plasmonic catalysts should no doubt focus on small, highly absorptive, multicomponent nanostructures. Capabilities to operate techniques under environmental or in operando conditions is also an important consideration to gain true insight into the dynamic chemical processes that occur at and within plasmonic photocatalysts.

Acknowledgments

D.F.S. gratefully acknowledges the Arnold O. and Mabel Beckman Foundation for their support with a postdoctoral fellowship in the chemical sciences.

This material is based upon work supported by the National Science Foundation Graduate Research Fellowship under Grant No. (2020291039).

References

1 Li, K., Hogan, N.J., Kale, M.J. et al. (2017). Balancing near-field enhancement, absorption, and scattering for effective antenna – reactor plasmonic photocatalysis. *Nano Lett.* 17: 3710–3717.

2 Engelbrekt, C., Crampton, K.T., Fishman, D.A. et al. (2020). Efficient plasmon-mediated energy funneling to the surface of Au@Pt core-shell nanocrystals. *ACS Nano* 14 (4): 5061–5074.

3 Baffou, G., Berto, P., Bermúdez Ureña, E. et al. (2013). Photoinduced heating of nanoparticle arrays. *ACS Nano* 7 (8): 6478–6488.

4 Zhang, X., Li, X., Reish, M.E. et al. (2018). Plasmon-enhanced catalysis: distinguishing thermal and nonthermal effects. *Nano Lett.* 18: 1714–1723.

5 Li, X., Zhang, X., Everitt, H.O., and Liu, J. (2019). Light-induced thermal gradients in ruthenium catalysts significantly enhance ammonia production. *Nano Lett.* 19 (3): 1706–1711.

6 Wang, P., Huang, B., Dai, Y., and Whangbo, M.-H.H. (2012). Plasmonic photocatalysts: harvesting visible light with noble metal nanoparticles. *Phys. Chem. Chem. Phys.* 14 (28): 9813–9825.

7 Zheng, Z., Tachikawa, T., and Majima, T. (2015). Plasmon-enhanced formic acid dehydrogenation using anisotropic Pd-Au nanorods studied at the single-particle level. *J. Am. Chem. Soc.* 137 (2): 948–957.

8 Awazu, K., Fujimaki, M., Rockstuhl, C. et al. (2008). A plasmonic photocatalyst consisting of silver nanoparticles embedded in titanium dioxide. *J. Am. Chem. Soc.* 130 (5): 1676–1680.

9 Fujishim, A. and Honda, K. (1972). Electrochemical photolysis of water at a semiconductor electrode. *Nature* 238: 37–39.

10 Zhang, X., Chen, Y.L., Liu, R.-S., and Tsai, D.P. (2013). Plasmonic photocatalysis. *Rep. Prog. Phys.* 76 (4): 046401.

11 Karger, B.L. (1997). HPLC: early and recent perspectives. *J. Chem. Educ.* 74 (1): 45–48.

12 Sparkman, O.D., Penton, Z.E., and Kitson, F.G. (2011). *Gas Chromatography and Mass Spectrometry*. Oxford, UK: Academic Press.

13 Hübschmann, H. (2009). *Handbook of GC/MS: Fundamentals and Applications*. Weinheim: Wiley-VCH.

14 Swearer, D.F., Zhao, H., Zhou, L. et al. (2016). Heterometallic antenna–reactor complexes for photocatalysis. *Proc. Natl. Acad. Sci. U. S. A.* 113 (32): 8916–8920.

15 Zhou, L., Martirez, J.M.P., Finzel, J. et al. (2020). Light-driven methane dry reforming with single atomic site antenna-reactor plasmonic photocatalysts. *Nat. Energy* 5: 61–70.

16 Zhou, L., Swearer, D.F., Zhang, C. et al. (2018). Quantifying hot carrier and thermal contributions in plasmonic photocatalysis. *Science (80-)* 362 (October): 69–72.

17 Gargiulo, J., Berté, R., Li, Y. et al. (2019). From optical to chemical hot spots in plasmonics. *Acc. Chem. Res.* 52 (9): 2525–2535.

18 Christopher, P., Xin, H., and Linic, S. (2011). Visible-light-enhanced catalytic oxidation reactions on plasmonic silver nanostructures. *Nat. Chem.* 3 (6): 467–472.

19 Mukherjee, S., Libisch, F., Large, N. et al. (2013). Hot electrons do the impossible: plasmon-induced dissociation of H_2 on Au. *Nano Lett.* 13 (1): 240–247.

20 Novo, C., Funston, A.M., and Mulvaney, P. (2008). Direct observation of chemical reactions on single gold nanocrystals using surface plasmon spectroscopy. *Nat. Nanotechnol.* 3 (10): 598–602.

21 Christopher, P., Xin, H., Marimuthu, A., and Linic, S. (2012). Singular characteristics and unique chemical bond activation mechanisms of photocatalytic reactions on plasmonic nanostructures. *Nat. Mater.* 11 (12): 1044–1050.

22 Linic, S., Christopher, P., and Ingram, D.B. (2011). Plasmonic-metal nanostructures for efficient conversion of solar to chemical energy. *Nat. Mater.* 10 (12): 911–921.

23 Zheng, Z., Tachikawa, T., and Majima, T. (2015). Plasmon-enhanced formic acid dehydrogenation using anisotropic Pd-Au nanorods studied at the single-particle level. *J. Am. Chem. Soc.* 137 (2): 948–957.

24 Baffou, G., Bordacchini, I., Baldi, A. et al. (2020). Simple experimental procedures to distinguish photothermal from hot-carrier processes in plasmonics. Light Sci Appl 9, 108 https://doi-org.stanford.idm.oclc.org/10.1038/s41377-020-00345-0

25 Denzler, D.N., Frischkorn, C., Hess, C. et al. (2003). Electronic excitation and dynamic promotion of a surface reaction. *Phys. Rev. Lett.* 91 (22): 1–4.

26 Menzel, D. (1975). Electron stimulated desorption: principles and recent developments. *Surf. Sci.* 47 (1): 370–383.

27 Sivan, Y., Baraban, J., Un, I.W., and Dubi, Y. (2019). Comment on "Quantifying hot carrier and thermal contributions in plasmonic photocatalysis". *Science* 364 (6439): 3–5.

28 Baumberg, J.J. (2019). Hot electron science in plasmonics and catalysis: what we argue about. *Faraday Discuss.* 214 (I): 501–511.

29 Sivan, Y., Un, I.W., and Dubi, Y. (2019). Assistance of metal nanoparticles in photocatalysis-nothing more than a classical heat source. *Faraday Discuss.* 214: 215–233.

30 Zhou, L., Swearer, D.F., Robatjazi, H. et al. (2019). Response to comment on "quantifying hot carrier and thermal contributions in plasmonic photocatalysis". *Science (80-)* 364 (6439).

31 Bonn, M., Funk, S., Hess, C. et al. (1999). Phonon- versus electron-mediated desorption and oxidation of CO on Ru(0001). *Science (80-)* 285 (5430): 1042–1045.

32 Funk, S., Bonn, M., Denzler, D.N. et al. (2000). Desorption of CO from Ru(001) induced by near-infrared femtosecond laser pulses. *J. Chem. Phys.* 112 (22): 9888–9897.

33 Kim, Y., Dumett Torres, D., and Jain, P.K. (2016). Activation energies of plasmonic catalysts. *Nano Lett.* 16 (5): 3399–3407.

34 Baffou, G. and Quidant, R. (2013). Thermo-plasmonics: using metallic nanostructures as nano-sources of heat. *Laser Photonics Rev.* 7 (2): 171–187.

35 Imelik, B. and Vedrine, J.C. (eds.) (1994). *Catalyst Characterization*. Boston, MA: Springer US.

36 Halas, N.J., Lal, S., Chang, W.S. et al. (2011). Plasmons in strongly coupled metallic nanostructures. *Chem. Rev.* 111: 3913–3961.

37 Olson, J., Dominguez-Medina, S., Hoggard, A. et al. (2015). Optical characterization of single plasmonic nanoparticles. *Chem. Soc. Rev.* 44 (1): 40–57.

38 Chen, Y., Munechika, K., and Ginger, D.S. (2007). Dependence of fluorescence intensity on the spectral overlap between fluorophores and plasmon resonant single silver nanoparticles. *Nano Lett.* 7 (3): 690–696.

39 Pestryakov, A.N., Petranovskii, V.P., Kryazhov, A. et al. (2004). Study of copper nanoparticles formation on supports of different nature by UV-Vis diffuse reflectance spectroscopy. *Chem. Phys. Lett.* 385 (3–4): 173–176.

40 Evanoff, D.D. and Chumanov, G. (2004). Size-controlled synthesis of nanoparticles. 2. Measurement of extinction, scattering, and absorption cross sections. *J. Phys. Chem. B* 108 (37): 13957–13962.

41 Avanesian, T., Dai, S., Kale, M.J. et al. (2017). Quantitative and atomic-scale view of CO-induced Pt nanoparticle surface reconstruction at saturation coverage via DFT calculations coupled with in situ TEM and IR. *J. Am. Chem. Soc.* 139 (12): 4551–4558.

42 Resasco, J., Dai, S., Graham, G. et al. (2018). Combining in-situ transmission electron microscopy and infrared spectroscopy for understanding dynamic and atomic-scale features of supported metal catalysts. *J. Phys. Chem. C* 122 (44): 25143–25157.

43 Hu, M., Novo, C., Funston, A. et al. (2008). Dark-field microscopy studies of single metal nanoparticles: understanding the factors that influence the linewidth of the localized surface plasmon resonance. *J. Mater. Chem.* 18 (17): 1949–1960.

44 Foerster, B., Joplin, A., Kaefer, K. et al. (2017). Chemical interface damping depends on electrons reaching the surface. *ACS Nano* 11 (3): 2886–2893.

45 Foerster, B., Spata, V.A., Carter, E.A. et al. (2019). Plasmon damping depends on the chemical nature of the nanoparticle interface. *Sci. Adv.* 5 (3).

46 Zamora-Perez, P., Tsoutsi, D., Xu, R., and Rivera-Gil, P. (2018). Hyperspectral-enhanced dark field microscopy for single and collective nanoparticle characterization in biological environments. *Materials (Basel).* 11 (243): 1–13.

47 Kirchner, S.R., Smith, K.W., Hoener, B.S. et al. (2018). Snapshot hyperspectral imaging (SHI) for revealing irreversible and heterogeneous plasmonic processes. *J. Phys. Chem. C* 122 (12): 6865–6875.

48 Willets, K.A., Wilson, A.J., Sundaresan, V., and Joshi, P.B. (2017). Super-resolution imaging and plasmonics. *Chem. Rev.* 117 (11): 7538–7582.

49 Chen, G., Zou, N., Chen, B. et al. (2017). Bimetallic effect of single nanocatalysts visualized by super-resolution catalysis imaging. *ACS Cent. Sci.* 3 (11): 1189–1197.

50 Ha, J.W., Ruberu, T.P.A., Han, R. et al. (2014). Super-resolution mapping of photogenerated electron and hole separation in single metal-semiconductor nanocatalysts. *J. Am. Chem. Soc.* 136 (4): 1398–1408.

51 Chen, T., Dong, B., Chen, K. et al. (2017). Optical super-resolution imaging of surface reactions. *Chem. Rev.* 117 (11): 7510–7537.

52 Zou, N., Chen, G., Mao, X. et al. (2018). Imaging catalytic hotspots on single plasmonic nanostructures via correlated super-resolution and electron microscopy. *ACS Nano* 12 (6): 5570–5579.

53 Simoncelli, S., Li, Y., Cortés, E., and Maier, S.A. (2018). Nanoscale control of molecular self-assembly induced by plasmonic hot-electron dynamics. *ACS Nano* 12 (3): 2184–2192.

54 Baffou, G., Girard, C., and Quidant, R. (2010). Mapping heat origin in plasmonic structures. *Phys. Rev. Lett.* 104 (13): 1–4.

55 Baffou, G. and Quidant, R. (2014). Nanoplasmonics for chemistry. *Chem. Soc. Rev.* 43: 3898–3907.

56 Sambur, J.B. and Chen, P. (2014). Approaches to single-nanoparticle catalysis. *Annu. Rev. Phys. Chem.* 65 (1): 395–422.

57 Nicoletti, O., De La Peña, F., Leary, R.K. et al. (2013). Three-dimensional imaging of localized surface plasmon resonances of metal nanoparticles. *Nature* 502 (7469): 80–84.

58 Aslam, U., Chavez, S., and Linic, S. (2017). Controlling energy flow in multimetallic nanostructures for plasmonic catalysis. *Nat. Nanotechnol.* 12 (10): 1000–1005.

59 Wu, Y., Li, G., and Camden, J.P. (2018). Probing nanoparticle plasmons with electron energy loss spectroscopy. *Chem. Rev.* 118 (6): 2994–3031.

60 Campos, A., Arbouet, A., Martin, J. et al. (2017). Plasmonic breathing and edge modes in aluminum nanotriangles. *ACS Photonics* 4 (5): 1257–1263.

61 Nelayah, J., Kociak, M., Stéphan, O. et al. (2007). Mapping surface plasmons on a single metallic nanoparticle. *Nat. Phys.* 3 (5): 348–353.

62 García De Abajo, F.J. (2010). Optical excitations in electron microscopy. *Rev. Mod. Phys.* 82 (1): 209–275.

63 Vadai, M., Angell, D.K., Hayee, F. et al. (2018). In-situ observation of plasmon-controlled photocatalytic dehydrogenation of individual palladium nanoparticles. *Nat. Commun.* 9 (1): 4658.

64 Yang, W.-C.D., Wang, C., Fredin, L.A. et al. (2019). Site-selective CO disproportionation mediated by localized surface plasmon resonance excited by electron beam. *Nat. Mater.* 18 (6): 614–619.

65 Mecklenburg, M., Hubbard, W.A., White, E.R. et al. (2015). Nanoscale temperature mapping in operating microelectronic devices. *Science (80-)* 347 (6222): 629–632.

66 Myroshnychenko, V., Rodríguez-Fernández, J., Pastoriza-Santos, I. et al. (2008). Modelling the optical response of gold nanoparticles. *Chem. Soc. Rev.* 37: 1792–1805.

67 Nørskov, J.K., Abild-Pedersen, F., Studt, F., and Bligaard, T. (2011). Density functional theory in surface chemistry and catalysis. *Proc. Natl. Acad. Sci. U. S. A.* 108 (3): 937–943.

68 Libisch, F., Huang, C., and Carter, E.A. (2014). Embedded correlated wavefunction schemes: theory and applications. *Acc. Chem. Res.* 47 (9): 2768–2775.

69 Martirez, J.M.P. and Carter, E.A. (2017). Prediction of a low-temperature N_2 dissociation catalyst exploiting near-IR–to–visible light nanoplasmonics. *Sci. Adv.* 2: 1–11.

70 Martirez, J.M.P. and Carter, E.A. (2017). Excited-state N_2 dissociation pathway on Fe-functionalized Au. *J. Am. Chem. Soc.* 139: 4390–4398.

71 Furube, A., Du, L., Hara, K. et al. (2007). Ultrafast plasmon-induced electron transfer from gold nanodots into TiO_2 nanoparticles. *J. Am. Chem. Soc.* 129 (48): 14852–14853.

72 Du, L., Furube, A., Yamamoto, K. et al. (2009). Plasmon-induced charge separation and recombination dynamics in gold-TiO_2 nanoparticle systems: dependence on TiO_2 particle size. *J. Phys. Chem. C* 113 (16): 6454–6462.

73 Brandt, N.C., Keller, E.L., and Frontiera, R.R. (2016). Ultrafast surface-enhanced Raman probing of the role of hot electrons in plasmon-driven chemistry. *J. Phys. Chem. Lett.* 7 (16): 3178–3185.

74 Keller, E.L. and Frontiera, R.R. (2018). Ultrafast nanoscale Raman thermometry proves heating is not a primary mechanism for plasmon-driven photocatalysis. *ACS Nano* 12 (6): 5848–5855.

75 Keller, E.L., Kang, H., Haynes, C.L., and Frontiera, R.R. (2018). Effect of silica supports on plasmonic heating of molecular adsorbates as measured by ultrafast surface-enhanced Raman thermometry. *ACS Appl. Mater. Interfaces* 10: 40577–40584.

76 Zhang, Y., Zhen, Y.R., Neumann, O. et al. (2014). Coherent anti-Stokes Raman scattering with single-molecule sensitivity using a plasmonic Fano resonance. *Nat. Commun.* 5: 1–7.

77 Stöckle, R.M., Suh, Y.D., Deckert, V., and Zenobi, R. (2000). Nanoscale chemical analysis by tip-enhanced Raman spectroscopy. *Chem. Phys. Lett.* 318 (1–3): 131–136.

78 Klingsporn, J.M., Sonntag, M.D., Seideman, T., and Van Duyne, R.P. (2014). Tip-enhanced Raman spectroscopy with picosecond pulses. *J. Phys. Chem. Lett.* 5 (1): 106–110.

79 Verma, P. (2017). Tip-enhanced Raman spectroscopy: technique and recent advances. *Chem. Rev.* 117 (9): 6447–6466.

80 Szczerbiński, J., Gyr, L., Kaeslin, J., and Zenobi, R. (2018). Plasmon-driven photocatalysis leads to products known from E-beam and X-ray-induced surface chemistry. *Nano Lett.* 18 (11): 6740–6749.

81 Chen, X., Liu, P., Hu, Z., and Jensen, L. (2019). High-resolution tip-enhanced Raman scattering probes sub-molecular density changes. *Nat. Commun.* 10 (1): 1–9.

82 Szczerbiński, J., Metternich, J.B., Goubert, G., and Zenobi, R. (2019). How peptides dissociate in plasmonic hot spots. *Small* 16 (4): 1905197.

83 Hoggard, A., Wang, L.Y., Ma, L. et al. (2013). Using the plasmon linewidth to calculate the time and efficiency of electron transfer between gold nanorods and graphene. *ACS Nano* 7 (12): 11209–11217.

84 Biswas, S., Husek, J., and Baker, L.R. (2018). Elucidating ultrafast electron dynamics at surfaces using extreme ultraviolet (XUV) reflection-absorption spectroscopy. *Chem. Commun.* 54 (34): 4216–4230.

85 Valley, D.T., Ferry, V.E., and Flannigan, D.J. (2016). Imaging intra- and interparticle acousto-plasmonic vibrational dynamics with ultrafast electron microscopy. *Nano Lett.* 16 (11): 7302–7308.

3

Synthesis of Plasmonic Nanoparticles for Photo- and Electrocatalysis

Wei Xie[1], Kaifu Zhang[1], Roland Grzeschik[2], and Sebastian Schlücker[2]

[1] *Key Lab of Advanced Energy Materials Chemistry (Ministry of Education), Renewable Energy Conversion and Storage Center, College of Chemistry, Nankai University, Tianjin, China*
[2] *Department of Chemistry and Center for Nanointegration Duisburg-Essen (CENIDE), University of Duisburg-Essen, Essen, Germany*

3.1 Introduction

In the preceding chapters the following key questions have been addressed:

1. What are the key ideas of plasmonic catalysis and what distinguishes it from conventional heterogeneous catalysis and conventional photocatalysis with molecular photocatalysts? Why is the subject highly complex and therefore requires a multidisciplinary approach?
2. What are the underlying physical processes by which plasmonic particles can contribute to catalysis and energy conversion? How can theory contribute to a fundamental understanding of dynamics and mechanisms?
3. What are the outstanding optical properties of plasmonic nanoparticles (NPs) which make them so attractive for use in catalysis and chemical energy conversion? How can we tailor them for fully exploiting their potential?

Since plasmonic NPs are an integral component of plasmonic catalysis, we focus on the chemistry and material science perspective in this chapter and specifically address the following two questions: (i) How can plasmonic NPs for photo- and electrocatalysis be synthesized in a material-, size-, and shape-selective manner and (ii) How have they been exploited in SERS-based studies of photo- and electrocatalysis?

We restrict ourselves to conventional wet-chemical approaches to synthesis because of their huge potential to precisely control the geometrical parameters' size and, in particular, shape. Other approaches such as pulsed laser ablation in liquids, although promising and offering many degrees of freedom in terms of selecting a huge variety of metals and combinations thereof, are not covered here and the reader is referred to the corresponding literature [1–3].

We first cover single monometallic NPs comprising different metals: gold (Section 3.2.1), silver (Section 3.2.2), copper (Section 3.2.3), as well as aluminum (Section 3.2.4). In contrast to the coinage metals Au, Ag, and Cu with localized surface

Plasmonic Catalysis: From Fundamentals to Applications, First Edition.
Edited by Pedro H.C. Camargo and Emiliano Cortés.

plasmon resonance (LSPR) in the visible (Vis), Al has a much higher abundancy and therefore lower prize, but an LSPR in the ultraviolet (UV). For each of these metals we summarize methods to synthesize isotropic (quasispherical/faceted) and, in particular, anisotropic particles (rods, cubes, stars, etc.). The assembly of two or more particles (Section 3.3) into defined structures (dimers, trimers and higher assemblies/superstructures, films) offers many additional degrees of freedom for tailoring the optical and/or chemical properties since the whole is more than the sum of its parts. For example, in dimers of two noble metal NPs, new plasmon modes and very high field enhancements in the particle gap arise from plasmonic coupling/hybridization.

Bimetallic NPs with both catalytic and plasmonic activity significantly expand the functional repertoire by integrating two metals with distinct optical and/or chemical properties into a new functional unit [4]. A prominent example is the combination of catalytically active Pt with plasmonically active Au cores/protuberances into bifunctional Au/Pt/Au nanoraspberries for label-free monitoring of chemical reactions by SERS [5]. By combining the best of two worlds – Pt for catalysis and Au for plasmonics – the Pt-catalyzed $NaBH_4$ reduction of 4-NTP to 4-ATP could be monitored directly in suspension, i.e. without the necessity of depositing the particles onto a solid substrate. This model reaction has subsequently been employed by many other groups in the SERS community [6–9]. A more generic approach to bifunctional NPs are 3D superstructures comprising a large plasmonically active core (e.g. 80 nm Au NPs) and small catalytically active satellites (e.g. 5 nm Au NPs) [10]. Due to the modular approach of the assembly scheme of the satellites onto the core, this approach to bifunctional NPs has been extended to combinations of many other metals [11–13].

Due to the central role of SERS in both conventional label-free reaction monitoring and in plasmonic chemistry, we decided to also include a subchapter on SERS applications in photo- and electrocatalysis (Sections 3.4.1 and 3.4.2). The first approach by conventional SERS for label-free reaction monitoring is passive: the plasmonic nanostructure is used only for Raman signal enhancement but does not participate in the actual chemical conversion. In contrast, the second approach – plasmonic chemistry, the topic of this book – is active: the plasmonic nanostructure is involved in charge transfer of hot carriers (electrons/holes) to/from (adsorbed) molecules.

In this chapter, we restrict ourselves to giving a brief overview on chemical reaction studies by SERS since specific classes of reactions in plasmonic chemistry are the subject of subsequent chapters, for example, on oxidation (Chapter 5) and hydrogenation (Chapter 6) reactions. The same applies to the use of SERS in electrocatalysis since a separate chapter is devoted exclusively to plasmonic chemistry in electrocatalysis (Chapter 12).

3.2 Monometallic Plasmonic Nanoparticles

3.2.1 Au Nanoparticles

Gold NPs are considered as the most inert metallic nanostructures and are employed in various fields due to their unique physical and chemical properties. In 1857, Michael Faraday published his pioneering work about gold nanosols [14].

Compared with bulk gold, gold nanosols show distinctive properties, particularly optical properties. In the past decades they have been widely used in the fields of enhanced spectroscopies and sensing, photodynamic therapy, and catalysis. Since the optical properties of metallic gold NPs are closely related to their morphology, it is very important to effectively control the shape by choosing proper reaction condition during synthesis.

3.2.1.1 Au Quasispheres

Among all gold NPs of various shapes, gold nanospheres have attracted wide attention due to their relatively simple structure and high thermodynamic stability. Faceted Au NPs are quasispherical and thus they typically exhibit only a single plasmon resonance peak. As the size of gold NPs increases, the LSPR extinction gradually red shifts and the ratio of scattering to absorption cross-sections increases.

The Turkevich method has been extensively used to synthesize Au NPs [16]. This method is easy to perform and only requires the addition of sodium citrate into boiling $HAuCl_4$ solution. The Au NPs size can be controlled by changing the molar ratio of the reducing agent sodium citrate and $HAuCl_4$. The crystal nuclei formed in the initial stage of the reduction have a high surface energy and $AuCl_4^-$ ions from the solution adsorb onto their surface. As the adsorbed $AuCl_4^-$ ions are further reduced, the crystal nuclei gradually become larger. Through further growth and maturation, the initial smaller particles become the final NPs with desired plasmonic and catalytic properties. TEM images of quasispherical Au NPs with different diameters are shown in Figure 3.1.

The polyhydroxy reduction approach has also been widely used in the preparation of noble metal NPs because it has good controllability and does not require seeds or templates, which is simple and convenient. The main process is the reduction of the inorganic salt with a polyhydric alcohol (usually ethylene glycol [EG]) at high temperature and the addition of a polymer (usually polyvinylpyrrolidone [PVP]) as the shape-directing stabilizer to prevent colloids aggregation.

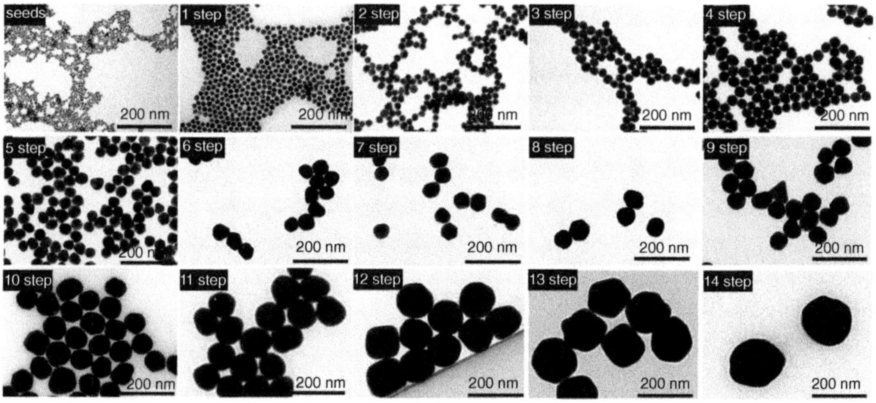

Figure 3.1 TEM images of quasispherical gold NPs with different diameters obtained after different growth steps [15].

The reagents used in these methods have a relatively weak affinity to the Au surface of the NPs, so quasispherical Au NPs can be prepared. If strongly binding agents such as hexadecyltrimethylammonium bromide (CTAB) are used as protective agents, their crystal facet-specific adsorption will affect the growth of the NPs. Thereby, anisotropic NPs with other morphologies can be obtained.

3.2.1.2 Au Nanorods

Rod-shaped Au NPs with long and short axes have physicochemical properties different from those of spherical Au NPs, such as higher electromagnetic enhancement and catalytic activity. The extinction spectra of Au nanorods (NRs) usually show two characteristic peaks, namely the long-axis and short-axis plasmonic band generated by the longitudinal and transverse oscillation of electrons, respectively. The electromagnetic field enhancement localized at the rod ends is more intense than that on the spherical counterparts. The ratio of the lengths of long and short axes plays a significant role in determining the position of the plasmon peak of Au NRs. Due to their high-index facets containing low-coordinated atomic steps and kinks, they exhibit a high catalytic activity.

A seeded growth method is usually employed for synthesizing Au NRs. This method originally established by Murphy's group comprises two steps: preparation of spherical gold NP seeds (3–4 nm in size) and their growth into rods [17]. However, the rod-shaped products prepared by this method have always been accompanied by more spherical Au NP byproducts. El-Sayed's group improved the seeded growth method through replacing the protective agent on the seeds: sodium citrate was replaced by CTAB to form a brown-yellow Au NP seed solution [18]. After the seed crystals are prepared, a certain amount of seed crystal suspension is added to the prepared mixture containing $HAuCl_4$, $AgNO_3$, CTAB, and ascorbic acid (AA), and then the conditions are controlled to grow them into short Au NRs. The aspect ratio (AR) of the NRs during the growth process can be controlled by changing the ratio of seed crystal to metal salt. In addition, by controlling the concentration of CTAB, Au NRs with high AR can be obtained (Figure 3.2).

The main influencing factors in this synthesis process are as follows. Halide ions such as Br^-, Cl^- have different priorities for adsorption on each crystal plane. The selective adsorption of halide ions limits the growth of Au NRs along the short axis, making the relative growth rate of the short axis and the long axis different, thereby controlling the directional growth of the crystal. In the synthesis of Au NRs, bromide in the surfactant is a key factor [19, 20]. It preferentially and selectively adsorbs on certain surfaces of the seed particles and then further attracts the surfactant CTAB. In addition, other factors also affect the synthesis: (i) pH value, (ii) seed concentration, and (iii) concentration of $AgNO_3$. Therefore, Au NRs with different ARs can be achieved by manipulating different conditions in the synthesis process.

3.2.1.3 Au Nanocubes

Among the many symmetrical gold nanostructures, the formation of gold nanocubes is particularly interesting because they are completely surrounded by {100} facets and are more difficult to synthesize. Huang and colleagues developed

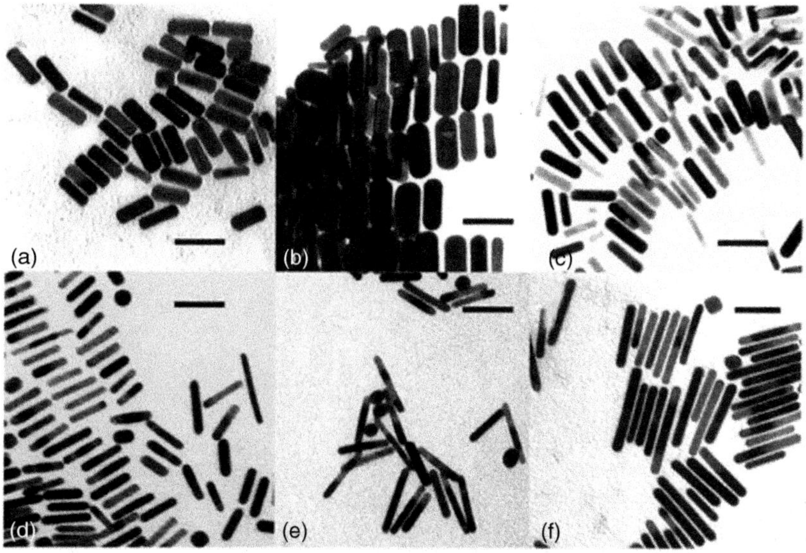

Figure 3.2 TEM images of gold NRs with increased average AR and resonance wavelengths of the plasmon peak at (a) 700 nm, (b) 760 nm, (c) 790 nm, (d) 880 nm, (e) 1130 nm, and (f) 1250 nm, respectively. Scale bar: 50 nm [18].

a facile seed-mediated synthesis for the preparation of relatively monodisperse Au nanocubes [21]. Au seeds were first synthesized and then added to a specific amount of mixed growth solution of hexadecyltrimethylammonium chloride, NaBr, $HAuCl_4$, and AA to further grow into Au nanocubes. Their research shows that a sufficiently high surfactant concentration and a very low bromide concentration are the key to the formation of cubes in high-yield cubes. Therefore, they used a small amount of NaBr and CTAC instead of CTAB as surfactant to synthesize nanocubes. Also, other nanostructures of different shapes such as the trisoctahedron and the rhombic dodecahedron can be obtained by adjusting the proportion of AA (Figure 3.3).

3.2.1.4 Au Nanotriangles

Changes in NP symmetry lead to changes in physical and chemical properties (structure–activity relationship) and further affect the application of nanostructures. Among many anisotropic NPs, Au nanotriangles are particularly outstanding for its unique structure, anisotropic surface energy, and controllable synthesis of side length and thickness.

Generally, Au nanotriangles have a side length of 40–1000 nm and a thickness of 5–50 nm. The ratio of the side length to the thickness is the AR, which is a parameter that measures the degree of anisotropy of the nanotriangles. The larger the value, the greater the degree of anisotropy. For example, for isotropic spherical NPs, the AR is one because the dimensions are the same in all directions. In contrast, the AR of gold nanotriangles is c. 5–40. In addition, the three sharp corners of the nanotriangular plate significantly affect its optical properties. In general, Au nanotriangles synthesized by the liquid phase method are single crystals with a face-centered cubic crystal

Figure 3.3 SEM images of the gold nanocrystals synthesized with shape evolution from truncated cubic to rhombic dodecahedral structures by increasing the amount of ascorbic acid added. The nanocrystals are (a) truncated cubes, (b) cubes, (c) type I transitional product, (d) trisoctahedra, (e) type II transitional product, and (f) rhombic dodecahedra [21].

structure. The triangular plane is a {111} facet and the three sides exposed are {100} or {110} facets. The structure of the material determines its physical and chemical properties. By controlling the side length, thickness, and sharp morphology of the Au nanotriangles, the surface plasmon resonance band can be continuously tuned across a wide range in the visible to near-infrared region (NIR) (Figure 3.4) [22–24].

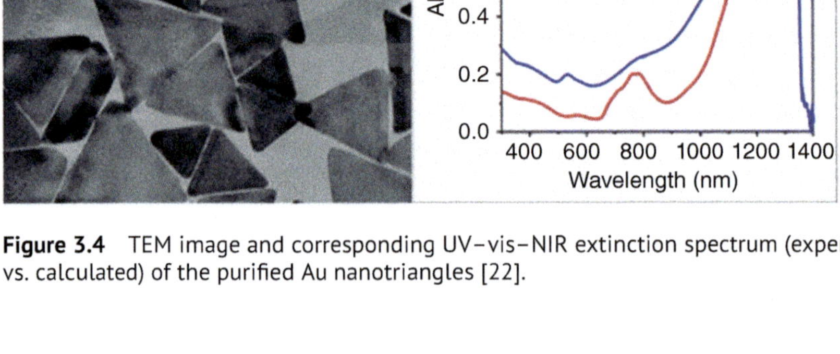

Figure 3.4 TEM image and corresponding UV–vis–NIR extinction spectrum (experimental vs. calculated) of the purified Au nanotriangles [22].

The synthesis of Au nanotriangles involves rapid reduction of $HAuCl_4$ in the presence of sodium citrate to obtain gold seeds with a diameter of 4–6 nm, which are then added to a growth solution containing CTAB, $HAuCl_4$, NaOH, and AA for growth into triangular particles with larger sizes [22]. The CTAB surfactant forms an electric double layer on the surface of the NPs to prevent the reduction of metal ions on the surface. The morphology and size of the Au nanotriangles depend not only on the concentration of CTAB, but also on the crystal structure of the seed particles and the concentration of Au ions and reducing agents [25]. These factors are interdependent and together determine the final morphology of the Au nanotriangles.

3.2.1.5 Au Nanostars

Au nanostars are multibranched NPs with pointed structures. Because the coupled plasmon modes in Au nanostars contain contributions from both the central nucleus and the nanotips, they have unique and interesting optical properties. The nanotip structure, like an antenna, generates a strong electromagnetic field at the end of the tip upon resonant excitation. At the same time, both tip shape and number affect the energy and intensity of the LSPR.

Since the tips of Au nanostars contain higher-order crystal planes, a strong protective agent needs to be added during the synthesis. Many methods for synthesizing Au nanostars are based on the chemical reduction of Au salts [26, 27]. This method requires the assistance of surfactants such as PVP or CTAB. Vo-Dinh and coworkers studied the growth mechanism of Au nanostars and their application in SERS [27]. By controlling the volume of the added seed solution, they obtained Au nanostars of different sizes and shapes (Figure 3.5). The results showed that the growth of Au nanostars comprises two steps: the rapid nucleation on the seed surface and the slow deposition in irregular voids.

Au nanostars exhibit strong LSPR peaks in the red to near-infrared range which makes them attractive for the field of optical bioimaging exploiting the "biological window" (Figure 3.6) [28, 29]. By adjusting the sharpness of the tips, the scattering

Figure 3.5 TEM images of Au nanostars in order of increasing size (a–f). Scale bars correspond to 200 nm for main panels and 100 nm for insets [26].

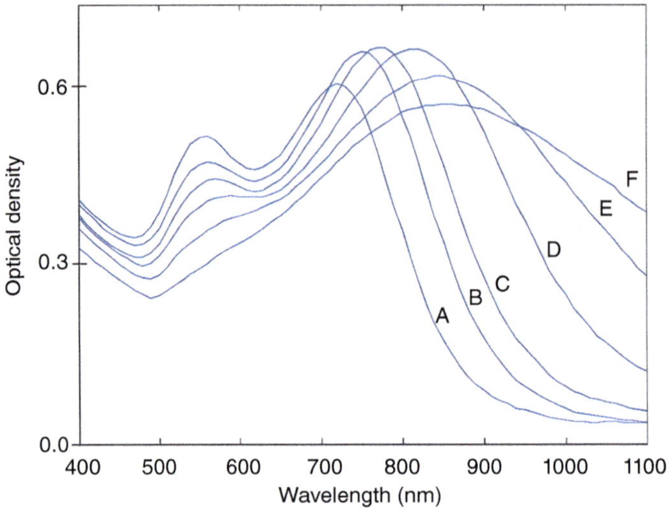

Figure 3.6 Vis/NIR extinction spectra of as-synthesized gold nanostars of increasing size from A to F [26].

cross-section can be enhanced. Therefore, Au nanostars can also be applied to SERS detection [30–32].

3.2.2 Ag Nanoparticles

Compared with gold, silver is cheaper and plasmonically more active. In addition, silver nanostructures show tunable plasmon bands across the entire visible light region, from blue to red, by controlling their size and shape. Therefore, the surface plasmons of silver NPs have been extensively studied. Xia and coworkers found that the SERS enhancement factor of a single nanocube is on the order of 10^8 at 514 nm excitation [33]. Similar to gold, various geometries of Ag nanostructures have different optical properties and chemical stability [34].

3.2.2.1 Ag Quasispheres

In the visible region, quasispherical Ag NPs are predicted to be the best materials with a very strong LSPR, which in general is governed by size, shape, crystallinity, and surface structure. In the synthesis of Ag quasispherical NPs, using sodium citrate as a reducing agent and a stabilizer, a certain amount of sodium citrate solution is added to the boiling $AgNO_3$ aqueous solution, and the mixture is mechanically stirred for one hour under boiling and then cooled to room temperature to form silver NPs [35]. However, due to the relatively weak reducing power of sodium citrate, the morphology and size of the prepared silver NPs are usually not uniform. Thus, Dennis and colleagues successfully synthesized silver NPs with uniform size by using the stronger reducing agent AA and PVP as a protective agent in a mixed system of glycerol and water [36]. Figure 3.7 shows the corresponding LSPR spectra and TEM images of the silver colloid. Therefore, the preparation of nanostructures can be accomplished by the selective adsorption of PVP.

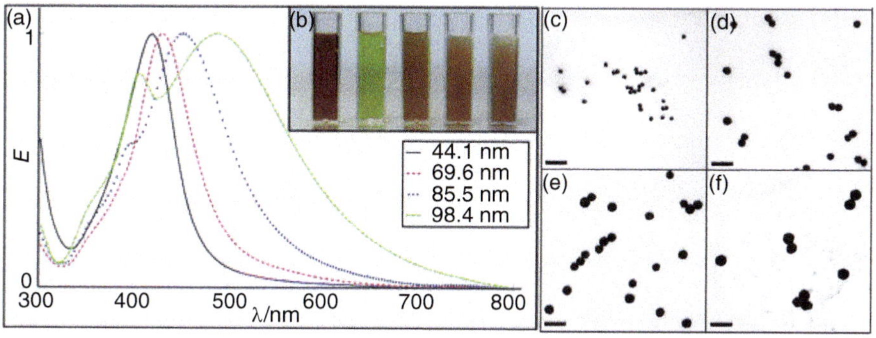

Figure 3.7 (a) LSPR spectrum of silver colloids in the size range of c. 40–100 nm. The LSPR peak occurs at 420 nm (44.1 nm diameter), 431 nm (69.6 nm), 453 nm (85.5 nm), and 490 nm (98.4 mm), respectively; (b) Photographs of the colloidal samples (from left to right: 30.0, 44.1, 69.6, 85.5, and 98.4 nm); (c–f) TEM images: 44.1 nm (c), 69.6 nm (d), 85.5 nm (e), and 98.4 nm (f); Scale bar 200 nm [36].

3.2.2.2 Ag Nanowires and Nanorods

One-dimensional silver nanostructures mainly include silver nanowires, rods, and tubes and the synthesis methods are also diverse. Xia and colleagues used a liquid phase method to synthesize silver nanowires with a uniform diameter and length on a large scale (Figure 3.8a, b) [37]. The results show that the simultaneous dropwise addition of silver nitrate and PVP is a key step in the generation of silver nanowires. At the same time, because PVP might interact more strongly with the {100} facets of the silver seeds, the silver seeds mainly grow along the {111} crystal plane to form

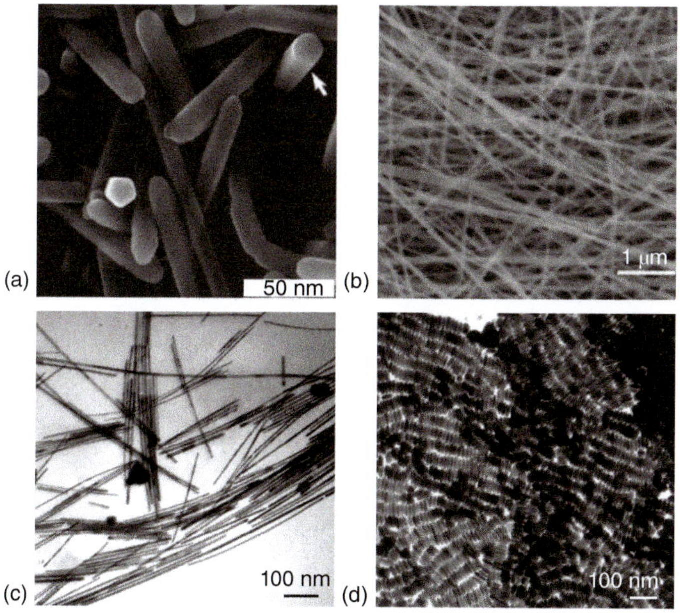

Figure 3.8 SEM image of Ag nanowires (a, b) [37, 38] and TEM image of Ag NPs (c, d) [39].

silver nanowires. Zelaski and colleagues used a microwave-assisted method to add a protective agent to an EG solution with a ratio of silver hexadecanoate and sodium chloride of 6 : 1 to 3 : 1 [38]. Silver nanowires with a diameter of 45 nm and a length of 4–12 μm were prepared by microwave reaction (300 W). The results also show that reaction time, microwave power, and the ratio of sodium chloride to silver nitrate are important factors affecting the synthesis of silver nanowires.

Murphy and colleagues used silver NPs with an average size of 4 nm as seed crystals, sodium hydroxide and CTAB as modifiers, and reduced silver nitrate with AA to prepare silver NRs [39]. By changing the amount of sodium hydroxide in the reaction solution to adjust the pH value, silver nanostructures with different morphologies were prepared. The experimental results show that when the pH value of the solution is close to 11.8, the AA dianion plays a leading role in morphology control, and it is easy to form silver NRs (Figure 3.8c). When the solution pH is lower than this value, the AA monoanion plays a major role, which is beneficial to the formation of silver nanowires (Figure 3.8d).

3.2.2.3 Ag Nanocubes

If single-crystalline silver nanosquares are used as seeds during the synthesis of silver nanostructures, the seed particles mainly grow epitaxially along two crystal directions, namely the {111} and/or {100} directions. If it is fully developed along the {111} direction, the single crystal silver nanosquares will maintain their square shape during the growth process and gradually grow up to yield a series of silver nanosquares with controllable size (Figure 3.9b). If they grow completely along the {100} direction, the single crystalline silver nanocubes will gradually transform into cut-off cubes, cubic octahedron, and finally octahedra. Therefore, it is particularly important to add structure-directing agents to control the growth of nanostructures during this process.

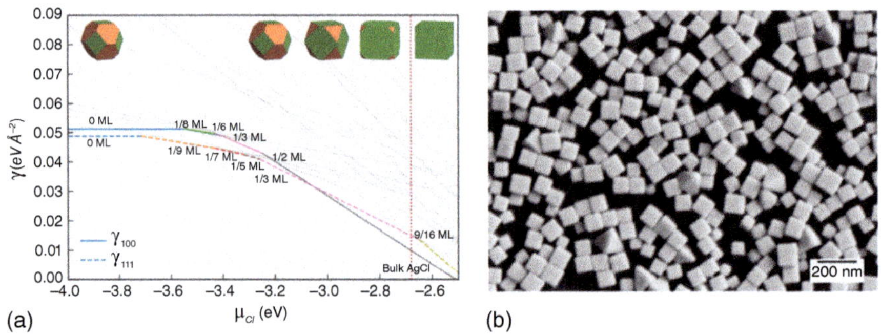

Figure 3.9 (a) Surface energies (γ) of Ag {100} and Ag {111} for different surface coverages of chloride as a function of the solution-phase chemical potential of chloride anions (μ_{Cl^-}). Each line represents one of the chloride coverages considered. Colored regions of the plot represent the minimum γ on each surface: minimum surface energies on Ag {100} are shown with solid lines and those for Ag {111} are shown with dashed lines. Wulff shapes are shown for several values of μ_{Cl^-}. The vertical red line at $\mu_{Cl^-} = -2.68$ eV denotes where the formation of bulk AgCl is thermodynamically favored over chloride-adsorbed Ag surfaces. (b) SEM image of Ag nanocubes [40].

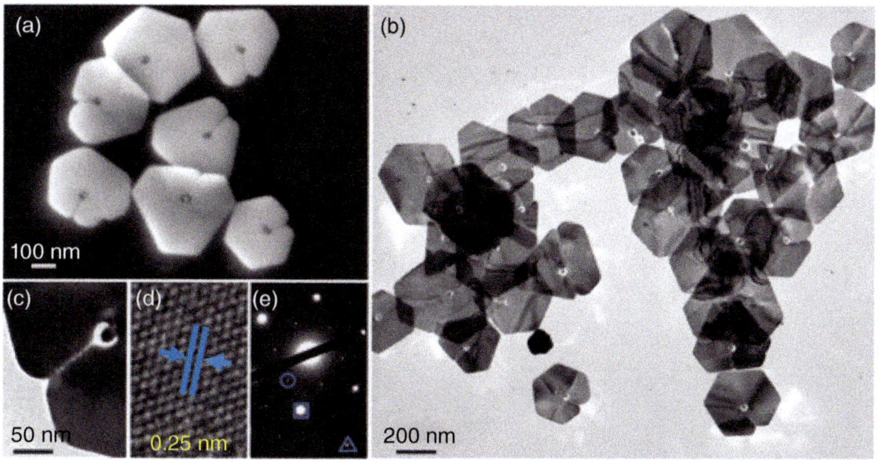

Figure 3.10 SEM (a), TEM (b, c), HRTEM images (d), and SAED pattern (e) of the hexagonal Ag nanoplates. The spots highlighted by a square, a triangle, and a circle correspond to {220}, {422}, and (1/3){422} reflections [42].

Chloride ions are a shape control agent used for the synthesis of Ag nanocubes exploiting thermodynamic control. Kristin and colleagues carried out calculations to confirm the role of the chloride anion in controlling the morphology of Ag NPs [40]. They found that with the increasing chemical potential of chloride, the surface free energy trend of Ag {100} and Ag {111} is reversed. Later, the surface free energy of Ag {100} is lower than that of Ag {111}, resulting in the formation of more Ag {100} facets. PVP as the most widely used structure-directing agent in the synthesis of Ag nanocubes, has also been widely studied and its dynamic control mechanism has been confirmed [41]. Generally, the synthesis route controlled by kinetics is efficient but hard to control compared with thermodynamic control (Figure 3.9a).

3.2.2.4 Ag Nanoplates with Long Narrow Gaps

In addition to the synthesis of Ag NPs with basic shapes (including Platonic bodies), there are also reports in the literature on direct synthesis of silver nanostructures with built-in hot spots. Chen and colleagues synthesized hexagonal and triangular Ag nanoplates with long and ultranarrow gaps via a seed-mediated growth method. In the synthesis process, Ag nanoplates could be formed without completely etching the seeds and the addition of NH_2OH tailored the shape of the nanoplates (Figure 3.10) [42]. It is intriguing that the formed nanogaps do not merge, which is addressed to a ligand effect. The ligand effect of polystyrene-block-poly(acrylic acid) is negligible between the two lobes with similar growth rate, but becomes inhibitive once the gap distance is reduced to its limit. These Ag nanoplates with ultranarrow nanogaps make the probe molecules accessible and exhibit high sensitivity in SERS detection, with a LOD of 10^{-9} M for 2-naphthalenethiol.

3.2.3 Cu Nanoparticles

From the technological, application-oriented and economic perspective, copper (Cu) has received much interest due to its low cost compared to the noble metals and its significant catalytic activity for reaction of carbon dioxide (CO_2) and nitrate (NO_3^-) [43]. Among the coinage metals Cu displays the largest onset wavelength of interband transitions in the extinction spectrum (about 600 nm) [44]. As mentioned above, considerable achievements have been made in the controlled synthesis of Au and Ag NPs with various morphologies and sizes with high monodispersity, yet comparable success with copper remains elusive because of its poor stability under ambient conditions. To address this issue, many synthesis routes have been reported in recent years and some Cu nanostructures have been readily prepared. Considering that Cu is easily oxidized, most of the schemes were carried out with an inert gas atmosphere or in a high vacuum environment.

3.2.3.1 Cu Quasispheres

The synthesis of spherical NPs has been of interest to researchers for a long time because of their isotropy. However, the high temperature required for copper ion reduction makes it difficult to control the morphology and size of the generated Cu NPs. Peng and colleagues demonstrated a simple one-pot method based on disproportionation reaction in oleylamine solvent, which successfully reduces the high temperature required for reduction reaction, and thus obtained copper nanospheres with excellent monodispersity [45]. Since $CuBr_2$ reduction required a higher temperature, they chose CuBr as the precursor and dissolved it in oleylamine, forming a complex, then under the promotion of trioctylphosphine (TOP) on the disproportionation reaction, monovalent Cu ions were reduced to Cu nanospheres (Figure 3.11).

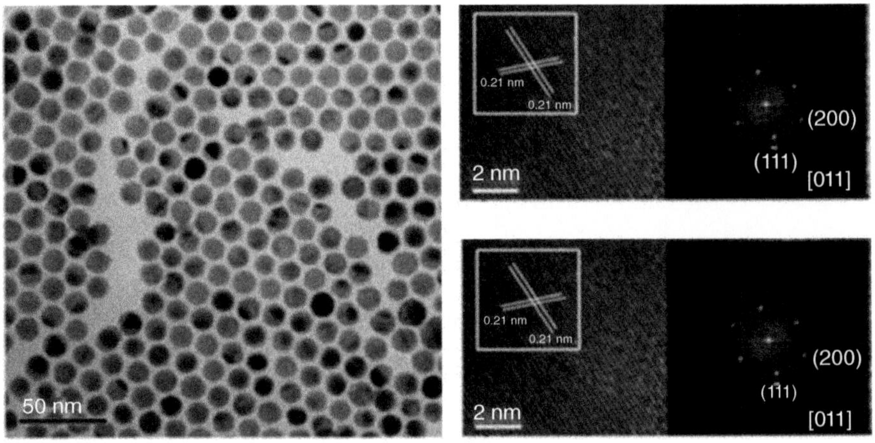

(a) (b)

Figure 3.11 (a) TEM image of monodisperse Cu nanospheres with an average diameter of 12.6 nm. HRTEM images of a single Cu nanosphere viewed along (b) {011} and (c) {001} zone axes [45].

3.2.3.2 Cu Nanorods

The synthesis of Cu NRs is different from the other metal NRs previously mentioned. So far, there is no efficient method for the synthesis of stable Cu NRs. Therefore, other metals (Au, Pd etc.) are usually used as seeds or templates to synthesize Cu NRs.

Yin and colleagues demonstrated a sample to induce twinning in metal nanostructures for anisotropic growth by taking advantage of the large lattice mismatch between two metals [46]. In a typical synthesis, $CuCl_2$ and hexadecylamine (HDA) were mixed together with glucose, followed by the addition of single-crystalline Au seeds. The mixture was heated to obtain Cu NRs with a gold seed inside.

3.2.3.3 Cu Nanocubes

Similar to the synthesis of Cu nanospheres, Peng and colleagues successfully prepared Cu nanocubes by using trioctylphosphine oxide (TOPO) instead of TOP [45]. Although both TOP and TOPO can promote the disproportionation reaction of Cu^+, the morphology of the obtained Cu NPs is very different. This can be attributed to the more stable complex formed by TOP and Cu^+, which prevents the selective adsorption of Br^- and thus leads to the formation of isotropic Cu NPs. When TOPO and Cu^+ form a complex, Br^- is more likely to selectively adsorb on the {100} facet, and a Cu nanocube is formed finally due to the dynamic differences of each facets.

3.2.4 Al Nanoparticles

In general, Au and Ag are excellent plasmonic materials with LSPR in the visible region, but especially their high-price limits the applications at large scale [47]. To extend the LSPR to the UV, many novel plasmonic materials have been reported, such as aluminum nanostructures, which are much cheaper than Au and Ag and exhibit a LSPR in the UV [48, 49].

3.2.4.1 Al Nanosheets

Sun and colleagues reported the wet-chemical synthesis of ultrathin Al nanosheets in the presence of a gas mixture of N_2 and O_2 with an appropriate ratio (10–80 vol%), in which $AlCl_3$ was first dissolved and $LiAlH_4$ was added [50]. The reaction occurred at 140 °C for 4 hours, resulting in gray products through centrifugation and washing with acetone and methanol in order to remove other impurities. Oxygen showed a major effect on the formation of Al nanosheets. Only a certain O_2 ratio resulted in the sheet-like structures, in which the selective oxygen adsorption on the {111} facet of the face-centered cubic (fcc) Al_{41} was believed to be the key factor to control the oriented growth of metallic Al nanosheets and reduce the thickness. With the preferential surface passivation, the nanosheets were endowed with improved stability in ambient environment. By changing the size and dimensions on the nanoscale, the thickness-dependent LSPR of the nanosheets was tunable (Figure 3.12).

3.2.4.2 Al Nanocrystals

Halas and coworkers reported a facile method for the synthesis of highly regular, faceted aluminum nanocrystals with controllable nanocrystal diameters

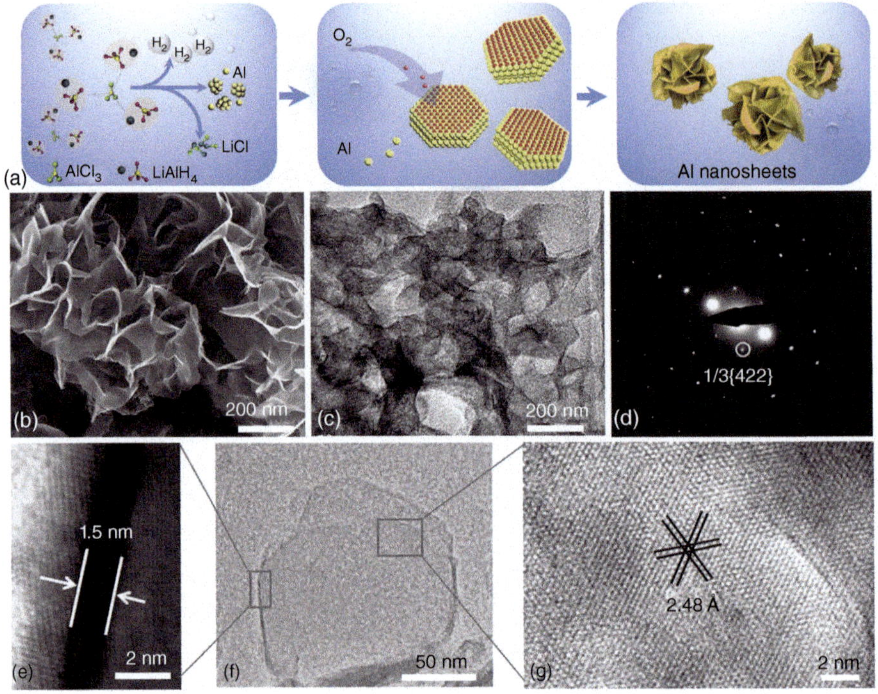

Figure 3.12 Schematic illustrations of the synthesis process and characterization of Al nanosheets [50]. (a) The wet-chemical synthesis of Al nanosheets involves three stages: nucleation, growth and assembly. SEM (b) and TEM (c) images of the flexible ultrathin Al nanosheets. (d) The electron diffraction pattern of a single Al sheet. (e–g) TEM analysis of the single Al sheet.

ranging from 70 nm to greater than 200 nm [51]. By adding a capping agent, *N*, *N*-dimethylethylamine alane was reduced in the mixed solution of 1,4-dioxane, tetrahydrofuran (THF) and titanium (IV) isopropoxide. Particles were centrifuged and cleaned to remove any unreacted reagents after reaction.

It was reported that the ratio of THF and dioxane was vital in controlling the sizes of the aluminum nanocrystals by this approach despite of their elusive roles. When dioxane was replaced by toluene which showed a similar dielectric constant, the particle sizes and shapes were the same. Results suggested that particle sizes were more directly related to the dielectric properties of the solution in this process. However, the role of titanium (IV) isopropoxide was not explicitly clear yet. The LSPR of Al nanocrystals showed a wide tunability from the UV to the NIR region of the spectrum as their size increased [51] (Figure 3.13).

3.2.4.3 Al Nanorods

In the synthesis of Al NRs, the reduction of alane (aluminum hydride) to metallic Al is accomplished by the hydride ligands, which produce thermodynamically governed isotropic nanocrystals owing to the presence of strong reducing agent.

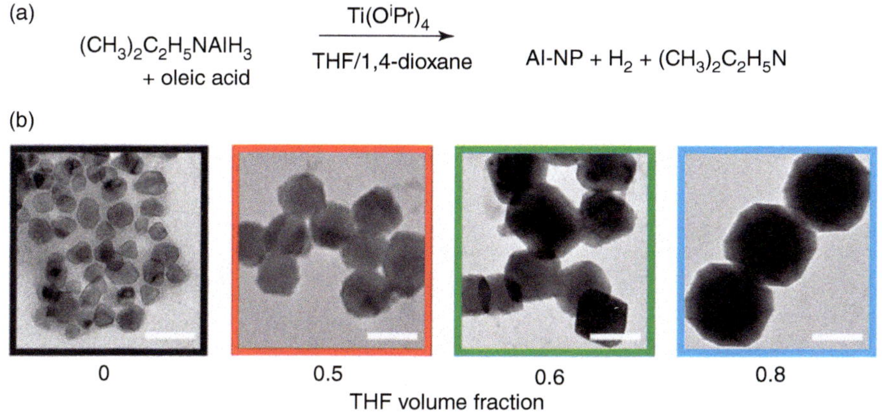

Figure 3.13 Size control of aluminum nanocrystals. (a) Reaction scheme for the synthesis of aluminum nanocrystals. (b) Representative TEM images from synthesis with different THF volume fractions in a THF/1,4-dioxane solution (from left to right) (Scale bar: 100 nm) [51].

Seed-mediated approaches to prepare anisotropic nanocrystals also require ligands, nevertheless common ligands such as CTAB, PVP, and sodium citrate are incompatible with alane. High-temperature decomposition of molecular precursors in solution phase has enabled the synthesis of monodispersed semiconductor nanostructures and similar thermal decomposition-based approaches may be applicable to synthesize Al nanocrystals.

Halas and colleagues mixed triisobutyl aluminum (TIBA) and trioctylamine in a three-neck round-bottom flask in an inert nitrogen atmosphere, heating the colorless reaction solution from room temperature to 250 °C. After 10 minutes at 250 °C, the solution turned opaque grey then the Al nanocrystal solution was cooled to room temperature [52].

Several studies showed that the shape of Al nanostructures could be optimized to create strong light-harvesting nanosystems, resulting from an improvement in the absorption efficiency combined with a suppression of the scattering efficiency [53] (Figure 3.14).

3.3 From Monometallic NP Films to Composite NP Architectures

Single isotropic plasmonic NPs generally do not have sufficiently high scattering cross-sections for SERS monitoring or cannot satisfy the need of both plasmonic activity and catalytic performance. Therefore, the appearance of plasmonic NPs with composite morphology has attracted much attention by researchers. Examples are the assembly of monomers into larger structures including NP monolayers, supra-particles/superstructures, and others, which exhibit significantly higher plasmonic activity than monomers.

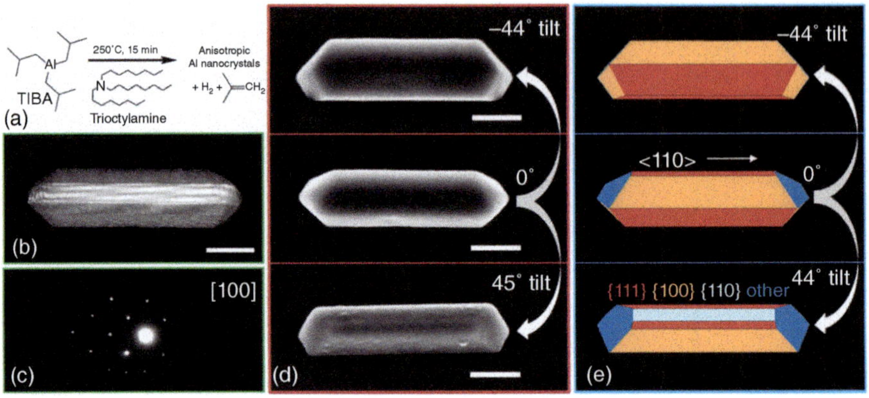

Figure 3.14 Synthesis and characterization of anisotropic aluminum nanocrystals. (a) Reaction scheme for the production of anisotropic Al nanorods. (b) Dark-field TEM micrograph of a representative Al nanorod. Scale bar is 250 nm. (c) Corresponding electron diffractogram. (d) Tilt series of the same Al nanorod, beginning with the orientation of the micrograph and tilted by $\pm\sim 44°$. All scale bars are 250 nm. (e) Model indicating the corresponding orientation of the Al nanorods [52].

3.3.1 Nanoparticle Monolayers

Yogev and colleagues reported first that silver films at dichloromethane–water interfaces could be obtained under certain conditions [54]. Since then such self-assembled NP monolayers have attracted extensive interest. In 2006, Sun and colleagues reported a universal approach for the self-assembly of hydrophilic NPs into ordered monolayer films at toluene–water interfaces, which has become a general method for the preparation of SERS substrates [55]. The individual preparation steps are shown in Figure 3.15. In this method, ethanol is an important inducer and the NPs are finally trapped at the interface.

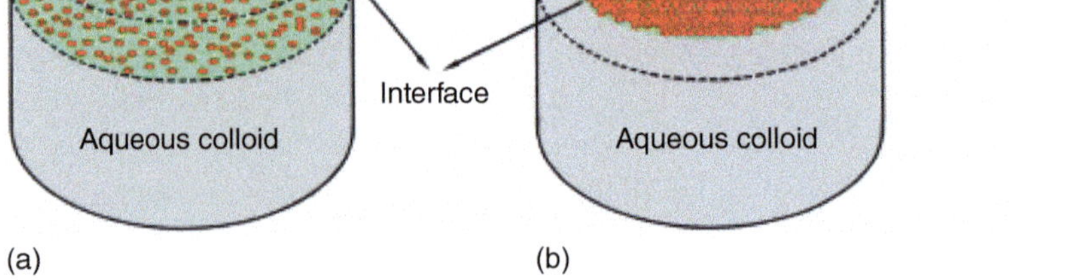

Figure 3.15 Schematic representation of the formation and transfer of an interfacial monolayer film to a solid substrate [55].

Xie and coworkers and other groups have applied this method for preparing Ag or Au NP monolayers as substrates to monitor reactions induced by energetic hot electrons on the Ag or Au NPs or to detect trace samples (see the Section 3.4 for specific applications) [56–58].

3.3.2 Superstructures

Superstructures or supraparticles, including core–shell–satellite and core–satellite structures, integrate high plasmonic and catalytic activity into a single bifunctional structure which provides us with a platform for label-free monitoring of chemical reactions by SERS. Au/Pt/Au core–shell nanoraspberries are the first reported type of such bifunctional NPs that have been used for *in situ* SERS monitoring of the 4-NTP reduction by hydride [5]. Large Au core particles were used to provide high SERS activity. A Pt shell was used because of its high catalytical activity. Au protuberances, obtained by galvanic replacement of a silver shell on top of the Pt shell, are present for adsorption of 4-NTP and provide sufficient SERS enhancement via plasmonic coupling with the large Au core. Therefore, this structure can both provide high SERS activity and catalytic activity. Figure 3.16 shows the morphological and structural changes involved in the synthesis of Au/Pt/Au core/shell nanoraspberries together with a SEM image of the final structure.

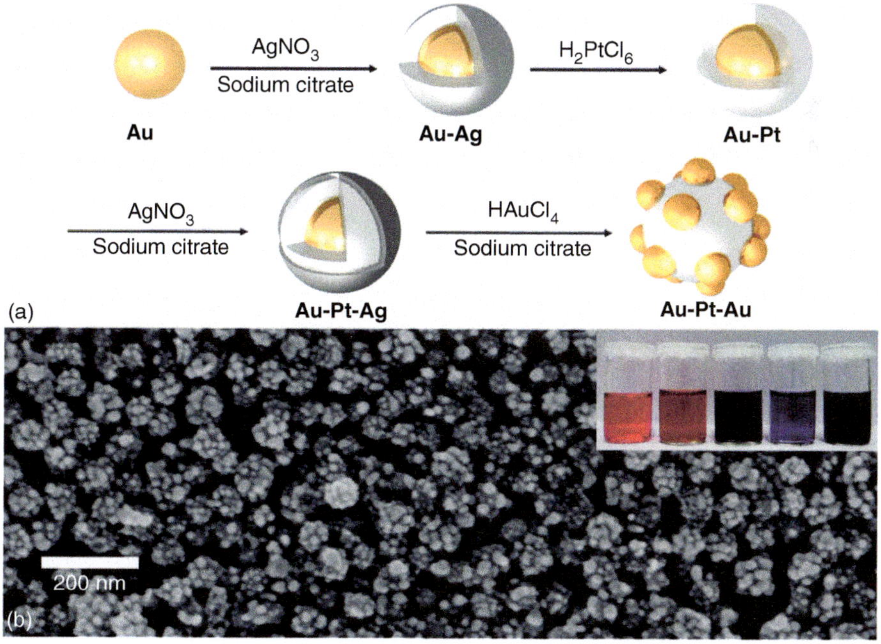

Figure 3.16 (a) Scheme showing morphological and structural changes involved in the synthesis of Au/Pt/Au core/shell nanoraspberries and (b) SEM image of Au/Pt/Au (inset: photos of the colloids during different synthesis steps) [5].

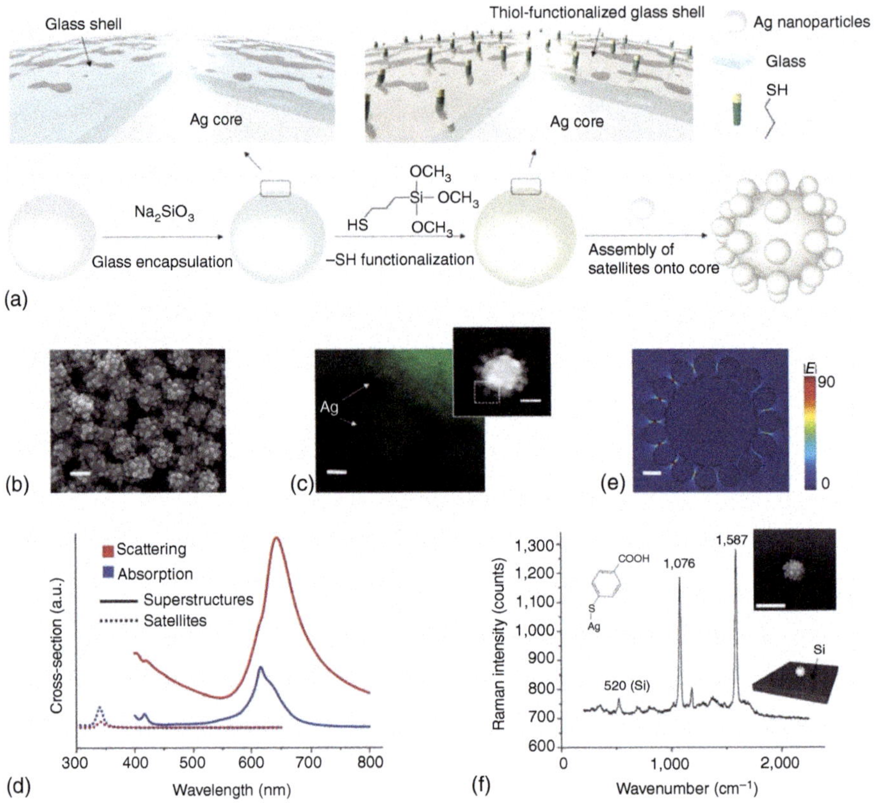

Figure 3.17 (a) Scheme of the synthesis of Ag core-satellite superstructures. (b) SEM image of the Ag superstructures (scale bar, 100 nm). (c) EDS element map (Ag) of the superstructure (scale bar, 10 nm) and the corresponding STEM image (scale bar, 50 nm). (d) Calculated absorption and scattering cross-sections of a small Ag satellite compared with the Ag superstructure. (e) Finite element method simulation of the incident electric field amplitude |E| distribution on resonant excitation at 632.8 nm (scale bar, 20 nm). (f) SERS spectrum of 4-MBA from a single Ag superstructure on a Si wafer shown in the SEM image (top right; scale bar, 200 nm).

In addition, Xie and Schlücker rationally designed core/satellite superstructures exhibiting significantly larger absorption and scattering cross-sections through plasmonic coupling between core and satellites. Mercaptosiloxane covalently bound to an ultrathin silica shell on the core was employed as a molecular linker for connecting core (silane) and satellite particles (thiol) [10, 59]. It is worth noting that the inert ultrathin silica shell on the Au core protects the metal surface from direct contact with the chemical species involved in the catalytic reaction and eliminates unwanted photocatalytic side reactions and realizes a noninvasive SERS investigation. A general synthesis sequence is shown in Figure 3.17a. From Figure 3.17b and c it can be seen that uniform and nearly perfect superstructure of Ag or Au NPs have been synthesized. Computer simulations predict a local electric field enhancement of $|E| \sim 90$ upon resonant excitation of the Ag superstructure. This corresponds

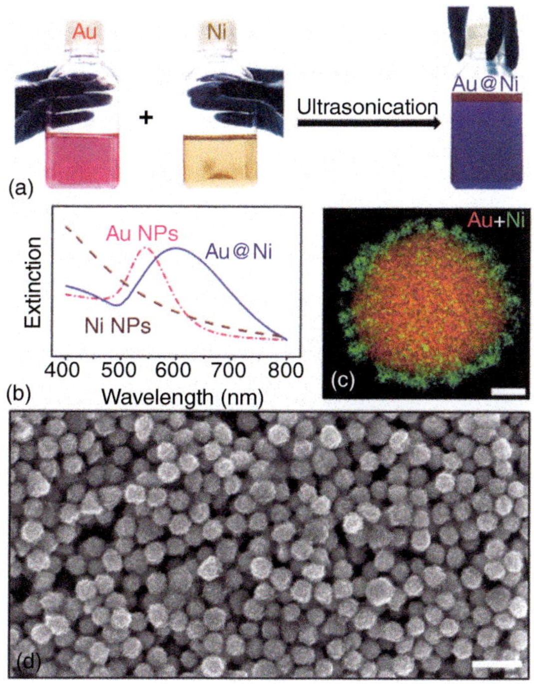

Figure 3.18 (a) One-step synthesis of Au@Ni superstructures. (b) Extinction spectra of Ni NPs, Au NPs, and Au-Ni superstructures. (c) Elemental map of a single superstructure. Scale bar: 20 nm. (d) Scanning electron microscopy (SEM) image of many Au-Ni superstructures. Scale bar: 200 nm [61].

to a SERS enhancement factor (EF) of $|E|^4 \sim 6.6 \times 10^7$. The SERS activity at the single-particle level enables quantitative and label-free reaction monitoring. Until now, this method has been successfully utilized in preparing superstructures with various kinds of metals and their oxides including Au-Cu, Au-CuO, and Au-NiO [60], in addition to the original reports on Au-Au and Ag-Ag structures.

In this context, for example, Xie and coworkers presented a one-step assembly strategy to obtain Au-Ni superstructures [61]. Compared to previous sophisticated synthetic methods, this strategy is very simple and just requires mixing Ni NPs suspended in hexane with Au NPs suspended in isopropanol under vigorous ultrasonication (Figure 3.18a). Up to 2 L of superstructure colloid can be synthesized in a single experiment. In the assembly process, hexane as the solvent induces a destabilization of ligand molecules on the Au surface, which leads to an assembly of Au and Ni NPs into core–satellite structures via both van der Waals (vdW) and electrostatic forces. The yield of Au-Ni superstructures is almost 100% and no further separation step is required. Self-assembly of the Ni satellites onto the Au core leads to a ~50 nm redshift of the plasmon band (Figure 3.18b). Figure 3.18c shows the elemental map of a single superstructure with an 80 nm Au core and many 7 nm Ni NPs as satellites. The average satellite number for each Au core is 72 ± 7 (Figure 3.18d).

3.3.3 Other Structures

In addition to the popular composite structures described above, there are increasing numbers of novel structure reported [62–64]. For instance, submonolayer Pt-coated

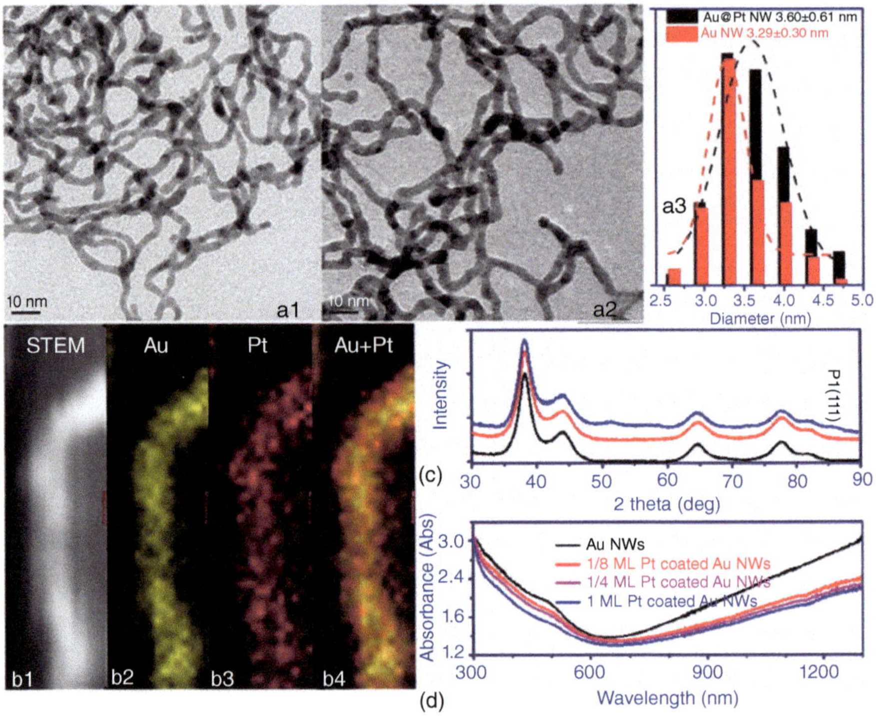

Figure 3.19 Characterization of Au and Au@Pt NWs by (a) TEM, (b) HAADF-STEM-EDS mapping, (c) XRD, and (d) UV–vis–NIR extinction spectra [64].

ultrathin Au nanowires (Au@Pt NWs) have been synthesized and used in catalysis [64]. Besides providing a strong local electric field for Raman signal enhancement, the Au NWs remarkably enhance the catalytic activity of Pt atoms through improving their dispersity and altering their electronic structure. As shown in Figure 3.19, ultrathin Au NWs were synthesized first and then ice-cold H_2PtCl_6 solution was added dropwise under vigorous magnetic stirring. According to the theory of heterogeneous nucleation and growth, in a solution with a low concentration of metal precursor, the generated reduced metal atoms prefer adsorbing on the existing dispersed Au NWs rather than forming separated nuclei. The integration of excellent SERS and high catalytic activity within Au@Pt NWs make them a good platform for studying catalyzed reaction. As a proof of principle, the self-organized Au@Pt NWs thin films were employed for in-operando SERS monitoring of the 4-NTP reduction process [64].

Generally, the plasmon resonances of most plasmonic metal nanostructures are in the visible region and NIR below 1000 nm, which correspondingly limits their applications to this spectral range [65, 66]. Extending this concept further to the NIR, i.e. the rational design of metal nanostructures with plasmon peak wavelengths beyond ~1000 nm paves the way to a potential broad range of applications in biomedical imaging [67], medical diagnostics, and solar energy harvesting [68]. To this end, Wang and colleagues reported a relatively general method for the synthesis

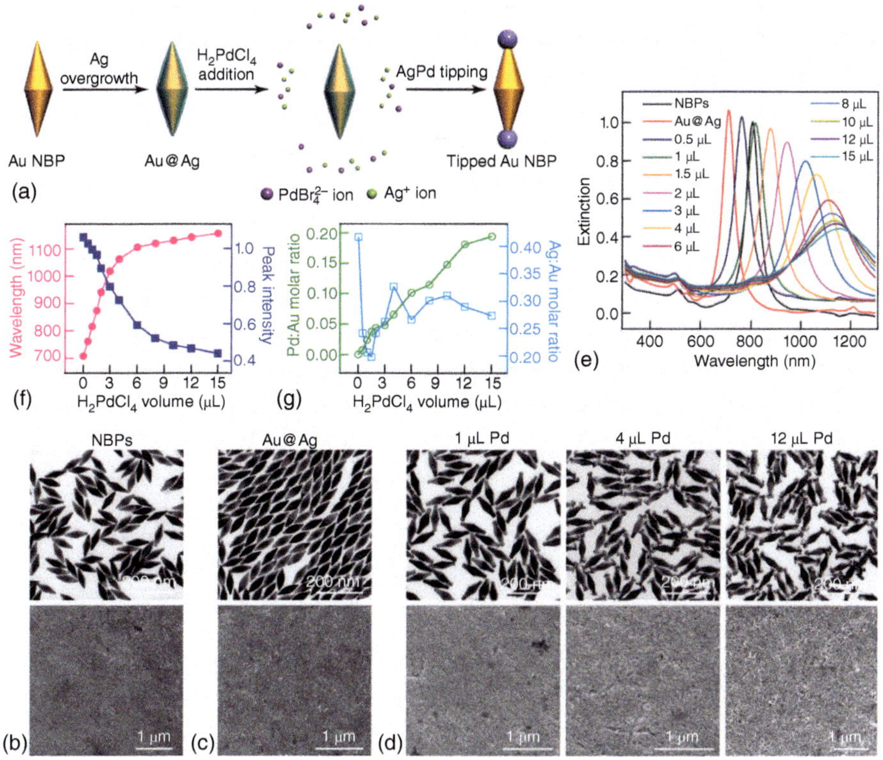

Figure 3.20 AgPd tipping on the 804 nm Au NBPs. (a) Schematic illustration of the synthesis. (b, c) TEM (top) and SEM (bottom) images of Au NBPs and the Au NBP@Ag samples, respectively. (d) TEM (top) and SEM (bottom) images of the Pd-induced metal-tipped Au NBPs produced with different volumes of aqueous H_2PdCl_4solution, respectively. (e) Vis/NIR extinction spectra of the Au NBPs, Au NBPs@Ag sample, and the Pd-induced metal-tipped Au NBPs produced with varying amounts of H_2PdCl_4 [63]. (f) Variations of the longitudinal dipolar plasmon wavelength and peak intensity as functions of the added volume of H_2PdCl_4. (g) Dependences of the Pd:Au and Ag:Au molar ratios of the AgPd-tipped Au NBP samples on the added volume of H_2PdCl_4. The point at 0 μL represents the Au NBP@Ag nanostructure sample [63].

of the metal-tipped Au nanobipyramids (NBPs) with tunable plasmon peaks also beyond 1000 nm [63]. As shown in Figure 3.20a, the first step is the overgrowth of Ag on the Au NBPs or NRs, yielding Au NBP@Ag or Au NRs@Ag nanostructures, respectively. This is a crucial step for achieving the anisotropic deposition of a second metal on these pregrown nanocrystals since Ag atoms are mainly deposited on the side surface. The tipped Au NBP is then formed by the sequential addition of metal precursors and AA as reductant. The amount of the added metal precursor determines the size of the tipped Au NBP as well as the position of its longitudinal plasmon peak, as shown in Figure 3.20d and e. Overall, the method is capable to obtain Au NBPs and NRs tipped by different kinds of metals or their alloys.

3.4 SERS Studies of Photo- and Electrocatalysis

3.4.1 Photocatalysis

Plasmonic nanostructures have unique optical properties and provide interesting alternative approaches to conventional photocatalysis with molecular photocatalysts. At present, gold, silver, and copper nanomaterials play a key role in plasmonic photocatalysis and SERS due to their pronounced LSPR. The nonradiative LSPR decay channel results in the generation of energetic hot electrons and holes that can actively participate in photochemical reactions [69]. Researchers have shown that hot electrons can be generated in two ways: plasmon-mediated electron-transfer (PMET) pathway and chemical interface damping (CID) [70, 71]. Studies have shown that the hot electrons or holes generated after LSPR excitation can induce the reaction directly or the reaction can be induced indirectly by activating H_2 or O_2 in the system. Because SERS and plasmonic catalysis both involve the interactions of photons, molecules, and metals, SERS is an excellent spectroscopic tool for studying plasmonic photocatalysis [72]. Two model reactions that have been widely investigated over the past years are the oxidative coupling of 4-ATP as well as the reductive coupling of 4-NTP both to the same azo product, 4,4′-dimercaptoazobenzene (4,4′-DMAB) [73–75].

3.4.1.1 Oxidation of Aniline

Ren and coworkers used normal quasispherical gold and silver NP and reported that Au or Ag oxides or hydroxides on the surface are the catalytically active centers to oxidize the surface species under LSPR assistance [76]. Catalytic oxidation of 4-ATP involves the interaction of O_2, light, and the metal surface. In negative control experiments, 4-ATP was also converted to DMAB even without O_2, which suggests that oxygen-containing components play a key role. DFT calculations reveal that 4-ATP oxidation by Au or Ag oxides/hydroxides is easier than that on bare Au or Ag NPs. Overall, this reaction is a prominent example of a LSPR-excited photocatalytic oxidation reaction (Figure 3.21).

Camargo and colleagues employed label-free SERS for revealing the mechanism of aniline oxidation with plasmonic Au@AgAu nanorattles [77]. Plasmonic Au@AgAu nanorattles possess two advantages for plasmonic photocatalysis: (i) nanorattles consist of an Au nanosphere inside of an AgAu nanoshell. The plasmon hybridization between Au nanosphere and AgAu nanoshell results in 14-fold higher electric field enhancements than bare Au or Ag NP; (ii) electromagnetic hot spots can be generated in individual nanostructures instead of uncontrolled aggregation. Laser power- and irradiation time-dependent SERS spectra of 4-ATP oxidation on Au@AgAu nanorattles demonstrated that the reaction is LSPR-induced rather than thermally driven. The generated hot electrons can directly activate the nearby O_2 in air into the active O_2^- species for further electron transfer to the absorbed molecules. Compared with AgAu nanoshells and Au NPs, the Au@AgAu nanorattles also exhibited an excellent activity for aniline coupling in real catalytic system under mild conditions.

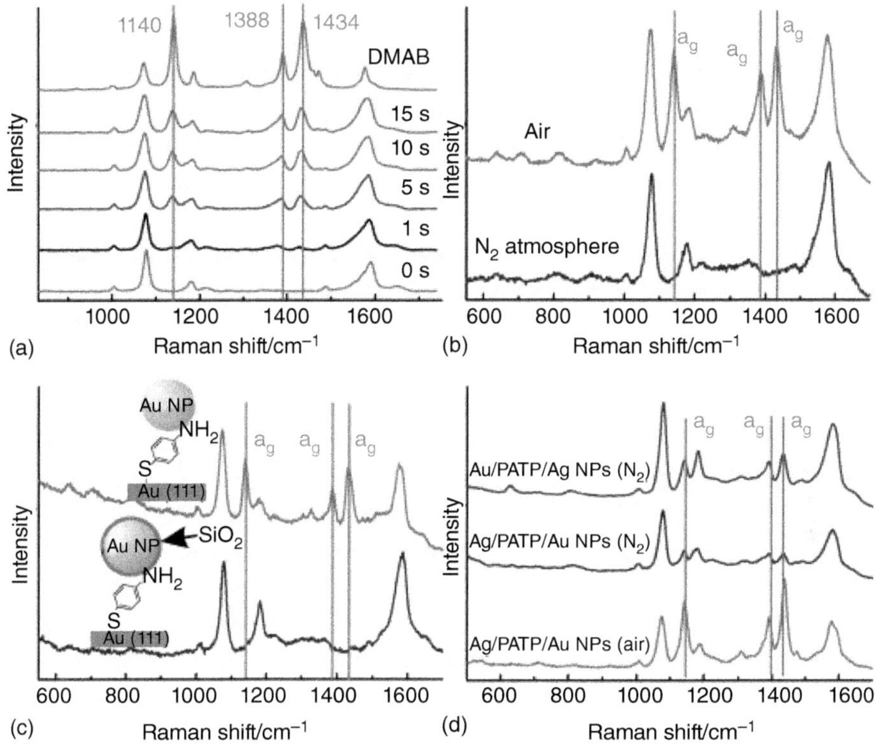

Figure 3.21 4-ATP oxidation induced by hot electrons. (a) Raman spectra of Au/4-ATP/Au in air under continuous laser irradiation. (b) Raman spectra of Au/4-ATP/Au in air and N_2 under the corresponding conditions. (c) Au {111}/4-ATP/Au NP (top) and Au {111}/4-ATP/Au@SiO_2 NP (bottom) junctions in the Raman spectra in air. (d) Ag film/4-ATP/Au NP and Au film/4-ATP/Ag NP junctions Raman spectra in air and N_2 [76].

Based on the above analysis, *in situ* SERS could not only provide fingerprint information about surface species, but also identify the catalytically active species in the LSPR-induced reaction.

3.4.1.2 Reduction of Nitroarenes

Hot electrons generated by LSPR excitation of Au can also drive the reductive coupling of the 4-ATP to 4,4′-DMAB monitored by SERS [78]. Xie and Schlücker observed the signal of 4,4′-DMAB during SERS studies of 4-NTP reduction on Au-Pt-Au nanoraspberries, which was later demonstrated to be induced by hot electrons [5]. They further found that the hot electron-induced coupling can be avoided when the 80 nm Au NPs cores in Au-Au superstructures were coated with an ultrathin silica shell [10]. Frontiera and colleagues utilized ultrafast time-resolved Raman thermometry for detecting the effective temperature of molecules attached to spherical Au NP aggregates in order to probe the role of hot electrons in plasmon-driven chemistry. Their ultrafast measurements indicate that this process is not primarily driven by heat [79]. Xie and Schlücker have also investigated the photorecycling mechanism on Ag NPs by promoting the catalytic

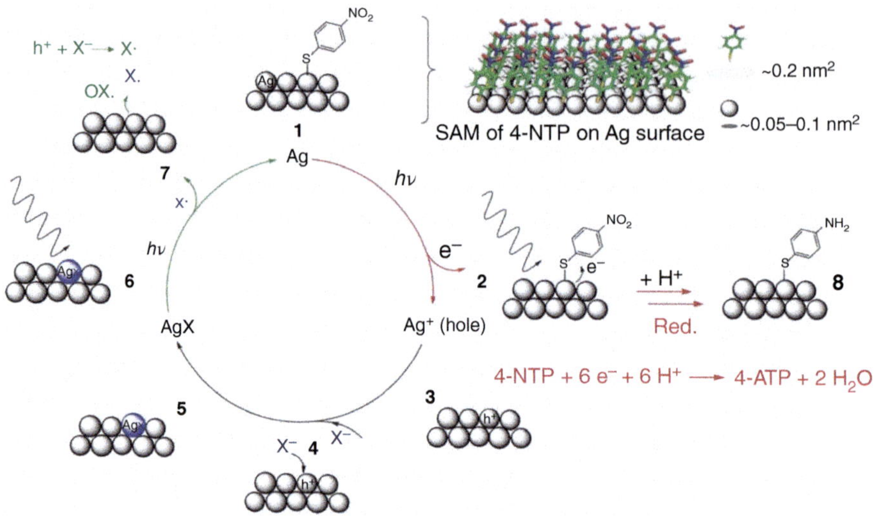

Figure 3.22 Counter-half-reaction-promoted hot electron photorecycling [59].

recovery of light in a plasmon-mediated oxidation reaction. In Figure 3.22, the six-electron reduction of 4-NTP to 4-ATP occurs even in the absence of hydride reagents [59]. Hot electrons generated from silver can be transferred to adsorbed molecules on the metal surface. Protons serve as the hydrogen source. Halide ions are required for photorecycling of the electron-donating silver atoms from hot holes after photodissociation of the insoluble silver halides present on the Ag surface. A series of control experiments demonstrate that without this photorecycling counter-half reaction, the six-hot electron reduction cannot proceed. The discovery was enabled by the use of *in situ* SERS spectroscopy, which enabled the direct identification of the reaction product online and in a label-free approach.

3.4.1.3 Dehalogenation

The activation of the carbon–halogen (C–X) bond is of great importance in organic synthesis. With the help of *in situ* SERS, Xie and colleagues demonstrated that plasmon-generated hot electrons on noble metal (Au or Ag) surfaces can contribute to an effective C–X bond activation (Figure 3.23) [56]. Using Au or Ag NP monolayers as SERS substrates, they found that the dissociation of the C–X bond on 80 nm Ag NP monolayers is much faster than that on 40 nm Ag and 80 nm Au under similar tested conditions. They demonstrated that the difference is closely related to the LSPR excitation of the metal NPs.

Based on the above results, the same group designed and synthesized an Au/CdS plasmonic catalyst and achieved the efficient dehalogenated deuteration under mild conditions [80]. SERS was again a useful tool to understand the mechanism. By using 2,6-dimethylphenyl isocyanide (2,6-DMPI) as the SERS probe molecule, they revealed an electron transfer from Au NPs to CdS. Since the wavenumber

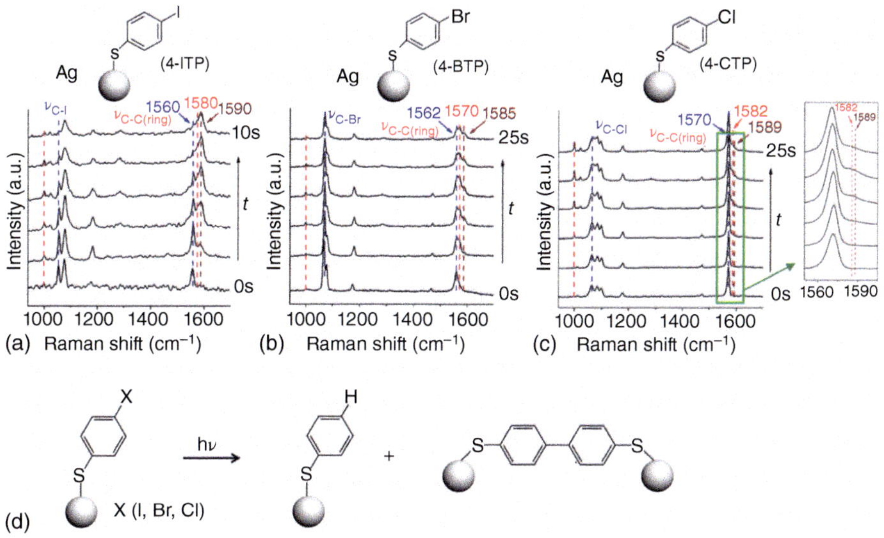

Figure 3.23 *In situ* SERS spectra acquired during the dehalogenation of 4-iodothiophenol (4-ITP) (a), 4-bromothiophenol (4-BTP) (b), and 4-chlorothiophenol (4-CTP) (c) on an 80 nm Ag NP monolayer. (d) Scheme of the aryl halide transformation on the metal NP monolayer under laser illumination [56].

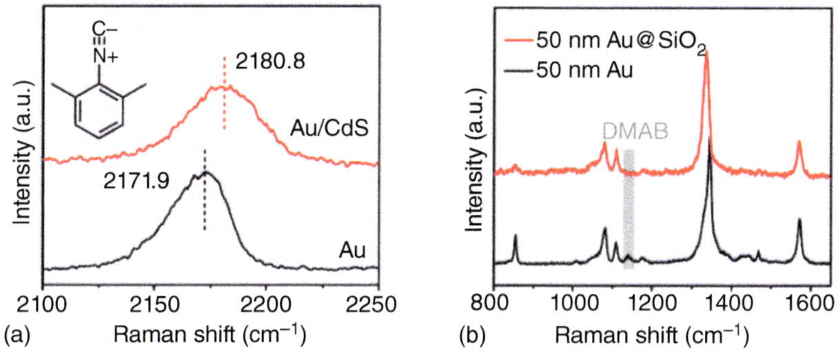

Figure 3.24 (a) SERS for revealing the electron transfer between Au and CdS by using 2,6-DMPI as the probe molecule. (b) SERS for revealing the blocked electron transfer between Au and Au@SiO_2 [80].

position of the N≡C stretching vibration is very sensitive to the electronic structure, the blue shift of the ν_{NC} stretching band of 2,6-DMPI from 2171.9 cm^{-1} (Au) to 2180.8 cm^{-1} (Au/CdS) (Figure 3.24) indicated an enhanced δ-donation from the N≡C to the d-band of Au on Au/CdS. This is likely due to the decreased d-electron density of Au induced by electron transfer from Au to CdS. The blocked electron transfer at the Au@SiO_2/CdS interface was also confirmed by SERS using the hot electron-induced reaction of 4-NTP to DMAB as a model reaction (Figure 3.24).

3.4.2 Electrocatalysis

3.4.2.1 Hydrogen Evolution Reaction

With the advantages of easy-obtainable water sources and their environmental benignity, water splitting reactions including both the hydrogen evolution reaction (HER) and the oxygen evolution reaction (OER) are regarded as one of the best choices for replacing the currently dominant carbon economy [81]. However, the costly and short-lived catalysts used in the water splitting process are limiting the large-scale implementation for hydrogen production. As a classical catalytic reaction, the mechanism of the water splitting reaction has received attention for a long time. Although researchers have tried many experimental and theoretical studies to reveal the mechanism, the detailed surface conversion process is still undisclosed. Clarifying the configuration of solvation-type interactions between the catalysts surface and reaction solutions could significantly improve our understanding of the electrode/electrolyte interface. SERS was applied to investigate the electrode surface reaction mechanism in the water splitting reactions. SERS offers several advantages over other vibrational spectroscopic probes of electrode surfaces. In particular, the enhancement is strongest for molecules very near at or on the surface of the substrate, which makes SERS particularly useful in the study of water splitting reactions [82].

Important progress in extending SERS studies of interfacial water to solutions free of (pseudo) halide ions was reported by Funtikov and colleagues in 1987 [83]. To control and improve HER, it is required to detect and observe the reaction system at the molecular level. Murakoshi and colleagues provided a method to investigate the electrode surface in HER with molecular specificity by using SERS-active Au dimer nanostructures prepared by angle-resolved nanosphere lithography (NSL) [84]. They designed an *in situ* electrochemical SERS device to evaluate the behavior of 4,4-bipyridine (4,4-bpy) on the electrode surface during the HER process. Based on both interpretation of SERS observations and theoretical simulations, the catalytic mechanism of 4,4-bpy on the metal electrode in HER was studied.

To gain a deep insight into the interfacial HER, Tian and colleagues proposed a spectroelectrochemical cell configuration which ensures the observation of both stretching and bending modes of water [85]. Cu and Ag were used as electrode material in the HER to achieve a significantly enhanced Raman signal.

3.4.2.2 Oxygen Evolution Reaction

Herein, some SERS studies about studying electrode surface reactions are summarized. However, direct evidence is still required to elucidate the mechanism of the OER. The involved species are well suited to be identified by SERS owing to their Raman activity.

The designed SERS strategy, which combined a SERS-active material and reactive substrate, was presented by Koper's group to investigate the OER on nickel oxyhydroxide (NiOOH) surface [86, 87]. For the SERS experiments they used electrodes comprising NiOOH electrodeposited on gold. They presented an *in situ* SERS study on NiOOH catalyst for the OER, providing spectroelectrochemical evidence for the active species and also demonstrating the influence of different reaction parameters.

3.4.2.3 Oxygen Reduction Reaction

The rapid development of our modern global society with high fossil fuels consumption has unavoidably resulted in the environmental deterioration and energy crisis [88]. The emerging demand for next generation renewable and sustainable energy conversion technologies is a strong motivation for studies on new energy conversion and storage devices. With the advantages of a high energy density and being environmentally friendly, fuel cells and alkali metal-oxygen batteries, which can effectively convert chemical energy to electric energy via electrochemical catalytic processes, are two of the most promising candidates for energy conversion and storage.

As the fundamental reaction in fuel cells and alkali metal-oxygen batteries, the oxygen reduction reaction (ORR) has attracted much attention for its sluggish kinetics. A comprehensive understanding of its mechanism is necessary for designing high-performance catalysts. SERS with high chemical specificity, high sensitivity, and surface selectivity is a powerful tool for monitoring the molecular transformations, identifying intermediates and elucidating the mechanism of the ORR.

3.4.2.3.1 SERS Study in Fuel Cells

In fuel cells, oxygen gas is reduced in aqueous electrolytes. At present, two kinds of mechanisms have been proposed for the ORR process in fuel cells, namely the four-electron and the two-electron pathway [88]. However, some essential questions and uncertainties for the ORR processes are still challenges, including slow kinetics, the origin of high over-potentials and the rate-determining step. As a classical catalyst in ORR, platinum-based catalysts have been in the focus in ORR research [89]. Many research groups have carried out experimental and theoretical studies to reveal the ORR mechanism.

Li and co-workers have employed *in situ* electrochemical (EC) SERS to examine the ORR intermediates at single-crystalline platinum surfaces [90–92]. They employed shell-isolated NPs (SHINs) on different Pt-surfaces for avoiding direct contact of the plasmonic material inside of the SHIN with the molecules on the Pt surface.

Under alkaline conditions, the O-O stretching vibration of the superoxide ion O_2^- ($1150\,cm^{-1}$) could be detected. Based on the EC-SHINERS experiments and in combination with results from theory as well as from previous research, the mechanism of ORR on the Pt {*hkl*} electrode surface was proposed based on SERS results. The key ORR intermediates, OH* and HO_2*, were detected for the first time at these high-index Pt {*hkl*} surfaces by SERS. With these SERS results it is possible to design new catalysts with different crystal surface structures to achieve higher ORR catalytic activities.

In addition to single metallic Pt catalysts, ORR processes on bimetallic catalysts have been investigated by Wang and colleagues, using assembled Pt_3Co nanocatalyst on SHINs and monitoring ORR processes by SERS [12].

Besides investigating the reaction mechanism, studying the failure mechanism of ORR for Pt-based catalysts is also critical to further enhance the activity and durability of catalysts. The current consensus is that the repetitive formation and reduction of Pt oxides play a crucial role in the dissolution of surface Pt atoms, which degrades the Pt-based catalysts. Tong and colleagues have employed *in situ* SERS to study the surface Pt oxidation effect on the ORR process [93]. Pt was used as SERS-active substrate directly and SERS measurements were taken with Pt-based catalysts with or without sulfide adsorption. During that SERS study they concluded that sulfide adsorption enhances the Pt surface's resistance to oxidation and makes the Pt surface nobler, which helps the Pt surface to become more ORR active and more structurally stable.

3.4.2.3.2 SERS Study in Alkali Metal-Oxygen Batteries

Compared with the fuel cells, the ORR in alkali metal-oxygen batteries occurs in nonaqueous aprotic electrolytes instead of an aqueous electrolyte. Aside from the four-electron and two-electron pathways, O_2 can also be reduced to superoxide (O_2^-) involving one electron transfer in nonaqueous aprotic electrolytes [94].

Diverse pathways lead to different intermediates and products with distinct stability and reversibility, which greatly influences the battery performance. Thus, the characterization of the intermediates and products for alkali metal-oxygen batteries are indispensable. As a powerful tool, SERS has also been utilized to study ORR in alkali metal-oxygen batteries [95].

Li-O_2 Batteries Most SERS studies on Li-O_2 batteries use a SERS-active gold substrate as a model cathode material because of its convenience and reproducibility. Bruce and colleagues used a nanoporous gold (NPG) cathode for SERS observations [96]. The NPG was prepared by chemical etching of a white gold leaf (Au-Ag alloy, 1 : 1 by weight) by concentrated nitric acid, thus one was able to obtain a free-standing gold film with a nanoporous structure. Peng and colleagues conducted a SERS study to demonstrate the formation of LiO_2 and Li_2O_2 during the discharge process [97]. They used a normal Au electrode as SERS substrate. The obtained Raman spectra provide direct spectroscopic evidence that the reduction of O_2, in a nonaqueous electrolyte in the presence of Li ions forms O_2^- which then forms LiO_2 electrode.

Aside from gold substrates, also SHINs were applied to SERS studies on Li-O_2 batteries. SHINERS can overcome the limitations of gold electrode and ORR study on other electrodes including glassy carbon, palladium, and platinum disk electrodes [98]. The use of SHINERS in the presence of Li^+ has shown that both surface and solvent can be harnessed to influence ORR pathways, which may be critical in designing electrode/electrolyte interfaces that can minimize side reactions within Li-O_2 cells.

Na-O_2 and K-O_2 Batteries Intensive research on Li-O_2 batteries has led to alternative alkali metal-oxygen batteries, such as Na-O_2 and K-O_2 batteries. The oxygen discharge products for Na-O_2 and K-O_2 electrochemistry have also been investigated by SERS [99, 100].

SERS data of Na-O_2 batteries provided by Hardwick and colleagues indicated that the choice of solvent can strongly affect the overall surface discharge products on planar roughened Au electrodes. *In situ* SERS signals of batteries with different electrolytes including dimethyl sulfoxide (DMSO), dimethylacetamide (DME), diethylene glycol dimethyl ether (DEGDME), and acetonitrile (MeCN) were collected during discharge process. Upon discharge, the high donor number solvents, DMSO and DMA, produced signals in the region for O_2^- and NaO_2. For DEGDME and MeCN-based electrolytes the obtained SERS data showed that the main discharge product was Na_2O_2. The absence of a NaO_2 signal indicated preferential Na_2O_2 formation, suggesting that any initially formed NaO_2 is short-lived or that superoxide was solely present before a second electron transfer.

For K-O_2 batteries, *in situ* SERS was also conducted to confirm that the product of the cathode reaction is a combination of KO_2 and K_2O_2, depending on the applied potential. The use of the low donor number solvent, MeCN enabled to directly probe the surface route [100].

Part Conclusion SERS has been widely used in ORR studies by many researchers. Much valuable information about the reaction intermediates and products has been obtained: influences of potential, solvents, and substrates as well as on redox mediators and reaction mechanisms were obtained based on the molecular specificity of the SERS technique. These observations have significantly improved our understanding of the reaction mechanism and the kinetics of the ORR process. Nevertheless, in most cases simple gold electrodes were used for SERS monitoring. From our point of view, the more complex SHINERS give better results due to their broader applicability to also nonplasmonically active electrodes.

3.4.2.4 Electrocatalytic CO_2 Reduction

The development of clean and efficient renewable new energy sources has become increasingly important. On the one hand, the rapid development of industry has further exacerbated the energy shortage. On the other hand, harmful gases such as carbon dioxide (CO_2) released by burning fossil fuels mainly composed of coal and petroleum has triggered the "greenhouse effect," environmental pollution, and global environmental safety issues. Through the reduction reaction with H_2O, CO_2 is converted into hydrocarbons or carbohydrates to truly realize the recycling of carbon materials. This could partially solve the problem of energy shortage and alleviate the greenhouse effect caused by the continuous accumulation of CO_2.

Among many methods for reducing CO_2, electrocatalytic CO_2 reduction is one of the most effective methods. The reaction rate is not only controllable, but also the product selectivity is high. In the past two decades, with the increasing attention of the international community to energy and environmental issues as well as the rapid development of nanomaterial chemistry, a large number of reports on the electrocatalytic reduction of CO_2 have emerged and become one of the most popular research topics at present. However, compared with the equally popular HER, the electrocatalytic reduction of CO_2 is a more complicated system. Under the conditions of an aqueous solution, room temperature, standard atmospheric pressure, neutral pH, and a standard hydrogen electrode (NHE) as a reference electrode, a variety of reactions can occur [101]. However, so far there is no clear explanation of the reaction mechanism which describes exactly what occurs during the electrocatalytic CO_2 reduction process. Research on the reaction mechanism is still relying largely on theory. The groups of Norskov [102], Koper [103], Cater [104], and others have worked on the CO_2 photocatalytic reduction; however, most of the theoretical results are not supported by experimental evidence. A detailed understanding of the electrocatalytic reduction mechanism requires *in situ* studies with molecular specificity, for example, by using SERS.

Copper-based catalysts are widely used in electrocatalytic reduction of carbon dioxide. Sargent and colleagues successfully prepared AgCu alloys, achieving 41% Faradaic efficiency and 25% energy efficiency for ethanol [105]. To confirm the mechanism, they used *in situ* SERS to investigate the energetic fingerprints of the binding sites on Cu and Ag/Cu samples, without any additional SERS-active substrate by exploiting the SERS-activity of Ag and Cu. Thus, with many different binding configurations on the Ag/Cu surface, each binding configuration has a different electron back-donating ability and shows up at a distinct wavenumber position. Therefore, the proposed mechanism is supported based on these Raman results. Introducing Ag to the Cu surface increases the diversity of binding sites. As a consequence, the coordination environment of intermediates in the ethylene path is reduced, making the catalyst highly selective for ethanol production.

3.4.2.4.1 Outlook

Through the past decades, well-defined and controllable noble metal nanocrystals have been synthesized via various methods and applied to sensing, imaging, biomedicine, environmental science, materials, energy, and catalysis due to their LSPR properties. Despite the incredible accomplishments, the synthesis of metal nanostructures still faces several challenges that require a tremendous effort to be addressed. In terms of monomer synthesis, for example, it remains difficult to produce monodispersed NPs at large scale. Also theory lacks behind in terms of quantitatively describing or even predicting the formation of various NP architectures.

In order to fully exploit their optical and catalytic properties, it is necessary to synthesize and assemble multifunctional and multimetallic nanoscale architectures for more extensive applications. Core-shell NPs, shell-isolated NPs, bifunctional superstructures, and dimers have demonstrated their performance in photo- and electrocatalysis and *in situ* SERS studies. In recent years, monitoring catalytic reactions at the single-particle level and exploring their mechanisms have attracted much attention. However, there is still no routine approach to synthesize highly uniform and stabilized multimetallic configurations for obtaining structural homogeneity and plasmonic uniformity at the single-particle level. In particular, novel nanostructures that can improve the efficient utilization of the hot electrons for better catalytic performance are highly desired. Overall, a lot of research remains to be done in terms of understanding all the complexities involved in the nanosynthesis of plasmonic particles in order to fully exploit their potential for highly efficient photo- and electrocatalysis.

References

1 Hussein, H.E.M., Amari, H., and Macpherson, J.V. (2017). Electrochemical synthesis of nanoporous platinum nanoparticles using laser pulse heating: application to methanol oxidation. *ACS Catalysis* 7 (10): 7388–7398.

2 Tomko, J., Naddeo, J.J., Jimenez, R. et al. (2015). Size and polydispersity trends found in gold nanoparticles synthesized by laser ablation in liquids. *Physical Chemistry Chemical Physics* 17 (25): 16327–16333.

3 Mendivil, M.I., Krishnan, B., Castillo, G.A., and Shaji, S. (2015). Synthesis and properties of palladium nanoparticles by pulsed laser ablation in liquid. *Applied Surface Science* 348: 45–53.

4 Heck, K.N., Janesko, B.G., Scuseria, G.E. et al. (2008). Observing metal-catalyzed chemical reactions in situ using surface-enhanced Raman spectroscopy on Pd-Au nanoshells. *Journal of the American Chemical Society* 130 (49): 16592–16600.

5 Xie, W., Herrmann, C., Kömpe, K. et al. (2011). Synthesis of bifunctional Au/Pt/Au core/shell nanoraspberries for in situ SERS monitoring of platinum-catalyzed reactions. *Journal of the American Chemical Society* 133 (48): 19302–19305.

6 Huang, J., Zhu, Y., Lin, M. et al. (2013). Site-specific growth of Au-Pd alloy horns on Au nanorods: a platform for highly sensitive monitoring of catalytic reactions by surface enhancement Raman spectroscopy. *Journal of the American Chemical Society* 135 (23): 8552–8561.

7 Li, J., Liu, J., Yang, Y., and Qin, D. (2015). Bifunctional Ag@Pd-Ag nanocubes for highly sensitive monitoring of catalytic reactions by surface-enhanced Raman spectroscopy. *Journal of the American Chemical Society* 137 (22): 7039–7042.

8 Zhang, J., Winget, S.A., Wu, Y. et al. (2016). Ag@Au concave cuboctahedra: a unique probe for monitoring Au-catalyzed reduction and oxidation reactions by surface-enhanced Raman spectroscopy. *ACS Nano* 10 (2): 2607–2616.

9 Joseph, V., Engelbrekt, C., Zhang, J. et al. (2012). Characterizing the kinetics of nanoparticle-catalyzed reactions by surface-enhanced Raman scattering. *Angewandte Chemie International Edition* 51 (30): 7592–7596.

10 Xie, W., Walkenfort, B., and Schlücker, S. (2013). Label-free SERS monitoring of chemical reactions catalyzed by small gold nanoparticles using 3D plasmonic superstructures. *Journal of the American Chemical Society* 135 (5): 1657–1660.

11 Zhang, H., Wang, C., Sun, H.L. et al. (2017). In situ dynamic tracking of heterogeneous nanocatalytic processes by shell-isolated nanoparticle-enhanced Raman spectroscopy. *Nature Communications* 8: 15447.

12 Wang, Y.H., Le, J.B., LI, W.Q. et al. (2019). In situ spectroscopic insight into the origin of the enhanced performance of bimetallic nanocatalysts towards ORR. *Angewandte Chemie International Edition* 58 (45): 16062–16066.

13 Zhao, Y., Du, L., Li, H. et al. (2019). Is the Suzuki–Miyaura cross-coupling reaction in the presence of Pd nanoparticles heterogeneously or homogeneously catalyzed? An interfacial surface-enhanced Raman spectroscopy study. *The Journal of Physical Chemistry Letters* 10 (6): 1286–1291.

14 Faraday, M. (1857). The bakerian lecture: experimental relations of gold (and other metals) to light. *Philosophical Transactions of the Royal Society of London* 147: 145–181.

15 Bastús, N.G., Comenge, J., and Puntes, V. (2011). Kinetically controlled seeded growth synthesis of citrate-stabilized gold nanoparticles of up to 200 nm: size focusing versus Ostwald ripening. *Langmuir* 27 (17): 11098–11105.

16 Enustun, B.V. and Turkevich, J. (1963). Coagulation of colloidal gold. *Journal of the American Chemical Society* 85 (21): 3317–3328.

17 Jana, N.R., Gearheart, L., and Murphy, C. (2001). Seed-mediated growth approach for shape-controlled synthesis of spheroidal and rod-like gold nanoparticles using a surfactant template. *Advanced Materials* 13 (18): 1389–1393.

18 Nikoobakht, B. and El Sayed, M.A. (2003). Preparation and growth mechanism of gold nanorods (NRs) using seed-mediated growth method. *Chemistry of Materials* 15 (10): 1957–1962.

19 Murphy, C.J., Thompson, L.B., Alkilany, A.M. et al. (2010). The many faces of gold nanorods. *The Journal of Physical Chemistry Letters* 1 (19): 2867–2875.

20 Caswell, K.K., Wilson, J.N., Bunz, U.H.F., and Murphy, C.J. (2003). Preferential end-to-end assembly of gold nanorods by Biotin–Streptavidin connectors. *Journal of the American Chemical Society* 125 (46): 13914–13915.

21 Wu, H.L., Kuo, C.H., and Huang, M.H. (2010). Seed-mediated synthesis of gold nanocrystals with systematic shape evolution from cubic to trisoctahedral and rhombic dodecahedral structures. *Langmuir* 26 (14): 12307–12313.

22 Millstone, J.E., Park, S., Shuford, K.L. et al. (2005). Observation of a quadrupole plasmon mode for a colloidal solution of gold nanoprisms. *Journal of the American Chemical Society* 127 (15): 5312–5313.

23 Millstone, J.E., Métraux, G.S., and Mirkin, C.A. (2006). Controlling the edge length of gold nanoprisms via a seed-mediated approach. *Advanced Functional Materials* 16 (9): 1209–1214.

24 Métraux, G.S. and Mirkin, C.A. (2005). Rapid thermal synthesis of silver nanoprisms with chemically tailorable thickness. *Advanced Materials* 17 (4): 412–415.

25 Sau, T.K. and Murphy, C.J. (2004). Room temperature, high-yield synthesis of multiple shapes of gold nanoparticles in aqueous solution. *Journal of the American Chemical Society* 126 (28): 8648–8649.

26 König, M., Radojcic, A., Schlücker, S., and Xie, W. (2016). Label-free SERS monitoring of hydride reduction catalyzed by Au nanostars. *Journal of Raman Spectroscopy* 47 (9): 1024–1028.

27 Khoury, C.G. and Vo Dinh, T. (2008). Gold nanostars for surface-enhanced Raman scattering: synthesis, characterization and optimization. *The Journal of Physical Chemistry C* 112 (48): 18849–18859.

28 Hao, F., Nehl, C.L., Hafner, J.H., and Nordlander, P. (2007). Plasmon resonances of a gold nanostar. *Nano Letters* 7 (3): 729–732.

29 Kim, C., Song, H.M., Cai, X. et al. (2011). In vivo photoacoustic mapping of lymphatic systems with plasmon-resonant nanostars. *Journal of Materials Chemistry* 21 (9): 2841–2844.

30 Fales, A.M., Yuan, H., and Vo Dinh, T. (2011). Silica-coated gold nanostars for combined surface-enhanced Raman scattering (sers) detection and singlet-oxygen generation: a potential nanoplatform for theranostics. *Langmuir* 27 (19): 12186–12190.

31 Yuan, H., Fales, A.M., Khoury, C.G. et al. (2013). Spectral characterization and intracellular detection of surface-enhanced Raman scattering (SERS)-encoded plasmonic gold nanostars. *Journal of Raman Spectroscopy* 44 (2): 234–239.

32 Abalde Cela, S., Aldeanueva Potel, P., Mateo Mateo, C. et al. (2010). Surface-enhanced Raman scattering biomedical applications of plasmonic colloidal particles. *Journal of the Royal Society Interface* 7 (suppl_4): S435–S450.

33 Rycenga, M., Xia, X., Moran, C.H. et al. (2011). Generation of hot spots with silver nanocubes for single-molecule detection by surface-enhanced Raman scattering. *Angewandte Chemie International Edition* 50 (24): 5473–5477.

34 Yang, Y., Zhang, Q., Fu, Z.W., and Qin, D. (2014). Transformation of Ag nanocubes into Ag–Au hollow nanostructures with enriched Ag contents to improve SERS activity and chemical stability. *ACS Applied Materials & Interfaces* 6 (5): 3750–3757.

35 Lee, P.C. and Meisel, D. (1982). Adsorption and surface-enhanced Raman of dyes on silver and gold sols. *The Journal of Physical Chemistry* 86 (17): 3391–3395.

36 Steinigeweg, D. and Schlücker, S. (2012). Monodispersity and size control in the synthesis of 20–100 nm quasi-spherical silver nanoparticles by citrate and ascorbic acid reduction in glycerol–water mixtures. *Chemical Communications* 48: 8682–8684.

37 Sun, Y., Mayers, B., Herricks, T., and Xia, Y. (2003). Polyol synthesis of uniform silver nanowires: a plausible growth mechanism and the supporting evidence. *Nano Letters* 3 (7): 955–960.

38 Gou, L., Chipara, M., and Zaleski, J.M. (2007). Convenient, rapid synthesis of Ag nanowires. *Chemistry of Materials* 19 (7): 1755–1760.

39 Jana, N.R., Gearheart, L., and Murphy, C.J. (2001). Wet chemical synthesis of silver nanorods and nanowires of controllable aspect ratio. *Chemical Communications* (7): 617–618.

40 Chen, Z., Balankura, T., Fichthorn, K.A., and Rioux, R.M. (2019). Revisiting the polyol synthesis of silver nanostructures: role of chloride in nanocube formation. *ACS Nano* 13 (2): 1849–1860.

41 Chen, Z., Chang, J.W., Balasanthiran, C. et al. (2019). Anisotropic growth of silver nanoparticles is kinetically controlled by polyvinylpyrrolidone binding. *Journal of the American Chemical Society* 141 (10): 4328–4337.

42 Jiang, T., Chen, G., Tian, X. et al. (2018). Construction of long narrow gaps in Ag nanoplates. *Journal of the American Chemical Society* 140 (46): 15560–15563.

43 Loiudice, A., Lobaccaro, P., Kamali, E.A. et al. (2016). Tailoring copper nanocrystals towards C_2 products in electrochemical CO_2 reduction. *Angewandte Chemie International Edition* 55 (19): 5789–5792.

44 Wang, H., Tam, F., Grady, N.K., and Halas, N.J. (2005). Cu nanoshells: effects of interband transitions on the nanoparticle plasmon resonance. *The Journal of Physical Chemistry B* 109 (39): 18218–18222.

45 Guo, H., Chen, Y., Cortie, M.B. et al. (2014). Shape-selective formation of monodisperse copper nanospheres and nanocubes via disproportionation reaction route and their optical properties. *The Journal of Physical Chemistry C* 118 (18): 9801–9808.

46 Wang, Z., Chen, Z., Zhang, H. et al. (2015). Lattice-mismatch-induced twinning for seeded growth of anisotropic nanostructures. *ACS Nano* 9 (3): 3307–3313.

47 Knight, M.W., King, N.S., Liu, L. et al. (2014). Aluminum for plasmonics. *ACS Nano* 8 (1): 834–840.

48 Hwang, E.H. and Das Sarma, S. (2007). Dielectric function, screening, and plasmons in two-dimensional graphene. *Physical Review B* 75 (20): 205418.

49 Ju, L., Geng, B., Horng, J. et al. (2011). Graphene plasmonics for tunable terahertz metamaterials. *Nature Nanotechnology* 6 (10): 630–634.

50 Luo, L., Li, Y., Sun, X. et al. (2020). Synthesis and properties of stable sub-2-nm-thick aluminum nanosheets: oxygen passivation and two-photon luminescence. *Chemistry* 6 (2): 448–459.

51 McClain, M.J., Schlather, A., Ringe, E. et al. (2015). Aluminum nanocrystals. *Nano Letters* 15 (4): 2751–2755.

52 Clark, B.D., Jacobson, C.R., Lou, M. et al. (2018). Aluminum nanorods. *Nano Letters* 18 (2): 1234–1240.

53 Lecarme, O., Sun, Q., Ueno, K., and Misawa, H. (2014). Robust and versatile light absorption at near-infrared wavelengths by plasmonic aluminum nanorods. *ACS Photonics* 1 (6): 538–546.

54 Yogev, D. and Efrima, S. (1988). Novel silver metal liquidlike films. *The Journal of Physical Chemistry* 92 (20): 5754–5760.

55 Li, Y.J., Huang, W.J., and Sun, S.G. (2006). A universal approach for the self-assembly of hydrophilic nanoparticles into ordered monolayer films at a toluene/water interface. *Angewandte Chemie, International Edition* 45 (16): 2537–2539.

56 Jiang, P., Dong, Y.Y., Yang, L. et al. (2019). Hot electron-induced carbon–halogen bond cleavage monitored by in situ surface-enhanced Raman spectroscopy. *The Journal of Physical Chemistry C* 123 (27): 16741–16746.

57 Zhang, Y.J., Chen, S., Radjenovic, P. et al. (2019). Probing the location of 3D hot spots in gold nanoparticle films using surface-enhanced raman spectroscopy. *Analytical Chemistry* 91 (8): 5316–5322.

58 Zong, Y., Guo, Q., Xu, M. et al. (2014). Plasmon-induced decarboxylation of mercaptobenzoic acid on nanoparticle film monitored by surface-enhanced Raman spectroscopy. *RSC Advances* 4 (60): 31810–31816.

59 Xie, W. and Schlücker, S. (2015). Hot electron-induced reduction of small molecules on photorecycling metal surfaces. *Nature Communications* 6 (1): 7570.

60 Zhang, K., Yang, L., Hu, Y. et al. (2020). Synthesis of Au-metal oxide core-satellite nanostructure for in situ SERS study of CuO-catalyzed photooxidation. *Angewandte Chemie International Edition* https://doi.org/10.1002/anie.202007462 59 (41): 18003–18009.

61 Li, Y., Hu, Y., Shi, F. et al. (2019). C−H arylation on nickel nanoparticles monitored by in situ surface-enhanced Raman spectroscopy. *Angewandte Chemie International Edition* 58 (27): 9049–9053.

62 Huang, Z., Liu, Y., Zhang, Q. et al. (2016). Collapsed polymer-directed synthesis of multicomponent coaxial-like nanostructures. *Nature Communications* 7 (1): 12147.

63 Zhu, X., Yip, H.K., Zhuo, X. et al. (2017). Realization of red plasmon shifts up to ~900 nm by AgPd-tipping elongated Au nanocrystals. *Journal of the American Chemical Society* 139 (39): 13837–13846.

64 Liu, R., Liu, J.F., Zhang, Z.M. et al. (2014). Submonolayer-Pt-coated ultrathin Au nanowires and their self-organized nanoporous film: SERS and catalysis active substrates for operando SERS monitoring of catalytic reactions. *The Journal of Physical Chemistry Letters* 5 (6): 969–975.

65 Chen, H., Shao, L., Li, Q., and Wang, J. (2013). Gold nanorods and their plasmonic properties. *Chemical Society Reviews* 42 (7): 2679–2724.

66 Lin, M., Kim, G.H., Kim, J.H. et al. (2017). Transformative heterointerface evolution and plasmonic tuning of anisotropic trimetallic nanoparticles. *Journal of the American Chemical Society* 139 (30): 10180–10183.

67 Tsai, M.F., Chang, S.H.G., Cheng, F.Y. et al. (2013). Au nanorod design as light-absorber in the first and second biological near-infrared windows for in vivo photothermal therapy. *ACS Nano* 7 (6): 5330–5342.

68 Cui, J., Li, Y., Liu, L. et al. (2015). Near-infrared plasmonic-enhanced solar energy harvest for highly efficient photocatalytic reactions. *Nano Letters* 15 (10): 6295–6301.

69 Linic, S., Aslam, U., Boerigter, C., and Morabito, M. (2015). Photochemical transformations on plasmonic metal nanoparticles. *Nature Materials* 14 (6): 567–576.

70 Kale, M.J., Avanesian, T., and Christopher, P. (2014). Direct photocatalysis by plasmonic nanostructures. *ACS Catalysis* 4 (1): 116–128.

71 Boerigter, C., Campana, R., Morabito, M., and Linic, S. (2016). Evidence and implications of direct charge excitation as the dominant mechanism in plasmon-mediated photocatalysis. *Nature Communications* 7 (1): 10545.

72 Gieseking, R., Ratner, M.A., and Schatz, G.C. (2016). Review of plasmon-induced hot-electron dynamics and related SERS chemical effects. *Frontiers of Plasmon Enhanced Spectroscopy* 1: 1–22.

73 Huang, Y.F., Zhu, H.P., Liu, G.K. et al. (2010). When the signal is not from the original molecule to be detected: chemical transformation of para-aminothiophenol on Ag during the SERS measurement. *Journal of the American Chemical Society* 132 (27): 9244–9246.

74 Zhan, C., Chen, X.J., Huang, Y.F. et al. (2019). Plasmon-mediated chemical reactions on nanostructures unveiled by surface-enhanced Raman spectroscopy. *Accounts of Chemical Research* 52 (10): 2784–2792.

75 Wu, D.Y., Zhao, L.B., Liu, X.M. et al. (2011). Photon-driven charge transfer and photocatalysis of p-aminothiophenol in metal nanogaps: a DFT study of SERS. *Chemical Communications* 47 (9): 2520–2522.

76 Huang, Y.F., Zhang, M., Zhao, L.B. et al. (2014). Activation of oxygen on gold and silver nanoparticles assisted by surface plasmon resonances. *Angewandte Chemie International Edition* 53 (9): 2353–2357.

77 da Silva, A.G.M., Rodrigues, T.S., Correia, V.G. et al. (2016). Plasmoninanorattles as next-generation catalysts for surface plasmon resonance-mediated oxidations promoted by activated oxygen. *Angewandte Chemie International Edition* 55 (25): 7111–7115.

78 Ren, X., Cao, E., Lin, W.H. et al. (2017). Recent advances in surface plasmon-driven catalytic reactions. *RSC Advances* 7 (50): 31189–31203.

79 Brandt, N.C., Keller, E.L., and Frontiera, R.R. (2016). Ultrafast surface-enhanced Raman probing of the role of hot electrons in plasmon-driven chemistry. *The Journal of Physical Chemistry Letters* 7 (16): 3179–3185.

80 Dong, Y.Y., Su, Y.L., Du, L.L. et al. (2019). Plasmon-enhanced deuteration under visible-light irradiation. *ACS Nano* 13 (9): 10754–10760.

81 Wu, W., Niu, C., Wei, C. et al. (2019). Activation of MoS_2 basal planes for hydrogen evolution by zinc. *Angewandte Chemie International Edition* 58 (7): 2029–2033.

82 Keeler, A.J., Salazar-Banda, G.R., and Russell, A.E. (2019). Mechanistic insights into electrocatalytic reactions provided by SERS. *Current Opinion in Electrochemistry* 17: 90–96.

83 Funtikov, A.M., Sigalaev, S.K., and Kazarinov, V.E. (1987). Surface enhanced Raman scattering and local photoemission currents on the freshly prepared surface of a silver electrode. *Journal of Electroanalytical Chemistry and Interfacial Electrochemistry* 228 (1-2): 197–218.

84 Yonezawa, Y., Minamimoto, H., Nagasawa, F. et al. (2017). In-situ electrochemical surface-enhanced Raman scattering observation of molecules accelerating the hydrogen evolution reaction. *Journal of Electroanalytical Chemistry* 800: 7–12.

85 Chen, Y.X., Zou, S.Z., Huang, K.Q., and Tian, Z.Q. (1998). SERS studies of electrode/electrolyte interfacial water part II—librations of water correlated to hydrogen evolution reaction. *Journal of Raman Spectroscopy* 29 (8): 749–756.

86 Garcia, A.C., Touzalin, T., Nieuwland, C. et al. (2019). Enhancement of oxygen evolution activity of nickel oxyhydroxide by electrolyte alkali cations. *Angewandte Chemie International Edition* 58 (37): 12999–13003.

87 Diaz Morales, O., Ferrus Suspedra, D., and Koper, M.T.M. (2016). The importance of nickel oxyhydroxide deprotonation on its activity towards electrochemical water oxidation. *Chemical Science* 7 (4): 2639–2645.

88 Wang, D.W. and Su, D. (2014). Heterogeneous nanocarbon materials for oxygen reduction reaction. *Energy & Environmental Science* 7 (2): 576–591.

89 Li, J., Yin, H.M., Li, X.B., and Okunishi, E. (2017). Surface evolution of a Pt–Pd–Au electrocatalyst for stable oxygen reduction. *Nature Energy* 2: 17111.

90 Dong, J.C., Zhang, X.-G., Briega-Martos, V. et al. (2019). In situ Raman spectroscopic evidence for oxygen reduction reaction intermediates at platinum single-crystal surfaces. *Nature Energy* 4 (1): 60–67.

91 Dong, J.C., Su, M., Briega-Martos, V. et al. (2020). Direct in situ Raman spectroscopic evidence of oxygen reduction reaction intermediates at high-index Pt(hkl) surfaces. *Journal of the American Chemical Society* 142 (2): 715–719.

92 Gómez-Marín, A.M. and Feliu, J.M. (2013). New insights into the oxygen reduction reaction mechanism on Pt(111): a detailed electrochemical study. *ChemSusChem* 6 (6): 1091–1100.

93 Wang, Y.Y., Chen, D.J., and Tong, Y.J. (2016). Mechanistic insight into sulfide-enhanced oxygen reduction reaction activity and stability of commercial Pt black: an in situ Raman spectroscopic study. *ACS Catalysis* 6 (8): 5000–5004.

94 Ivanov, I. (2013). Double-edged sword for catalysis. *Nature Chemistry* 5 (1): 6–7.

95 Han, X.B., Kannari, K., and Ye, S. (2019). In situ surface-enhanced Raman spectroscopy in Li–O_2 battery research. *Current Opinion in Electrochemistry* 17: 174–183.

96 Peng, Z., Freunberger, S.A., Chen, Y., and Bruce, P.G. (2012). A reversible and higher-rate Li-O_2 battery. *Science* 337 (6094): 563–566.

97 Peng, Z., Freunberger, S.A., Hardwick, L.J. et al. (2011). Oxygen reactions in a non-aqueous Li^+ electrolyte. *Angewandte Chemie International Edition* 50 (28): 6351–6355.

98 Galloway, T.A. and Hardwick, L.J. (2016). Utilizing in situ electrochemical SHINERs for oxygen reduction reaction studies in aprotic electrolytes. *The Journal of Physical Chemistry Letters* 7 (11): 2119–2124.

99 Aldous, I.M. and Hardwick, L.J. (2016). Solvent-mediated control of the electrochemical discharge products of non-aqueous sodium–oxygen electrochemistry. *Angewandte Chemie International Edition* 55 (29): 8254–8257.

100 Chen, Y., Jovanov, Z.P., Gao, X. et al. (2018). High capacity surface route discharge at the potassium-O_2 electrode. *Journal of Electroanalytical Chemistry* 819: 542–546.

101 Chang, X., Wang, T., and Gong, J. (2016). CO_2 photo-reduction: insights into CO_2 activation and reaction on surfaces of photocatalysts. *Energy & Environmental Science* 9 (7): 2177–2196.

102 Peterson, A.A., Abild Pedersen, F., Studt, F. et al. (2010). How copper catalyzes the electroreduction of carbon dioxide into hydrocarbon fuels. *Energy & Environmental Science* 3 (9): 1311–1315.

103 Koper, M.T.M. (2013). Theory of multiple proton–electron transfer reactions and its implications for electrocatalysis. *Chemical Science* 4 (7): 2710–2723.

104 Keith, J.A. and Carter, E.A. (2012). Theoretical insights into pyridinium-based photoelectrocatalytic reduction of CO_2. *Journal of the American Chemical Society* 134 (18): 7580–7583.

105 Li, Y.C., Wang, Z., Yuan, T. et al. (2019). Binding site diversity promotes CO_2 electroreduction to ethanol. *Journal of the American Chemical Society* 141 (21): 8584–8591.

4

Plasmonic Catalysis Toward Hydrogenation Reactions

Gareth D. Price[1,2], *Alexandra Gellé*[1] *and Audrey Moores*[1,2]

[1] *Centre for Green Chemistry and Catalysis, Department of Chemistry, McGill University, Montreal, QC, Canada*
[2] *Department of Materials Engineering, McGill University, Montreal, QC, Canada*

4.1 Introduction

Hydrogenation reactions are critical chemical transformations with significant industrial importance [1]. In particular, these reactions are applied to a variety of industries, from very large volume ones, such as the production of ethylene from acetylene [2], or the food relevant hydrogenated oil industry, to smaller ones, such as the pharma or perfume markets [3, 4]. Transition metals have been the catalysts of choice for over 100 years in both their homogeneous and heterogeneous forms [5]. The heterogeneous form has been favored for large scale reactions, yet they require harsher reaction conditions to function than their homogeneous counterparts. Conversely, homogenous versions are better suited for small volume applications, with better performances, selectivity, and milder conditions, but their easy and effective recyclability have been a clear limitation [6]. Hydrogenation reactions are also considered as green transformations: in the direct hydrogenation of substrates with the H_2 molecule, they proceed with perfect atom economy [7]. Furthermore, researchers are looking into replacing the current fossil sourcing of H_2 with the more sustainable water splitting process [8]. Transfer hydrogenation, whereby another substrate in the medium acts as a hydrogen source, such as $NaBH_4$ or an alcohol, are also exciting avenues, the latter being made particularly appealing by the availability of renewable sourcing of alcohols, or by reagent recycling [9]. In this context, plasmonic nanoparticles (PNPs) and hydride structures containing PNPs have been highlighted as promising candidates to boost the reactivity of nanosized catalysts to afford improved reactivity and selectivity, under mild conditions, for hydrogenation reactions. PNPs feature exciting properties, based on the ability of their conductive electron cloud to interact with incoming light and be displaced while causing the creation of restoring charges [10]. There exists a wavelength of light for which this system enters resonance, called the localized surface plasmon resonance (LSPR). In PNPs, LSPR leads to measurable thermal, optical, and/or electronic properties, which can be exploited for many applications, including in

Plasmonic Catalysis: From Fundamentals to Applications, First Edition.
Edited by Pedro H.C. Camargo and Emiliano Cortés.

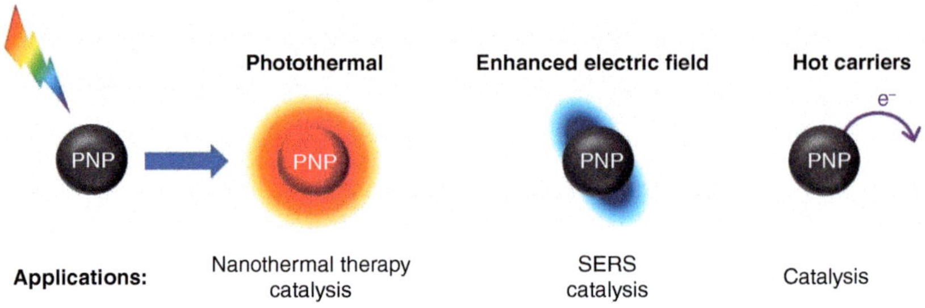

Figure 4.1 Three mechanisms are at play in plasmonic catalysis. Each mechanism leads to properties which have been exploited in noncatalysis-related applications. *Source*: Gellé and Moores [11].

catalysis (Figure 4.1) [12–15]. Specifically, thermal effects can induce temperature increase locally around the PNPs, which have led, for instance, to applications in nanomedicine, for the targeted burning of tumors, or in catalysis to thermally accelerate reactions. Another interesting property of the LSPR is the enhanced local electric field in the close proximity of PNPs, which has been exploited for surface-enhanced Raman spectroscopy (SERS) and catalysis. Finally, the property the most exploited in the context of catalysis, is the creation of hot carriers. At LSPR, PNPs see their electrons and holes energy schemes altered in the way that can trigger reaction. Gold, silver, copper, and aluminum are the most common metals featuring measurable LSPR properties at the nanoscale to have been exploited in the context of plasmonic catalysis.

In the context of hydrogenation, several catalytic systems have been envisaged and developed, as shown in Scheme 4.1 [15]: (i) PNPs may act both as light absorbers and catalysts; (ii) PNPs are used as a support for catalytically active but nonplasmon active metals, either as decorated or core–shell nanohybrids; (iii) both plasmon-active and catalytically active metals are alloyed together.

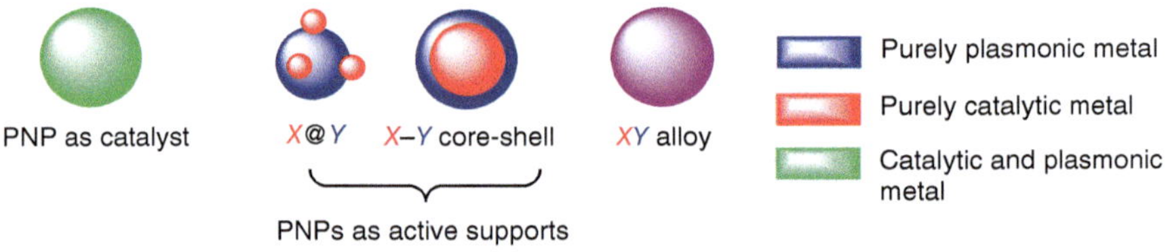

Scheme 4.1 Schematic representation of the different catalyst designs for PNP-activated hydrogenation catalysis. *Source*: Gellé et al. [15].

In this chapter, we focus on the use of PNPs for the direct hydrogenation of organic molecules. Examples from the literature are presented per family of substrates.

4.2 Hydrogenation of Alkenes and Alkynes

The hydrogenation of unsaturated carbon–carbon bonds is a fundamental reaction useful for the production of commodity products and fine chemicals alike [3]. A series of examples were reported with monometallic systems for such transformations. The Xiong group demonstrated that Pd concave nanocubes (NC) catalyzed the hydrogenation of styrene to ethylbenzene under mild conditions, while being under visible light irradiation (Scheme 4.2). Pd is usually an ineffective plasmonic metal due to its UV-range plasmonic band and low plasmonic cross-section. Yet, careful NP engineering allowed to tweak its properties in order to exploit them in catalysis. A Ru^{3+}-mediated process enabled the synthesis of Pd NC of about 40 nm in edge length, which featured a strong plasmonic band around 400 nm [16]. Interestingly, the concave Pd NC were capable of driving the reduction of styrene at room temperature with irradiation of visible light (>400 nm), providing an activation comparable to thermally driven reactions at 70 °C. They also noted that thermostatic reactions under irradiation were hardly affected in terms of activity, so they concluded that the hot electron pathway is the most likely mechanism. The authors suggested the concave structure allowed for enhanced LSPR properties at the edges and corners. Several experimental and theoretical studies have proven that hot spots are concentrated on the extremities of nonspherical structures [17–19].

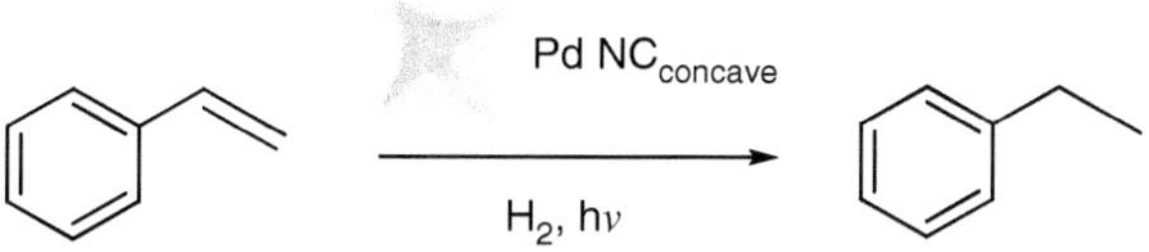

Scheme 4.2 Styrene to ethylbenzene hydrogenation with concave Pd NC [16]. *Source*: Based on Long et al. [16].

Monometallic Au NPs made by Nguyen and coworkers were shown to also catalyze styrene hydrogenation using $NaBH_4$ as a reducing agent (Scheme 4.3), and were used to determine the impact of hot electron diffusion on catalytic activity [20]. They showed the photocatalytic activity had an inverse relationship with particle size, when studying Au NPs between 10 and 40 nm in diameter. This effect was attributed to the increased probability of diffusion of the hot carriers to the surface in particles with a higher surface to volume ratio.

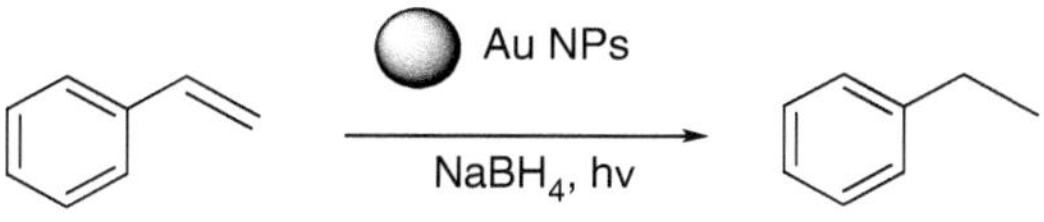

Scheme 4.3 Styrene to ethylbenzene hydrogenation with Au NPs [20]. *Source*: Based on Mao et al. [20].

While most hydrogenations of unsaturated carbon bonds have been shown to take place on the plasmonic surface, Halas and coworkers were able to demonstrate the application of oblique Pd nanocones on a large surface area substrate to induce hydrogenation in adjacent layers of graphene tented above the catalyst. [21] To create the structure, hole-mask colloidal lithography was used to grow the Pd cones on a silicon substrate which was then covered by a monolayer graphene sheet *via* wet transfer. When these samples were exposed to a 450 nm laser in an H_2 environment, Raman spectroscopy indicated the formation of hydrogenated graphene, with both fluorescence quenching microscopy (FQM) and Raman mapping corroborating the initial results. The authors proposed that the graphene hydrogenation was due solely to hot electrons from the plasmonic Pd nanocones, which activated and then desorbed hydrogen species into the adjacent graphene monolayer, catalyzing its hydrogenation. Any thermal effects were discounted as theoretical calculations as IR measurements of catalyst temperatures indicated a maximum heating of only 6 K, while an increased temperature of even 43 K in dark conditions failed to initiate graphene hydrogenation. The authors suggested that this novel design showcases the idea of using PNPs to functionalize a material in proximity via hydrogenation, or even extending this reactivity beyond the scope of hydrogenation with 2D materials.

The Halas group has also pioneered the use of Al nanoislands as plasmonically active supports to affect the catalytic properties of Pd nanoparticles deposited onto them. Using these Pd@Al NPs, they studied the partial hydrogenation of acetylene so as to obtain the ubiquitous commodity monomer ethylene, a key starting material for the production of polyethylene (Scheme 4.4). [22] Pd NPs are excellent acetylene hydrogenation catalysts but are inferior at selective partial hydrogenation. Solutions such as Pd doping to improve selectivity, the strategy used with the Pb-doped Lindlar catalyst, have been developed but compromise overall activity [23]. PNPs offer an alternative strategy. Specifically, Al nanoislands acted as a plasmonic "antennae", absorbing incident visible light and altering the reactivity on the Pd "reactor". [24]

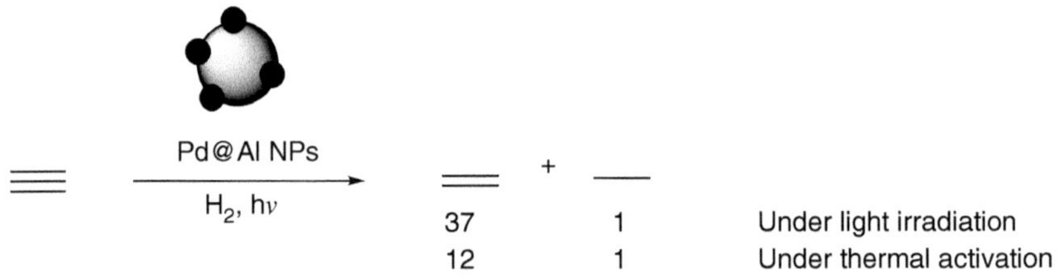

Scheme 4.4 Acetylene to ethylene partial hydrogenation with Pd@Al NPs [22]. *Source:* Based on Swearer et al. [22].

Upon exposure of this antenna–reactor complex to acetylene in a mild H_2 environment under 492 nm irradiation, the Pd@Al NPs were shown to be more selective toward the partial hydrogenation of acetylene to ethylene than the same system under thermal activation. At the maximum intensity tested at 14.3 W cm^{-1}, a ~37:1 product ratio of ethylene:ethane was observed in contrast to an upper limit of ~12:1 under thermal activation. This increased selectivity was attributed to the more facile desorption of H_2 from the Pd surface in the presence of light activated Al antenna, limiting the availability of hydrogen for complete reduction to ethane. [22] A computational, mechanistic study of this hydrogen desorption on heterometallic antenna reactor complexes by Spata and coworkers provided a quantitative description of this phenomenon. [25] Using embedded correlated wave function (ECW) theory, these researchers were able to model the interaction of H_2 and Pd(111) in these Pd@Al NPs antenna–reactor complexes. Their results indicated that the rate limiting photodesorption mechanism caused an increase in hybridized H/Pd excited states, rather than the generation of negative ion species from hot-electron transfer into absorbates. The photocatalytic mechanism of these Pd@Al NPs was determined to involve a predominant direct absorption mechanism with the rare, and possibly competitive, fluorescence resonance energy (FRET) to generate the excited H/Pd states.

This hybrid nanostructure serves as an illustration of the advantages of combining the light harvesting capabilities of the PNPs with the appealing chemistry of traditional catalysts, allowing for a more selective system which is greater than the sum of its parts to be conceived for hydrogenation under mild conditions.

These advantages were reinforced in the recent exploration of two catalyst-PNP heterostructures by the Camargo group for the partial hydrogenation of phenylacetylene to styrene, another commercially valuable commodity monomer used for the production of polystyrene (PS) (Scheme 4.5). [26] The first PNP was a spherical Au–Ag core–shell NP structure with an ultrathin Pt catalytic surface (Au-Ag-Pt NPs), and the other was an Au–AgPt nanorattle structure composed of a hollow interior section which separated the pure Au core from the AgPt alloy shell. By increasing the concentration of Pt precursor introduced to the initial Au–Ag core–shell, Pt was shown to form a PtAg alloy shell *via* galvanic replacement, which encapsulated the Au core and left behind the hollow interior layer caused by the partial dissolution of Ag. Both catalysts demonstrated an ~85% selectivity toward styrene formation under visible light irradiation and an H_2 pressure of six bars. This selectivity was attributed to the higher stability of carbon–carbon triple bonds onto the Pt surface compared to the styrene double bond, which was congruent with DTF calculations and experimental results. However, the Au-AgPt nanorattle showed an enhanced activity compared to the Au-Ag-Pt core–shell particle, with a nearly ninefold increase in phenylacetylene conversion. This remarkable activity was attributed to the plasmonic hybridization between the plasmonic Au core and the surrounding plasmonic-catalytic AgPt shell, which featured an increased field enhancement *via* LSPR excitement compared to a core–shell nanostructure, as further explored by the same group for other reactions [27, 28].

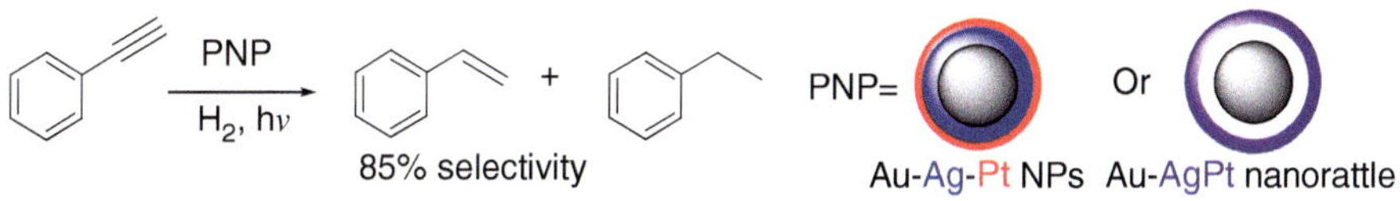

Scheme 4.5 Phenylacetylene to styrene partial hydrogenation with Au-Ag-Pt NPs and Au-AgPt nano-rattle [26]. *Source*: Modified from Quiroz et al. [26].

In another example, the same group reported the use of an Ag-Pd alloy in order to perform plasmon-enhanced styrene hydrogenation. Galvanic replacement was used to generate PdAg nanocages, hollow cubes featuring a high surface area of reactive Pd sites [29]. By tailoring the Pd/Au ratio of alloy and illumination intensity, the researchers were able to obtain a 97% conversion of styrene after only one hour under mild conditions using visible light ($\lambda > 400$ nm), nearly four times that of pure Pd concave nanostructure under similar conditions. The same group continued the development of bimetallic structure by synthesizing Au-Pd core–shell nanorods (NRs) with facile and atom-level control over the thickness of the Pd shell, and used ultrafast absorption spectroscopy characterizations to rationalize the reaction mechanisms [30]. The precise control over the Au-Pd NRs structures provided another tool for the independent tuning of the photothermal and hot-electron transfer effects to optimize the systems catalytic properties, in this case applied to the reduction of styrene to ethylbenzene.

The Scott group also reported the use of a plasmonic bimetallic system for the hydrogenation of 2-methyl-3-buten-2-ol catalyzed by Pd decorated Au nanotriangles *via* thermal activation of the Pd surface from the photothermal effect induced by the Au core (Scheme 4.6) [17]. An 82% yield was obtained under ambient conditions, which is similar to results obtained in dark conditions at 30 °C, corroborating the claim that thermal effects were the dominant influence on catalytic activity rather than hot electron generation.

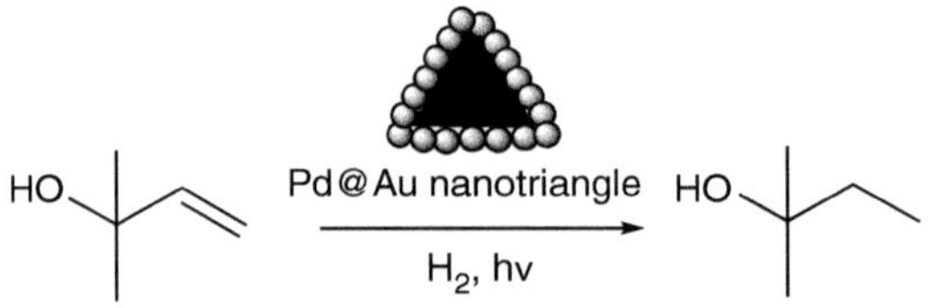

Scheme 4.6 Hydrogenation of 2-methyl-3-buten-2-ol catalyzed by Pd@Au nanotriangles [17]. *Source*: Based on Gangishetty et al. [17].

Jin and coworkers developed a macroscopic nanoassembly (NA) composed of a layer of PdAg nanosheets (NSs) surrounding a PS core for the hydrogenation of styrene to ethylbenzene [31]. Hexagonal PdAg NSs were first synthesized and then attached to amine-modified PS microspheres which formed the PdAg-PS NAs by coordination of the metal to the amine. The photothermal effects were again shown to be responsible for the increased activity under full illumination which was comparable to 70 °C without illumination, especially between the narrow gaps separating NSs. Using finite-difference time-dependent (FDTD) simulations, a nearly 12-fold field enhancement between certain particle edges was determined. PdAg-PS NAs also feature an increased photostability compared to solitary PdAg NSs, leading to a drop from 63.3 to 22.5% yield after five cycles under full illumination, whereas PdAg-PS NAs decreased from 80.8 to only 72.5%.

To explore the impact of bimetallic composition on catalytic properties, Guselnikova and coworkers selectively reduced the alkyne moiety of 4-ethynylbenzenediazonium tosylate by covalently bonding the substrate, a phenyl acetylene moiety, to the surface of an undulating Au-Pt bimetallic sheet supported on glass, leaving the alkyne group in the para position relative to the covalent bond between the substrate and the catalysts surface (Scheme 4.7) [32]. Under IR laser illumination (λ = 785 nm), the alkyne groups were partially reduced to alkenes when the Pt layers were thin (<9.1 nm) and fully reduced to alkanes when thick Pt layers were thicker (>9.1 nm) without the detection of an alkene group first. The IR signal

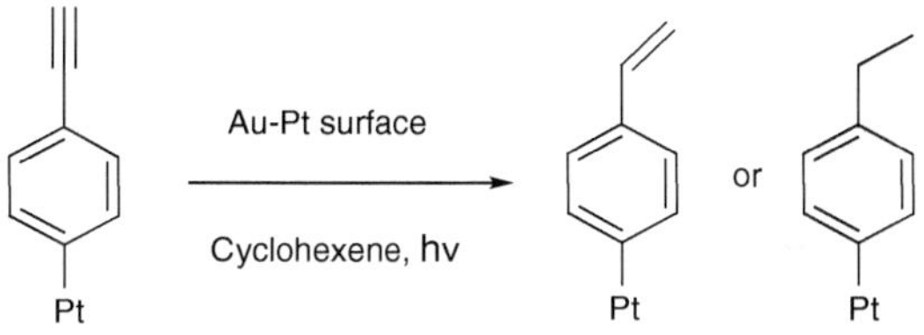

Scheme 4.7 Hydrogenation of anchored phenylacetylene over an Au-Pt bimetallic sheet supported on glass [32]. *Source*: Based on Guselnikova et al. [32].

served for both plasmonic activation in the Au subsurface and as a means to monitor the evolution of the reaction in real time via SERS, which is discussed further in Section 4.5. A hot-electron mechanism was proposed, whereby hydrogens from cyclohexenes activated, and subsequently catalyzed hydrogenation on the Pt surface. While this methodology does not provide a viable method of chemical transformation, it does grant meaningful insight into the dependence of bimetallic PNP's performance on their design, in particular their spatial composition.

4.3 Hydrogenation of Aldehydes and Ketones

The transformation of carbonyl containing molecules is another critical reaction for the production of industrially relevant chemicals. Interestingly, this reaction features both examples of direct hydrogenation and transfer hydrogenation, where polar alcohols are an ideal source of H_2. Looking at monometallic systems, our group reported that Ag nanocubes (Ag NCs) were efficient to selectively reduce carbonyls over C=C bonds using H_2 as reducing reagent (Scheme 4.8) [33].

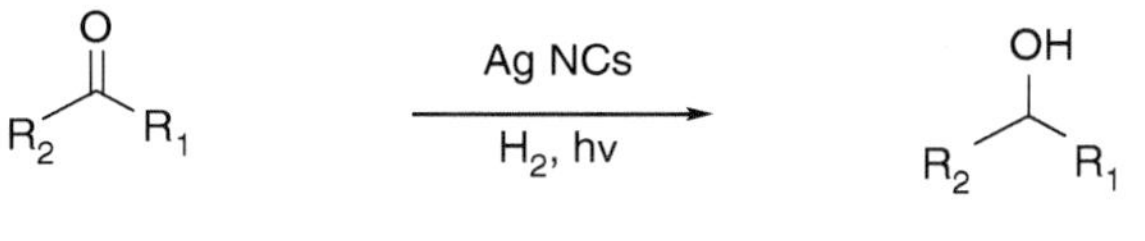

Scheme 4.8 Carbonyl reduction under light irradiation catalyzed by Ag NCs [33]. *Source*: Modified from Landry et al. [33].

Under one H_2 atmosphere at 80 °C with 405 nm irradiation, the hot electrons generated by the plasmonic Ag NCs lead to a homoleptic cleavage of H_2 to form reactive species on the surface of the catalyst. Absorbed carbonyls then reacted with the active hydrogen to form the corresponding alcohol, followed by the subsequent desorption of the product from the catalyst surface (Figure 4.2). These Ag NCs were the first examples of plasmonic Ag catalyst able to perform carbonyl reduction at atmospheric pressure, overcoming the typically limited ability of Ag to activate molecular hydrogen by clever exploitation of its plasmonic properties. When tested with 12 ketones and aldehydes, yields of the corresponding alcohol ranged between 32 and 92% without the reduction of other moieties. Their scope included molecules such as citronellol, which are relevant for the perfume industry. Lower activity were reported for more sterically hindered carbonyls.

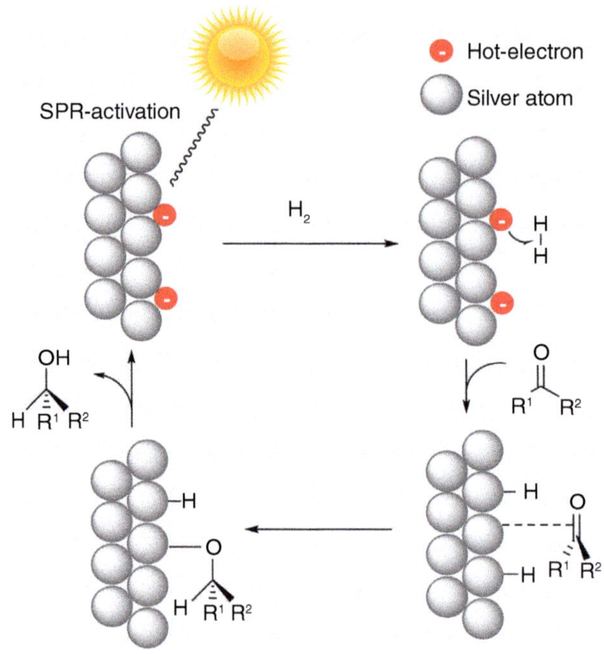

Figure 4.2 Proposed mechanism for the hydrogenation of carbonyl compounds with H_2 catalyzed by Ag NCs. *Source*: Landry et al. [33].

While Ag NCs were shown to be effective catalysts independently, plasmonic particles often benefit greatly from synergistic interactions with other materials. In contrast to the previously discussed bimetallic combinations for alkene and alkyne hydrogenation, often the interactions between plasmonic metals and semiconductors produced structures with the exceptional ability to catalyze transfer hydrogenations of carbonyls. This may be explained by the ability of such systems to separate charges, an appealing feature when trying to activate a polar bond such as C=O.

Yang, Guo, and coworkers reported that Au@SiC NPs could transfer hydrogenate cinnamaldehyde in an isopropanol solution at room temperature (Scheme 4.9) when excited by full spectrum visible light (400 nm < λ < 800 nm). [34] This catalyst showed remarkable activity for the reduction of cinnamaldehyde to cinnamyl alcohol progressed with a 100% selectivity and conversion, with a turnover frequency (TOF) of 487 h^{-1}. The 11 other α,β-unsaturated aldehydes tested also showed remarkable selectivity, conversion, and activity, with the majority having both above 80% and all with a TOF greater than 100 h^{-1}. The authors proposed that the

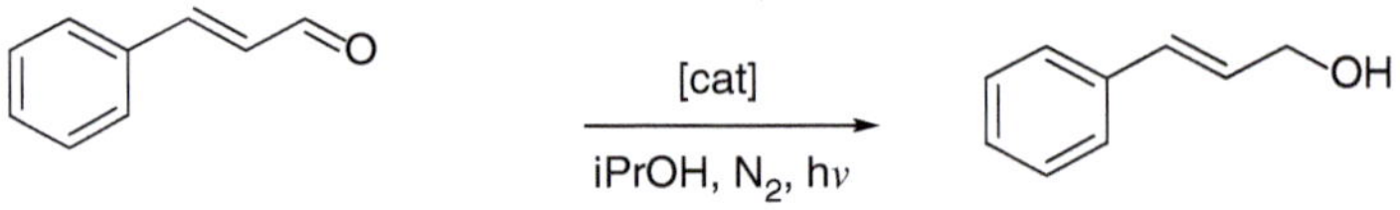

Scheme 4.9 Selective C—O reduction of cinnamaldehyde with plasmon-enhanced catalyst. [cat]=Au@SiC (with visible light) [34], Au@TiO_2 (with visible light) or Pt@TiO_2 (with UV light) [35]. *Source*: Hao et al. [34]; Ma and Li [35].

exceptional chemoselectivity and activity observed were derived from the interplay between the plasmonic metal (Au) and semiconductor (SiC) by promoting charge separation, which accounted for the superior activity. The proposed mechanism begins with the LSPR excitation of Au that causes hot electrons, which then migrate to the conduction band of the SiC, inducing local positive sites on the surface of the Au and an electron rich site at the material interface. The positive Au sites catalyzed the oxidation of 2-propanol to acetone, generating an active hydrogen species which was free to combine with C=O bonds absorbed at the electron-rich interface, and therefore finally forming the corresponding alcohol. The high selectivity was attributed to the significant absorption preference toward C=O bonds over C=C, as well as steric effects which disfavored unsaturated bonds close bulky rings, such as the C=C bonds in cinnamaldehyde.

Studies by Li and coworkers demonstrated that Au@TiO_2 and Pt@TiO_2 NPs could also enable the hydrogenation of cinnamaldehyde under plasmonic activation (Scheme 4.9) [35]. Because of the respective position of their plasmon bands, Au@TiO_2 was active under visible irradiation, while Pt@TiO_2 NPs functioned with UV light. The proposed mechanism is similar to that of Au@SiC NPs: the LSPR of the plasmonic metal generated hot electrons, which were transferred to the interface of the plasmonic metal and TiO_2, activating both the isopropanol and the substrate carbonyl group. Expanding on the previous mechanism, the authors proposed that proton dissociation forms the corresponding alkoxide which then participates in a transition state 6-member ring with the aldehyde to finally obtain the corresponding alcohol, following the well know Meerwein–Ponndorf–Verley (MPV) mechanism [36]. Even after five cycles, a ~100% yield was reported for cinnamaldehyde transfer hydrogenation catalyzed by Pt@TiO_2 under UV irradiation, with a reported TOF of up to 197 h^{-1}. Au@TiO_2 NPs were shown to be perfectly selective under visible light, but with a lower activity leading to a 70% conversion and a TOF of 56 h^{-1}.

Another report by Ke et al. demonstrated the potential of Au@CeO_2 NPs to reduce styrene oxide to styrene and acetophenone to benzyl alcohol (Scheme 4.10) under mild conditions in a KOH isopropanol solution [37]. While the selectivities were above 85%, the catalyst's low activity resulted in conversions of only up to 30%. This limitation was offset to some degree due to a range of substrates,

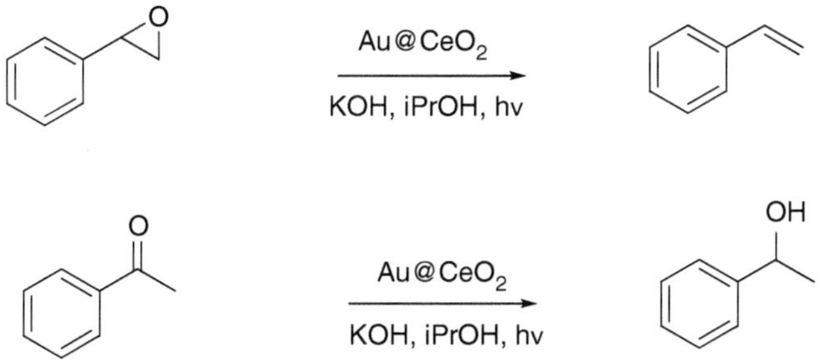

Scheme 4.10 Styrene oxide and acetophenone reduction under light irradiation catalyzed by Au@CeO_2 [37]. *Source*: Based on Ke et al. [37].

including azobenzene and nitrobenzene, for which the catalyst proved to be active (Section 4.5).

The PNP-catalyzed hydrogenation of carbonyl group has also garnered significant interest recently due to their applications in biomass upgrading to produce fine chemicals and fuels from sustainable feedstocks [38]. The hydrogenation of furfural (FAL) to furfuryl alcohol (FOL) is a critical step in biomass upgrading, which PNPs have recently been shown to effectively catalyze under mild conditions.

Li and coworkers developed a hybrid structure capable of hydrogenating FAL by exploiting the plasmonic abilities of Cu, a low-cost, earth abundant alternative to other noble plasmonic metals supported on mesoporous carbon (MC) [39]. Using pyrolysis, the group deposited ~1 nm nanoparticles of Cu and Cu_2O onto ~15 nm MC which proved to be effective for transfer hydrogenation FAL to FOL (94% conversion, 91% selectivity, TOF = 4.65 h^{-1}) in a basic, K_2CO_3 isopropanol solution under visible light irradiation (Scheme 4.11). The charge separation between the semiconducting MC and Cu_2O, and the plasmonic Cu was again proposed as the reason for the high catalytic activity. The excitation of Cu_2O and plasmonic Cu generate hot electrons which are transferred to the surface of the MC followed by the same PMV 6-member ring formation seen for Pt@TiO_2 (Figure 4.3). Both Cu and Cu_2O were excited under irradiation which catalyzed the hydrogenation of furfural. Yet, when paired, Cu acted as a mediator to facilitate the transfer of electrons from the conduction band of CuO_2 to the MC, and increase the catalysts' activity. The low cost of Cu, compared to other plasmonic metals, such as Au, provides an appealing avenue for industrial applications of such catalysts which could economically compete with traditional hydrogenation Ni- based catalysts. The durability of these catalysts also contributes to their potential, with no degradation in catalytic activity even after three cycles.

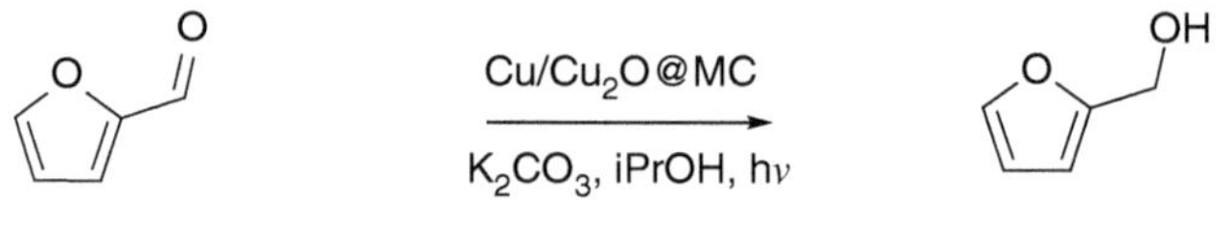

Scheme 4.11 Furfural to furfuryl alcohol reduction under light irradiation catalyzed by Cu/Cu_2O@MC [39]. *Source*: Modified from Zhang and Li [39]. © 2019, American Chemical Society.

The potential of Cu as a plasmonic catalyst was demonstrated by Gu, Zheng, and coworkers who were able to hydrogenate FAL using Cu NPs with a thin carbon coating (Cu-C NPs) derived from a metal organic frameworks (MOF) [40]. The Cu-based MOF HKUST-1 ($[Cu_3(btc)_2]$) was pyrolyzed at 400–800 °C in an H_2/Ar atmosphere to generate the Cu-C NPs. Using DFT calculations, it was determined that the e^- transfer between the metallic Cu core and thin C shell stabilized the Cu-C NPs, which preserved catalytic activity and prevented metal oxidation even after six reaction cycles. Cu-C NPs annealed at 600 °C were shown to hydrogenate FAL to FOL in a 70 °C isopropanol solution with an H_2 atmosphere under visible light (λ = 400–730 nm) with a ~99% yield (Scheme 4.12). These researchers proposed that the LSPR-generated hot electrons dissociate the molecular H_2, creating an

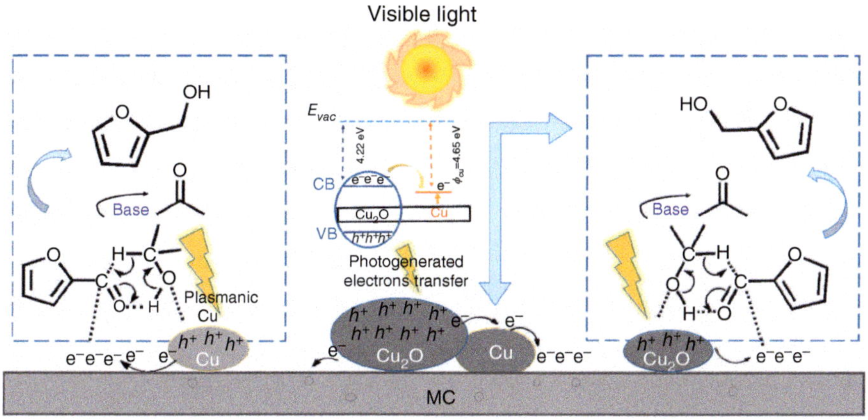

Figure 4.3 Proposed mechanism for light initiated selective hydrogenation of FAL to produce FOL over MC supported Cu-based catalysts. *Source*: Zhang and Li [39].

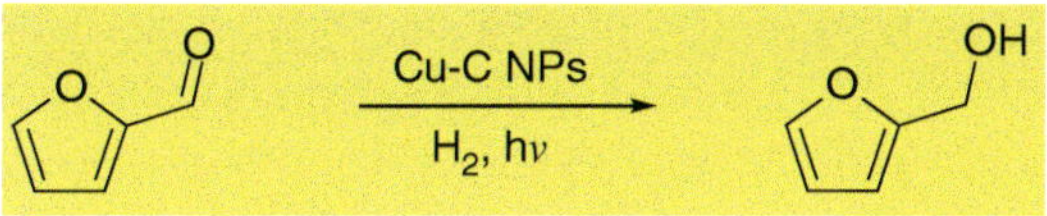

Scheme 4.12 Furfural to furfuryl alcohol reduction under light irradiation catalyzed by Cu-C NPs [40]. *Source*: Based on Wang et al. [40].

active Cu-H species on the metal's surface which hydrogenated the carbonyl group of absorbed FAL, generating solely FOL as end product. While other plasmonic catalysts for carbonyl reduction have been reported with similar yields, stability, the use of comparably an inexpensive plasmonic metal, and the need for only H_2 as a reductant made these Cu-C NPs an exciting example of the possible advantages of novel PNPs.

Research by Wang, Zhang, and coworkers exploited a similar synergy for previously discussed carbonyl hydrogenation between plasmonic and semiconducting materials. They attached Au nanoparticles to $CuCo_2O_4$ nanotubes (Au@$CuCo_2O_4$ NTs) to obtain a system able to catalyze the hydrogenation of α,β-unsaturated aldehydes including FAL to FOL (99% yield, TOF = 2902.8 h^{-1}). [41] An isopropanol solution of KOH and Au@Cu $CuCo_2O_4$ NTs catalyzed the hydrogenation of a range of unsaturated aldehyde substrates with yields between 72 and 99% and TOFs between 2111.1 and 2902.8 h^{-1} while at room temperature under visible light (Scheme 4.13). The hydrogenation mechanism proposed is similar to that of other

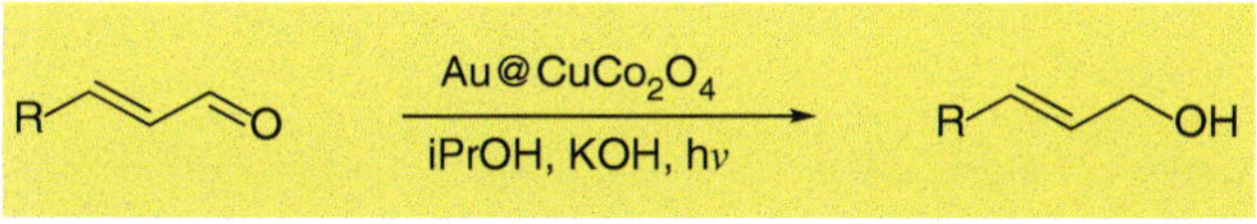

Scheme 4.13 Selective C=O hydrogenation of α,β-unsaturated aldehydes under visible light catalyzed [41]. *Source*: Based on Hu et al. [41].

plasmonic-semiconductor catalysts, where LSPR-generated hot electrons from the Au excited valence band electrons of the $CuCo_2O_4$ to migrate to the conduction band of the semiconducting support. The resulting charge separation catalyzes isopropanol oxidation, generating active H species on the Au surface which can further react with adsorbed unsaturated aldehyde to produce the corresponding alcohol. In addition to high yields and TOF, catalyst stability was also exceptional, with no reported decrease in activity after six cycles or change in morphology when the particles were analyzed using TEM.

As discussed previously, the catalytic performance of PNPs relies not only on the nature of the particle itself, but also on its architecture and the presence of other metal or support. While isolated PNPs have shown promising photoactive activity on their own, high efficiencies for the hydrogenation of some carbonyls were reported using bimetallic nanomaterials. In reported systems, it was observed that plasmonic enhancement may facilitate reactivity for both the direct and transfer hydrogenation toward carbonyl species. The recent development of such systems with cheaper plasmonic metals such as Ag [33] and even better, Cu, opens opportunities for potentially industrially relevant applications, despite the inherent complexity associated with the specific design developed so far [39]. Combined with advances in upgrading biobased feedstocks, these catalysts may prove to have a key role in promoting green chemistry as a viable solution to climate change concerns and as a replacement to unsustainable chemical sources.

4.4 Reduction of Nitro Compounds

4.4.1 Hydrogenation of Nitro Groups

A lot of work in the last two decades developing our understanding of plasmonic nanoparticle for hydrogenation catalysis has focused on nitroaromatic compounds. In particular, the use of sodium borohydride ($NaBH_4$) or H_2 to reduce nitroaromatic compounds in the presence of plasmonic catalysts has become a quintessential model reaction to test plasmonic efficacy. These reactions are typically the conversion of nitrobenzene (NB) to aniline, or 4-nitrophenol (4-NPh) to 4-aminophenol (4-AP) (Scheme 4.14).

The first use of plasmonic NPs for nitroarene reduction was reported by the Pal group in 2000, where they demonstrated the use of Au, Ag, and Cu plasmonic NPs in both aqueous and micellar mediums to reduce 4-NPh to 4-AP, as well as several other similar nitroaromatics [42]. From this initial report, dozens of unique plasmonic

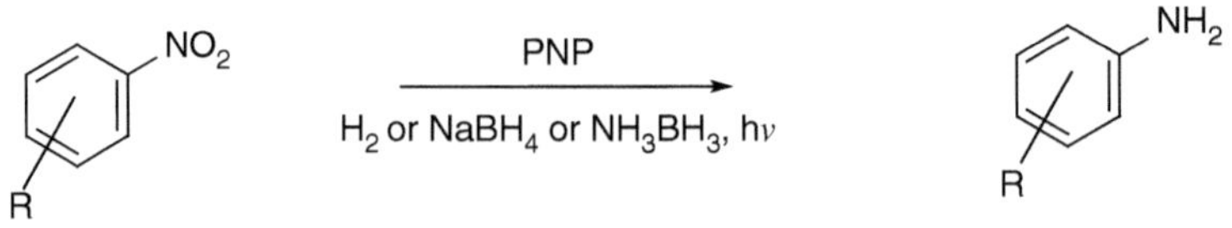

Scheme 4.14 Nitroarene reduction.

catalyst designs for nitroaromatic reduction have been developed, but a fundamental understanding of the mechanism which underpinned this reaction was reported only recently.

In 2018, the Camargo group shed light onto the mechanism of these reactions with both computational and experimental results that showed reaction-pathway-dependent behavior of the catalysts [43]. By studying the hydrogenation of 4-NPh on plasmonic Au NPs supported on either TiO_2 or SiO_2 using $NaBH_4$ or H_2 as reducing agents, Camargo and coworkers were able to demonstrate that the interaction between the reductant and the charge-transfer processes resulting from the plasmonic metal–support interactions had drastic impacts on the performance of the catalyst. For example, when comparing the pseudo-first order rate constants, Au@TiO_2 was observed to be three times more active than Au@SiO_2 when using H_2, but in the presence of $NaBH_4$ the rates were similar, with Au@SiO_2 even shown to be slightly faster. This work also proposed a detailed mechanism for PNP catalyzed 4-NPh hydrogenation with $NaBH_4$ (Figure 4.4).

The efforts to understand and apply PNPs for nitroaromatic reduction have generated a vast body of work which spans a range of approaches with varied benefits and

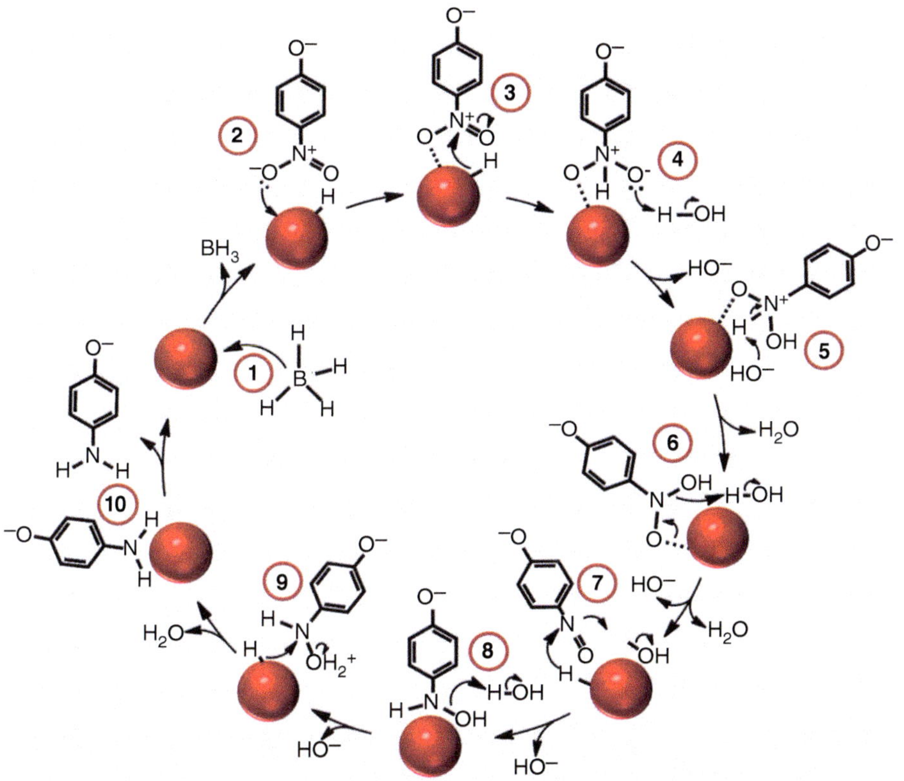

Figure 4.4 Proposed mechanism for the 4-nitrophenol hydrogenation by $BH_4^-{}_{(aq)}$ catalyzed by metal NPs. *Source*: Barbosa et al. [43].

trade-offs based on the nanoarchitecture desired to exploit a particular plasmonic property. The Pal group pursued their exploration of such systems with colloidal Ag NPs, [44] either supported or supported by silica gel [45]. Both catalysts were active for the conversion of the familiar 4-NPh hydrogenation with $NaBH_4$ under light activation, while neither worked with other reducing agents such as hydrazine.

Other studies focused on enhancing the activity of these monometallic PNPs, often adjusting their morphology to increase their photoactivity by tuning the surface area and formation of hot spots on complex structures. Chakraborty, Parikh, and coworkers developed hollow colloidal Ag NPs that showed a slight increase in activity compared to similar spherical Ag PNPs for the reduction of 4-NPh [46]. Carefully engineered geometries such as porous multishell Au NPs, composed of concentric Au spheres modified using poly(sodium-p-styrenesulfonate) (PSS), have also been shown to be effective and selective catalysts. [47] Under mild conditions, the reduction of 4-nitrostyrene to 4-amineostyrene using NH_3BH_3 as a reducing agent was observed with >99% conversion and 97% selectivity under 420 nm light irradiation. This study also focused on the effects of surface modification using PSS, which increased selectivity toward nitro reduction when compared to un-modified particles. Cai and coworkers also developed NPs which exploited surface functionalization to enhance catalytic activity and stability [48]. They reported AuNPs coated with polyphenol–Fe^{3+} as active nanocatalyst for 4-NP hydrogenation which could be recycled up to four times without significant loss of activity.

Deviating from the typical spherical PNPs, Au nanostars have been shown to exhibit increased activity due to their enhanced plasmonic properties resulting from the several sharp points of their unique morphology. Au nanostars catalyzed the hydrogenation of 4-NPh and 4-nitroanailine using $NaBH_4$ as a reductant which was monitored by UV–Vis absorption spectroscopy but were ineffective when using 4-nitrothiophenol (4-NTP) as a substrate [49]. This was attributed to the preferential adsorption of the thiol moiety on the Au surface, rather than the nitro functional group. This observation reinforces the importance of deliberate catalyst design to truly exploit the vast potential of PNPs when applied to a particular process.

Other reported plasmonic catalysts have employed metallic particles immobilized on an inert support to increase particle stability and ease of catalyst recyclability. Au NPs trapped in resin were shown to be efficient for the hydrogenation of 4-NPh and other polysubstituted nitro aromatics using $NaBH_4$ under mild conditions with yields up to 95% [50]. Similar Au NPs have also been synthesized *in situ* inside a chitosan hydrogel which increased recyclability and stability, preventing any meaningful change in catalyst activity after either five cycles or a year of storage [51]. The hydrogenation of 4-NPh with $NaBH_4$ resulted in ~100% conversion after less than 15 minutes and the catalyst was also shown to be active for a range of other nitro aromatic compounds.

Using microwave assisted synthesis, Mori et al. were able to grow size-controlled Ag PNPs directly onto SBA-15 mesoporous silica, and were used to hydrogenate 4-NP reduction using NH_3BH_3 as reducing agent [52]. This synthesis method provided a useful tool for exploring the impact of Ag NPs size on the property of the material in detail, while in tandem developing a highly size controllable method for Ag NPs synthesis. In general, they concluded that larger (48 × 7.9 nm) blue tinted Ag NRs showed the greatest catalytic activity owing to their increased LSPR excitation at those sizes.

While monometallic plasmonic particles have shown to be effective catalysts, a range of synergistic interactions and complimentary properties open up when considering the use of bimetallic plasmonic systems, either hybrids made of combined plasmonic metals or other nonplasmonic metals. Zhu and coworkers compared the hydrogenation of NB in a KOH, isopropyl alcohol solution on pure Au@ZrO_2 NPs and alloyed $Au_{2.6}Cu_{0.4}$@ZrO_2 NPs, under visible light irradiation [53]. Through experimental results and DFT calculations, these researchers demonstrated that this alloying affected the reaction pathway. Specifically, stronger interaction between the nitro group and the Cu atoms, compared to the pure Au surface, caused the adsorbed NB to convert directly to aniline, whereas a complex series of intermediates following the classic mechanism occurred on pure Au [53]. This hybrid catalyst showed remarkable yields (>90%) for the selective hydrogenation of several nitroaromatic by facilitating the reduction of the desired nitro group and preventing competing reactions such as dimerization caused by unwanted intermediates.

The benefits of AuCu alloys were also exploited by the Hou group. AuCu pentacle NPs showed increased activity compared to spherical Au NPs and other AuCu geometries for 4-NP hydrogenation using $NaBH_4$, under visible light irradiation [54]. Another study considered immobilized AuCu triangular NPs and reported enhanced catalytic activity compared to hemispherical AuCu NPs and truncated octahedrons Au NPs for the hydrogenation of 4-NP, showing similar results to the AuCu pentacle NPs [55]. In both cases, the careful combination of a favorable heterometallic composition and increased catalytic cites caused by their sharp geometry outperformed simple spherical NPs with a monometallic composition.

These AuCu heterometallic structures can also bring added economic benefit by increasing the content of inexpensive materials such as Cu. Further developing these economic factors for PNPs will be a critical step for potential industrial applications to compete with currently available catalysts. This aspect was further explored by Wu et al. who developed a method for the single-pot synthesis of AgCu NPs aimed at producing an efficient catalyst with a less material and energy intensive synthesis [56]. This scalable method of AgCu NPs production employs inexpensive glucose to produce the NPs while still maintaining the superior catalytic capabilities compared to than monometallic Ag NPs for the reduction of 4-NPh with visible light.

In a similar vein, the Baker group synthesized Au, Ag, and AuAg NPs for 4-NPh reduction using a less complex, time-consuming, and hazardous material intensive process [57]. Bio-based and inexpensive citric acid-derived carbon nanodots were shown by the authors to be an effective replacement for laser ablation or hazardous reductants such as hydrazine for the synthesis of AuAg NP, all while still maintaining their catalytic property.

Core–shell or decorated structures, predominantly combinations of Au and Ag, have been investigated for nitro-aromatic hydrogenation due to their synergistic effects. Kanta et al. reported that Au-Ag core shell NPs were 12 times more active when compared to pure Au NPs for the hydrogenation of 4-NPh using $NaBH_4$ with a reported 96.5% conversion under visible light [58]. Yin et al. also studied Au-Ag core–shell NPs and attributed the increased activity of these heterometallic PNPs to the couplings of the Au and Ag plasmons [59]. In these PNPs, the Au core acts as a light harvesting antenna which induces a near-field enhancement at the substrate-Ag interface causing a direct interfacial electronic transition. The LSPR coupling between the two metals decays into hot electrons which can transfer directly into the absorbate, catalyzing chemical transformation. By combining similar Au-Ag NP hybrid particles into self-assembling, 1D nanochains of either Au-Ag or Ag@Au spherical NPs were used to hydrogenate o-chloronitrobenzene under mild conditions in an H_2 atmosphere. The most promising Au-Ag nanochains, composed of 25 nm Au cores with a thin 4 nm Ag shell, could obtain a TOF of 16.5 h^{-1} and a yield of ~77%, which was better than individual Au-Ag NPs and an order of magnitude greater than monometallic Au NPs. The nanochain activity compared to individual NPs was determined to arise primarily from the hot-spot areas between the individual NPs where the electromagnetic enhancement is known to be greatly enhanced [60]. Enhanced generation of hot electrons there, as well as preferential adsorption properties were proposed as the cause of these enhancements.

Focusing on the practical application of these PNPs into industrially relevant catalysts, Yilmaz et al. looked at overcoming the issues related to aggregation, low recyclability, and low surface area often associated with suspended PNPs [61]. They deposited an array of 3-D Au-Ag NRs onto an inert support, which had catalytic properties similar to other bimetallic PNPs used for 4-NPh hydrogenation using $NaBH_4$, but with the notable benefit of simplified extraction. Unfortunately, this proposed design has some significant usability drawbacks, in particular it quickly decreased in efficacy after the first cycle, potentially due to poisoning on the catalysts surface.

While the combination of plasmonic metals to produce a hybrid PNP with superior properties has been thoroughly explored, another useful development of plasmonic catalysts for nitroaromatic reduction is the generation of a magnetic core surrounded by a catalytically active shell. The Peng group reported a magnetically recoverable Fe_3O_4-Au core shell structure recoverable and usable for up to six hydrogenations of 4-NPh, while maintaining 100% conversion in just over an hour [62]. Another catalyst developed by Shin et al. implemented an Fe_3O_4-SiO_2-Ag multishell structure which was also effective in the hydrogenation of 2-NPh, 3-NPh, and 4-NPh and could also be easily extracted from the reaction mixture after use [63]. The focus of these PNPs on utility and ease of recovery are another example of a potentially useful development to create PNPs for nitro reductions which are appealing in an industrial setting, and also offer greener and less costly alternatives.

The previously discussed benefits derived from a synergistic combination of a plasmonic metal with a traditional catalyst and/or a semiconductors support have also show great promise for nitroarene reduction. By combining a SiC support with Au and Pd nanoislands (Au/Pd@SiC), Hao et al. reported the hydrogenation of NB under 0.8 W/cm [2] intensity visible light (400–800 nm) with a TOF of up to 1715 h^{-1} and a 100% yield [64]. Under mild conditions under one H_2 atmosphere, the hot electrons generated by the LSPR of the Au NPs transferred into the adjacent SiC, activating the adsorbed NB or transferring to the Pd islands. These adsorbed NB then readily reacted with active hydrogen present on the surface of the SiC that migrated from the electron rich Pd sites due to the rapid dissociation of H_2 on this highly catalytic metal. This synergistic catalyst demonstrated a broad applicability which exploited this interaction, hydrogenating several other nitroarene to their corresponding anilines with similar, or even enhanced, TOF and yields compared to NB hydrogenation.

Another approach by the Yamashita group used Ag@TiO_2 NPs for the selective hydrogenation of 4-nitrostyrene (4-NS) to 4-aminostyrene (4-AS) with NH_3BH_3 as a reductant. For the tested metallic oxides (TiO_2, ZrO_2, Al_2O_3, and CeO_2) TiO_2 showed the greatest promise with an ~81% yield under light irradiation [65]. To determine the impact of irradiation wavelength on the catalyst, the reaction was activated with blue (470 nm), green (530 nm), and red (627 nm) irradiation. Green light demonstrated the greatest yield of ~83%, consistent with the 516 nm plasmon peak of the Ag@TiO_2 NPs. Further developments by this group employed CeO_2 coated mesoporous silica supporting Ag NPs (Ag@CeO_2-SiO_2) to obtain a yield of 90% for the same 4-NS hydrogenation [66]. Kaur et al. also used a similar Ag@TiO_2 plasmonic catalyst to hydrogenate 4-nitrobenzoic acid to 4-aminobenzoic acid in isopropanol by directly irradiating these samples with sunlight to achieve a 100% yield [67]. In addition to their enhanced catalytic properties, these supported NPs also benefited from easier extraction and reuse compared to their unsupported counterparts, a notable quality which has been thoroughly discussed.

One of the few reported heterometallic plasmonic catalysts for nitroaromatic reduction was reported by Kim et al. It combined plasmonic Ag and catalytic Ni to produce a coupled binary system composed of a large Ni particle and a slightly smaller Ag decoration (Ag@Ni NPs) [68]. When these Ag@Ni NPs were compared to monometallic Ni NPs and core–shell Ag-Ni NPs for the hydrogenation of 4-NPh in a $NaBH_4$, isopropanol solution under visible light, the pseudo first order rate constant of Ag@Ni NPs were the highest reported and showed a sixfold increase compared to hydrogenation preformed in the dark. The exact mechanism requires further exploration, but it was suggested that the binary morphology allowed for the Ag surface plasmon to enhance the Ni reactivity without allowing the Ni to quench the plasmon, as seen in the Ag-Ni core–shell structures.

In 2020, Wang et al. reported one of the first plasmonic catalysts which was capable of reducing nitrates rather than an aromatic nitro groups with the familiar synergy between plasmonic metals and semiconducting supports [69]. The reduction of aqueous nitrates, preferably to N_2, is of critical importance for treating anthropogenic water pollution which may otherwise lead to eutrophication if left unchecked. These catalysts were composed of an oxygen deficient P25 TiO_2 center (R-TiO_2) synthesized via lithium thermal reduction that were then decorated with Ag and Cu metallic NPs to yield Ag/Cu@R-TiO_2 NPs. When these catalysts were combined with an aqueous nitrate solution under broad-spectrum light, 93% of the nitrate was reduced to N_2 and NH_4^+ (Scheme 4.15) with a selectivity of 68% toward environmentally benign N_2 without the need for additional reactants requires in

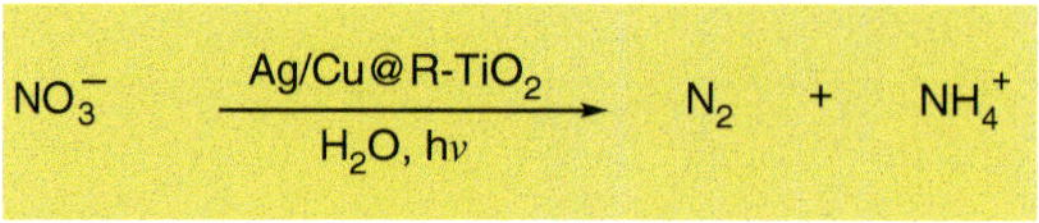

Scheme 4.15 Nitrate Reduction under light irradiation catalyzed by Ag/Cu@R-TiO_2 NPs [69]. *Source*: Based on Wang et al. [69].

many other traditional photocatalytic nitrate reductions. These researchers proposed that primarily a multistep mechanism initially produces hot electrons from the plasmonic metals which converts NO_3^- to NO_2^-, followed by the competitive reduction of NO_2^- to either N_2 or NH_4^+. These reactions consume protons, which are generated by a water oxidation process taking place at the valence band of R-TiO_2. There, holes are present and result from the migration of electrons to the valance band, and eventually the metals, thus closing the catalytic cycle. The water oxidation reaction caused the formation of O_2 as a benign byproduct. The selectivity toward the favorable N_2 over NH_4^+ correlated with a higher concentration of absorbed NO_2^- which was promoted by the protonated surface of R-TiO_2 in the acidic environment. Further, the R-TiO_2 increased the ability of photogenerated electrons to reduce nitrate intermediates compared to typical TiO_2 or SiO_2 due to the narrowed bandgap caused by defects states available below the typical conduction band energy.

4.4.2 Reductive Coupling of Nitroaromatics Compounds

While the formation of anilines from nitroarenes has been the primary focus of plasmon-mediated nitro reduction, PNPs also proved to be active for the reductive coupling of nitroaromatics compounds.

Early work by Zhu et al. showed that plasmonic activation could enhance selectivity for this reaction. Indeed Au@ZrO_2 NPs could effectively couple NB (~100% yield) to azobenzene under mild conditions in a KOH isopropanol solution and was also effective toward coupling of other nitroarenes under visible light irradiation (Scheme 4.16). [70] The proposed mechanism includes the oxidation of isopropanol at the active Au surface initiating the coupling of adsorbed NB which was sustained by an azoxybenzene intermediate to produce azobenzene (Figure 4.5).

In contrast, azobenzene was mainly an unstable intermediate of NB hydrogenation under thermal activation at 100 °C and five bars of H_2 using similar oxide

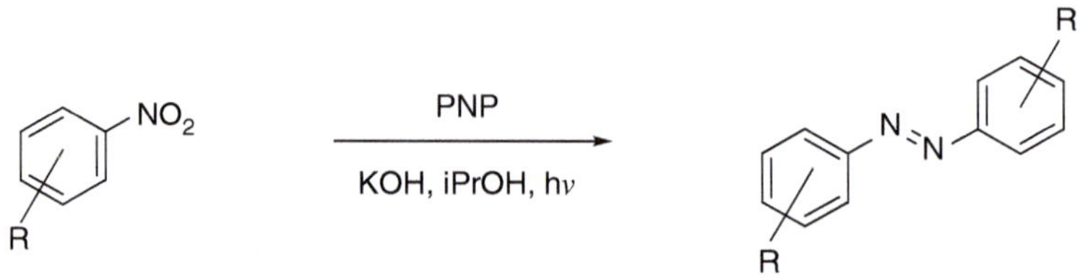

Scheme 4.16 Nitroaromatic coupling. *Source*: Based on Zhu et al. [70].

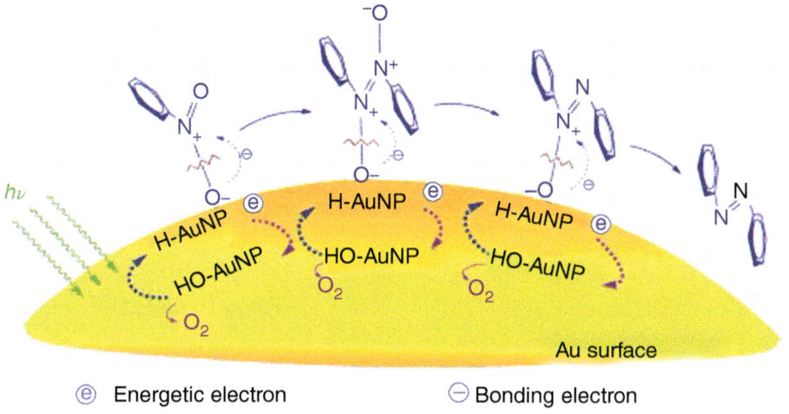

Figure 4.5 Mechanism for the photocatalytic reduction of nitroaromatic compounds. H–AuNP reacts with the N–O bonds to produce HO–AuNP species, which subsequently decompose to produce oxygen molecules and H–AuNP species. *Source*: Zhu et al. [70].

supported Au NPs [71]. The lower energy environment of the LSPR-mediated NB reduction permitted the unstable azobenzene to become the terminal product, illustrating the possible chemo selectivity of PNPs which can be obtained in greener, lower energy conditions [70].

The Tada group was able to dramatically increase the yield of these coupling reactions by replacing the inert support with a synergistic semi-conductor and decorating its surface with Au NPs with a 1:14 ratio of large (9 nm) and small (2 nm) bimodally distributed particles (BM Au@TiO_2) [72]. In a KOH, isopropanol solution under visible light, the smaller 2 nm Au NPs oxidized isopropanol which produced active hydrogen and hot electrons that could be transported to the larger 9 nm Au NPs, where the reductive coupling catalytic activity of the sites was enhanced. The familiar plasmonic metal and semiconductor interaction, under light irradiation, combined with the bimodal Au NPs which created active oxidative and reductive sites depending on the Au NPs size, caused a twofold enhancement of catalytic activity for several nitroaromatics compared to mono-sized Au on an inert surface [72].

Other products can also be obtained by nitroarene coupling which expands the possible utility for chemical manufacturing. Using AgCu@ZrO_2 NPs, Liu et al. could selectively produce azoxybenzene from NB in a visible light illuminated KOH isopropanol solution (Scheme 4.17) with a yield of ~80% and a TOF of 2.0 h^{-1} [73] without the need for stoichoimetric amounts of activated reducing agents required

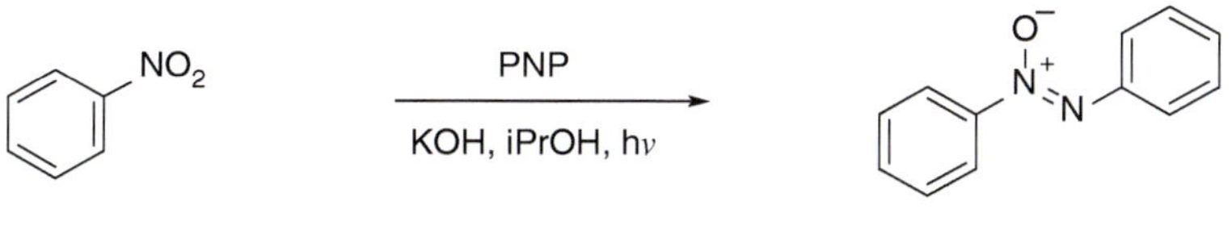

Scheme 4.17 Selective coupling of nitrobenzene to azoxybenzene under visible light irradiation. *Source*: Based on Liu et al. [73].

by other catalysts [74]. Azoxybenzene is often reported as an unstable intermediate of NB coupling (Figure 4.5) and it was noted by the authors that, contrary to monometallic Ag NPs, AgCu@ZrO_2 NPs were uniquely able to selectively produce azoxybenzene. Reaction selectivity was controlled by tuning the wavelength of the visible irradiating light and a similar control over conversion was demonstrated by increasing light intensity and reaction temperature.

Au@CeO_2 NPs were also shown to be moderately effective at producing azoxybenzene from NB, but with only half the yield obtained by AgCu@ZrO_2 NPs [37]. Au@CeO_2 NPs were also shown to be moderately effective toward the hydrogenation of azobenzene to hydrazobenzene (Scheme 4.18), another industrially relevant compound.

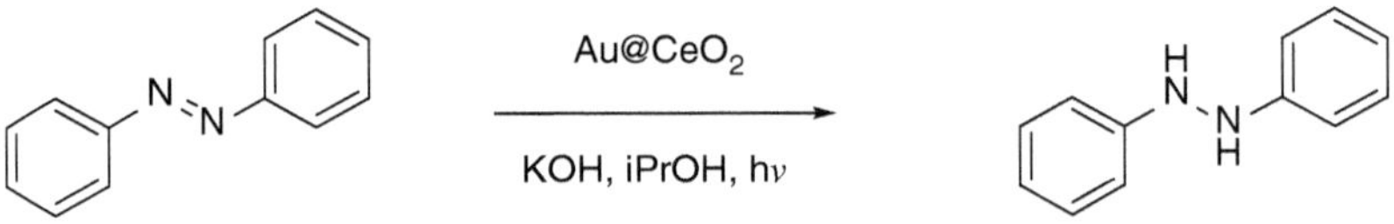

Scheme 4.18 Selective hydrogenation of azobenzene under light irradiation catalyzed by Au@CeO_2 [37]. *Source*: Based on Ke et al. [37].

Other reported systems afford accurate control over the desired product based on reaction conditions and catalyst design. The Huang group showed that Au@TiO_2 NPs could couple NB selectively at 90 °C under visible light irradiation to either hydrazobenzene or azobenzene by simply varying the reacting time (4.16 and 4.19) [75]. This gives a useful flexibility to the catalysts which already showed exceptional catalytic properties, with yields over 95% and a TOF up to 400 h^{-1}.

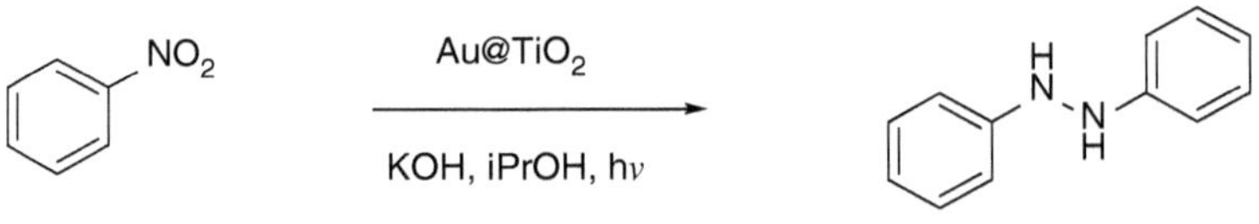

Scheme 4.19 Selective coupling of nitrobenzene to hydrazobenzene under visible light catalyzed by Au@TiO_2 at 90 °C [75]. *Source*: Based on Liu et al. [75].

By tuning the reaction temperature, Guo et al. were able to obtain either azobenzene or azoxybenzene (4.16 and 4.17) via NB coupling, using Cu NPs stabilized on graphene under visible light irradiation (400–800 nm) [76]. This selectivity was contributed to the lower activation energy required to convert NB to azoxybenzene compared to the further reduction of azoxybenzene to azobenzene, which allowed the authors to control the formation of the product by adjusting temperature. By using a series of cut off filters to block portions of the light, it was also demonstrated that 43% of the total light-induced activity was initiated by the 530–600 nm range which overlapped with the catalyst's plasmon peak, compared to the 28% activity for the higher energy and larger wavelength range of 450–530 nm.

It has recently been reported that SERS is also capable of initiating catalytic reduction of nitroarenes. SERS is a powerful technique relying on plasmonic field enhancement to detect the presence of very small concentration of Raman active molecules. As it uses a plasmonic surface for detection, plasmon-induced transformation can also happen at the surface of the photoactive metal. While transformations happening in SERS apparatus may not correspond to traditional in batch catalytic reactions, they provide important insights into the potential of plasmonic metals. One of the first plasmon-induced transformations observed *via* SERS was reported by Tian and coworkers in 2010. The authors noticed the appearance of a new signal, which lead them to realize a plasmon-mediated transformation had occured, and Ag had therefore converted the para-aminothiophenol into 4,4′-dimercaptoazobenzene. [77] Since this study demonstrated plasmon-induced oxidative coupling on Ag surface, the same group conducted extensive studies on SERS-mediated amine and nitro couplings, both experimental [77, 78] and theoretical. [79, 80]

Reductive coupling of nitrophenol was also observed *via* SERS using Au, as reported by Ren et al. [81] Zhang et al. explored metal–metal, metal–semiconductor, and metal–insulator interaction used in the dimerization of 4-nitrothiophenol [82]. The authors increased the activity of the plasmonic metal by the formation of heterogeneous nanomaterials. They used Au cores encapsulated by shells composed of metal, semiconductor, or insulator (Au@Ag, Au@Ag_2S, and Au@SiO_2) core–shell to study the photoactivity for the formation of p,p′-dimercaptoazobenzene from 4-NPh. The Schlücker group studied 3D superstructures, gold satellites self-assembled onto a large shell-isolated gold core, for the reduction of 4-NPh using $NaBH_4$ [83]. SERS was also used to observe interesting selectivity for plasmon-driven diazo coupling reaction of p-nitroaniline. Ding et al. reported the formation of 4,4′-diaminoazobenzene through a selective reduction of the nitro groups, rather than oxidization of the amines into 4,4′-dinitroazobenzene, using supported Ag NPs [84].

4.5 Outlook

This chapter has provided an overview of catalytic reductions catalyzed by PNPs across a range of substrates. The plasmonic enhancement caused by the use of such catalysts under light irradiation is typically associated with high yields, improved selectivity, and reduced energy use. Throughout the chapter, an emphasis was placed on the variety of PNP compositions and morphologies which dramatically influenced their optical properties and efficacy. A number of groups focused on nanoparticle engineering in order to create structures able to feature LSPR hot spots, and demonstrated that these usually enhanced the desired catalytic effect. This was achieved either by the shape of the nanoparticle, or by the assembly of multiple particles. Furthermore, the combination of plasmonic metals with traditional catalysts, semiconductors, and supports provide another dimension to the tunability of these particles for a variety of applications. While industrial instances of these catalyst are still on the horizon, researchers have continued to expand their feasibility through thorough, ongoing analysis of their properties and novel particle designs to provide a novel avenue for effective, green catalysis.

Future work will undoubtedly focus on the use of cheaper and more earth abundant metals, such as Cu and Al. The hydrogenation reaction features specific challenges that PNP-based catalysts can help solve. On the reducing agent side, moving away from hazardous reductants, used in large excess and causing large amounts of waste such as $NaBH_4$ is still a relevant topic of research. Several examples covered in this chapter highlight the possibility to use LSPR-generated hot electrons to activate H_2 or isopropanol and turn them into more active reducing agents under the used conditions. Another classic feature of hydrogenation focused research lies in the ability to precisely control the selectivity of these processes. This is of tremendous importance in the fields of pharma and perfume, but is becoming hot as well in the context of biomass conversion. Very recent works covered in this chapter illustrate this trend and there is no doubt future work will dig more in this direction. Also, as more active and selective PNP-based reduction catalysts are developed, we expect to see reactions with less activated substrates, such as aliphatic compound expanding their applicability beyond the predominant focus on aromatic ones. On the catalysts side, we also anticipate that new, emerging plasmonic nanomaterials, such as group-4 nitride [85] are also of great interest and should be explored in the future. More investigation of reaction mechanisms is also anticipated, in an effort to better rationalize the role of LSPR enhancement. To that effect, we want to point the reader to a recent article which revisited several seminal plasmonic catalysis papers and argued that hot electron mechanisms claims may in fact be inaccurate and that simple photothermal effects may explain observed activity [86]. Lastly, the reader is pointed to a recent article by our group showcasing for the first time the use of plasmonic catalysis for arene hydrogenation [87]. As these articles appeared during the production of this chapter, they were not discussed in length herein. In this context, hydrogenation reactions which have been comparatively less explored than their oxidation counterparts are anticipated to be the topic of future mechanistic studies.

References

1 Irfan, M., Glasnov, T.N., and Kappe, C.O. (2011). Heterogeneous catalytic hydrogenation reactions in continuous-flow reactors. *ChemSusChem* 4 (3): 300–316.

2 Tiedtke, D. B.; Cheung, T. T. P.; Leger, J.; Zisman, S. A.; Bergmeister, J. J.; Delzer, G. A. In *Chemicals influencing the activity of palladium-based catalysts for the selective hydrogenation of acetylene to etylene in acetylene converters*, 13th Ethylene Producers' Conference, 2001; American Institute of Chemical Engineers: 2001.

3 Behr, A. and Neubert, P. (2012). *Applied homogeneous catalysis*. John Wiley & Sons.

4 Nishimura, S. (2001). *Handbook of heterogeneous catalytic hydrogenation for organic synthesis*. New York: Wiley.

5 de Vries, J.G. and Elsevier, C.J. (2007). *Handbook of homogeneous hydrogenation.* WeinheimWiley-VCH.

6 Cole-Hamilton, D.J. and Tooze, R.P. (2006). *Catalyst separation, recovery and recycling: chemistry and process design*, vol. 30. Springer Science & Business Media.

7 Trost, B.M. (1995). Atom economy—a challenge for organic synthesis: homogeneous catalysis leads the way. *Angewandte Chemie International Edition in English* 34 (3): 259–281.

8 Ahmad, H., Kamarudin, S., Minggu, L., and Kassim, M. (2015). Hydrogen from photo-catalytic water splitting process: a review. *Renewable and Sustainable Energy Reviews* 43: 599–610.

9 Moores, A. (2010). Atom economy–principles and some examples. *Handbook of Green Chemistry: Online*: 1–15.

10 Moores, A. and Goettmann, F. (2006). The plasmon band in noble metal nanoparticles: an introduction to theory and applications. *New Journal of Chemistry* 30 (8): 1121–1132.

11 Gellé, A. and Moores, A. (2019). Plasmonic nanoparticles: photocatalysts with a bright future. *Current Opinion in Green and Sustainable Chemistry* 15: 60–66.

12 Long, R., Li, Y., Song, L., and Xiong, Y. (2015). Coupling solar energy into reactions: materials design for surface plasmon-mediated catalysis. *Small* 11 (32): 3873–3889.

13 Wang, C. and Astruc, D. (2014). Nanogold plasmonic photocatalysis for organic synthesis and clean energy conversion. *Chemical Society Reviews* 43 (20): 7188–7216.

14 Linic, S., Aslam, U., Boerigter, C., and Morabito, M. (2015). Photochemical transformations on plasmonic metal nanoparticles. *Nature Materials* 14 (6): 567–576.

15 Gellé, A., Jin, T., de la Garza, L. et al. (2019). Applications of plasmon-enhanced nanocatalysis to organic transformations. *Chemical Reviews* 120 (2): 986–1041.

16 Long, R., Rao, Z., Mao, K. et al. (2015). Efficient coupling of solar energy to catalytic hydrogenation by using well-designed palladium nanostructures. *Angewandte Chemie International Edition* 54 (8): 2425–2430.

17 Gangishetty, M.K., Fontes, A.M., Malta, M. et al. (2017). Improving the rates of Pd-catalyzed reactions by exciting the surface plasmons of AuPd bimetallic nanotriangles. *RSC Advances* 7 (64): 40218–40226.

18 Hrelescu, C., Sau, T.K., Rogach, A.L. et al. (2011). Selective excitation of individual plasmonic hotspots at the tips of single gold nanostars. *Nano Letters* 11 (2): 402–407.

19 Lu, X., Rycenga, M., Skrabalak, S.E. et al. (2009). Chemical synthesis of novel plasmonic nanoparticles. *Annual Review of Physical Chemistry* 60 (1): 167–192.

20 Mao, Z., Vang, H., Garcia, A. et al. (2019). Carrier diffusion—the main contribution to size-dependent photocatalytic activity of colloidal gold nanoparticles. *ACS Catalysis* 9 (5): 4211–4217.

21 Yuan, L., Zhang, C., Zhang, X. et al. (2019). Photocatalytic hydrogenation of graphene using Pd nanocones. *Nano Letters* 19 (7): 4413–4419.

22 Swearer, D.F., Zhao, H., Zhou, L. et al. (2016). Heterometallic antenna–reactor complexes for photocatalysis. *Proceedings of the National Academy of Sciences* 113 (32): 8916–8920.

23 Dubuis, H. L. a. R., Palladium catalyst for partial reduction of acetylenes. *Organic Syntheses* 1966, 46.

24 Zhang, C., Zhao, H., Zhou, L. et al. (2016). Al–Pd nanodisk heterodimers as antenna–reactor photocatalysts. *Nano Letters* 16 (10): 6677–6682.

25 Spata, V.A. and Carter, E.A. (2018). Mechanistic insights into photocatalyzed hydrogen desorption from palladium surfaces assisted by localized surface plasmon resonances. *ACS Nano* 12 (4): 3512–3522.

26 Quiroz, J., Barbosa, E.C.M., Araujo, T.P. et al. (2018). Controlling reaction selectivity over hybrid plasmonic nanocatalysts. *Nano Letters* 18 (11): 7289–7297.

27 Schnepf, M.J., Mayer, M., Kuttner, C. et al. (2017). Nanorattles with tailored electric field enhancement. *Nanoscale* 9 (27): 9376–9385.

28 da Silva, A.G., Rodrigues, T.S., Correia, V.G. et al. (2016). Plasmonic nanorattles as next-generation catalysts for surface plasmon resonance-mediated oxidations promoted by activated oxygen. *Angewandte Chemie International Edition* 55 (25): 7111–7115.

29 Zhao, X., Long, R., Liu, D. et al. (2015). Pd–Ag alloy nanocages: integration of Ag plasmonic properties with Pd active sites for light-driven catalytic hydrogenation. *Journal of Materials Chemistry A* 3 (18): 9390–9394.

30 Huang, H., Zhang, L., Lv, Z. et al. (2016). Unraveling surface plasmon decay in core–shell nanostructures toward broadband light-driven catalytic organic synthesis. *Journal of the American Chemical Society* 138 (21): 6822–6828.

31 Zhang, J., Li, C., Xu, C. et al. (2019). Enhancing photothermal effect and stability of plasmonic Pd/Ag-nanosheet by nanoassembly for efficient light-driven catalytic organic hydrogenation. *ChemistrySelect* 4 (45): 13173–13181.

32 Guselnikova, O., Olshtrem, A., Kalachyova, Y. et al. (2018). Plasmon catalysis on bimetallic surface—selective hydrogenation of alkynes to alkanes or alkenes. *The Journal of Physical Chemistry C* 122 (46): 26613–26622.

33 Landry, M.J., Gellé, A., Meng, B.Y. et al. (2017). Surface-plasmon-mediated hydrogenation of carbonyls catalyzed by silver nanocubes under visible light. *ACS Catalysis* 7 (9): 6128–6133.

34 Hao, C.-H., Guo, X.-N., Pan, Y.-T. et al. (2016). Visible-light-driven selective photocatalytic hydrogenation of cinnamaldehyde over Au/SiC catalysts. *Journal of the American Chemical Society* 138 (30): 9361–9364.

35 Ma, Y. and Li, Z. (2018). Coupling plasmonic noble metal with TiO_2 for efficient photocatalytic transfer hydrogenation: M/TiO_2 (M = Au and Pt) for chemoselective transformation of cinnamaldehyde to cinnamyl alcohol under visible and 365 nm UV light. *Applied Surface Science* 452: 279–285.

36 Cohen, R., Graves, C.R., Nguyen, S.T. et al. (2004). The mechanism of aluminum-catalyzed Meerwein–Schmidt–Ponndorf–Verley reduction of carbonyls to alcohols. *Journal of the American Chemical Society* 126 (45): 14796–14803.

37 Ke, X., Zhang, X., Zhao, J. et al. (2013). Selective reductions using visible light photocatalysts of supported gold nanoparticles. *Green Chemistry* 15 (1): 236–244.
38 Lange, J.-P., van der Heide, E., van Buijtenen, J., and Price, R. (2012). Furfural—a promising platform for lignocellulosic biofuels. *ChemSusChem* 5 (1): 150–166.
39 Zhang, M. and Li, Z. (2019). Cu/Cu_2O-MC (MC = mesoporous carbon) for highly efficient hydrogenation of furfural to furfuryl alcohol under visible light. *ACS Sustainable Chemistry & Engineering* 7 (13): 11485–11492.
40 Wang, R., Liu, H., Wang, X. et al. (2020). Plasmonic enhanced furfural hydrogenation catalyzed by stable carbon coated copper nanoparticles driven from metal-organic frameworks. *Catalysis Science & Technology*.
41 Hu, G., Huang, Z.-P., Hu, C.-X. et al. (2020). Selective photocatalytic hydrogenation of α,β-unsaturated aldehydes on $Au/CuCo_2O_4$ nanotubes under visible-light irradiation. *ACS Sustainable Chemistry & Engineering* 8 (22): 8288–8294.
42 Pradhan, N., Pal, A., and Pal, T. (2001). Catalytic reduction of aromatic nitro compounds by coinage metal nanoparticles. *Langmuir* 17 (5): 1800–1802.
43 Barbosa, E.C.M., Fiorio, J.L., Mou, T. et al. (2018). Reaction pathway dependence in plasmonic catalysis: hydrogenation as a model molecular transformation. *Chemistry – A European Journal* 24 (47): 12330–12339.
44 Pradhan, N., Pal, A., and Pal, T. (2002). Silver nanoparticle catalyzed reduction of aromatic nitro compounds. *Colloids and Surfaces A: Physicochemical and Engineering Aspects* 196 (2–3): 247–257.
45 Kundu, S., Mandal, M., Ghosh, S.K., and Pal, T. (2004). Photochemical deposition of SERS active silver nanoparticles on silica gel and their application as catalysts for the reduction of aromatic nitro compounds. *Journal of Colloid and Interface Science* 272 (1): 134–144.
46 Vadakkekara, R., Chakraborty, M., and Parikh, P.A. (2012). Reduction of aromatic nitro compounds on colloidal hollow silver nanospheres. *Colloids and Surfaces A: Physicochemical and Engineering Aspects* 399: 11–17.
47 Song, S., Li, J., Liu, Y. et al. (2020). Robust synthesis of Au-based multishell structures as plasmonic catalysts for selective hydrogenation of 4-nitrostyrene. *Angewandte Chemie* 59 (3): 1103–1107.
48 Zeng, T., Zhang, X., Guo, Y. et al. (2014). Enhanced catalytic application of Au@polyphenol-metal nanocomposites synthesized by a facile and green method. *Journal of Materials Chemistry A* 2 (36): 14807–14811.
49 Cui, Q., Xia, B., Mitzscherling, S. et al. (2015). Preparation of gold nanostars and their study in selective catalytic reactions. *Colloids and Surfaces A: Physicochemical and Engineering Aspects* 465: 20–25.
50 Shah, D. and Kaur, H. (2014). Resin-trapped gold nanoparticles: An efficient catalyst for reduction of nitro compounds and Suzuki-Miyaura coupling. *Journal of Molecular Catalysis A: Chemical* 381: 70–76.
51 Wu, X.-Q., Wu, X.-W., Huang, Q. et al. (2015). In situ synthesized gold nanoparticles in hydrogels for catalytic reduction of nitroaromatic compounds. *Applied Surface Science* 331: 210–218.
52 Mori, K., Verma, P., Hayashi, R. et al. (2015). Color-controlled Ag nanoparticles and nanorods within confined mesopores: microwave-assisted rapid

synthesis and application in plasmonic catalysis under visible-light irradiation. *Chemistry--A European Journal* 21 (33): 11885–11893.

53 Xiao, Q., Sarina, S., Waclawik, E.R. et al. (2016). Alloying gold with copper makes for a highly selective visible-light photocatalyst for the reduction of nitroaromatics to anilines. *ACS Catalysis* 6 (3): 1744–1753.

54 He, R., Wang, Y.-C., Wang, X. et al. (2014). Facile synthesis of pentacle gold–copper alloy nanocrystals and their plasmonic and catalytic properties. *Nature Communications* 5: 4327.

55 Hajfathalian, M., Gilroy, K.D., Yaghoubzade, A. et al. (2015). Photocatalytic enhancements to the reduction of 4-nitrophenol by resonantly excited triangular gold–copper nanostructures. *The Journal of Physical Chemistry C* 119 (30): 17308–17315.

56 Wu, W., Lei, M., Yang, S. et al. (2015). A one-pot route to the synthesis of alloyed Cu/Ag bimetallic nanoparticles with different mass ratios for catalytic reduction of 4-nitrophenol. *Journal of Materials Chemistry A* 3 (7): 3450–3455.

57 Essner, J.B., Laber, C.H., and Baker, G.A. (2015). Carbon dot reduced bimetallic nanoparticles: size and surface plasmon resonance tunability for enhanced catalytic applications. *Journal of Materials Chemistry A* 3 (31): 16354–16360.

58 Haldar, K.K., Kundu, S., and Patra, A. (2014). Core-size-dependent catalytic properties of bimetallic Au/Ag core–shell nanoparticles. *ACS Applied Materials & Interfaces* 6 (24): 21946–21953.

59 Yin, Z., Wang, Y., Song, C. et al. (2018). Hybrid Au–Ag nanostructures for enhanced plasmon-driven catalytic selective hydrogenation through visible light irradiation and surface-enhanced Raman scattering. *Journal of the American Chemical Society* 140 (3): 864–867.

60 Besteiro, L.V. and Govorov, A.O. (2016). Amplified generation of hot electrons and quantum surface effects in nanoparticle dimers with plasmonic hot spots. *The Journal of Physical Chemistry C* 120 (34): 19329–19339.

61 Yilmaz, M., Demirel, G.B., and Demirel, G. (2020). Silver thin film deposited 3-dimensional gold nanorod arrays for plasmonic catalysis. *Journal of Vacuum Science & Technology A* 38 (2): 023407.

62 Wu, Y., Zhang, T., Zheng, Z. et al. (2010). A facile approach to Fe_3O_4@ Au nanoparticles with magnetic recyclable catalytic properties. *Materials Research Bulletin* 45 (4): 513–517.

63 Shin, K.S., Cho, Y.K., Choi, J.-Y., and Kim, K. (2012). Facile synthesis of silver-deposited silanized magnetite nanoparticles and their application for catalytic reduction of nitrophenols. *Applied Catalysis A: General* 413: 170–175.

64 Hao, C.-H., Guo, X.-N., Sankar, M. et al. (2018). Synergistic effect of segregated Pd and Au nanoparticles on semiconducting SiC for efficient photocatalytic hydrogenation of nitroarenes. *ACS Applied Materials & Interfaces* 10 (27): 23029–23036.

65 Verma, P., Kuwahara, Y., Mori, K., and Yamashita, H. (2020). Visible-light-driven reduction of nitrostyrene utilizing plasmonic silver nanoparticle catalysts immobilized on oxide supports. *Catalysis Today* 355: 620–626.

66 Verma, P., Kuwahara, Y., Mori, K., and Yamashita, H. (2019). Plasmonic catalysis of Ag nanoparticles deposited on CeO_2 modified mesoporous silica for the nitrostyrene reduction under light irradiation conditions. *Catalysis Today* 324: 83–89.

67 Kaur Aulakh, M., Sharma, R., Pal, B., and Prakash, R. (2020). Photo-induced oxidation and reduction by plasmonic Ag-TiO_2 nanocomposites under UV/sunlight. *Solar Energy* 196: 427–436.

68 Kim, C., Kwon, Y., and Lee, H. (2015). Shape effect of Ag–Ni binary nanoparticles on catalytic hydrogenation aided by surface plasmons. *Chemical Communications* 51 (61): 12316–12319.

69 Wang, C., Dong, S., Wang, Y. et al. (2020). Selective removal of nitrate via the synergistic effect of oxygen vacancies and plasmon-induced hot carriers. *Chemical Engineering Journal* 397: 125435.

70 Zhu, H., Ke, X., Yang, X. et al. (2010). Reduction of nitroaromatic compounds on supported gold nanoparticles by visible and ultraviolet light. *Angewandte Chemie International Edition* 49 (50): 9657–9661.

71 Grirrane, A., Corma, A., and García, H. (2008). Gold-catalyzed synthesis of aromatic azo compounds from anilines and nitroaromatics. *Science* 322 (5908): 1661–1664.

72 Naya, S.i., Niwa, T., Kume, T., and Tada, H. (2014). Visible-light-induced electron transport from small to large nanoparticles in bimodal gold nanoparticle-loaded titanium (IV) oxide. *Angewandte Chemie International Edition* 53 (28): 7305–7309.

73 Liu, Z., Huang, Y., Xiao, Q., and Zhu, H. (2016). Selective reduction of nitroaromatics to azoxy compounds on supported Ag–Cu alloy nanoparticles through visible light irradiation. *Green Chemistry* 18 (3): 817–825.

74 Ohe, K., Uemura, S., Sugita, N. et al. (1989). Sodium arenetellurolate-catalyzed selective conversion of nitro aromatics to aromatic azoxy or azo compounds and its application for facile preparation of 3,3′- and 4,4′-bis[.beta.-(aryltelluro)vinyl]azobenzenes from (3- and 4-nitrophenyl)acetylenes. *The Journal of Organic Chemistry* 54 (17): 4169–4174.

75 Liu, Q., Zhang, J., Xing, F. et al. (2020). Plasmon-enhanced and controllable synthesis of azobenzene and hydrazobenzene using Au/TiO_2 composite. *Applied Surface Science* 500: 144214.

76 Guo, X., Hao, C., Jin, G. et al. (2014). Copper nanoparticles on graphene support: an efficient photocatalyst for coupling of nitroaromatics in visible light. *Angewandte Chemie International Edition* 53 (7): 1973–1977.

77 Huang, Y.-F., Zhu, H.-P., Liu, G.-K. et al. (2010). When the signal is not from the original molecule to be detected: chemical transformation of para-aminothiophenol on Ag during the SERS measurement. *Journal of the American Chemical Society* 132 (27): 9244–9246.

78 Huang, Y.-F., Wu, D.-Y., Zhu, H.-P. et al. (2012). Surface-enhanced Raman spectroscopic study of p-aminothiophenol. *Physical Chemistry Chemical Physics* 14 (24): 8485–8497.

79 Zhao, L.-B., Huang, Y.-F., Liu, X.-M. et al. (2012). A DFT study on photoinduced surface catalytic coupling reactions on nanostructured silver: selective formation of azobenzene derivatives from para-substituted nitrobenzene and aniline. *Physical Chemistry Chemical Physics* 14 (37): 12919–12929.

80 Zhao, L.-B., Zhang, M., Huang, Y.-F. et al. (2014). Theoretical study of plasmon-enhanced surface catalytic coupling reactions of aromatic amines and nitro compounds. *Journal of Physical Chemistry Letters* 5 (7): 1259–1266.

81 Ren, X., Tan, E., Lang, X. et al. (2013). Observing reduction of 4-nitrobenzenthiol on gold nanoparticles in situ using surface-enhanced Raman spectroscopy. *Physical Chemistry Chemical Physics* 15 (34): 14196–14201.

82 Zhang, K., Xu, N., Jia, M. et al. (2019). In situ detection of hot-electron-induced photocatalytic reduction using Au@Ag, Au@Ag_2S, and Au@SiO_2 core-shell nanoparticles. *Journal of Applied Physics* 125 (18): 183101.

83 Xie, W., Walkenfort, B., and Schlücker, S. (2012). Label-free SERS monitoring of chemical reactions catalyzed by small gold nanoparticles using 3D plasmonic superstructures. *Journal of the American Chemical Society* 135 (5): 1657–1660.

84 Ding, Q., Chen, M., Fang, Y. et al. (2017). Plasmon-driven diazo coupling reactions of *p*-nitroaniline via– NH_2 or– NO_2 in atmosphere environment. *The Journal of Physical Chemistry C* 121 (9): 5225–5231.

85 Karaballi, R.A., Humagain, G., Fleischman, B.R., and Dasog, M. (2019). Synthesis of plasmonic group-4 nitride nanocrystals by solid-state metathesis. *Angewandte Chemie* 131 (10): 3179–3182.

86 Dubi, Y., Un, I.W., and Sivan, Y. (2020). Thermal effects – an alternative mechanism for plasmon-assisted photocatalysis. *Chemical Science* 11: 5017–5027.

87 de la Garza, L.C., Brodusch, N., Gauvin, R., Moores, A. (2021). Plasmon-Enhanced Hydrogenation of 1-Dodecene and Toluene Using Ruthenium-Coated Gold Nanoparticles. *ACS App. Nano Mater*. DOI: 10.1021/acsanm.0c03077.

5

Plasmonic Catalysis, Photoredox Chemistry, and Photosynthesis

Sungju Yu[1,#] *and Prashant K. Jain*[1,2,3,4]

[1] *Department of Chemistry, University of Illinois at Urbana-Champaign, Urbana, IL, USA*
[2] *Materials Research Laboratory, University of Illinois at Urbana-Champaign, Urbana, IL, USA*
[3] *Department of Physics, University of Illinois at Urbana-Champaign, Urbana, IL, USA*
[4] *Beckman Institute for Advanced Science and Technology, University of Illinois at Urbana-Champaign, Urbana, IL, USA*

5.1 Introduction

Great strides have been made in the exploitation of the optical characteristics of noble metal nanostructures in a variety of fields such as photovoltaics [1–5], heterogeneous reactions [6–18], photodetectors [19–23], and sensors [24–28]. These nanostructures interact strongly with light. This interaction takes the form of a coherent collective oscillation of free electrons (e^-) of the nanostructure resonantly induced by the alternating electromagnetic field of light of a specific frequency. This resonant electronic oscillation is confined by the boundaries of the nanostructure and is therefore known as the localized surface plasmon resonance (LSPR). The LSPRs of Au, Ag, and Cu nanoparticles (NPs) occur in the visible-frequency range and therefore, they have been used to enhance visible-light absorption and charge carrier generation in photocatalytic and photovoltaic processes. For example, TiO_2 photocathodes modified with Au NPs produce a photocurrent under the irradiation of visible light [29–33]. In this composite metal–semiconductor system, the TiO_2 does not absorb the visible light; rather, the light is absorbed by the Au NPs via the excitation of LSPRs. The LSPRs decay to hot e^- in the Au NPs, which are transferred to the TiO_2 and harvested in the form of a photocurrent in the external circuit. Instead of leading to photocurrent, energetic e^- and holes (h^+) produced by plasmonic excitation can be directly harvested as reaction equivalents for driving redox reactions. This is the central principle behind plasmon-driven photoredox chemistry, which is the focus of this chapter. The salient features and operating mechanisms of plasmonic photoredox chemistry are described along with important examples. In particular, the plasmonic excitation of Au NPs has been shown to drive the multi-e^-, multi-proton (H^+) redox transformations such as H_2O splitting [7, 30–32, 34–37]

[#] Department of Energy Systems Research, Department of Chemistry, Ajou University, Republic of Korea.

Plasmonic Catalysis: From Fundamentals to Applications, First Edition.
Edited by Pedro H.C. Camargo and Emiliano Cortés.

and CO_2 reduction [13, 38–45] for the renewable production of fuels. We highlight particular observations where plasmon excitation has been found to drive an otherwise nonpreferred chemical pathway or a thermodynamically uphill ($\Delta G > 0$) redox reaction. The chapter ends with an outlook, where the challenges and prospects of plasmonic photoredox chemistry and photosynthesis are discussed.

5.2 Energy Conversion Following Plasmonic Excitation

5.2.1 Plasmon-Induced Generation of Charge Carriers

The collective plasmon oscillation dephases on a timescale of a few femtoseconds. Following dephasing, on the 100 fs–1 ps timescale, LSPRs decay either by radiative damping emitting photons, or by nonradiative damping forming e^-–h^+ pairs. Nonradiative damping process, i.e. the generation of energetic charge carriers, occurs via two channels: interband excitation (Figure 5.1a) and intraband excitation (Figure 5.1b). Interband excitation involves the transition of an e^- from the d band of the metal to an empty sp band state above the Fermi level of energy, ε_f. Intraband excitation involves the transition of an e^- from an occupied state (below the Fermi level) in the sp band to an empty state (above the Fermi level) of the sp band.

According to a theoretical study by Atwater and co-workers [47], the energy distribution of energetic charge carriers produced by the decay of excited LSPRs is influenced by the electronic band structure of the metal, particularly the position of the d band relative to ε_f. In Ag, the energetic threshold of interband transitions is ~3.6 eV. The lowest energy transitions arise from the d band states at the X and L points as well as from sp band states at ε_f at the L point. These transitions, therefore, give rise to e^- and h^+, both of considerably high energy and narrow energy

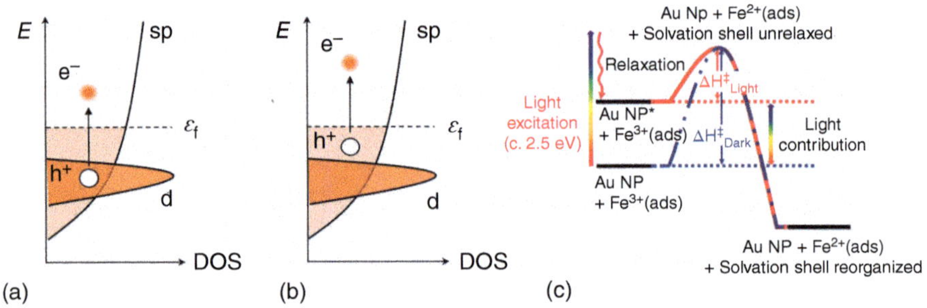

Figure 5.1 Energy (E) vs. density of states (DOS) diagrams illustrating the generation of energetic e^-–h^+ pairs by (a) interband transition and (b) intraband transition in a metal NP following the excitation of LSPRs. (c) Schematic showing the reaction activation barrier for the reduction of adsorbed ferricyanide, $Fe^{3+}{}_{(ads)}$, to ferrocyanide, $Fe^{2+}{}_{(ads)}$, on the surface of an Au NP under continuous LSPR excitation as compared to that for the reduction reaction on an Au NP under dark conditions. Under LSPR excitation, the Au NP is cathodically photocharged: the e^- supplied by such a photocharged Au NP (depicted as Au NP*) have a higher chemical potential as compared to an unexcited Au NP. As a result, the apparent activation enthalpy under light excitation, $\Delta H^{\ddagger}{}_{light}$, is lower than that under dark conditions, $\Delta H^{\ddagger}{}_{dark}$. *Source*: Panel (c) is reprinted with permission from Ref. [46]. Copyright American Chemical Society 2016.

distributions. Interband transitions in Cu and Au arise from the d band states near the X and L points, whereas, in the case of Cu, from d band states near the K point. These d band states are located ~2 eV below ε_f. As a result, the transitions result in e^-–h^+ pairs where the h^+ are ~2 eV more energetic than the e^-.

5.2.2 Extraction of Charge Carriers Generated by Plasmonic Excitation

The energetic charge carriers can be extracted from the plasmon-excited NP and utilized in redox reactions [46, 48, 49]. The e^- and h^+ generated in colloidal Au NPs by LSPR excitation are transferred to acceptor species (e.g. ferricyanide and ethanol) that adsorb from the solution phase to the NP surface. Interband damping of LSPRs or direct interband excitation produces h^+ in the d band of the Au NPs and e^- in sp states just above ε_f. These energetic h^+ are readily accepted by surface-adsorbed ethanol, which undergoes oxidation as a result [46, 48, 49]. The h^+ transfer events effectively amount to charge separation: the excess e^- build up on the Au NP. Under continuous LSPR excitation, an appreciable steady-state excess e^- density builds up on the NP. The cathodically photocharged NP has a raised quasi-Fermi level (dotted red line in Figure 5.1c) as compared to the Fermi level in the dark (dotted blue line in Figure 5.1c). A photoinduced Fermi-energy rise as high as 240 meV has been measured under 488 nm light excitation of an intensity of $0.9\,\mathrm{W\,cm^{-2}}$. On a timescale slower than the timescale of h^+ transfer, the excited e^- are transferred to surface-adsorbed ferricyanide complexes, triggering their reduction to ferrocyanide. Because of the higher chemical potential of the e^- supplied by a photocharged NP, as compared to those from an unexcited NP, the apparent activation barrier for the 1e^--reduction of Fe^{3+} to Fe^{2+} is lower under LSPR excitation ($\Delta H^{\ddagger}{}_{\mathrm{light}}$) as compared to that in the dark ($\Delta H^{\ddagger}{}_{\mathrm{dark}}$), as depicted in Figure 5.1c. The magnitude of the decrease ($\Delta H^{\ddagger}{}_{\mathrm{dark}} - \Delta H^{\ddagger}{}_{\mathrm{light}}$) corresponds to the Fermi-energy rise brought about by light excitation. However, the Fermi-energy rise is not of the same magnitude as the excitation photon energy, $h\nu$, because the energetic relaxation and recombination of charge carriers (which is rather fast in metals) competes with the carrier accumulation process. To achieve a higher ε_f under LSPR excitation, it is necessary to separate or extract the charge carriers at a rate much faster than their energetic relaxation.

5.2.3 Mechanisms of Charge Transfer

As made clear by the previous section, the transfer of excited charge carriers to adsorbates is a critical step in plasmon-excitation-driven photoredox chemistry. Some understanding of this step can be obtained from the analogous plasmon-excitation-driven e^- and h^+ transfer from a metal NP to a solid-state material – such as an oxide semiconductor, graphene [44], or MoS_2 [50, 51] – in electronic contact with the NP. There is a well-established model of charge transfer across a defect-free junction between a metal and an attached semiconductor [11, 19, 29–33, 52–54]. Consider a metal with a work function, φ_m, in contact with a semiconductor of a work function, φ_s. When $\varphi_m > \varphi_s$ (i.e. metal and *n*-type semiconductor), the e^- in the semiconductor are at a higher potential, which results in a transfer of e^- from the semiconductor to the metal until the two potentials equilibrate. As a result, a depletion region is formed in the semiconductor near the junction, and the bands in the semiconductor bend upward toward the junction. The upward band-bending by a magnitude, $\varphi_m - \varphi_s$, results in a barrier for the transfer of e^- from the metal to the conduction band (CB) of the semiconductor. This barrier, known as a Schottky barrier, has a height, φ_b, which is given by:

$$\varphi_b = \varphi_m - \chi, \tag{5.1}$$

where χ is the e^- affinity measured from the vacuum level to the CB edge of semiconductor. Sufficiently energetic e^- generated by LSPR excitation can transfer from the metal to the CB of the semiconductor by crossing over the barrier height. The band-bending enhances the injection of e^- from the junction into the bulk of the semiconductor and hinders back e^- transfer to the metal. Thus, Schottky junctions can enhance the efficiency of carrier harvesting in the form of photocurrent [29–33] or reaction equivalents [11, 30–32, 41, 42].

The interfacial e^- transfer can take place via one of two pathways. In one case (Figure 5.2a), energetic charge carriers are first generated in the metal NP by LSPR decay, following which the e^- are injected into the CB of semiconductor. In contrast to this so-called indirect pathway, recent studies have proposed a direct pathway (Figure 5.2b) where e^- transfer occurs concurrently with the decay of the collective excitation [55, 56]. The efficiency of carrier harvesting is expected to be higher in the direct pathway than in the indirect case; because in the latter case, energy relaxation by e^-–e^- scattering competes with the transfer process [57].

The role played by the semiconductor in a metal–semiconductor junction is played by an adsorbate in plasmon-driven photoredox chemistry. The chemisorption of a molecule to a metal surface often involves hybridization of the orbitals of the molecule and electronic states of the metal. The electronic admixture of the metal–adsorbate complex has empty states often of a lower energy than the lowest unoccupied molecular orbital (LUMO) of the free molecule. These states are accessible by plasmon-excitation-generated e^- through indirect or direct transfer pathways. Such e^- transfer to adsorbates on the metal surface (Figure 5.2c) has been detected in a number of surface-enhanced Raman scattering (SERS) studies [58–61]. Secondly, the frontier orbital energies of the electronic admixture have a reduced energy gap as compared to the gap of highest occupied molecular orbital (HOMO) and LUMO of the free molecule. Visible-region LSPR excitations can directly couple to electronic transitions from occupied to unoccupied states of the admixture (Figure 5.2d), which amount to transient charge transfer from the metal to the molecule [62]. This process is known as chemical interface damping (CID). The prevalence of CID on the surface of a plasmonic nanostructure is manifested as the broadening of the LSPR band of the nanostructure [57, 63–65].

CID is thought to be the operative mechanism in the LSPR-mediated bond dissociation in small molecules such as H_2 [14, 15] and O_2 [6, 8]. The transfer of an e^- from an LSPR-excited Au or Ag NP to the unoccupied antibonding states of the metal–adsorbate complex forms a transient negative ion (TNI) (e.g. $H_2^{\delta-}$ or $O_2^{\delta-}$). Subsequently, the TNI, possibly in a vibrationally excited state, travels on the excited

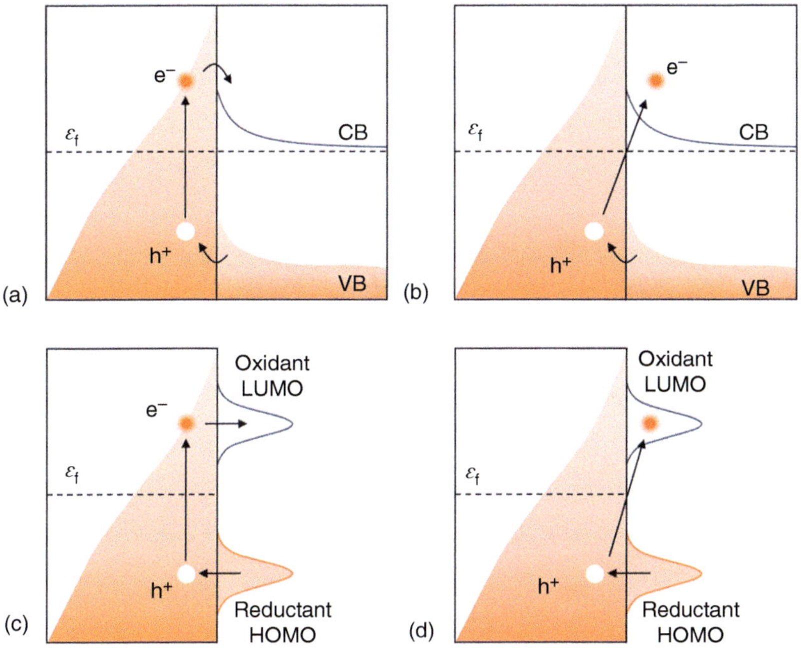

Figure 5.2 Schematics for LSPR-excitation-induced charge transfer in (a,b) a plasmonic metal–(*n*-type) semiconductor heterostructure and (c,d) a plasmonic metal–adsorbate complex by the (a,c) indirect and (b,d) direct pathways. ε_f refers to the Fermi level of the plasmonic metal. The conduction band (CB) and valence band (VB) of the semiconductor are indicated. Likewise, the highest occupied molecular orbital (HOMO) and lowest unoccupied molecular orbital (LUMO) of the adsorbate are also indicated.

potential energy surface, extending its bond length and ultimately resulting in bond dissociation. Such a mechanism was previously found to be operative in ultrashort pulse excitation-induced bond dissociation on a plasmonic NP surface [66]; but its role in photochemistry induced by continuous-wave (CW) excitation is of immense practical interest. The kinetics and efficiency of plasmon-excitation-induced charge transfer in metal–adsorbate complexes depend on the strength of electronic hybridization between the metal and molecule and the presence of resonant overlap between electronic transitions of the admixture and the LSPR excitation photon energy.

5.2.4 Energetics and Kinetics of Carrier Harvesting

Experimental and theoretical studies have been devoted to the elucidation of photoinduced carrier generation on plasmonic nanostructures [52, 53, 59, 61, 67–71]. Brus and co-workers measured the photoelectrochemical response of electrodes comprising Ag (Figure 5.3a,b) [67–69, 71] and Au (Figure 5.3c,d) [70] NPs under CW light excitation. Ag NPs in a citrate solution under 488 nm excitation of an intensity of 17 mW cm^{-2} exhibit a steady-state potential shifted by as much as −50 mV relative to its dark value (Figure 5.3a). Au NPs in the presence of citrate exhibit a

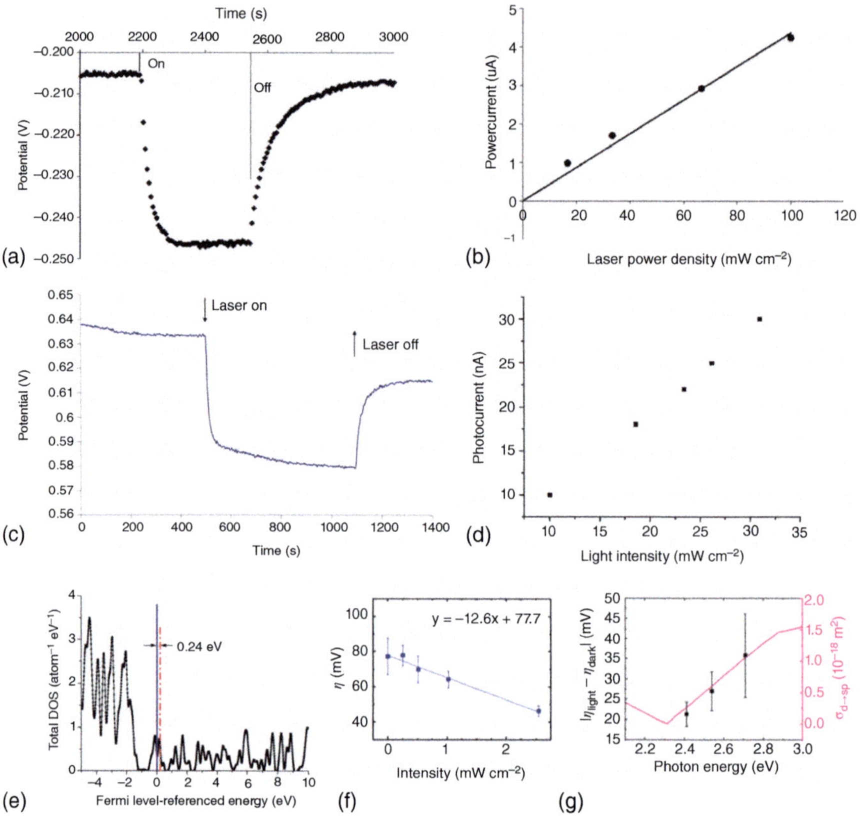

Figure 5.3 Plasmon-excitation-induced photocharging and photopotential of metal nanostructures. (a) The photoresponse of an electrode consisting of Ag NPs on an indium tin oxide (ITO) substrate in a 0.5 mM sodium citrate and 0.1 M potassium nitrate solution. The time-profile of the open-circuit potential, V_{oc}, of the electrode is shown under 488 nm laser excitation of an intensity of 17 mW cm^{-2} being turned on and off. (b) Photocurrent measured as a function of light intensity on the Ag NP/ITO electrode under 488 nm laser excitation in a 0.5 mM sodium citrate and 0.1 M potassium nitrate solution. (c) The photoresponse of an electrode consisting of Au NPs on an ITO substrate in a 0.5 mM sodium citrate and 0.1 M potassium nitrate solution. The time-profile of the V_{oc} of the electrode is shown under 514 nm laser excitation of an intensity of 18 mW cm^{-2} being turned on and off. (d) Photocurrent measured as a function of light intensity on the Au NP/ITO electrode under 514 nm laser excitation in a 0.5 mM sodium citrate and 0.1 M potassium nitrate solution. (e) The total density of states (DOS) calculated by DFT for bulk Au as a function of the energy of electronic states referenced to the Fermi energy in the dark. The vertical blue line shows the Fermi energy, $\epsilon_{f,dark}$, of Au in the dark, while the vertical red line indicates the quasi-Fermi energy, $\epsilon_{f,light}$, under 488 nm laser excitation of a power of 500 mW. The photoinduced rise in the Fermi energy is 0.24 eV, as indicated. (f) The overpotential (blue data points) for the hydrogen evolution reaction (HER) on Au NPs supported on a glassy carbon electrode under 488 nm laser excitation of different intensities. Each data point is an average of measurements from four different electrodes and the error bar represents the standard deviation. A linear fit to the data points is shown by the blue line, the fit equation of which is indicated. (g) The absolute difference (black data points) in the overpotential under light excitation relative to that in the dark for the HER on Au NPs supported on a glassy carbon electrode under laser excitation of different photon energies but the same intensity of ~1 W cm^{-2}. Each data point is an average of measurements from three different electrodes and the error bar represents the standard deviation. The measured trend matches the photon-energy-dependence of the d → sp interband absorption cross-section, $\sigma_{d\to sp}$ (pink curve), calculated by Mie theory for 12 nm-diameter Au NPs in water. *Source*: Panels (a) and (b) are reprinted with permission from Ref. [67]. Copyright American Chemical Society 2007. Panels (c) and (d) are reprinted with permission from Ref. [70]. Copyright American Chemical Society 2010. Panel (e) is reprinted with permission from Ref. [46]. Copyright American Chemical Society 2016. Panels (f) and (g) are reprinted with permission from Ref. [72]. Copyright American Chemical Society 2019.

similar effect under 514 nm excitation of an intensity of 18 mW cm^{-2} (Figure 5.3c). The negative shift of the potential results from the cathodic polarization of the NPs under plasmonic excitation. Similar to the process described in Section 5.2.2, h^+ generated by plasmonic excitation are irreversibly scavenged by adsorbed citrate molecules, which undergo oxidation. The excess e^- are stored on the NP, which is registered as the negative potential shift. The closed circuit currents measured on Ag and Au NP electrodes (Figure 5.3b,d) are enhanced under light excitation. The photocurrent has a linear dependence on the light intensity. The current generation is 50-fold higher for Ag NPs than that for Au NPs. Accounting for the absorbance cross-section of the metal, the quantum yield (QY), i.e. the ratio of the number of e^- harvested in the form of photocurrent to the number of absorbed photons, was found to be 6.3×10^{-5} e^- per photon for Ag NPs [67] and 3.3×10^{-6} e^- per photon for Au NPs [70]. In other words, at the same rate of photoexcitation, the rate of h^+-mediated oxidation of citrate on Ag NPs is about 20-fold higher than that on Au NPs. The higher QY of Ag NPs possibly originates from the symmetric, high energy profiles of both photoexcited charge carriers, as stated in Section 5.2.1. A photoexcitation-induced potential has been measured in plasmonic NP and nanohole arrays by Sheldon and co-workers [73]. A negative (or positive) potential was measured under excitation at a frequency that is blue-shifted (or red-shifted) with respect to the LSPR frequency. The magnitude of the measured potential was as high as 100 mV at an intensity of 100 mW cm^{-2} and found to increase with an increase in the light excitation power.

Photoinduced cathodic charging of plasmonic NPs and resulting photopotentials are responsible for the kinetic enhancement of redox reactions under LSPR excitation, as shown by Jain and co-workers. As described in Section 5.2.2, the Fe^{3+}-to-Fe^{2+} redox conversion catalyzed by Au NPs has a lower activation barrier under LSPR or interband excitation [46]. The decrease in the apparent activation barrier is a direct manifestation of the steady-state photopotential on the photoexcited NP that results from the accumulation of excess e^- on the NP. The photopotential increases with increasing light intensity until it plateaus to a magnitude of 240 mV at an intensity of 0.9 W cm^{-2} of 488 nm light excitation (Figure 5.3e). The excess e^- accumulated on an NP occupy sp states above the Fermi level of the metal NP. In essence, CW light-excitation causes a Fermi-energy rise. From the energy of the quasi-Fermi level at a given light intensity, $\varepsilon_{\mathrm{f,light}}$, the steady-state concentration of e^- on the NP at that light intensity can be estimated. The number of e^- per NP, $[e^-]_{NP}$, is given by:

$$[e^-]_{NP} = N_A \cdot V_{NP} \cdot \rho_{Au} \cdot {M_{Au}}^{-1} \cdot [e^-]_{atom}, \tag{5.2}$$

where N_A is the Avogadro number, V_{NP} is the volume of an NP, ρ_{Au} is the bulk density of metallic Au, and M_{Au} is the molar mass of metallic Au. $[e^-]_{atom}$ is the number of excited e^- per atom, which is determined as:

$$[e^-]_{atom} = \int_{\varepsilon_{\mathrm{f,dark}}}^{\infty} \mathrm{DOS}(\varepsilon) \cdot f(\varepsilon, T)\, \mathrm{d}\varepsilon, \tag{5.3}$$

where DOS(ε) is the density of states ($atom^{-1}$ eV^{-1}) as a function of the e^- energy, ε. The DOS is determined by density functional theory (DFT) calculations of the band structure of Au. $f(\varepsilon,T)$ is the Fermi–Dirac distribution function:

$$f(\varepsilon, T) = \frac{1}{1 + e^{(\varepsilon - \varepsilon_{f,light})/k_B T}}, \tag{5.4}$$

where k_B is the Boltzmann constant and T is the temperature. The $[e^-]_{NP}$ estimated from Eqs. (5.2)–(5.4) increases with increasing light intensity, I, with an apparent scaling of $I^{1/2}$. A simple model shows the origin of this intensity-dependence. From a steady-state balance between the NP photoexcitation, e^-–h^+ recombination, and e^- and h^+ transfer steps, the excess e^- concentration, $[e^-]$, on the NP is estimated as:

$$[e^-] = \sqrt{\frac{k_g k'_{ht}}{k_{nr} k'_{et}}}, \tag{5.5}$$

where k'_{et} is the rate of e^- transfer, which is the product of the rate constant of e^- transfer and the concentration of the e^- acceptor. Analogously, k'_{ht} is the rate of h^+ transfer, which is the product of the rate constant of h^+ transfer and the concentration of the h^+ acceptor. k_{nr} is the rate constant for nonradiative charge recombination and k_g is the rate constant for charge carrier generation, which is a function of light intensity I:

$$k_g = \frac{\sigma \eta I}{h\nu}, \tag{5.6}$$

where σ is the absorption cross-section of the metal NP at the excitation photon energy $h\nu$, h is the Planck constant, and ν is the photon frequency. σ depends on the NP size and shape. η is the efficiency of conversion of LSPR excitations into e^-–h^+ pairs. Thus, Eqs. (5.5) and (5.6) show that $[e^-]$ has a light-intensity dependence of $I^{1/2}$ with other parameters – such as the excitation photon energy, the size, shape, and composition of the metal NP, and the identity and concentration of the acceptor species – being held constant.

Similar to redox reactions, electrochemical reactions on plasmonic NPs are also enhanced in kinetics under continuous LSPR and interband excitation. This effect has been found in the electrochemically driven hydrogen evolution reaction (HER) on Au NP electrodes [72]. As an outcome of charge accumulation and photopotential sustained on NPs under CW excitation, the overpotential of HER is lower (Figure 5.3f,g). As a corollary effect, the current density at a given applied potential is enhanced under light excitation. The overpotential decreases with increasing light intensity in a linear fashion (Figure 5.3f). The overpotential also decreases with increasing photon energy (Figure 5.3g), which is due to the increase in the interband absorption cross-section of Au at higher photon energies.

5.2.5 Chemical Potential of Plasmonic Excitations

As described in the previous section, the potential of charge carriers generated by LSPR excitation results in a kinetic enhancement of charge transfer reactions. In addition, this photopotential can also be harvested as free energy for driving thermodynamically uphill ($\Delta G > 0$) chemical reactions [74]. Such a reaction is nonspontaneous in the absence of an external source of free energy: it does not proceed forward at any appreciable rate, regardless of the presence of catalytic metal NPs. However, under plasmonic excitation, the reaction proceeds, which can be attributed to the free energy contribution of the excited charge carriers.

The Gibbs free energy, ΔG, of a reversible reaction at a constant temperature and pressure is given as:

$$\Delta G = -RT \ln \left(R_\mathrm{f}/R_\mathrm{b} \right), \tag{5.7}$$

where R is the gas constant and T is the reaction temperature. R_f and R_b are rates of the forward and backward reaction, respectively. Under LSPR excitation, the free energy of the reaction is modified due to the contribution of e^-–h^+ pairs generated by interband damping of LSPR excitations:

$$RT \ln \left(R_\mathrm{f}/R_\mathrm{b} \right) = -\Delta G_\mathrm{dark} + \frac{z\mu_\mathrm{e-h}\sigma\eta I}{k_\mathrm{nr}h\nu}. \tag{5.8}$$

Here, ΔG_dark is the Gibbs free energy of the reaction under dark conditions and the last term on the right-hand side represents the free energy contribution of e^-–h^+ pairs, wherein z is the number of e^-–h^+ pairs involved in the initiation step of the photoreaction and $\mu_\mathrm{e\text{-}h}$ is the chemical potential of an e^-–h^+ pair, which is equal to the interband gap energy, E_ib.

For an endoergic reaction, $\Delta G_\mathrm{dark} > 0$. Thus, in the dark ($I = 0$), $R_\mathrm{f} < R_\mathrm{b}$, i.e. there is no net forward progress of the reaction. However, under light excitation of a sufficiently large intensity, the last term on the right-hand side in Eq. (5.8) makes the net free energy negative and $R_\mathrm{f} > R_\mathrm{b}$, i.e. the reaction can progress spontaneously in the forward reaction. Higher the light intensity, higher is the net forward rate of the reaction as shown by Eq. (5.8). The increase in the quantity $RT\ln(R_\mathrm{f}/R_\mathrm{b})$ provides a measure of the free energy contribution of e^-–h^+ pairs generated by LSPR excitation. As shown by a plot of this quantity vs. the light intensity (Figure 5.4), this free energy contribution to the reaction increases linearly with increasing intensity, provided all other parameters are constant.

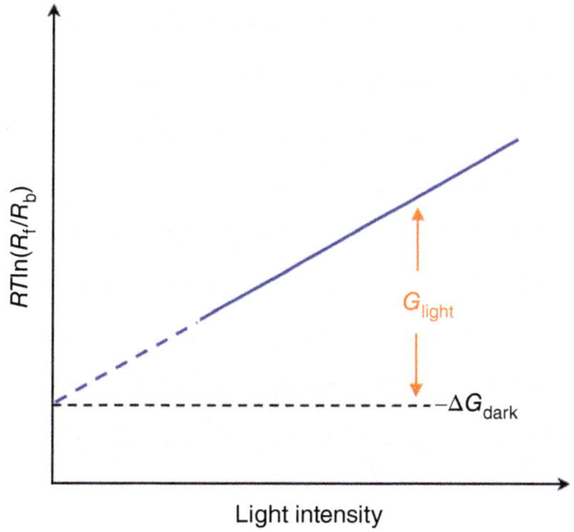

Figure 5.4 A plot of $RT\ln(R_\mathrm{f}/R_\mathrm{b})$ for a LSPR-excitation-driven reaction as a function of the light intensity, I. As per Eq. (5.8), the change in $RT\ln(R_\mathrm{f}/R_\mathrm{b})$ relative to its dark value, is a measure of the free energy contribution of e^-–h^+ pairs generated by LSPR excitation, $G_\mathrm{light} = \frac{z\mu_\mathrm{e-h}\sigma\eta I}{k_\mathrm{nr}h\nu}$. As depicted by this plot, this free energy contribution to the reaction increases linearly with increasing intensity, provided all other parameters are constant.

5.3 Plasmon-Excitation-Assisted Charge Transfer Reactions

5.3.1 Photo-Driven Growth of Ag and Au NPs

Among the earliest plasmon-excitation-driven photoredox reactions is the photo-driven colloidal growth of plasmonic metal NPs by Brus et al. [68, 69] and Mirkin et al. [75]. In this reaction, plasmonic NPs (e.g. citrate-capped Ag NPs) are subject to CW visible-light excitation in the presence of a metal ion precursor, e.g. Ag^+ salt. The e^- generated by the LSPR excitation are transferred from the Ag NP to adsorbed Ag^+, resulting in the reduction of the Ag^+ to Ag: $*Ag^+ + e^- \rightarrow Ag$, where the asterisk refers to a surface-bound species. The Ag atoms formed by reduction are deposited onto the seed Ag NP, resulting in NP growth. Alongside, h^+ are transferred via the indirect (Figure 5.5a) or direct (Figure 5.5b) pathways to adsorbed citrate, triggering the oxidation of citrate: $*C_6H_5O_7{}^{3-} + 2h^+ \rightarrow *C_5H_5O_5{}^- + CO_2$.

LSPR-excitation-induced metal deposition and growth has also been demonstrated for colloidal Au nanostructures [76]. The basic principle behind the photo-driven growth of Au NPs is similar to that described for the Ag system above. In a scheme developed by Wei and co-workers, Au NP seeds were stabilized by capping their surfaces with polyvinylpyrrolidone (PVP) ligands. The e^- generated in the Au NP by LSPR excitation are transferred to the PVP/Au interface, which serves as an intermediate site for the transfer of these e^- to Au^{3+} adsorbed from solution. The e^--driven photodeposition and growth of the Au NP occurs preferentially on the perimeter sites where there is a relatively high-density of PVP. The outcome is the anisotropic growth of the Au NP to form a nanoprism. The counter half-reaction involves the scavenging of the h^+ by methanol, which undergoes oxidation. PVP has a wide HOMO–LUMO gap relative to a ligand such as sodium citrate, allowing PVP to be inert against photoredox conversion [76].

5.3.2 Switching of Redox States

LSPR-excitation-generated carriers can be used to switch the redox state of a metal complex, as demonstrated for the Fe^{3+}/Fe^{2+} system. In the presence of a reducing agent such as $NaBH_4$, Fe^{3+} undergoes $1e^-$-reduction to Fe^{2+}. However, instead of a conventional reducing agent, a plasmonically excited Au NP can be used as a source of e^- for this reduction [48]. In the counter reaction, the h^+ is sacrificially consumed by an easily oxidizable scavenger such as ethanol. This redox reaction has a small residual rate in the dark, possibly due to the direct, but slow reaction of ethanol with Fe^{3+}; but the reaction rate is considerably higher under LSPR excitation at the same temperature, which demonstrates the action of charge carriers generated by LSPR excitation. The reaction rate is dependent on the light intensity, the photon energy, and the concentration of the h^+ scavenger. The higher the concentration of ethanol (within the range where the colloid is stable), the larger is the rate of h^+ removal, the higher are the steady-state concentration and Fermi energy of e^- on the NP, and consequently the larger are the rates of the e^- transfer and the overall redox reaction.

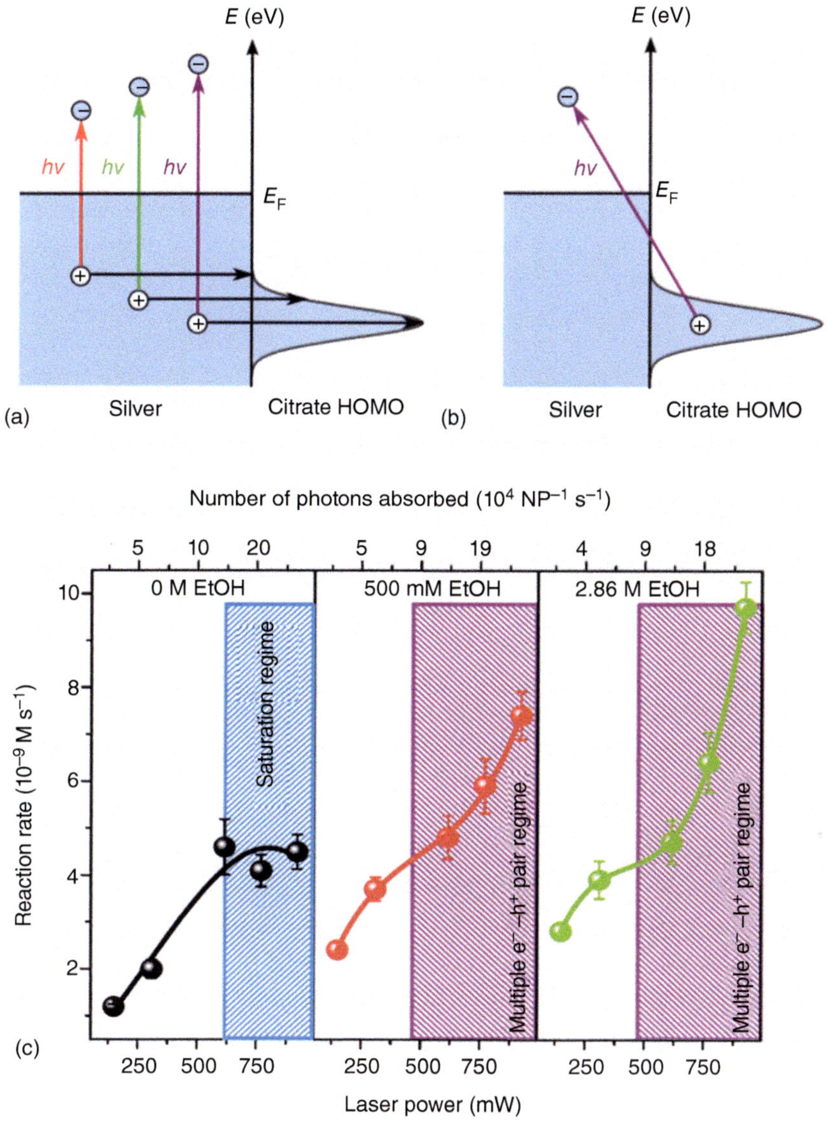

Figure 5.5 Plasmon-excitation-induced photoredox reactions. Depiction of (a) indirect and (b) direct pathways for the transfer of a h^+ from a photoexcited Ag NP to the HOMO of adsorbed citrate. (c) Rate of the Au NP-photocatalyzed Fe^{3+}-to-Fe^{2+} redox conversion as a function of the laser power at different concentrations of ethanol, the h^+ scavenger. *Source*: Panels (a) and (b) are reprinted with permission from Ref. [71]. Copyright American Chemical Society 2013. Panel (c) is reprinted with permission from Ref. [48]. Copyright Springer Nature 2018.

Analogously, the higher the intensity or photon energy of light, the larger is the rate of e^-–h^+ pair generation, the higher are the steady-state concentration and Fermi energy of e^- on the NP, and consequently the larger are the rates of the e^- transfer and the overall redox reaction.

It has also been found that at high enough light intensities, multiple e^- or h^+ can be transferred collectively, rather than sequentially, to an adsorbate. Such a phenomenon has been observed in the aforementioned photoredox conversion involving the reduction of Fe^{3+} alongside the oxidation of ethanol. When ethanol is absent, the only source of h^+ removal is slow water oxidation and h^+ removal is limiting. With increasing light intensity, the rate of e^- transfer to Fe^{3+} increases linearly, but the rate of e^-–h^+ recombination increases quadratically. As a result, the reaction rate increases with increasing light intensity in the low-intensity regime; but as higher intensities are approached, e^-–h^+ recombination becomes progressively more competitive than e^- transfer to Fe^{3+}, which results in a plateauing of the reaction rate (Figure 5.5c, black curve). In the presence of ethanol, a different trend is observed: although the reaction rate begins to plateau, a further increase in the light intensity results in a rise in the reaction rate (Figure 5.5c, red and green curves). This rise is a signature of a multicarrier redox conversion: specifically, ethanol can acquire $2h^+$ – collectively rather than sequentially – to undergo oxidation to acetaldehyde. The rate of such a $2h^+$-transfer process scales quadratically as the light intensity; therefore, this process becomes prevalent at high intensities and is not outcompeted by e^-–h^+ recombination. One can infer that as the photoexcitation rate (i.e. the light intensity) is progressively increased, two-carrier-mediated processes, then three-carrier-mediated processes, and so on become likely.

5.4 Plasmon-Excitation-Driven Processes Relevant for Fuel Generation

5.4.1 H_2O Splitting

The splitting of H_2O yields O_2 and H_2, a clean storable fuel. However, this process needs both an external source of free energy (such as an electric potential or light excitation) and a catalytic enhancement of its inherently sluggish rate. Plasmon-excitation-generated charge carriers can be used to serve both purposes as shown by Moskovits and co-workers [31]. A plasmonic device was constructed to harvest charge carriers generated by LSPR excitation and utilize them for H_2O splitting. The device consists of a TiO_2-modified Au nanorod photocathode, a Pt anode, a circuit that connects the two electrodes, and a separator for H^+ exchange (Figure 5.6a). By visible-light excitation of LSPRs in the Au nanorods, photoexcited e^- are generated in the states above the Fermi level of Au in dark. A fraction of these e^- get injected into the CB of TiO_2 (see Section 5.2.3 for details). The Au–TiO_2 interface constitutes a Schottky junction that facilitates charge separation. The e^- diffuse to the surface of TiO_2 where they are trapped by a hydrogen evolution catalyst (HEC) and utilized for the reduction of adsorbed H^+ to H_2 ($2H^+ + 2e^- \rightarrow H_2$).

The h^+ left behind in the Au nanorod are shuttled via a conductive substrate, i.e. ITO, and an electrical circuit to the Pt anode. At the anode, the h^+ participate in H_2O oxidation, generating O_2 and H^+ ($2H_2O \rightarrow O_2 + 4H^+ + 4e^-$). The H^+ generated migrate through the separator to the surface of HEC, where they are adsorbed, thereby closing a catalytic cycle. A wireless version of such a device has also been developed [30], which splits H_2O powered by plasmonic excitation of the Au nanorod absorber, but without generating power in an external load (Figure 5.6b). The working mechanism (Figure 5.6c) is similar to that of the photoelectrochemical device shown in Figure 5.6a. The e^- photogenerated in the Au nanorod transfer to a Pt reduction catalyst by passing through the TiO_2 layer, which acts as a filter that blocks h^+ but allows e^- flow. At the Pt surface, the e^- participate in the reduction of H^+ to H_2. The h^+ are transferred to a Co oxidation catalyst coated on the Au nanorod. On the Co surface, the h^+ mediate the oxidation of H_2O to form O_2 and H^+. In this wireless device, H_2 is produced at the rate of 5×10^{13} molecules $cm^{-2}\,s^{-1}$ under AM1.5 100 mW cm^{-2} illumination. The total efficiency, η_t, of H_2O splitting is determined by the efficiencies of the elementary processes involved [77]:

$$\eta_t = \eta_{abs} \cdot \eta_g \cdot \eta_{et} \cdot \eta_{ht} \cdot \eta_{red} \cdot \eta_{ox}, \tag{5.9}$$

where η_{abs} is the efficiency of light absorption, η_g is the efficiency of energetic charge carrier generation, η_{et} is the efficiency of e^- transfer to the reduction catalyst, η_{ht} is the efficiency of h^+ transfer to the oxidation catalyst, η_{red} is the efficiency of the reduction reaction, and η_{ox} is the efficiency of the oxidation reaction.

5.4.2 CO_2 Reduction

Visible-light-driven reduction of CO_2 into higher-energy-value hydrocarbons, a process typically identified as artificial photosynthesis, is desirable for the conversion of intermittent solar energy into a high-density fuel that can be stored, transported, and used on demand. This conversion being a multi-e^-, multi-H^+ process is kinetically challenging. Secondly, when no H_2 is used as the reductant, H_2O oxidation is often the half-reaction that generates H^+ for the CO_2-to-hydrocarbon conversion. The overall redox conversion is thermodynamically uphill similar to carbon fixation in natural photosynthesis. Plasmon-excitation-generated charge carriers have been utilized for driving the CO_2 reduction reaction (CO_2RR) both kinetically and thermodynamically [39–41, 43, 45].

Visible-light excitation of colloidal Au NPs in a CO_2-saturated aqueous solution has been shown to trigger CO_2RR [39]. In addition to their strong visible-range LSPR absorption, Au NPs have the ability to adsorb CO_2. Moreover, they are stable against photocorrosion and with the aid of surface-passivating PVP ligands, they are colloidally stable under photoreaction conditions. Either by LSPRs excitation using green light or by direct interband excitation using blue light, e^-–h^+ pairs are generated in the Au NPs. Isopropanol dissolved in the aqueous medium is used as the sacrificial agent for scavenging h^+ from the d-band of Au. The e^- transfer to surface-adsorbed CO_2 and trigger their reduction. Quite unlike thermocatalytic and electrocatalytic CO_2RR on Au catalysts, where CO is the major product of

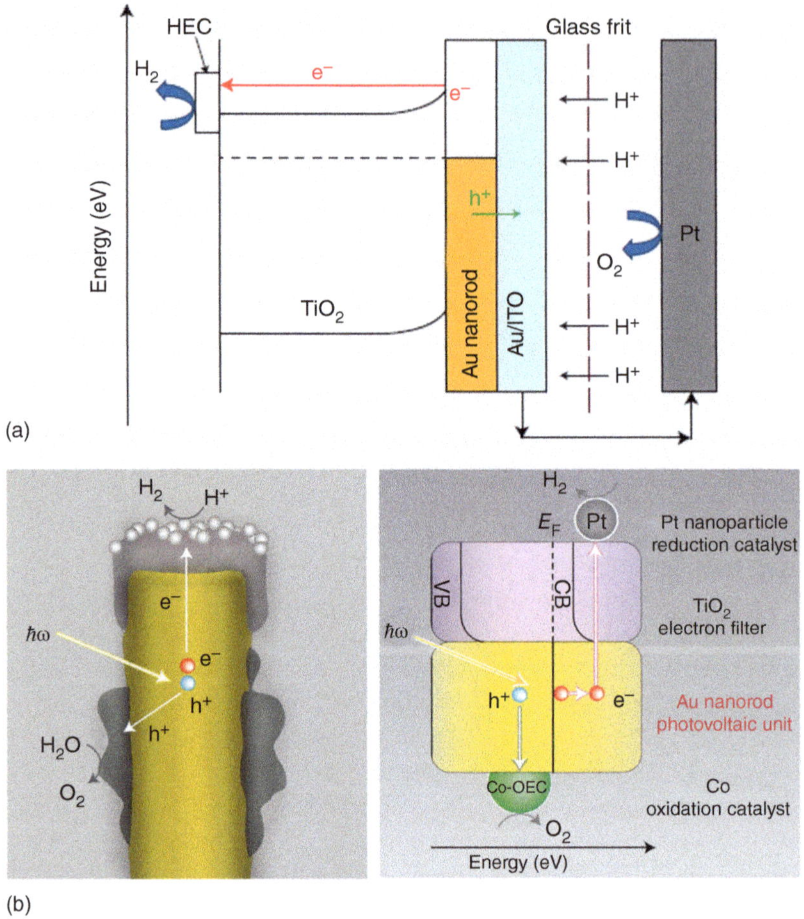

Figure 5.6 Plasmon-excitation-driven H_2O splitting. (a) A photoelectrochemical device consisting of TiO_2-modified Au nanorod photocathode and a Pt anode. The cathode and anode compartments are separated by a glass frit with a H^+ exchange membrane, Nafion. An HEC is deposited on the surface of the TiO_2 for catalyzing e^--mediated reduction of H^+ to form H_2 at the cathode. O_2 is generated by H_2O oxidation at the anode. (b) Illustration and (c) energy band diagram of a wireless device for plasmon-excitation-driven H_2O splitting. The device consists of an array of Au nanorods. Each nanorod is coated with a Co oxygen evolution catalyst (OEC) and a TiO_2 layer on which is deposited a Pt reduction catalyst. *Source*: All the panels are reprinted with permission from Ref. [30]. Copyright Springer Nature 2013.

reduction, conversion of CO_2 to CH_4 and C_2H_6 is observed. CH_4 formation is an $8e^-$–$8H^+$ process. C_2H_6 formation requires a $14e^-$–$14H^+$ reduction and a C–C bond formation step. These multi-e^-, multi-H^+ processes appear to be favored by the large cathodic photocharge on Au NPs under light excitation. The H^+ are sourced either by the oxidation of isopropanol ($C_3H_7OH + 2h^+ \rightarrow C_3H_6O + 2H^+$) or from the aqueous medium. In this light-driven CO_2RR process, the product distribution was modulated by varying attributes of incident light, such as the photon energy and intensity (Figure 5.7a). CH_4 production was observed under at all photon energies ranging from 532 nm light (which is overlapped with the LSPR) to 488 nm light (which primarily excites interband transitions in Au). C_2H_6 production was observed only at the highest photon energy, i.e. it required interband excitation. As another key observation, under 488 nm light excitation, CH_4 was produced at a rate that increases near-linearly with increasing light intensity. However, C_2H_6 was only produced above a threshold light intensity. At intensities higher than this threshold, the C_2H_6 production rate increased linearly with increasing intensity. Thus, the selectivity of C_2H_6 production was higher at higher intensities.

All these trends are attributed to a critical difference between CH_4 and C_2H_6 production kinetics. The rate of CH_4 production is limited by the transfer of one photogenerated e^- to a CO_2 adsorbate; whereas the rate-limiting step in C_2H_6 formation is a $2e^-$ transfer, one each to two CO_2 adsorbates, which must then undergo C–C coupling. The relative likelihood of these events is estimated by Poisson statistics, according to which the probability $P(n; <N>)$, of harvesting ne^- ($n = 1, 2, \ldots$) from a NP is:

$$P(n; <N>) = \frac{e^{-<N>} <N>^n}{n!}, \tag{5.10}$$

where $<N>$ is the average number of e^- harvested from an NP in the residence time, τ_r, of a CO_2 adsorbate on the surface of NP. $<N>$ increases linearly with increasing light intensity. The calculated probabilities of harvesting $1e^-$, $P(1; <N>)$, and $2e^-$, $P(2; <N>)$ (Figure 5.7b), showed intensity-dependent trends that matched the trends of the experimentally measured turnover frequencies (TOFs) of CH_4 and C_2H_6, respectively. The net rate for e^- harvesting was determined from the production rates of CH_4 and C_2H_6 by taking into account the fact that the formation of each CH_4 requires $8e^-$ and that of each C_2H_6 molecule requires $14e^-$. Under LSPR (i.e. 532 nm) excitation, the rate of e^- harvesting increases with increasing light intensity, until it reaches a plateau (Figure 5.7c, red curve). This trend, which is similar to that observed for plasmon-excitation-assisted Fe^{3+} reduction in the absence of ethanol (Figure 5.5c, black curve), results from the e^-–h^+ recombination rate dominating over the $1e^-$ transfer rate at high intensities (see Section 5.3.2 for details). Under interband (i.e. 488 nm) excitation, the e^- harvesting rate approaches a plateau at intermediate intensities, but it rises again steeply above an intensity (Figure 5.7c, blue curve) where C_2H_6 formation onsets. In other words, $2e^-$ harvesting becomes prevalent in the intermediate-to-high intensity regime, similar to what has been observed for Fe^{3+} reduction in the presence of ethanol (Figure 5.5c, red and green curves, also see Section 5.3.2 for details). Thus, light-excitation-driven multicarrier processes underlie this scheme of CO_2RR.

The LSPR excitation of metal NPs produces intense electric fields at the NP surface, which amplifies the Raman scattering cross-sections of molecules located on or near the surface by several orders of magnitude. SERS is particularly intense in the case of Ag NP dimers [78, 79]. This intense SERS has been exploited for in situ tracking of elementary steps involved in CO_2RR on a single Ag nanostructure under LSPR excitation (514.5 nm CW laser) [62]. Raman scattering spectra collected continuously under reaction conditions captured adsorbates and species formed on the Ag

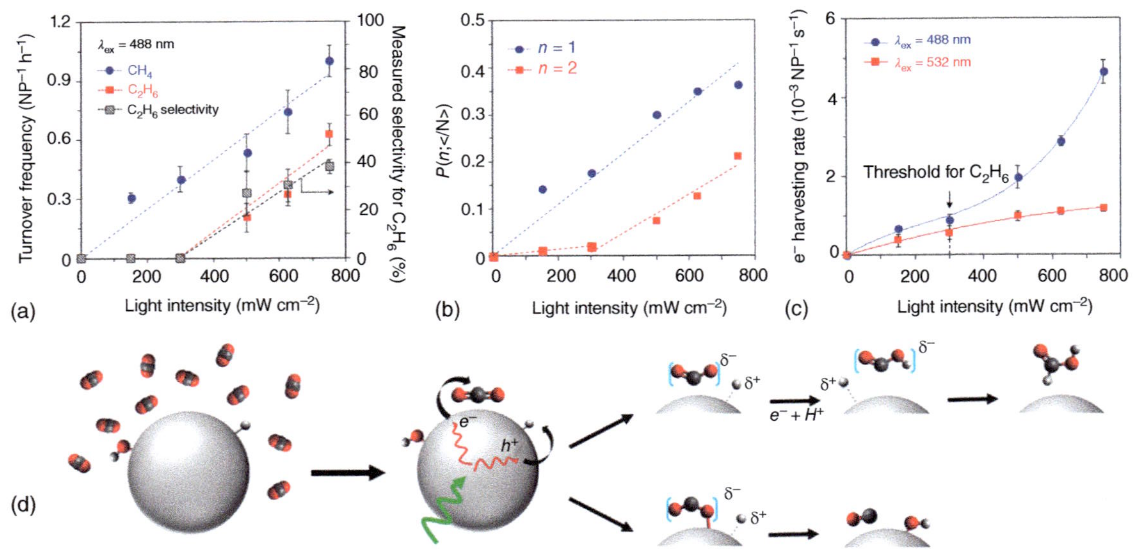

Figure 5.7 Visible-light-driven CO_2 reduction on Au NPs. (a) TOFs for CH_4 and C_2H_6 formation and measured selectivity (%) of C_2H_6 formation as a function of light intensity under interband excitation, i.e. excitation wavelength, λ_{ex}, of 488 nm. (b) Probabilities, $P(n; <N>)$, of 1e$^-$ and 2e$^-$ harvesting as a function of light intensity. (c) Rate of overall e$^-$ harvesting as a function of light intensity under interband excitation (λ_{ex} = 488 nm, blue curve) and LSPR excitation (λ_{ex} = 532 nm, red curve). The threshold intensity for C_2H_6 formation, determined to be 300 mW cm^{-2} from the plot in panel (a), is labeled by the arrow. Each data point in panels (a) and (c) is an average of three separate trials performed under the same conditions, and the standard deviation of these measurements is indicated as the error bar in each case. (d) Proposed reaction pathway for photocatalytic CO_2 reduction on the surface of an Ag NP under CW 514.5 nm excitation. This pathway is based on combined results of in situ SERS spectroscopy and DFT simulations. C atoms are indicated by dark gray spheres, O atoms by red spheres, H atoms by white spheres, and Ag NPs by large gray spheres. *Source*: Panels (a)–(c) are reprinted with permission from Ref. [39]. Copyright American Chemical Society 2017. Panel (d) is reprinted with permission from Ref. [62]. Copyright American Chemical Society 2018.

surface in the course of the reaction. Spectra, assigned with the aid of DFT calculated Raman scattering modes, showed CO_2RR products such as CO and formate formed on the surface. One other surface species detected was hydroxycarbonyl, HOCO*. DFT simulations showed that HOCO* is a critical intermediate in the CO_2RR formed by light-driven transfer of $1e^-$–$1H^+$ to an adsorbed CO_2. This intermediate then undergoes further conversion to various hydrocarbons by additional e^-–H^+ transfer processes. On the basis of the insight gained from these findings, a reaction pathway for plasmon-excitation-driven CO_2RR has been proposed (Figure 5.7d).

5.4.3 CO_2 Reduction with a Reaction Promoter

In CO_2-to-fuel conversion, longer-chain hydrocarbons are preferred as products due to their higher energy density as well as ease of transport in liquid form. However, most CO_2RR schemes largely yield C_1 products such as CO, HCHO, and CH_4 [41, 43, 45, 80–82]; C_{2+} hydrocarbons, which need additional e^- and H^+ transfers and C–C coupling steps [83, 84], are more kinetically challenging to produce. Even C_1 formation can be challenging in aqueous media because H^+ outcompete CO_2 for harvestable e^- resulting in the evolution of H_2 and a reduced selectivity of CO_2-to-hydrocarbon conversion. Thus, if charge transfer to CO_2 and C–C coupling steps can be sped up, the selectivity of hydrocarbon formation can be increased and the generation of larger hydrocarbons can become feasible.

In a plasmon-excitation-driven scheme, the kinetics of charge transfer can be optimized by proper choice of the excitation intensity and photon energy. In addition to these factors, proper engineering of the reaction environment can allow enhancement of charge transfer rates. For instance, e^- transfer to CO_2 is kinetically difficult due to the large energetic cost of reorganizing the linear, sp-bonded structure of CO_2 to the bent, sp^2-bonded structure of $CO_2^{\bullet-}$. If a solution-phase species – or in some cases, the catalyst surface itself – were to stabilize $CO_2^{\bullet-}$ by complexation or solvation, the free energy of formation of $CO_2^{\bullet-}$ would be reduced and the rate of e^- transfer to CO_2 would be enhanced. As an outcome, if e^- transfer to CO_2 is rate-limiting, the CO_2RR rate would be enhanced. The solution-phase species is thus a promoter of CO_2RR. This concept has been realized in a plasmon-excitation-driven scheme [38] by the use of an imidazolium salt, 1-ethyl-3-methylimidazolium tetrafluoroborate (EMIM–BF_4), as a solution-phase promoter. EMIM–BF_4 has been shown to reduce the overpotential for electrochemical CO_2RR [85, 86]: it is thought that the imidazolium cation complexes with polar intermediates or transition states of CO_2RR such as $CO_2^{\bullet-}$, lowering their free energy of formation, and enhancing the rate of CO_2RR [87–89]. Continuous LSPR excitation (λ_{ex} = 532 nm and intensity of 1 W cm^{-2}) of Au NPs in pure water without an h^+ scavenger did not drive CO_2RR: no hydrocarbon products were observed. Under the same conditions, but with the addition of EMIM–BF_4 to the aqueous medium, C_1 (CH_4), C_2 (C_2H_4 and C_2H_2), and C_3 (C_3H_6 and C_3H_8) hydrocarbons were produced by CO_2RR (Figure 5.8a) [38].

It is important to confirm that the precursor for the hydrocarbon products is indeed CO_2 rather than another carbon source in the reaction medium, such as leftover organic reagents used in the synthesis of NP photocatalyst or the

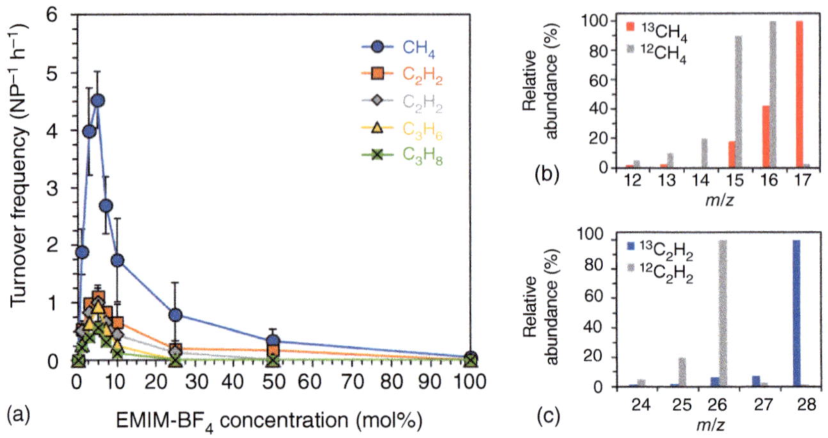

Figure 5.8 Plasmon-excitation-driven CO_2RR on Au NPs promoted by the imidazolium salt, EMIM–BF_4. (a) TOFs of hydrocarbon formation as a function of the EMIM–BF_4 concentration under continuous LSPR excitation with CW light of a wavelength of 532 nm and an intensity of 1 W cm^{-2}. (b, c) Mass fragmentation patterns of hydrocarbon products of plasmon-excitation-driven CO_2RR on Au NPs in an EMIM–BF_4 medium when $^{13}CO_2$ is used as the reactant. The products are identified by GC-MS to be (b) $^{13}CH_4$ (red bars) and (c) $^{13}C_2H_2$ (blue bars) by comparison of their mass fragmentation patterns with reference fragmentation patterns (gray bars) of (b) $^{12}CH_4$ and (c) $^{12}C_2H_2$. The fragmentation patterns of the ^{13}C-isotopologues are shifted to a higher *m/z* relative to those of the corresponding ^{12}C-isotopologues. *Source*: All the panels are reprinted with permission from Ref. [38]. Copyright Springer Nature 2019.

EMIM–BF_4 promoter. Such confirmation is obtained by performing a control photoreaction without CO_2, i.e. in an atmosphere of an inert gas such as Ar or He, with all other conditions being kept the same. The lack of any hydrocarbon products in this control photoreaction indicates that CO_2 is the source of hydrocarbon products. For stronger confirmation, it is recommended to perform a test with the heavier ^{13}C-isotopologue of CO_2. If CO_2 is indeed the precursor of hydrocarbons formed in the reaction, then the hydrocarbons must contain ^{13}C, which is possible to determine by gas chromatography-mass spectrometry (GC-MS) or nuclear magnetic resonance (NMR) measurement. When $^{13}CO_2$ was used in plasmon-excitation-driven CO_2RR on Au NPs in an EMIM–BF_4 medium, while all other conditions were maintained the same, the hydrocarbons generated were found to be ^{13}C-labeled. Specifically, $^{13}CH_4$ (Figure 5.8b) and $^{13}C_2H_2$ (Figure 5.8c) were identified from their characteristic mass fragmentation patterns in GC-MS, confirming that CO_2 was the precursor of hydrocarbon products.

All the product TOFs showed a volcano-type dependence on the concentration of EMIM–BF_4 (Figure 5.8a): the TOFs were most optimal at an EMIM–BF_4 concentration of 5 mol%, above which the TOFs decrease with increasing EMIM–BF_4 concentration. No products were observed when 100 mol% EMIM–BF_4 was used, i.e. H_2O was absent in the medium, which indicates that EMIM–BF_4 and H_2O are both needed: the former acts as a promoter, while the latter is a source of H^+ for CO_2RR. In 1–10 mol% EMIM–BF_4, the product selectivity followed the order: $C_1 > C_2 > C_3$, but under the most optimal conditions, the C_{2+} selectivity approached ~50%. Thus, from the results it appears that the stabilization of $CO_2^{\bullet-}$ by EMIM–BF_4 enhances the rate of e^- transfer to CO_2 and increases the likelihood of C–C coupling between two $CO_2^{\bullet-}$ intermediates.

The increased e^- transfer rates have yet another favorable consequence that a sacrificial h^+ scavenger is no longer required to ensure that e^-–h^+ recombination does not mar e^- harvesting. The h^+ can accumulate on the NP, wherefrom they can be removed by H_2O albeit on a slower timescale because of the thermodynamic and kinetic difficulty of H_2O oxidation. The h^+ located in the d-band of Au are energetic enough to transfer to H_2O. Typically, H_2O undergoes h^+-mediated oxidation to form O_2 ($2H_2O \rightarrow O_2 + 4H^+ + 4e^-$), which is the case in natural photosynthesis. However, careful characterization of nonhydrocarbon gas-phase products by gas chromatography with a thermal conductivity detector (GC–TCD) showed no formation of O_2 in the photoreaction. H_2O_2 is another possible product of H_2O oxidation reaction. Being a $2h^+$-reaction ($2H_2O \rightarrow H_2O_2 + 2H^+ + 2e^-$), H_2O_2 formation is expected to be kinetically favored over the $4h^+$-mediated O_2 generation. The production of H_2O_2 in the photoreaction was determined by a fluorogenic test using the Amplex red reagent. In the presence of a horseradish peroxidase catalyst, the nonfluorescent Amplex red reacts with H_2O_2 in a 1:1 stoichiometry. The reaction forms highly fluorescent resorufin, which has a molar extinction coefficient of 58 000 cm^{-1} M^{-1} [90], an excitation maximum at ~570 nm, and a fluorescence emission maximum at ~585 nm. The Amplex red reagent subjected to the reaction mixture of a photoreaction showed an increase in the fluorescence emission at ~585 nm (relative to the background fluorescence of the reagent mixture). Thus, H_2O_2 was detected in the reaction mixture [38], indicating that the counter reaction in this EMIM–BF_4-promoted CO_2RR scheme is the oxidation of H_2O to H_2O_2. Along with the productive use of photogenerated e^-, the free energy of h^+ carriers is harvested by H_2O oxidation rather than being dissipated in a thermodynamically downhill sacrificial isopropanol oxidation.

DFT calculations suggest a possible mechanism by which the imidazolium salt stabilizes the intermediate in CO_2RR. The imidazolium cation can lose an H^+ to form an *N*-heterocyclic carbene, EMIM*, which in turn can form a complex with CO_2 (Figure 5.9a). The energy of intermolecular interaction, E_{m-m}, between EMIM* and CO_2 is calculated to be −0.36 eV, which is threefold higher than the E_{m-m} of −0.11 eV for complexation between CO_2 and H_2O (Figure 5.9b). Such a strong complexation between EMIM* and CO_2 leads to a structural reorganization of CO_2 from a linear structure to a bent structure with an O=C=O angle of 133.7°. The C=O bond length is extended from 1.17 to 1.24 Å. The structure of the CO_2 moiety in the complex is thus close to that of $CO_2^{\bullet-}$, which has a C=O bond length of 1.23 Å and an O=C=O angle of 135.8°. Mulliken charge analysis indicates that the CO_2 moiety in the [EMIM*–CO_2] complex has a net charge of −0.73. Thus, compared to free CO_2, the CO_2 moiety in the [EMIM*–CO_2] complex requires substantially reduced reorganization upon e^- acceptance. In other words, complexation of CO_2 and EMIM* reduces the free energy barrier for $1e^-$ transfer to CO_2 (Figure 5.9c,d). Besides, the lifetime of $CO_2^{\bullet-}$ can be lengthened by Coulombic stabilization by the

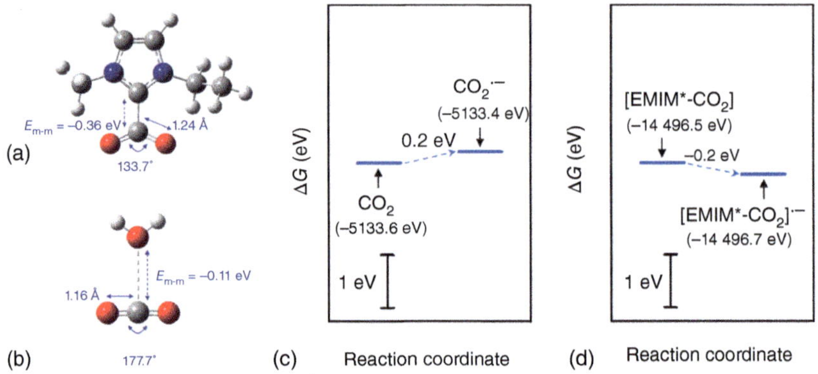

Figure 5.9 The role of EMIM−BF_4 in plasmon-excitation-driven CO_2RR. The lowest energy structures of (a) [EMIM*−CO_2] and (b) [H_2O−CO_2] complexes computed by DFT. The C=O bond length, O=C=O bond angle, and energy of intermolecular interaction, E_{m-m}, are provided for each structure. DFT-computed free energies of formation of (c) CO_2^- and (d) [EMIM*−CO_2]$^-$ by 1e$^-$ addition to CO_2 and [EMIM*−CO_2], respectively. *Source*: All the panels are reprinted with permission from Ref. [38]. Copyright Springer Nature 2019.

$EMIM^+$ cation. Based on this insight, it is expected that the higher the concentration of [EMIM*–CO_2], the higher is the rate of CO_2RR. The [EMIM*–CO_2] complex is postulated to form by the following reaction:

$$\mathrm{EMIM-BF_4 + CO_2} + x\mathrm{H_2O} \rightarrow \left[\mathrm{EMIM^* - CO_2}\right] + \left[\mathrm{BF}_{4-x}(\mathrm{OH})_x\right]^- \\ + (x+1)\,\mathrm{H^+} + x\mathrm{F^-} \qquad (5.11)$$

The [EMIM*–CO_2] concentration is expected to be most optimal at intermediate concentrations of EMIM–BF_4: in pure water or pure EMIM–BF_4, no [EMIM*–CO_2] complex is expected to be formed. This effect explains the observed volcano-type dependence of CO_2RR activity on the EMIM–BF_4 concentration.

As mentioned earlier, in aqueous media, the HER ($2H^+ + 2e^- \rightarrow H_2$) typically outcompetes CO_2RR at e$^-$ harvesting due to the more facile kinetics of the former. This is also the case in EMIM–BF_4-promoted CO_2RR, where H_2 generation dominates over hydrocarbon production. However, the relative kinetics of the CO_2RR and HER pathways can be modulated by variation of the H^+ concentration, [H^+], in the EMIM–BF_4 medium [91]. At the optimal concentration (5 mol%) of EMIM–BF_4, an increase in the [H^+] decreases the overall e$^-$ harvesting rate. However, the increase in the [H^+] enhances e$^-$ harvesting via the CO_2RR pathway at the expense of the HER pathway (Figure 5.10a). A kinetic model captures these [H^+]-dependent trends (Figure 5.10b). As per this model, h^+ capture by H_2O being considerably slower than the reduction half-reactions is the rate-limiting process in the overall carrier harvesting. With increasing [H^+], the oxidation of H_2O to H_2O_2 becomes increasingly difficult as seen from the law of mass action; consequently, the e$^-$ harvesting rate decreases. However, the rate of production of a CO_2RR intermediate such as HOCO [80] increases with increasing availability of H^+, because it involves the transfer of 1e$^-$ and 1H^+, possibly in a coupled manner, to CO_2:

$$\mathrm{CO_2 + e^- + H^+ \rightarrow HOCO} \qquad (5.12)$$

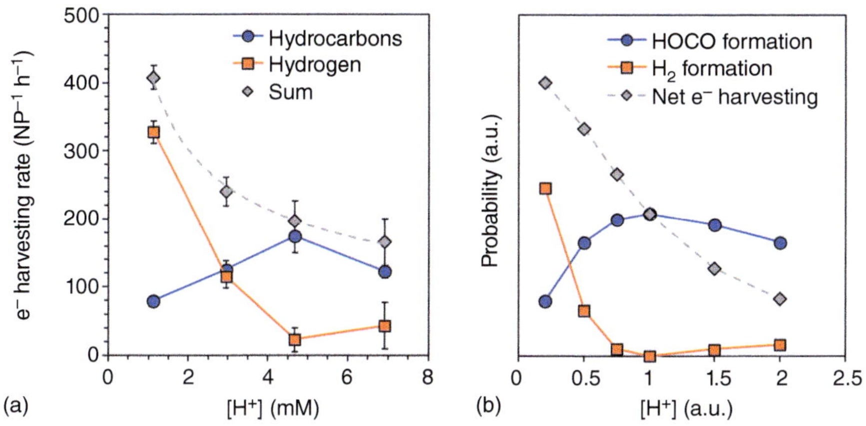

Figure 5.10 Competition between plasmon-excitation-driven CO_2RR and HER on Au NPs in an EMIM–BF_4-containing medium. (a) Measured rate of e^- harvesting by the hydrocarbon production channel, by the HER, and by both channels as a function of the $[H^+]$ in 5 mol% EMIM–BF_4. Each data-point is an average of three identical measurements, the standard deviation of which is represented by the error bar. (b) The calculated probability of overall e^- harvesting, HOCO formation, and H_2 formation as a function of $[H^+]$. *Source*: All the panels are reprinted with permission from Ref. [91]. Copyright American Chemical Society 2019.

As a result, the rate of e^- harvesting by the CO_2RR pathway increases with increasing $[H^+]$. The decrease in the rate of e^- harvesting by the HER pathway with increasing $[H^+]$ is counter-intuitive; but it is explicable if HOCO formation rather than HER is the dominant pathway for e^-–H^+ consumption. In other words, only e^-–H^+ pairs not consumed by HOCO formation are available for HER. This behavior is seen in the anticorrelated trends of the CO_2RR and HER as a function of $[H^+]$. At $[H^+]$ above 4.7 mM, the contribution of the CO_2RR pathway decreases, in anticorrelation with which that of the HER pathway increases. The CO_2RR selectivity is most optimal at ~5 mM H^+ where 84% of e^- were harvested in the form of hydrocarbon products and HER was largely suppressed.

It must also be noted that in plasmon-excitation-driven CO_2RR on Au NPs, there is no detectable production of CO, otherwise known to be the primary product of electrochemical CO_2RR on Au [92–95]. CO is either not produced or it is produced at levels too low to be detected by GC–TCD. The lack of dominance of CO production or HER demonstrates, in general, that selectivity of a heterogeneous catalyst can be overturned by plasmonic excitation coupled with the use of catalytic promoters.

5.4.4 Thermodynamic Insights into Plasmon-Excitation-Driven CO_2 Reduction

The standard reduction potentials, $E°$, for half-reactions involved in plasmon-excitation-driven CO_2RR are listed below [38]:

$$CO_2 + 8H^+ + 8e^- \rightarrow CH_4 + 2H_2O \qquad E^\circ = 0.17\,\text{V} \qquad (5.13)$$

$$2CO_2 + 10H^+ + 10e^- \rightarrow C_2H_2 + 4H_2O \qquad E^\circ = -0.05\,V \tag{5.14}$$

$$2CO_2 + 12H^+ + 12e^- \rightarrow C_2H_4 + 4H_2O \qquad E^\circ = 0.08\,V \tag{5.15}$$

$$3CO_2 + 18H^+ + 18e^- \rightarrow C_3H_6 + 6H_2O \qquad E^\circ = 0.10\,V \tag{5.16}$$

$$3CO_2 + 20H^+ + 20e^- \rightarrow C_3H_8 + 6H_2O \qquad E^\circ = 0.14\,V \tag{5.17}$$

$$2H^+ + 2e^- \rightarrow H_2 \qquad E^\circ = 0.00\,V \tag{5.18}$$

$$H_2O_2 + 2H^+ + 2e^- \rightarrow 2H_2O \qquad E^\circ = 1.83\,V \tag{5.19}$$

All potentials are referenced to the standard hydrogen electrode (SHE) at pH 0. The hydrocarbon formation reactions require the transfer of $8e^-$ to $20e^-$ and a corresponding number of H^+ to CO_2. The overall redox conversion of CO_2 to a hydrocarbon (e.g. CH_4) and H_2O to H_2O_2 is highly endoergic under standard conditions with a cell potential of −1.66 V:

$$CO_2 + 6H_2O \rightarrow CH_4 + 4H_2O_2 \tag{5.20}$$

This redox conversion is not expected to take place in the dark unless an external source of free energy is provided. However, this conversion can become thermodynamically feasible when energetic enough e^-–h^+ pairs generated by visible-light excitation of Au NPs are available to participate in the reaction. The ε_f of Au is at a potential of 0.66 V vs. SHE at pH 0. For 532 nm excitation, the photon energy, $h\nu$, is 2.33 eV, which is the maximum possible energy of an e^-–h^+ pair formed by photoexcitation. An h^+ formed by interband decay of LSPR excitations in Au is located in the d band. The top of the d band is 1.8 eV below ε_f. Thus, a d-band h^+ has a maximum potential of 2.50 V (vs. SHE at pH = 0), which is more positive than the E° for the H_2O oxidation half-reaction. The corresponding photogenerated e^- is located in the sp band just above ε_f and is estimated to have a maximum potential of 0.17 V, just energetic enough for the CH_4 generation half-reaction under standard conditions. Thus, the CO_2RR becomes feasible under visible-light excitation on account of the photochemical action of light.

Photothermal heating caused by photoexcitation of Au NPs alone is not sufficient to induce this endoergic reaction. In fact, a control reaction performed in the dark, with all other conditions the same, does not show any measurable activity [91]. For removing any ambiguity, this control reaction was performed at a temperature higher than the steady-state temperature of the reaction medium in the photoreaction. Since this redox conversion (Eq. (5.20)) is nonspontaneous under dark conditions, light excitation is not serving simply to provide kinetic enhancement of an otherwise slow reaction; rather it is a source of free energy for this endoergic conversion [96]. Thus, plasmon-excitation-driven CO_2RR on Au NPs in EMIM–BF_4 with no sacrificial h^+ scavenger constitutes a photosynthetic reaction.

The potential of energetic e^-–h^+ pairs generated by LSPR excitation drives the redox reaction. A signature of this effect is observed in the light-intensity dependence of the CO_2RR activity [74]. The hydrocarbon TOFs increases with increasing light intensity in a super-linear fashion (Figure 5.11a). The C_{2+} selectivity also increases as the light intensity increases, reaching 50% at an intensity of $0.6\,W\,cm^{-2}$ and slightly exceeding 50% above that intensity (Figure 5.11b). For all the products, hydrocarbons as well as hydrogen, $RT\ln(TOF)$ increases in a linear fashion with increasing light intensity (Figure 5.11c), which is in accordance with the behavior predicted by Eq. (5.8). Thus, the increase in the rate of CO_2RR at higher light intensities is attributed to the increased chemical potential of e^-–h^+ pairs generated under higher intensities of photoexcitation (see Section 5.2.5 and Figure 5.4).

5.5 Outlook

It is now clearly established that a range of photoredox reactions, including ones involving multicarrier processes, can be driven by LSPR excitation of noble metal NPs. While the central principle of plasmon-excitation-driven photoredox chemistry is fairly general, practical demonstration in a wider range of redox reactions remains to be accomplished. Can nearly any redox reaction be driven by plasmonic excitation possibly with some engineering of the nanostructure, reaction environment, and light excitation attributes?

Another major frontier for scientific exploration is the mechanistic understanding of the photochemistry, even if some aspects are well understood. There is a need to determine what role is played by plasmonic excitation other than the generation of energetic e^-–h^+ pairs. Potentially, electrodynamic fields and field gradients generated by plasmonic excitation influence charge transfer processes. Electrostatic potentials on the photoexcited NP can modulate adsorption processes. The effect, if any, of resonant coupling between the plasmonic excitation and electronic transitions in the adsorbate also needs to be elucidated.

The most illuminating insight can be obtained by deciphering the molecular mechanism, elementary steps, intermediates, and pathway of a plasmon-excitation-driven redox reaction and by exploring how they differ from the electrochemically driven counterpart. This understanding will aid the development of theoretical models of plasmon-excitation-driven chemistry. For instance, CO_2 conversion is a multistep process involving CO_2 activation, multiple e^- and H^+ transfers, C=O bond dissociation, and C–H and C–C bond formation steps. Plasmon-excitation-driven CO_2RR is therefore rich in mechanistic details. The precise pathway under photoexcitation and the manner in which light excitation triggers or modifies these elementary steps need to be understood.

Some crucial understanding can be obtained by the detection of the transient intermediates formed in the reaction under light excitation. In situ SERS spectroscopy performed with single-NP spatial resolution, so that it is free of ensemble-averaging, is a powerful tool for such molecular-level detection. The dynamic vibrational modes captured by SERS serve as signatures of adsorption processes and bond breaking, formation, and reorganization steps occurring on the nanostructure surface under plasmonic excitation (Figure 5.12a). SERS spectra can be interpreted with the aid of isotopologue studies and DFT calculations. In fact,

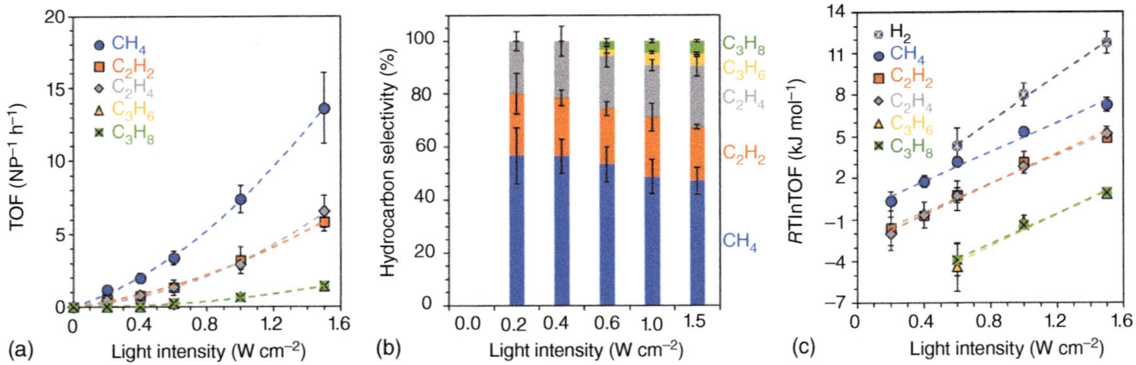

Figure 5.11 Effect of light intensity on the activity of plasmon-excitation-driven CO_2RR on Au NPs in an EMIM−BF_4-containing medium. (a) TOFs and (b) % selectivities of hydrocarbon products as a function of the intensity of 532 nm light excitation. (c) *RT*ln(TOF) plotted as a function of the light intensity for products of the photoreaction including hydrocarbons and H_2. The plots are linear as seen from the straight-line fits (dotted lines) to the data-points (symbols). *Source*: All the panels are reprinted with permission from Ref. [74]. Copyright Wiley-VCH Verlag GmbH & Co. KGaA, Weinheim 2020.

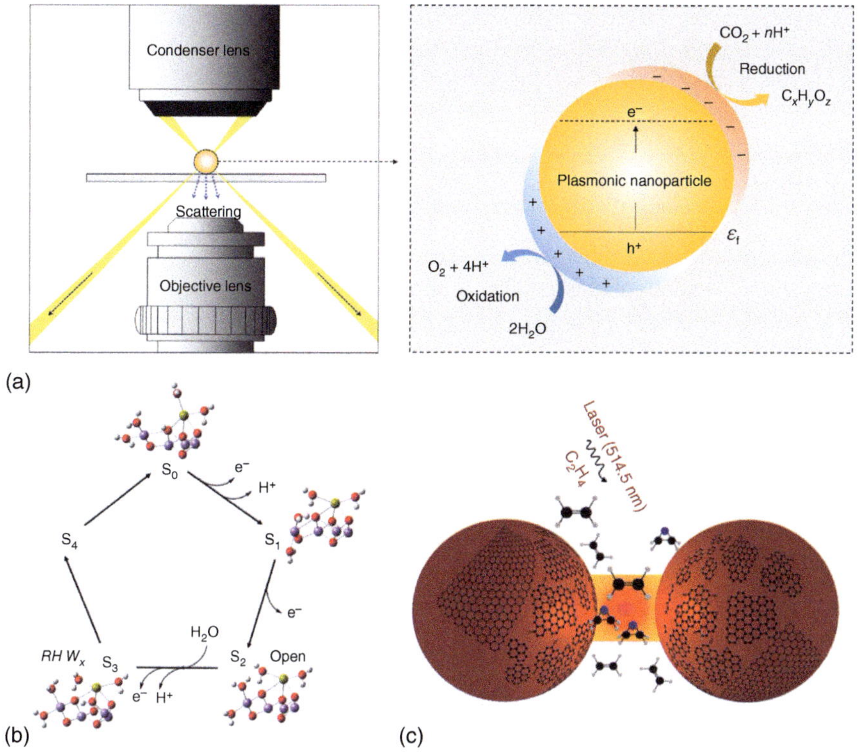

Figure 5.12 (a) Schematic of in situ single-NP-level SERS spectroscopy for the elucidation of the molecular mechanism of a plasmon-excitation-induced photoredox reaction, such as CO_2-to-hydrocarbon conversion, on the surface of a plasmonic NP. (b) Cycling of the oxygen-evolving complex through a series of oxidation states (S_0–S_4) in visible-light-driven H_2O splitting on plasmonic Ag NPs. In situ SERS spectroscopy in conjunction with DFT simulations was used to determine the structures of the intermediate states of the photocycle. Mn atoms are shown by purple spheres, Ca atoms by yellow spheres, O atoms by red spheres, and H atoms by white spheres. (c) Visible-light-induced formation of graphene fragments in ethylene epoxidation on an Ag NP dimer under CW 514 nm excitation, as revealed by in situ single-NP-level SERS spectroscopy. *Source*: Panel (b) is reprinted with permission from Ref. [97]. Copyright American Chemical Society 2018. Panel (c) is reprinted with permission from Ref. [98]. Copyright Springer Nature 2018.

the combination of in situ single-NP-level SERS spectroscopy and DFT has been employed for the elucidation of previously unknown or ambiguous mechanistic aspects of plasmon-excitation-induced CO_2RR [62] and a range of other photoredox reactions (Figure 5.12b,c) [97, 98].

There is one other central challenge in plasmonic photoredox chemistry which needs to be overcome. The ultrafast energy relaxation and recombination of energetic charge carriers in the metal limits the efficiency of conversion of photons to charge carriers harvested as reaction equivalents. For more efficient harvesting, nanoengineered Schottky junctions and/or inner-sphere charge acceptors should be explored in greater detail. For practical utility in solar energy harvesting, the plasmonic absorption characteristics should be tuned by nanostructure design to capture the solar spectrum. However, the greatest opportunities of plasmonic photoredox chemistry lay in the modulation of the chemical selectivities of noble metal catalysts and the discovery of new photochemical reactions.

Acknowledgments

Funding for this work was provided by the Energy & Biosciences Institute (EBI) through the EBI-Shell program. We acknowledge the contributions of former members of the Jain lab, Youngsoo Kim, Andrew J. Wilson, Gayatri Kumari, and Xueqiang Zhang described in this chapter. We also thank the funding agencies (EBI through the EBI-Shell program, National Science Foundation through Grant NSF CHE-1455011, and the Arnold and Mabel O. Beckman Foundation through the Young Investigator program) that supported the work of the Jain lab described in this chapter.

References

1 Mendes, M.J., Luque, A., Tobías, I., and Martí, A. (2009). *Appl. Phys. Lett.* 95: 071105.
2 Pala, R.A., White, J., Barnard, E. et al. (2009). *Adv. Mater.* 21: 3504.
3 Reineck, P., Lee, G.P., Brick, D. et al. (2012). *Adv. Mater.* 24: 4750.
4 Pala, R.A., Liu, J.S.Q., Barnard, E.S. et al. (2013). *Nat. Commun.* 4: 2095.
5 Wu, B., Wu, X., Guan, C. et al. (2013). *Nat. Commun.* 4: 2004.
6 Christopher, P., Xin, H., and Linic, S. (2011). *Nat. Chem.* 3: 467.
7 Ingram, D.B. and Linic, S. (2011). *J. Am. Chem. Soc.* 133: 5202.
8 Christopher, P., Xin, H., Marimuthu, A., and Linic, S. (2012). *Nat. Mater.* 11: 1044.
9 Marimuthu, A., Zhang, J., and Linic, S. (2013). *Science* 339: 1590.
10 Yu, S., Lee, S.Y., Yeo, J. et al. (2014). *J. Phys. Chem. C* 118: 29583.
11 Yu, S., Kim, Y.H., Lee, S.Y. et al. (2014). *Angew. Chem. Int. Ed.* 53: 11203.
12 Song, C.K., Baek, J., Kim, T.Y. et al. (2016). *Appl. Catal. B* 198: 91.
13 Dhiman, M., Maity, A., Das, A. et al. (2019). *Chem. Sci.* 10: 6594.
14 Mukherjee, S., Libisch, F., Large, N. et al. (2013). *Nano Lett.* 13: 240.
15 Mukherjee, S., Zhou, L., Goodman, A.M. et al. (2014). *J. Am. Chem. Soc.* 136: 64.
16 Zhao, J., Nguyen, S.C., Ye, R. et al. (2017). *ACS Cent. Sci.* 3: 482.
17 Huang, Y.-F., Zhu, H.-P., Liu, G.-K. et al. (2010). *J. Am. Chem. Soc.* 132: 9244.
18 Suh, J.S., Jang, N.H., Jeong, D.H., and Moskovits, M. (1996). *J. Phys. Chem.* 100: 805.
19 Knight, M.W., Sobhani, H., Nordlander, P., and Halas, N.J. (2011). *Science* 332: 702.
20 Liu, Y., Cheng, R., Liao, L. et al. (2011). *Nat. Commun.* 2: 579.
21 Tanzid, M., Ahmadivand, A., Zhang, R. et al. (2018). *ACS Photonics* 5: 3472.
22 Sobhani, A., Knight, M.W., Wang, Y. et al. (2013). *Nat. Commun.* 4: 1643.

23 Zheng, B., Zhao, H., Cerjan, B. et al. (2018). *Appl. Phys. Lett.* 113: 101105.
24 King, N.S., Liu, L., Yang, X. et al. (2015). *ACS Nano* 9: 10628.
25 Jain, P.K. and El-Sayed, M.A. (2008). *Nano Lett.* 8: 4347.
26 Lee, K.-S. and El-Sayed, M.A. (2006). *J. Phys. Chem. B* 110: 19220.
27 Sil, D., Gilroy, K.D., Niaux, A. et al. (2014). *ACS Nano* 8: 7755.
28 Zhang, S., Bao, K., Halas, N.J. et al. (2011). *Nano Lett.* 11: 1657.
29 Nishijima, Y., Ueno, K., Yokota, Y. et al. (2010). *J. Phys. Chem. Lett.* 1: 2031.
30 Mubeen, S., Lee, J., Singh, N. et al. (2013). *Nat. Nanotechnol.* 8: 247.
31 Lee, J., Mubeen, S., Ji, X. et al. (2012). *Nano Lett.* 12: 5014.
32 Liu, Z., Hou, W., Pavaskar, P. et al. (2011). *Nano Lett.* 11: 1111.
33 Tian, Y. and Tatsuma, T. (2005). *J. Am. Chem. Soc.* 127: 7632.
34 Hou, B., Shen, L., Shi, H. et al. (2017). *Phys. Chem. Chem. Phys.* 19: 2877.
35 Qin, L., Wang, G., and Tan, Y. (2018). *Sci. Rep.* 8: 16198.
36 Takakura, R., Oshikiri, T., Ueno, K. et al. (2017). *Green Chem.* 19: 2398.
37 Wang, X., Long, R., Liu, D. et al. (2016). *Nano Energy* 24: 87.
38 Yu, S. and Jain, P.K. (2019). *Nat. Commun.* 10: 2022.
39 Yu, S., Wilson, A.J., Heo, J., and Jain, P.K. (2018). *Nano Lett.* 18: 2189.
40 Hou, W., Hung, W.H., Pavaskar, P. et al. (2011). *ACS Catal.* 1: 929.
41 Neaţu, Ş., Maciá-Agulló, J.A., Concepción, P., and Garcia, H. (2014). *J. Am. Chem. Soc.* 136: 15969.
42 Kang, Q., Wang, T., Li, P. et al. (2015). *Angew. Chem. Int. Ed.* 54: 841.
43 Wang, C., Ranasingha, O., Natesakhawat, S. et al. (2013). *Nanoscale* 5: 6968.
44 Kumar, D., Lee, A., Lee, T. et al. (2016). *Nano Lett.* 16: 1760.
45 Cui, X., Wang, J., Liu, B. et al. (2018). *J. Am. Chem. Soc.* 140: 16514.
46 Kim, Y., Dumett Torres, D., and Jain, P.K. (2016). *Nano Lett.* 16: 3399.
47 Sundararaman, R., Narang, P., Jermyn, A.S. et al. (2014). *Nat. Commun.* 5: 5788.
48 Kim, Y., Smith, J.G., and Jain, P.K. (2018). *Nat. Chem.* 10: 763.
49 Kim, Y., Wilson, A.J., and Jain, P.K. (2017). *ACS Catal.* 7: 4360.
50 Kang, Y., Najmaei, S., Liu, Z. et al. (2014). *Adv. Mater.* 26: 6467.
51 Fang, Z., Wang, Y., Liu, Z. et al. (2012). *ACS Nano* 6: 10222.
52 Wu, K., Rodríguez-Córdoba, W.E., Yang, Y., and Lian, T. (2013). *Nano Lett.* 13: 5255.
53 Wu, K., Chen, J., McBride, J.R., and Lian, T. (2015). *Science* 349: 632.
54 Yu, S., Mohan, V., and Jain, P.K. (2020). *MRS Bull.* 45: 43.
55 Boerigter, C., Aslam, U., and Linic, S. (2016). *ACS Nano* 10: 6108.
56 Boerigter, C., Campana, R., Morabito, M., and Linic, S. (2016). *Nat. Commun.* 7: 10545.
57 Foerster, B., Joplin, A., Kaefer, K. et al. (2017). *ACS Nano* 11: 2886.
58 Osawa, M., Matsuda, N., Yoshii, K., and Uchida, I. (1994). *J. Phys. Chem.* 98: 12702.
59 Wu, D.-Y., Liu, X.-M., Huang, Y.-F. et al. (2009). *J. Phys. Chem. C* 113: 18212.
60 Dong, B., Fang, Y., Xia, L. et al. (2011). *J. Raman Spectrosc.* 42: 1205.
61 Wu, D.-Y., Zhao, L.-B., Liu, X.-M. et al. (2011). *Chem. Commun.* 47: 2520.
62 Kumari, G., Zhang, X., Devasia, D. et al. (2018). *ACS Nano* 12: 8330.
63 Hövel, H., Fritz, S., Hilger, A. et al. (1993). *Phys. Rev. B* 48: 18178.

64 Hendrich, C., Bosbach, J., Stietz, F. et al. (2003). *Appl. Phys. B* 76: 869.
65 Douglas-Gallardo, O.A., Berdakin, M., and Sánchez, C.G. (2016). *J. Phys. Chem. C* 120: 24389.
66 Jain, P.K., Qian, W., and El-Sayed, M.A. (2006). *J. Am. Chem. Soc.* 128: 2426.
67 Redmond, P.L. and Brus, L.E. (2007). *J. Phys. Chem. C* 111: 14849.
68 Redmond, P.L., Wu, X., and Brus, L. (2007). *J. Phys. Chem. C* 111: 8942.
69 Wu, X., Redmond, P.L., Liu, H. et al. (2008). *J. Am. Chem. Soc.* 130: 9500.
70 Wu, X., Thrall, E.S., Liu, H. et al. (2010). *J. Phys. Chem. C* 114: 12896.
71 Thrall, E.S., Preska Steinberg, A., Wu, X., and Brus, L.E. (2013). *J. Phys. Chem. C* 117: 26238.
72 Wilson, A.J., Mohan, V., and Jain, P.K. (2019). *J. Phys. Chem. C* 123: 29360.
73 Sheldon, M.T., van de Groep, J., Brown, A.M. et al. (2014). *Science* 346: 828.
74 Yu, S. and Jain, P.K. (2020). *Angew. Chem. Int. Ed.* 59: 2085.
75 Jin, R., Cao, Y., Mirkin, C.A. et al. (2001). *Science* 294: 1901.
76 Zhai, Y., DuChene, J.S., Wang, Y.-C. et al. (2016). *Nat. Mater.* 15: 889.
77 Moskovits, M. (2015). *Nat. Nanotechnol.* 10: 6.
78 Hao, E. and Schatz, G.C. (2004). *J. Chem. Phys.* 120: 357.
79 Kim, K. and Lee, H.S. (2005). *J. Phys. Chem. B* 109: 18929.
80 Yui, T., Kan, A., Saitoh, C. et al. (2011). *ACS Appl. Mater. Interfaces* 3: 2594.
81 Zhai, Q., Xie, S., Fan, W. et al. (2013). *Angew. Chem. Int. Ed.* 52: 5776.
82 Shown, I., Hsu, H.-C., Chang, Y.-C. et al. (2014). *Nano Lett.* 14: 6097.
83 Jiang, K., Sandberg, R.B., Akey, A.J. et al. (2018). *Nat. Catal.* 1: 111.
84 Yu, S., Wilson, A.J., Kumari, G. et al. (2017). *ACS Energy Lett.* 2: 2058.
85 Sun, L., Ramesha, G.K., Kamat, P.V., and Brennecke, J.F. (2014). *Langmuir* 30: 6302.
86 Wang, Y., Hatakeyama, M., Ogata, K. et al. (2015). *Phys. Chem. Chem. Phys.* 17: 23521.
87 Rosen, B.A., Salehi-Khojin, A., Thorson, M.R. et al. (2011). *Science* 334: 643.
88 Rosen, B.A., Haan, J.L., Mukherjee, P. et al. (2012). *J. Phys. Chem. C* 116: 15307.
89 Asadi, M., Kim, K., Liu, C. et al. (2016). *Science* 353: 467.
90 Zhou, M., Diwu, Z., Panchuk-Voloshina, N., and Haugland, R.P. (1997). *Anal. Biochem.* 253: 162.
91 Yu, S. and Jain, P.K. (2019). *ACS Energy Lett.* 4: 2295.
92 Wuttig, A., Yoon, Y., Ryu, J., and Surendranath, Y. (2017). *J. Am. Chem. Soc.* 139: 17109.
93 Mariano, R.G., McKelvey, K., White, H.S., and Kanan, M.W. (2017). *Science* 358: 1187.
94 Liu, M., Pang, Y., Zhang, B. et al. (2016). *Nature* 537: 382.
95 Manthiram, K., Surendranath, Y., and Alivisatos, A.P. (2014). *J. Am. Chem. Soc.* 136: 7237.
96 Jain, P.K. (2019). *J. Phys. Chem. C* 123: 24347.
97 Wilson, A.J. and Jain, P.K. (2018). *J. Am. Chem. Soc.* 140: 5853.
98 Zhang, X., Kumari, G., Heo, J., and Jain, P.K. (2018). *Nat. Commun.* 9: 3056.

6

Plasmonic Catalysis for N_2 Fixation

Tomoya Oshikiri[1] *and Hiroaki Misawa*[1,2]

[1] *Research Institute for Electronic Science, Hokkaido University, Sapporo, Japan*
[2] *Center for Emergent Functional Matter Science, National Yang Ming Chiao Tung University, Hsinchu, Taiwan*

6.1 Introduction

The synthesis of ammonia (NH_3) from dinitrogen (N_2) molecules developed by Haber and Bosch is one of the most important discoveries in the twentieth century, not only because of its application, through which synthetic fertilizers have contributed enormously to the survival of humankind by obtaining "bread from the air," but also from the viewpoint of fundamental science. Recently, NH_3 has also been considered a mobile energy carrier with a high energy density that does not cause pollution [1, 2]. The Haber–Bosch process is performed under harsh conditions, such as high pressure (10–25 MPa) and temperature (300–500 °C) [3, 4]. The energy required to maintain the harsh conditions for this reaction, which consumes more than 2% of the world's annual energy supply, is mainly derived from the reforming of fossil fuels [5]. Therefore, it is indispensable to explore carbon-free and sustainable methods for NH_3 synthesis.

Although numerous efforts have been devoted to NH_3 synthesis under milder conditions by improving thermochemical inorganic catalysts [6–9], molecular catalysts [10–12], and electrochemical catalysts [13–15], all of them require thermal or electric energy input. Therefore, these systems can only use renewable energy indirectly. To achieve NH_3 synthesis through an ultimate energy-saving process, it is essential to develop a system to directly utilize renewable energy, e.g. solar light.

NH_3 photosynthesis was first reported by Schologer [16, 17] using iron-doped titanium dioxide (TiO_2) following the discovery of water splitting by Honda and Fujishima [18, 19]. Although there were some reports about NH_3 photosynthesis after that [20–22], it was very difficult to accurately evaluate the NH_3 photosynthesis due to the small amount of product and contamination from the environment and nitrogen source until recent years [23]. Since the beginning of the twenty-first century, reliable NH_3 photosynthesis has started to be reported due to the improvement in the detection procedures for NH_3 [24–28]. "Artificial photosynthesis" is defined as a reaction that is driven by visible light, uses water as an electron donor, and

Plasmonic Catalysis: From Fundamentals to Applications, First Edition.
Edited by Pedro H.C. Camargo and Emiliano Cortés.

is energetically uphill to create energy-storage materials such as hydrogen (H_2), hydrocarbons, and NH_3. However, it is still far from meeting the requirements of practical applications, such as artificial photosynthesis of NH_3, because the photocatalysts work with sacrificial electron donors. Therefore, N_2 fixation methods with a downhill reaction are also introduced in this chapter.

In addition, it is difficult for a semiconductor photocatalyst to utilize a broadband solar spectrum because of the band structure limitations. In recent years, localized surface plasmon resonance (LSPR) has received much attention for solar energy conversion due to the light-harvesting properties. In this chapter, we discuss plasmonic catalysis for N_2 fixation from the viewpoint of the mechanisms. In Section 6.2, the mechanisms of plasmon-induced NH_3 photosynthesis are categorized as (i) N_2 fixation through near-field enhancement (NFE) of LSPR, (ii) direct hot-electron injection (DHEI) from a plasmonic metal into N_2 molecules, and (iii) hot-electron transfer (HET) from a plasmonic metal to a semiconductor. The evaluation procedures are also described in the section. In Sections 6.3–6.5, examples of plasmon-induced NH_3 photosynthesis via each mechanism categorized in Section 6.2 are introduced. We also introduce examples of plasmon-induced N_2 reduction to NH_3 using visible light as an energy source and water as an electron donor in the final part of Section 6.5.

6.2 Reaction Mechanism and Evaluation of N_2 Fixation

6.2.1 Principles of Plasmon-Enhanced NH_3 Photosynthesis

Metal nanoparticles (NPs), such as gold (Au), silver (Ag), and copper (Cu), show LSPR, which is the collective oscillation of conduction electrons and enhances the electromagnetic field in the vicinity of metal NPs, so-called NFE [29, 30]. As a characteristic feature of LSPR, the resonant wavelength can be easily controlled by changing the size and shape of the NPs [31, 32]. Therefore, metal NPs exhibiting LSPR have received considerable attention as optical antennas for light energy conversion systems such as photocatalysts. For example, efficient excitation of a semiconductor photocatalyst is expected based on the NFE effect of metal NPs.

A schematic illustration of the radiative and nonradiative decay of LSPR in metallic NPs is shown in Figure 6.1 [33]. The radiative decay corresponds to light scattering, which is a loss in LSPR from the viewpoint of energy conversion. On the other hand, nonradiative decay is induced through excitation of the electrons of the metal NP itself, not only within the conduction band (CB) (intraband transition) but also between the d bands and the sp CB (interband transition). The excitation of electron–hole pairs (hot carriers) can also be utilized as a mechanism to construct plasmon-induced light energy conversion systems such as solar cells and artificial photosynthesis devices [34–36]. Studies have shown that HET occurs when an *n*-type semiconductor such as TiO_2 is attached to metal NPs with a larger work function than that of the semiconductor to form a Schottky barrier [37]. The electrons transferred from the metal NPs to the semiconductor CB and the holes are separated by

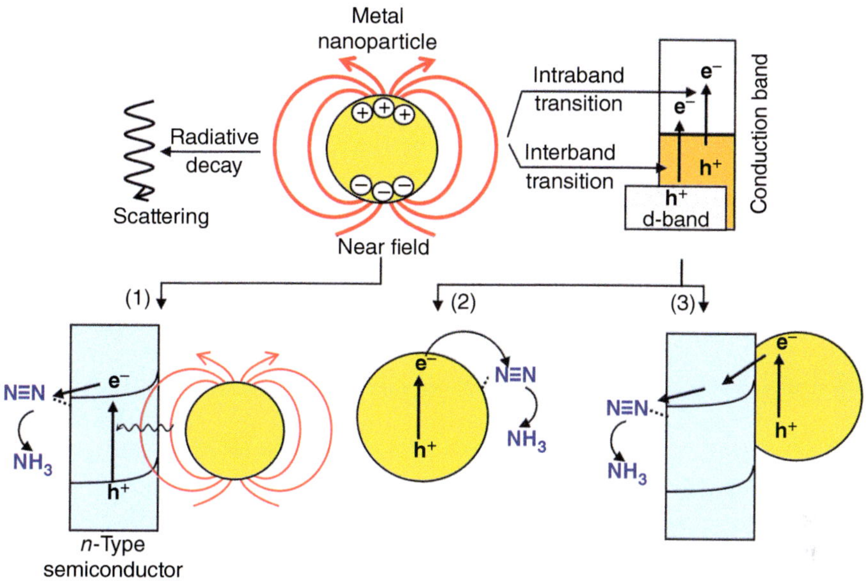

Figure 6.1 Schematic illustrations of the decay of LSPR in metallic NPs (upper), and the reaction scheme of plasmon-induced photocatalysis for N_2 fixation (lower). (1) N_2 fixation through NFE, (2) DHEI into N_2 molecules, and (3) HET between a plasmonic metal and a semiconductor.

the Schottky barrier. If the CB energy is more negative than the reduction potential from N_2 to NH_3 (0.096 V vs. RHE) and N_2 is adsorbed on the photocatalyst, then N_2 can be reduced by the electrons in the CB of the semiconductor transferred from the metal NPs.

Highly energetic hot electrons can be directly injected into molecular orbitals [38–40]. Multiple injection of hot electrons into the lowest unoccupied molecular orbital (LUMO) of N_2 cleaves the N≡N bond.

We have introduced three kinds of operational mechanisms for plasmon-induced NH_3 photosynthesis based on (1) N_2 fixation through NFE, (2) DHEI into N_2 molecules, and (3) HET from a plasmonic metal to a semiconductor, as shown in Figure 6.1.

The difference in the active wavelength of NH_3 photosynthesis among the reaction mechanisms should also be discussed. In the case of NFE, the resonant wavelength of LSPR has to basically overlap with the excitation wavelength of the base material because the base material itself is excited by the near field. In contrast, in the case of the HET mechanism, hot electrons are generated in the metal NPs independent of the semiconductor excitation. Therefore, the photocatalyst responds to the LSPR band wavelength in addition to the semiconductor absorption wavelength. However, hot electrons should have sufficient energy to reach the flat band of the semiconductor. Similarly, although the hot-electron excitation wavelength purely depends on the LSPR band in the case of DHEI, the hot electron should have sufficient energy to be directly injected into the N_2 molecular orbital.

Although plasmon-related NH_3 synthesis using other mechanisms, such as plasmonic photothermal conversion [41, 42] and storing electrons using metal NPs, [43] have also been reported, we do not discuss them in detail in this chapter because these mechanisms do not involve a plasmon-induced photochemical process.

6.2.2 Associative and Dissociative Pathways of N_2 Fixation

In principle, two types of reaction pathways are possible in the reduction of N_2 to NH_3 on a heterogeneous surface, as shown in Figure 6.2 [44, 45]. One is an associative pathway in which N_2 molecules on the catalyst are hydrogenated sequentially and NH_3 is then released at the same time as cleavage of the N–N bond [46]. The other is a dissociative pathway in which nitrogen and hydrogen react after cleavage of the strong N_2 triple bond [47]. In nature, enzymes such as nitrogenase follow the associative pathway, as the activation energy is lower. However, the Haber–Bosch process uses the dissociative pathway.

In both pathways, the overall reaction of N_2 with water in which NH_3 and O_2 are formed is an uphill reaction, as shown in Eq. (6.1). Therefore, N_2 fixation with water under sunlight is one of the most critical targets of artificial photosynthesis.

$$1/2N_2 + 3/2H_2O \rightarrow 3/4O_2 + NH_3 \quad (\Delta G^o = 339\ \text{kJ mol}^{-1}) \tag{6.1}$$

6.2.3 Analysis and Quantification of Plasmon-Induced NH_3 Evolution

The produced NH_3 is quantified by the colorimetric method using indicators such as indophenols and Nessler reagents. Ion chromatography is also a useful tool to quantify NH_3. The fatal problem in NH_3 quantification is contamination. NH_3 is very abundant in the atmosphere and instruments. In addition, the crude N_2 gas source includes a trace of NH_3. To distinguish the NH_3 evaluation target and contamination, a high purity isotopic $^{15}N^{15}N$ gas should be used as a N_2 source. The evolved $^{15}NH_3$

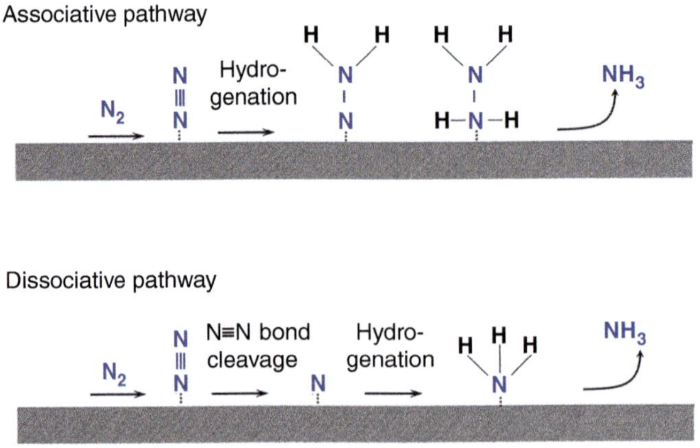

Figure 6.2 Generic reaction pathways for N_2 reduction to NH_3 on heterogeneous catalysts. *Source*: Oshikiri et al. [44].

and its derivatives can be detected by mass spectrometry (MS) or nuclear magnetic resonance (NMR) spectroscopy. In addition, a similar problem arises in proving that water works as an electron donor. The O_2 gas as a result of water oxidation cannot be distinguished from the O_2 in air. Therefore, water labeled with ^{18}O provides direct evidence of water oxidation [48]. The stoichiometry in the overall reaction should also be confirmed.

The quantum efficiency becomes essential and acceptable for evaluating the photocatalytic activity for N_2 fixation. The external quantum efficiency (EQE), which is the ratio of the number of electrons used for the product to the number of photons incident on the catalyst and is also called as the apparent quantum efficiency (AQE), is defined for N_2 fixation by Eq. (6.2). The internal quantum efficiency (IQE), which is the ratio of the number of electrons used for the product to the number of photons absorbed by the catalyst, is estimated to be smaller than the EQE because the number of absorbed photons is usually smaller than the number of photons in the incident light.

$$\mathrm{EQE} = \frac{[3 \times \text{Number of evolved } \mathrm{NH_3}]}{[\text{Number of incident photons}]} \times 100\ (\%) \tag{6.2}$$

The activities of photochemical catalysts are usually assessed by the turnover number (TON). TON is the number of moles of the substrate that a mole of catalyst can convert before becoming inactivated and is defined by Eq. (6.3). Therefore, when TON > 1, the reaction proceeds "catalytically."

$$\mathrm{TON} = \frac{[\text{Number of evolved } \mathrm{NH_3}]}{[\text{Number of catalytic species}]} \tag{6.3}$$

In the case of a photoelectrochemical reaction, the faradaic efficiency (FE), which is the proportion of the photocurrent contributing to N_2 fixation, is useful for evaluating the reaction selectivity. The FE for NH_3 evolution is calculated by Eq. (6.4).

$$\mathrm{FE} = \frac{n \cdot N\,(\mathrm{mol}) \cdot F}{Q\,(\mathrm{C})} \times 100\%, \tag{6.4}$$

where N is the amount of products generated in the process, F is the faradaic constant (96 485 C/mol), n is the number of electrons involved in the reaction, and Q is the total charge involved in the whole reaction.

An action spectrum, which is a plot of the rate of a photochemical reaction as a function of the wavelength of light, is essential to confirm the contribution of LSPR to the reaction. When the reaction is proceeded or enhanced by LSPR, the shape of the action spectrum should correspond to the plasmon resonance band. To discuss the contribution of LSPR to the reaction precisely, the near-field spectrum is obtained by electromagnetic field simulation and spectroscopy using special techniques. The finite-element method and finite-difference frequency-domain simulations are frequently used to calculate the near-field distribution and the spectrum near plasmonic NPs [49]. To obtain the experimental near-field spectrum, scanning near-field optical microscopy [50, 51], cathode luminescence [52, 53], electron energy-loss spectroscopy [54], and photoemission electron microscopy [55] are used.

However, because direct measurement of the near-field spectrum is experimentally complicated, the far-field absorption, scattering, and extinction spectra resulting from plasmon decay are usually used for comparison with the action spectrum.

6.3 N_2 Fixation Through NFE

As described in Section 6.2.1, plasmon resonances are the collective oscillations of conduction electrons coupled with NFE. Improving the energy transfer from the near field to a semiconductor photocatalyst is an approach to enhance the efficiency of NH_3 photosynthesis. NFE mechanisms have the advantages of increasing the absorption of a photocatalyst with a low absorption cross-section or decreasing the necessary volume of a photocatalyst as well as the weight and cost. The NFE process that occurs through the dipole–dipole interaction is also called plasmon-induced resonant energy transfer (PIRET).

Zhang and MacFarlane et al. reported N_2 fixation to NH_3 on black silicon (bSi) decorated with Au-NPs and a hole-sink layer of Cr (Figure 6.3) [56]. The bSi was fabricated from *p*-type boron-doped <100> silicon using a dry etching method. The photoelectrode works under the illumination of light in all visible regions, and the NH_3 yield is enhanced with the Au-NPs compared to that without the Au-NPs. The primary process of the enhancement should be the energy transfer arising from the near field of the Au-NPs. Because the authors used sulfite as a strong sacrificial electron donor, the total reaction is downhill.

Gong et al. loaded Au-NPs on rutile TiO_2 nanorod arrays, and then, the TiO_2/Au was covered with amorphous TiO_2 using atomic layer deposition techniques (Figure 6.4) [57]. Surface oxygen vacancies (OVs) in the outer amorphous TiO_2 thin layer promote the adsorption of N_2, which facilitates N_2 reduction to NH_3. Under the irradiation of visible light, the generation rate of electron–hole pairs within the TiO_2 could be improved via PIRET. The authors suggested that HET from Au to TiO_2 also participates in the reduction of N_2. Although they used pure water as a reaction solution, the oxidation reaction involving holes has not been investigated.

Near-field-enhanced N_2 fixation is beneficial for enhancing the excitation mode with a low absorption cross-section. In contrast, it does not have a strong effect on high absorption processes, such as the direct excitation of a semiconductor. In addition, it should be noted that the near-field energy cannot enhance the nonactive wavelength of the base material.

6.4 N_2 Fixation Through DHEI into N_2 Molecules

DHEI from a plasmonic metal into N_2 molecules is one of the primary mechanisms of plasmon-induced NH_3 photosynthesis. The plasmonic material plays two roles in the DHEI mechanism: generation of hot electrons through LSPR decay and adsorption of N_2 molecules.

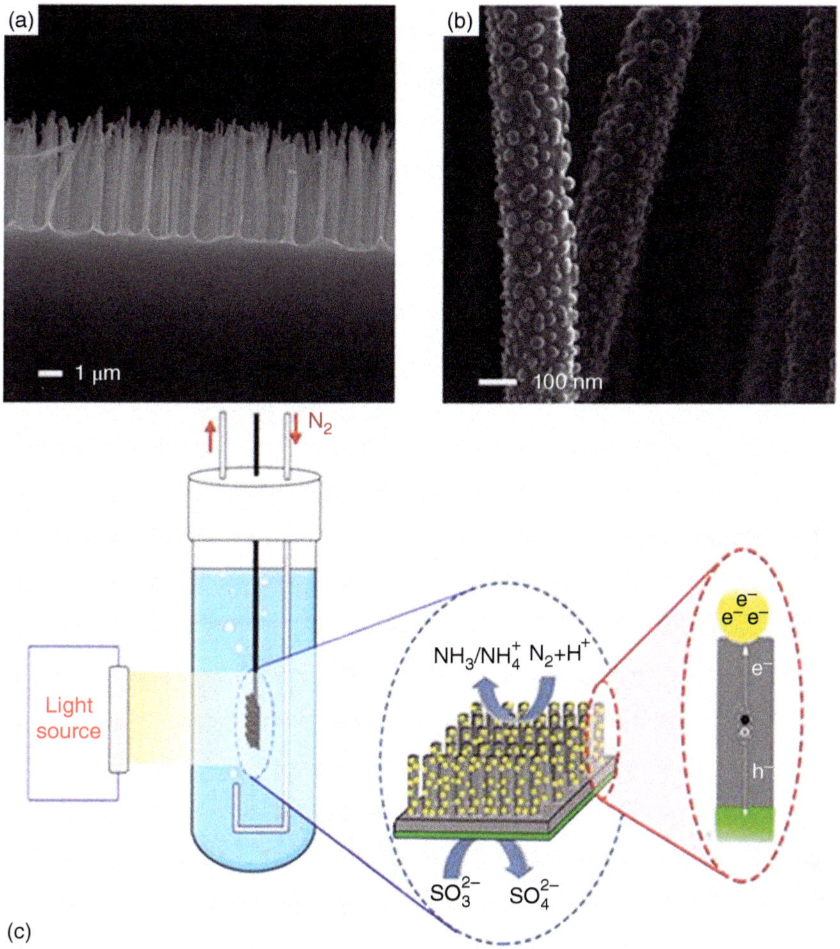

Figure 6.3 (a) Cross-sectional view and (b) magnified view SEM images of an Au-NP-coated black silicon nanostructure. (c) Schematic diagram of the cell. *Source*: Reproduced with permission from [56]. Licensed under CC BY 4.0.

Cater et al. theoretically calculated the DHEI and subsequent N_2 dissociation processes on Mo-doped Au using density functional theory [58]. Mo adsorbs N_2 and works as an active site. They speculated that hot electrons excited within the surface defects of the plasmonic metal are more likely to be injected into the molecular orbital because the direct charge-transfer excitation from the Fermi level into the molecular orbital requires high energy.

Long and Xiong et al. reported N_2 fixation on nanocomposite particles of Au and Ru through a dissociative pathway (Figure 6.5) [59]. The catalytic experiments were performed in water under Xe lamp irradiation at a N_2 pressure of 2 atm. $AuRu_{0.31}$, with a Ru ratio of 31%, achieves a higher NH_3 production rate than that of bare Au nanocrystals. In this work, Ru works as a catalytic active site. Figure 6.5c shows that N_2 fixation catalyzed by $AuRu_{0.31}$ nanostructures has a nearly linear rate dependence

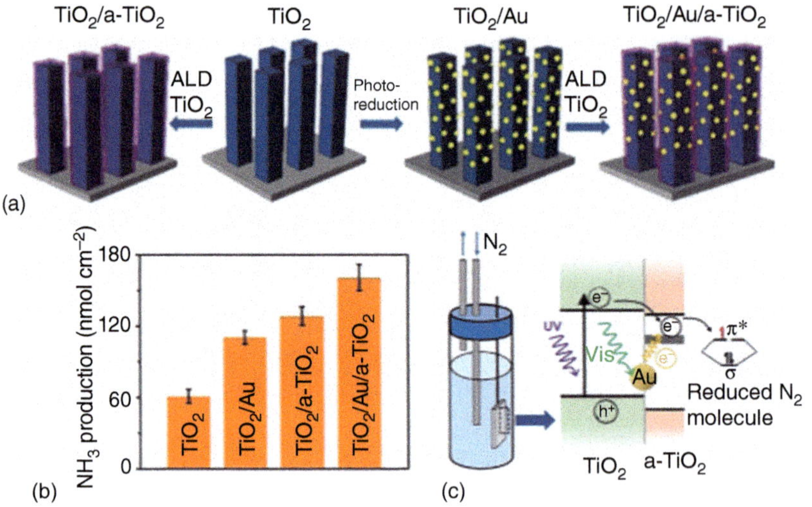

Figure 6.4 (a) Illustration of the experimental procedures for the preparation of bare TiO_2, TiO_2/Au, TiO_2/a-TiO_2, and TiO_2/Au/a-TiO_2 photoelectrodes. (b) NH_3 production over 12 h obtained from different photoelectrodes under AM 1.5G illumination and N_2 flow. (c) Illustration of the single-compartment cell (left) and the synergistic effect of surface O_{vac} and plasmonic Au-NPs for photocatalytic N_2 reduction on TiO_2/Au/a-TiO_2 (right). *Source*: Wang et al. [57] © 2018, John Wiley & Sons.

on the light intensity. The linear rate–intensity relationship is a typical indicator of the induction of a reaction by a single charge carrier. This result suggests that the reduction of N_2 to NH_3 is driven by plasmonic hot electrons. To evaluate the light utilization efficiency, the wavelength dependence of the AQE of $AuRu_{0.31}$ was determined by measuring NH_3 production rates under various monochromatic light irradiation conditions in pure water (Figure 6.5d). The AQEs match the full extinction spectral range of AuRu nanostructures, indicating a high utilization efficiency of the incident light. To verify the origin of the generated NH_3, catalytic N_2 reduction using isotopic $^{15}N_2$ as a nitrogen source was performed. As revealed by ^{1}H NMR spectroscopy (Figure 6.5e), the detected $^{15}NH_4^+$ identifies the role of $^{15}N_2$ as the nitrogen source. These results reaffirm that the N_2 fixation process indeed occurs through $AuRu_x$ plasmonic catalysts via photon energy coupling. Although no sacrificial electron donors were added into the reaction solution, the hydroxyl radical was detected as an oxidation product instead of the O_2 molecule.

Zeng et al. reported NH_3 synthesis using a composite of Os and Au loaded on Cs_2O in a mixture of N_2 and H_2 gases under visible light irradiation (Figure 6.6) [60]. Although the action spectrum of the NH_3 production rate conforms to the LSPR band, as shown in Figure 6.6b, it is difficult to distinguish the DHEI into N_2 and the reaction enhancement due to plasmonic heating. Another disadvantage of the bi-metal system is the damping of the LSPR. The LSPR band is broadened by the addition of Os, as shown in Figure 6.6c.

Another approach that achieves both high hot-electron energy and adsorption ability for N_2 was proposed by Wang et al. [61]. They fabricated semiconductor plasmonic MoO_{3-x} nanosheets that integrate plasmon resonance with rich active

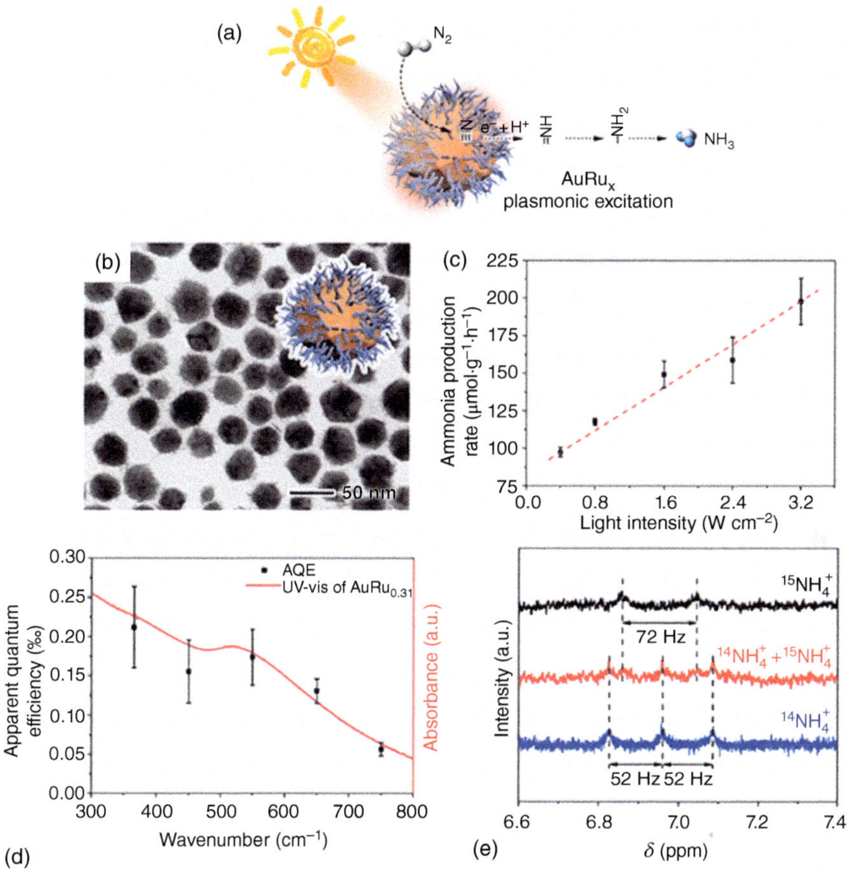

Figure 6.5 (a) Schematic of N_2 fixation on $AuRu_x$ (b) TEM image of $AuRu_{0.31}$ core-antenna nanostructures. (c) Photocatalytic NH_3 production rates by $AuRu_{0.31}$ in the first two hours under different light intensities. (d) Calculated AQEs (■) for N_2 fixation over $AuRu_{0.31}$ in pure water under 20 mW cm^{-2} monochromatic light irradiation, along with the UV–vis extinction spectrum. (e) ^{1}H NMR (400 MHz) spectra of the solution after N_2 fixation reaction using $AuRu_{0.31}$ as a catalyst in a $^{15}N_2$, $^{15}N_2 + {}^{14}N_2$, or $^{14}N_2$ atmosphere. *Source*: Reproduced with permission from [59]. Copyright 2019 American Chemical Society.

sites in one nanostructure. The intrinsic OVs introduce electrons into the CB of the MoO_{3-x} nanosheet and induce a broad LSPR. The hot electrons excited from the CB of MoO_{3-x} have high reduction potential, as shown in Figure 6.7a. Furthermore, the low-valence Mo moieties work as N_2 chemisorption active sites via electron back-donation. The action spectrum of the AQE for N_2 fixation is in accordance with the LSPR spectrum of the MoO_{3-x} nanosheets (Figure 6.7b). Remarkably, the MoO_{3-x} nanosheets are driven to act as a N_2 fixation photocatalyst under near-infrared (NIR) light with a wavelength of up to 905 nm. In this case, the energy of the hot holes is not sufficient for oxidation. Although the authors did not add any sacrificial reagent into the reaction solutions, hot holes are mainly consumed by the oxidation of the OVs in this system. Therefore, the MoO_{3-x} nanosheet does not work as a true catalyst.

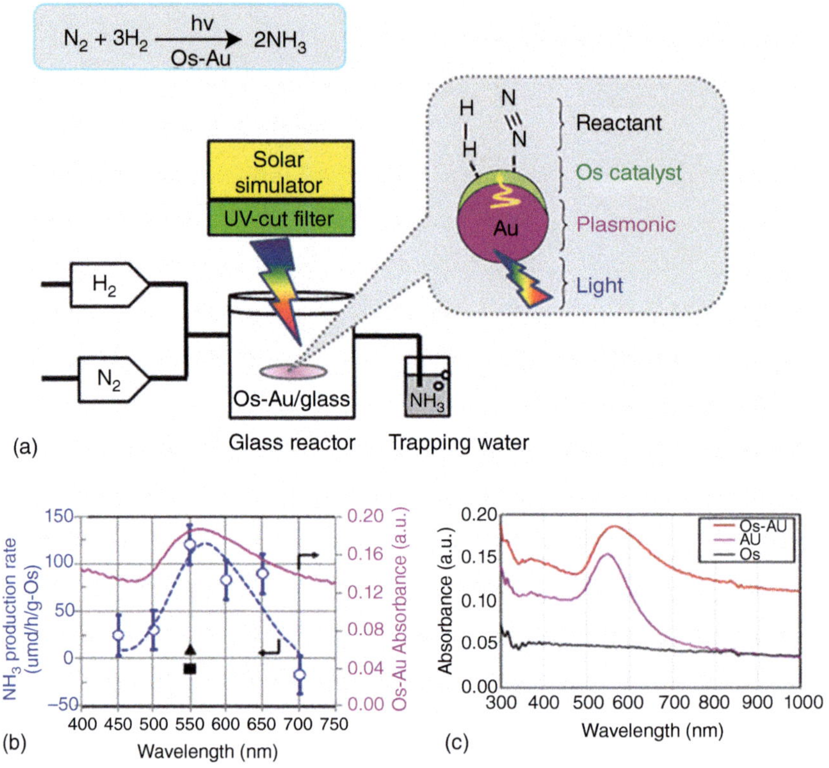

Figure 6.6 (a) Schematic diagram of an Os–Au composite NP for NH_3 synthesis. (b) Dependences of the NH_3 production rate and the LSPR absorbance of Cs_2O-promoted Os–Au (○) particles on the irradiation wavelength. NH_3 was not produced in the reaction using Cs_2O-promoted Os (▲) or Cs_2O-promoted Au (■) particles. (c) Absorbance spectra of the prepared NPs (Os–Au, Au, and Os). *Source*: Zeng et al. [60] © 2015, Elsevier.

The common problem of N_2 fixation using DHEI into N_2 is the low probability of the reaction of hot electrons and holes based on their very short lifetime [62–64]. An additional approach that elongates the lifetime of hot carriers is indispensable to improve the reaction efficiency for the practical use of DHEI.

6.5 HET from a Plasmonic Metal to a Semiconductor

HET from a plasmonic metal to an *n*-type semiconductor is an attractive strategy for extending the absorption range of wide-bandgap semiconductors. With the benefit of the plasmonic sensitization effect, UV-responsive semiconductors can become active under visible and even NIR light.

6.5.1 N_2 Fixation Through HET with Sacrificial Electron Donors

Most of the N_2 fixation processes through HET were performed with sacrificial electron donors. It is very difficult for water oxidation to proceed because this process

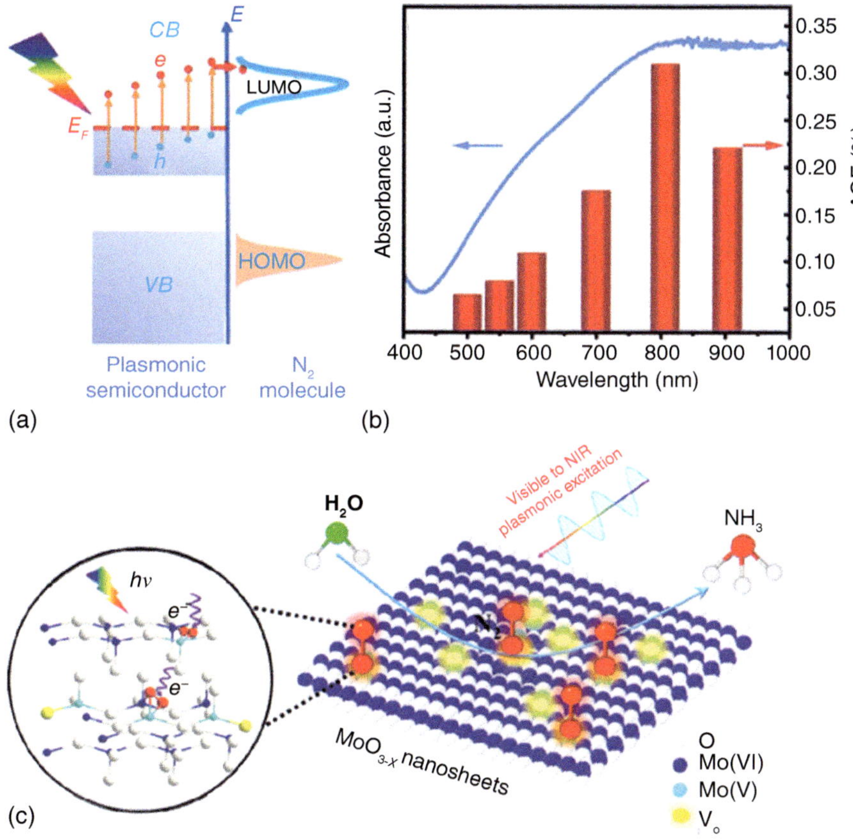

Figure 6.7 (a) Intraband excitation of a plasmonic semiconductor (CB to CB) to generate reductive electrons for molecular N_2 fixation. (b) The AQEs were calculated for N_2 fixation over MoO_{3x} nanosheets in pure water under various monochromatic light irradiation conditions in reference to their UV-vis-NIR diffuse reflectance spectra. (c) Schematic illustration of N_2 photofixation to NH_3 by MoO_{3-x} nanosheets under ambient conditions and visible to NIR plasmonic excitation. *Source*: Wu et al. [61] © 2020, Royal Society of Chemistry.

is kinetically and thermodynamically unfavorable due to the four-electron process and positive oxidation potential (1.23 V vs. RHE). Alcohols and amines are usually used as sacrificial electron donors.

For example, Zhao et al. applied the nanocomposite of $KNbO_3$ and Au to NH_3 photosynthesis (Figure 6.8) [65]. Because $KNbO_3$ has a negative CB potential and a wide bandgap [66], NH_3 was synthesized under UV irradiation of the $KNbO_3$ photocatalyst. After loading of Ag on $KNbO_3$, the reaction efficiency increased even under UV irradiation because Ag worked as an electron sink. The Ag/$KNbO_3$ composite also responded to visible light. However, the NH_3 synthesis efficiency under both UV and visible light irradiation was suppressed compared to that under only UV irradiation (Figure 6.8c,d). This phenomenon occurred because the electron injection from $KNbO_3$ into Ag was canceled by the opposite electron flow from Ag to $KNbO_3$

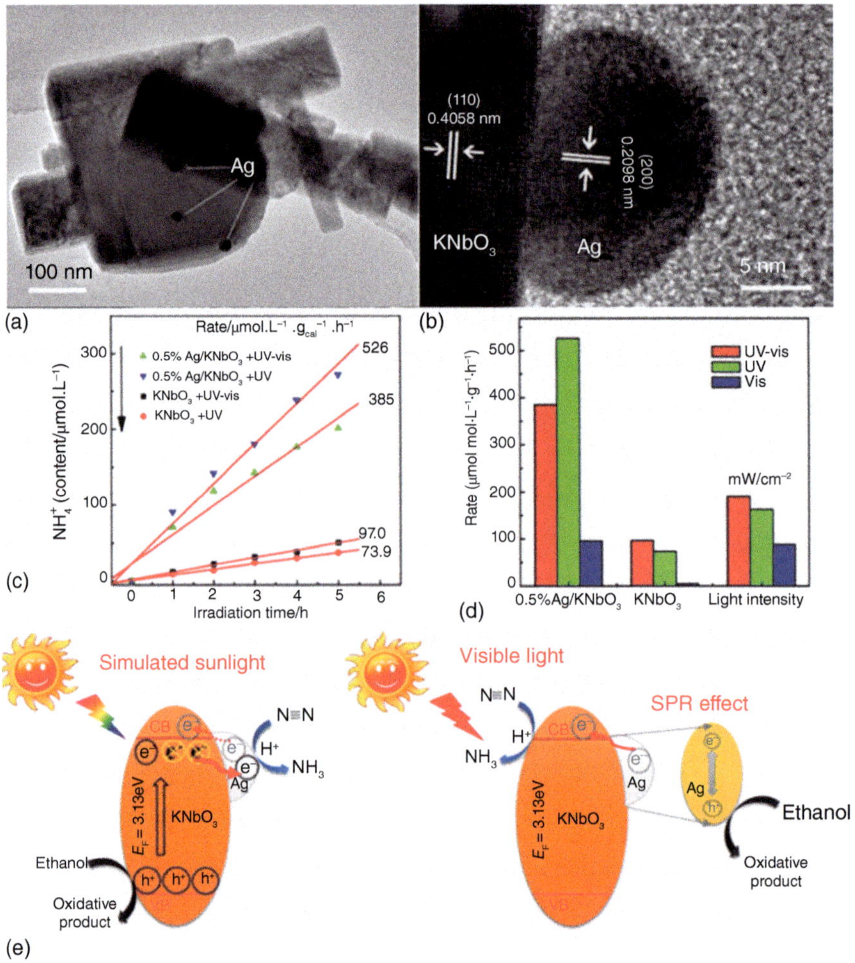

Figure 6.8 (a) TEM and (b) HRTEM images of the 0.5% $Ag/KNbO_3$ composite. (c,d) Performance of $KNbO_3$ and 0.5% $Ag/KNbO_3$ in photocatalytic N_2 fixation under different light irradiation conditions. (e) Possible mechanisms in the $Ag/KNbO_3$ composite under simulated sunlight and visible light. *Source*: Reproduced with permission from [65]. Copyright 2019 American Chemical Society.

(Figure 6.8e). This problem is based on the poor rectification properties because of the absence of an effective space charge layer derived from the small work function of Ag and the tiny particle size of $KNbO_3$. Ethanol was added to the reaction solution to prevent oxidation of Ag and to accelerate N_2 fixation.

Jiang et al. fabricated TiO_2 nanosheets with OVs loaded with Au-NPs (Au/TiO_2-OV, Figure 6.9) [67]. They clearly showed the contribution of LSPR to NH_3 photosynthesis by obtaining the action spectrum of the AQE for NH_3 synthesis, which well corresponds to the absorption spectrum of the LSPR band of Au-NPs (Figure 6.9b). In this case, TiO_2 works as the adsorption center of N_2 on the OVs in addition to the acceptor of the hot electrons from Au-NPs

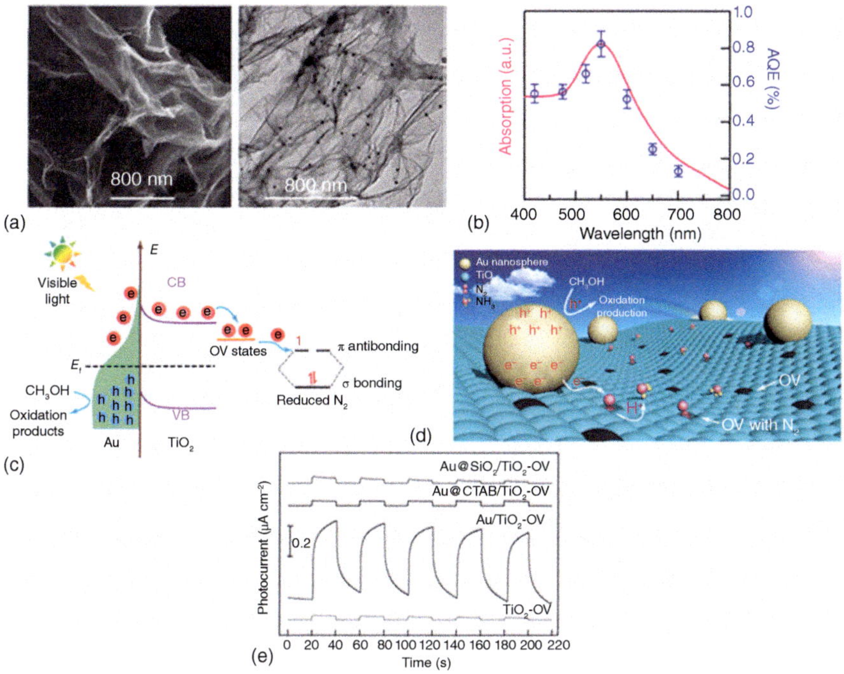

Figure 6.9 (a) SEM (left) and TEM (right) images of the Au/TiO_2-OV sample. (b) Absorption and NH_3 evolution action spectra of Au/TiO_2-OV (1.5 wt% Au). (c) Schematic illustrating the plasmonic hot-electron generation, injection, and N_2 reduction processes in N_2 photofixation with the Au/TiO_2-OV catalyst under visible light. (d) Artistic illustration of efficient plasmonic N_2 photofixation. (e) Photocurrents measured for the Au@SiO_2/TiO_2-OV, Au@CTAB/TiO_2-OV, Au/TiO_2-OV, and TiO_2-OV samples with the visible light irradiation switched on and off repeatedly. *Source*: Reproduced with permission from [67]. Copyright 2018 American Chemical Society.

(Figure 6.9c). The adsorption of N_2 was confirmed by an in situ IR measurement. These authors also performed control experiments in which Au-NPs were coated with hexadecyltrimethylammonium bromide (CTAB) or a SiO_2 thin layer to prevent HET. In these cases, the NH_3 production rate was significantly suppressed (Figure 6.9e), indicating that the HET mainly enhanced the visible response activity of Au/TiO_2-OV.

NH_3 photosynthesis with a similar concept was also reported by the same research group. Crystalline ceria was selectively grown at the ends of gold nanorods (Figure 6.10) [68]. As a result, only the ends of the Au NRs were coated with ceria (Au/end-CeO_2). OVs were also introduced into the CeO_2 to adsorb N_2 molecules. The research group also prepared Au NRs fully covered with CeO_2 (Au@CeO_2 core–shell structure). The spatial separation design of the Au/end-CeO_2 nanostructures offers reaction sites for both reduction and oxidation, with the reduction of N_2 occurring on the surface of CeO_2 and the consumption of hot holes occurring on the side surface of the Au NRs (Figure 6.10e). In contrast, in the core@shell nanostructures, the Au NRs are buried inside, making hot hole access with CH_3OH

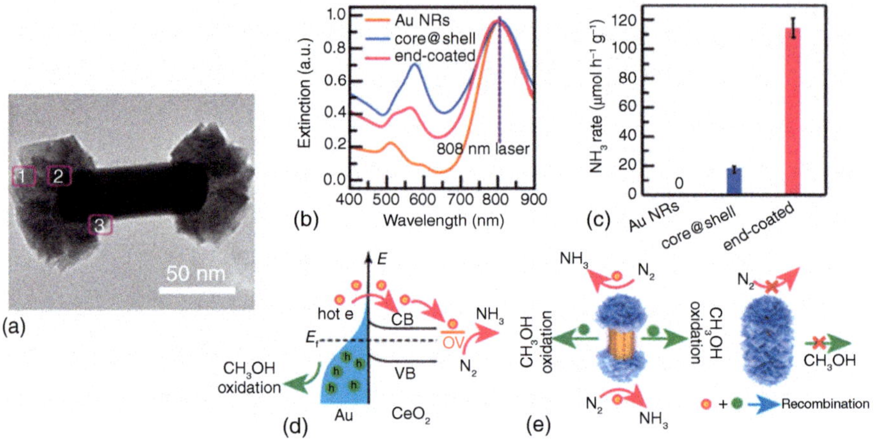

Figure 6.10 (a) HRTEM images of the single Au/end-CeO_2 nanostructure. (b) Extinction spectra of three photocatalyst nanostructures. (c) N_2 photofixation rates of three photocatalysts. (d) Mechanism of N_2 photofixation for the Au/end-CeO_2 nanostructure. E_f, Fermi level; CB, conduction band; VB, valence band; OV, oxygen vacancy state. (e) Comparison of the hot-carrier separation behaviors of the Au/end-CeO_2 nanostructure and the core@shell nanostructure. *Source*: Reproduced with permission from [68]. Copyright 2019 American Chemical Society.

difficult, which results in electron–hole recombination. Notably, the small activity obtained for the core@shell nanostructures may result from the presence of tiny gaps among the CeO_2 nanosheets. These results provide strong evidence that N_2 fixation is mainly driven by HET-induced charge separation rather than by near-field energy transfer from Au to CeO_2.

In both cases of Au/TiO_2-OV and Au/end-CeO_2, alcohols were needed as sacrificial electron donors to achieve N_2 fixation. This suggests that it is difficult to achieve efficient charge separation without rapid hole consumption because nanosized photocatalysts do not form an obvious space charge layer between the plasmonic metal and semiconductor to prevent back-electron transfer. Another possibility is that holes may oxidize NH_3 in the absence of a sacrificial donor.

Misawa et al. fabricated a semiconductor photoelectrode loaded with Au-NPs. In this study, 0.05 wt% niobium-doped strontium titanate (Nb-$SrTiO_3$) with a bandgap of 3.2 eV, which is comparable to that of TiO_2 anatase, was employed as a semiconductor photoelectrode for NH_3 formation (Figure 6.11) [69]. The NH_3 synthesis device contained two compartments to separate the reduced and oxidized products. To promote NH_3 formation, a chemical bias was applied by regulating the pH value of these compartments, rather than using an external electrochemical apparatus. The anodic chamber was filled with an aqueous solution of potassium hydroxide (KOH). Simultaneously, an aqueous solution of hydrogen chloride (HCl) was injected into the cathodic chamber with nitrogen gas. The HCl also served as a proton source for NH_3 synthesis. In addition, ethanol was added to the anodic compartment as a sacrificial donor. The bar chart in Figure 6.11d shows the action spectrum for the AQE of the NH_3 formation $\eta_{app, NH3}$. In the 410–800 nm region, the $\eta_{app, NH3}$

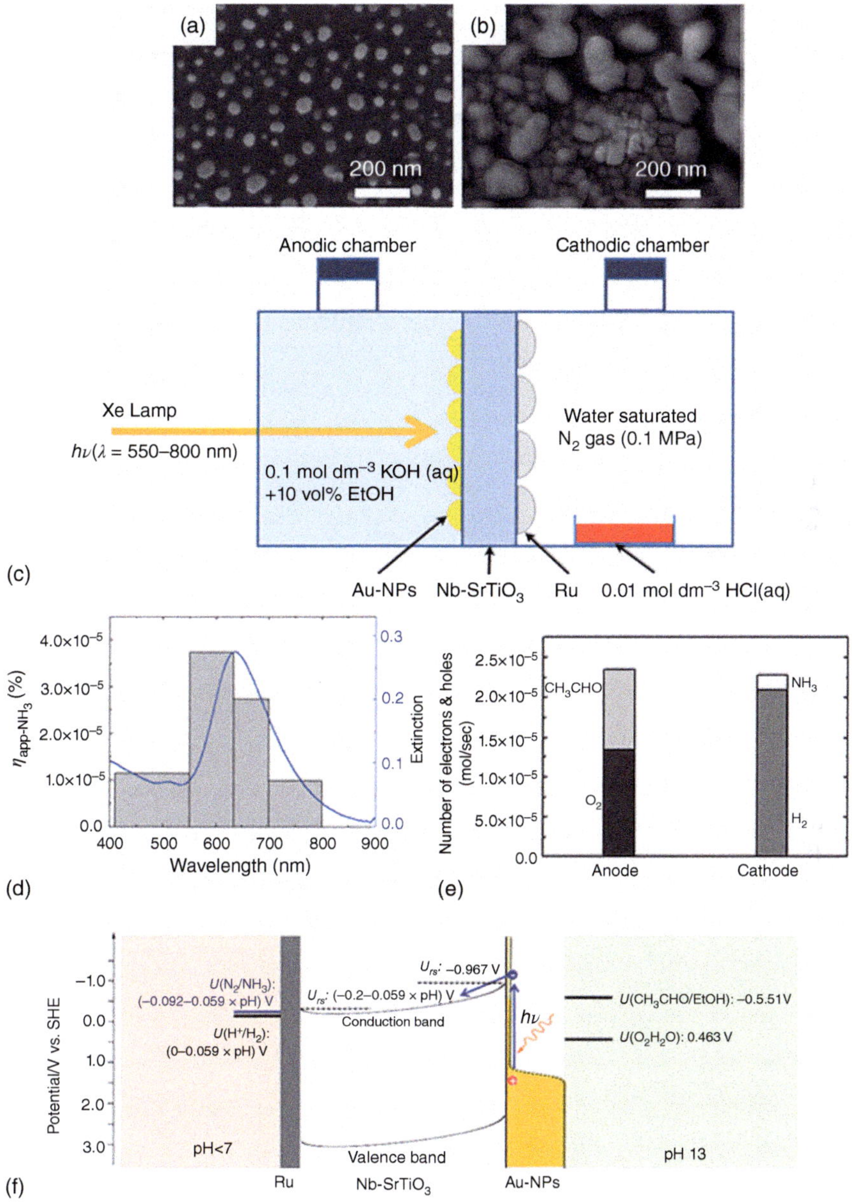

Figure 6.11 Scanning electron microscope image of Au-NPs (a) and Ru (b) on Nb-$SrTiO_3$. (c) Schematic illustration of the NH_3 synthesis device using a Nb-$SrTiO_3$ photoelectrode loaded with Au-NPs. (d) Bar chart of the action spectrum of the AQE of NH_3 formation for several wavelength regions. The solid line indicates the LSPR band. (e) Numbers of electrons and holes that react per second derived from the reduction and oxidization on the cathodic and anodic sides, respectively. (f) Energy diagram of the plasmon-induced NH_3 photosynthesis system using a $SrTiO_3$ substrate loaded with Au-NPs. *Source*: Reproduced with permission from [69]. Copyright 2014 Wiley-VCH Verlag GmbH & Co. KGaA.

value approximately correlates with the LSPR band, which is indicated by the solid line in Figure 6.11d. It is worth noting that the NH_3 formation was highly related to the LSPR excitation.

In this system, four primary types of reactions proceed on the cathode and anode. These redox reactions proceed with electrons and holes generated by LSPR excitation. The numbers of electrons on the cathodic side and holes on the anodic side are highly consistent (Figure 6.11e), indicating a stoichiometric reaction within the full redox system. Figure 6.11f illustrates an energy diagram for the plasmon-induced system for the photosynthesis of NH_3 using a $SrTiO_3$ photoelectrode loaded with Au-NPs. It was hypothesized that excited hot electrons are transferred to the CB of $SrTiO_3$ immediately following the Au-NP inter- or interband transition. These events yield holes trapped in the $SrTiO_3$ surface near the Au/$SrTiO_3$/water interface; subsequently, the trapped holes can efficiently induce oxidation of hydroxyl ions and ethanol through multielectron transfer. By contrast, the photogenerated electrons injected into the CB of $SrTiO_3$ induce nitrogen and proton reduction at the Ru surface, which was deposited on the backside of the $SrTiO_3$. The chemical bias formed by the alkaline solution in the cathodic chamber and acidic solution in the anodic chamber accelerates the reaction.

It is well known that Ru is a suitable catalyst not only for N_2 reduction but also for H_2 evolution because H_2 strongly adsorbs onto a ruthenium surface [70, 71]. Furthermore, the relatively slow rate of NH_3 formation may also represent the larger activation energy required to cleave the triple bond of the nitrogen molecule compared with the energy required to produce a hydrogen molecule from protons. Therefore, we focused on the binding affinity of H and N adatoms on the surface of a co-catalyst. Density functional theory calculations for a wide variety of transition-metal surfaces have predicted that the adsorption energy of H (ΔH^*) is smaller than that of N (ΔN^*) on the Ru surface. However, ΔN^* is smaller than ΔH^* on the Zr surface, indicating that this surface preferentially adsorbs N adatoms [72, 73]. Based on this prediction, Misawa et al. loaded Zr as a co-catalyst instead of Ru on the backside of an Au-NPs/Nb-$SrTiO_3$ photoelectrode (Au-NPs/Nb-$SrTiO_3$/Zr/ZrO_x, where the Zr surface was covered with a passivation thin film) [74]. As a result, NH_3 was selectively formed under visible light irradiation.

6.5.2 N_2 Fixation Through HET Using Water as an Electron Donor

As described in Section 6.2.2, water should be used as an electron donor for NH_3 photosynthesis to achieve uphill artificial photosynthesis. However, surprisingly, there are few reports that quantitatively confirm the oxygen evolution from water.

In light of this, Au-NPs/Nb-$SrTiO_3$/Zr/ZrO_x was applied for HET-induced NH_3 synthesis from N_2 and water without the use of any sacrificial donors (Figure 6.12a) [74]. As a result, the formation of both NH_3 and O_2 increased linearly with increasing irradiation time, and the production ratio between NH_3 and O_2 was approximately 4:3 ($NH_3/O_2 = 1.33$; Figure 6.12b). This result indicates that N_2 reduction and water oxidation proceed stoichiometrically. It has been reported that water might

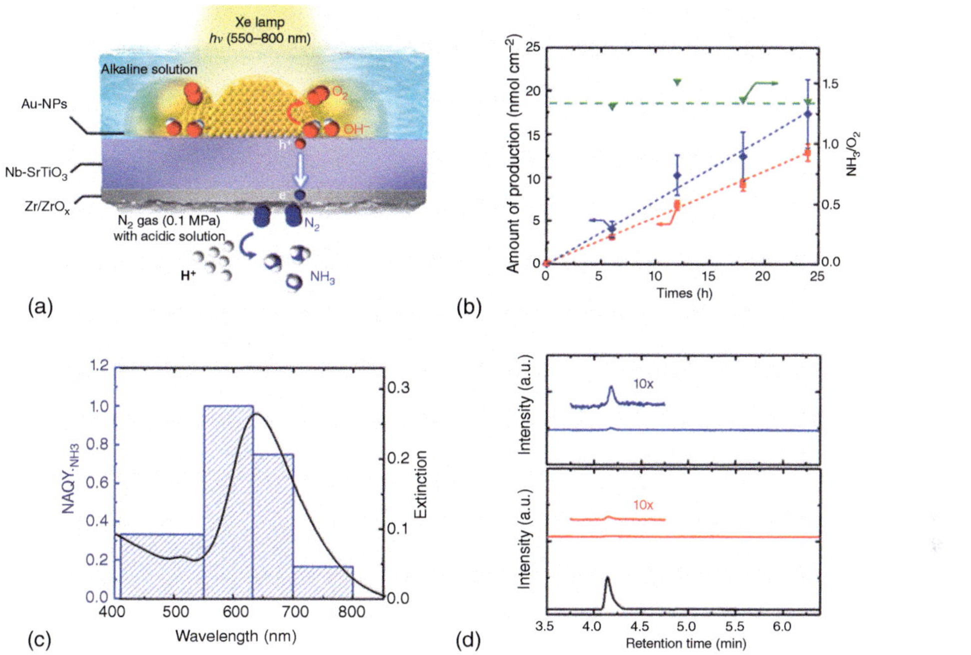

Figure 6.12 (a) Layout of the NH_3 synthesis device bearing the Nb-$SrTiO_3$ photoelectrode loaded with Au-NPs and a Zr/ZrO_x thin film using water as an electron donor. (b) Irradiation time dependence of NH_3 formation on the cathodic side of the chamber (diamond), O_2 evolution on the anodic side of the chamber (square), and the ratio of NH_3 and O_2 (triangle). The dashed green line indicates the ideal stoichiometric ratio (1.33). The irradiation conditions included xenon light spectrally filtered to wavelengths from 550 to 800 nm. (c) Bar chart of the action spectrum of the NAQY of NH_3 formation for several wavelength regions. The solid line indicates the LSPR band. (d) GC-MS chromatogram. Upper: sample obtained from the reaction solution after a 46-h irradiation time ($m/z = 18$, line in blue). Lower: 2.5 ppm NH_3 aqueous solution ($m/z = 17$, line in black; $m/z = 18$, line in red). The insets show the magnified view from 3.75 to 4.75 min. *Source*: Oshikiri et al. [74] © 2016, John Wiley & Sons.

be efficiently oxidized by multiple stored holes trapped in the surface states of the semiconductor near the hot site of the plasmon [48]. Therefore, both the efficient Zr co-catalyst and the Au-NPs are considered to contribute to the complete redox reaction. The bar chart in Figure 6.12c shows the action spectrum for the normalized AQE of NH_3 formation, $NAQY_{NH3}$. The NH_3 formation is strongly related to the LSPR excitation because the value of $NAQY_{NH3}$ is highly dependent on the position of the LSPR band (solid line). In addition, NH_3 formation was observed in all visible wavelength regions, indicating that the plasmon-induced NH_3 synthesis system can efficiently utilize visible light.

NH_3 synthesis on Au-NPs/Nb-$SrTiO_3$/Zr/ZrO_x from isotopic N_2 gas was investigated to obtain direct evidence of N_2 fixation. The photoelectrochemical reaction was promoted by the use of $^{15,15}N_2$ gas instead of $^{14,14}N_2$. Figure 6.12d shows the GC-MS chromatogram of the sample obtained from the reaction solution after irradiation with visible light and that of a standard NH_3 aqueous solution. A peak derived from

$^{15}NH_3$ ($m/z = 18$) is clearly observed at the same retention time as for the standard $^{14}NH_3$ aqueous solution ($m/z = 17$). Furthermore, the $^{15}NH_3$ peak intensity from the reaction solution is much larger than the natural abundance ratio of ^{15}N. These results provide direct evidence that N_2 gas and water were converted into NH_3 under visible-light irradiation.

Misawa et al. divided the integrated NH_3 synthesis device shown in Figure 6.12a into the Au-NPs/Nb-$SrTiO_3$ photoanode and Zr coil cathode. Then, the photoanode and cathode were applied to a two-electrode system connected to an external circuit to investigate the redox reaction by electrochemical analysis in addition to product analysis (Figure 6.13). Understanding the reaction mechanism of NH_3 formation on Zr facilitates the development of selective and efficient NH_3 synthesis techniques. The effects of the applied potential and concentration of reactants on the reaction are important. Figure 6.13b shows the FE of NH_3 and N_2H_4 formation as a function of the applied bias between two electrodes. When the applied bias was less than 0.7 V, no NH_3 was detected. The FE of NH_3 increased as the applied bias increased, reaching c. 80% when the applied bias was 1.0 V. In addition, N_2H_4 was observed when the applied bias ranged from 0.5 to 0.9 V. Figure 6.13c shows the FEs of NH_3 and N_2H_4 as a function of pH under an applied bias of 1.0 V. The FE of NH_3 was very poor when the pH was greater than 4.0. However, the FE of NH_3 dramatically increased as the pH decreased below 4.0. In contrast, N_2H_4 was observed when the pH was between 3.5 and 4.0. This result indicates that the proton concentration directly affects NH_3 production. Notably, both the bias and pH dependencies indicate that intermediate values of bias and pH produce N_2H_4. These results suggest that reaction acceleration is needed for NH_3 production, and N_2H_4 may be an intermediate in NH_3 production.

That plasmon-induced NH_3 synthesis on Zr proceeds via an associative pathway in which N_2 is hydrogenated by protons was proposed, as shown in Figure 6.13d. Initially, the N_2 molecule is preferentially adsorbed on the Zr surface (step i). Subsequently, the reduction of N_2 proceeds via proton addition (step ii). When the pH is large and the applied potential is low, H_2 is also produced as a byproduct (steps ii′ and iii′). In addition, when the applied potential and proton concentration are moderate (pH 3.5–4.0 at an applied bias of 1.0 V or an applied bias of 0.5–0.9 V at pH 3.0), N_2H_4 is produced as the major product via a four-electron transfer process (steps iii–iv). If the reaction rate is high enough with a large applied bias and a high proton concentration, then NH_3 is produced as the main product via a six-electron transfer process (step v). The authors conclude that the reason for the high selectivity in N_2 conversion on Zr is due to the association process without direct cleavage of the strong N_2 triple bond.

To enhance the NH_3 photosynthesis efficiency via HET from a plasmonic metal to a semiconductor, both the photoanode and cathode should be improved (Figure 6.14) [75]. A photoanode composed of an Au film, TiO_2, and Au-NP (ATA) structure with partially inlaid Au-NPs is a promising candidate to increase the number of hot electrons and holes. The ATA structure with modal strong coupling between Fabry–Pérot nanocavities and LSPR modes shows high photocurrent generation by optical hybrid modes across a broad range of wavelengths, followed by HET from the plasmonic metal to semiconductor [76]. In the absorption

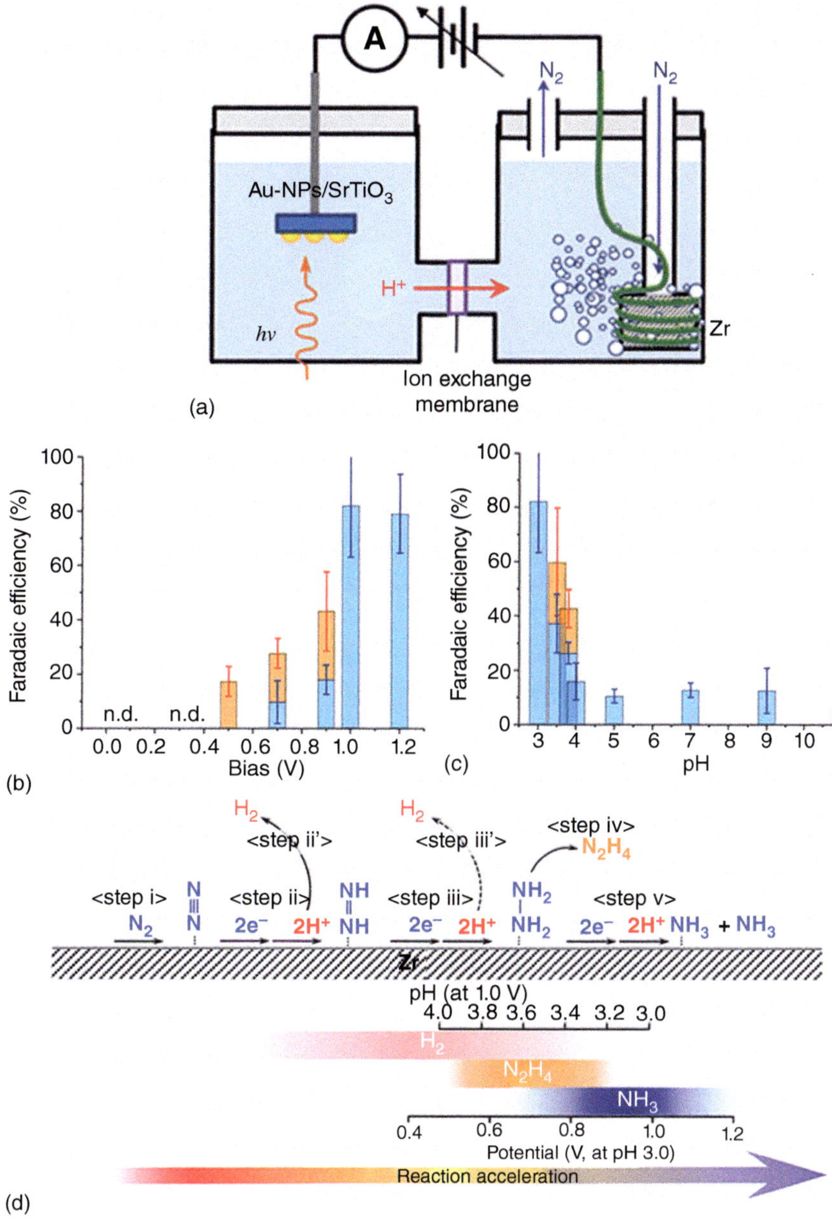

Figure 6.13 (a) Schematic of the reaction cell. (b,c) FEs for NH_3 and N_2H_4 in the two-electrode system using an Au-NPs/$SrTiO_3$|Zr coil as a function of the applied bias (b) and pH (c). The pH of the solution was fixed at 3 when determining the bias dependence. The applied bias was fixed at 1.0 V when determining the pH dependence. The irradiation wavelength was 410–800 nm. A value less than the detection limit is represented as not determined (n.d.). (d) Possible reaction pathway of NH_3 formation on Zr. *Source*: Oshikiri et al. [44].

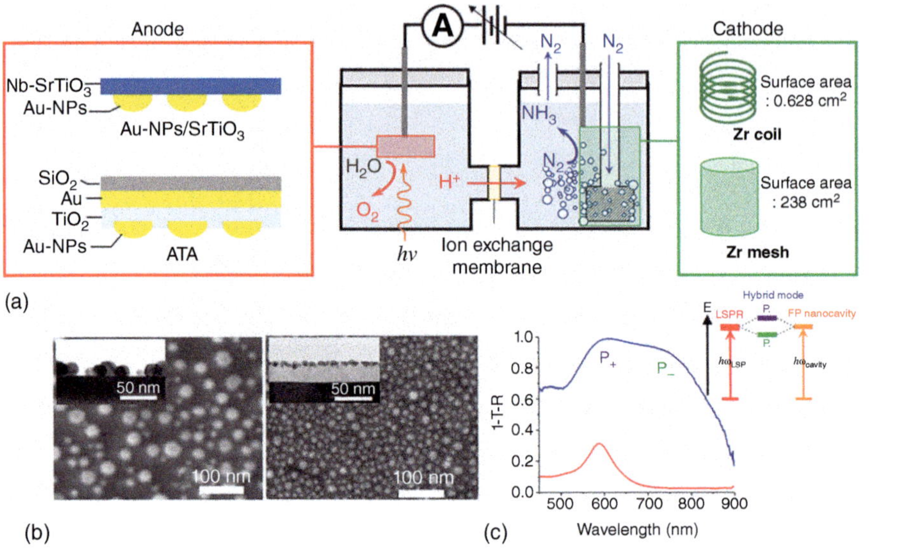

Figure 6.14 (a) Schematic of the reaction cell. (b) SEM images of Au-NPs/$SrTiO_3$ (left) and ATA (right). The inset figures show cross-sectional STEM images of the corresponding structures. (c) Absorption spectrum of ATA and extinction spectrum of Au-NPs on $SrTiO_3$. The references for the absorption spectra of ATA and Au-NPs on $SrTiO_3$ were air and $SrTiO_3$, respectively. The inset figure shows the energy level diagram of the hybridization obtained by modal strong coupling between the LSPR and FP nanocavity modes of ATA. *Source*: Reproduced with permission from [75]. Copyright 2020 Wiley-VCH Verlag GmbH & Co. KGaA.

spectrum shown in Figure 6.14c, dual absorption bands, corresponding to the strong-coupling-induced splitting of energy levels into upper and lower branches, were observed at wavelengths longer than 550 nm. In addition, approximately 80% of visible light with a wavelength of 400–800 nm was absorbed by ATA. This value is four times the absorption of visible light by Au-NPs/$SrTiO_3$ without modal strong coupling. In addition, to enhance the adsorption of N_2 molecules, increasing the surface area of the cathode is an effective method.

Figure 6.15 depicts the incident photon-to-current efficiency (IPCE) action spectra of the four types of two-electrode systems. In the case of the Au-NPs/$SrTiO_3$|Zr coil, the IPCE action spectrum showed a peak at a wavelength of 600 nm, approximately corresponding to the LSPR band shown in Figure 6.14c. Even when the cathode surface area was increased (Au-NPs/$SrTiO_3$|Zr mesh, Figure 6.15b), the IPCE spectral shape was similar to that in Figure 6.15a. This result indicates that the Zr coil has sufficient surface area for the electron supply from the Au-NPs/$SrTiO_3$ photoanode under visible light irradiation. When the anode was changed from Au-NPs/$SrTiO_3$ (Figure 6.15a) to ATA (ATA|Zr coil, Figure 6.15c), the IPCE action spectrum was broadened. However, contrary to expectations, the absolute value of the IPCE was not enhanced much even though the absorption was significantly increased. Finally, the AT anode and Zr mesh cathode were combined (ATA|Zr mesh, Figure 6.15d). The IPCE action spectrum was broad and higher than that of the other three systems, and the active wavelength corresponded to the absorption spectrum shown

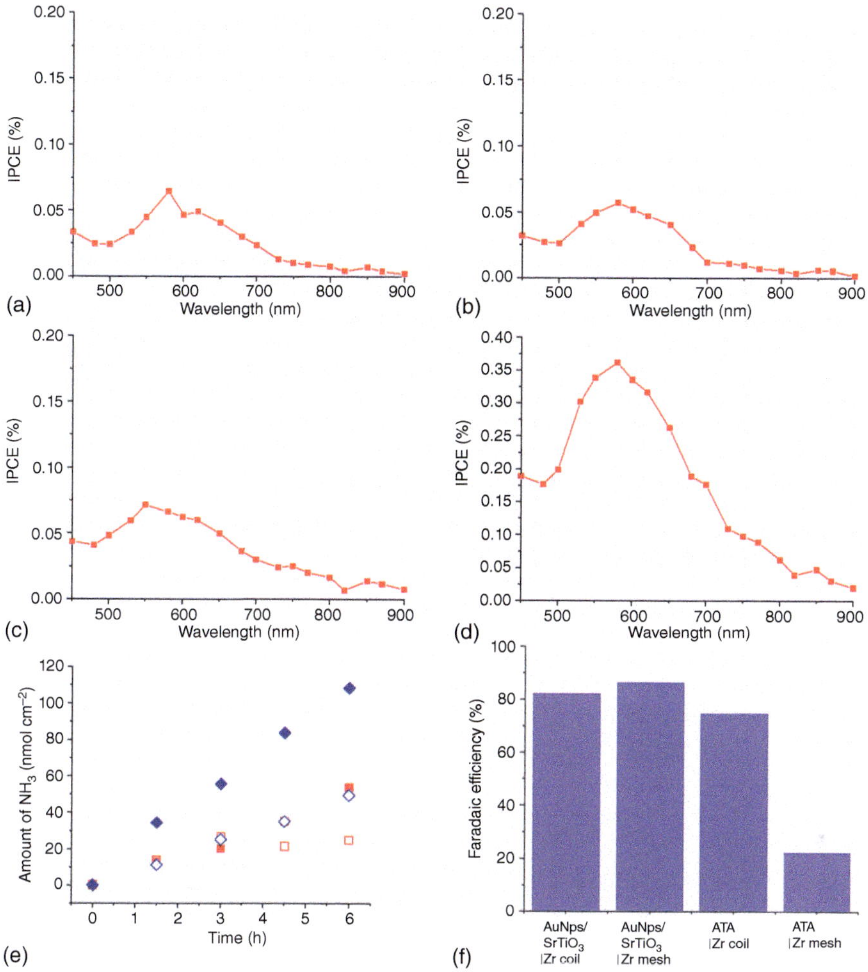

Figure 6.15 IPCE action spectra of the two-electrode systems using (a) Au-NPs/$SrTiO_3$|Zr coil, (b) Au-NPs/$SrTiO_3$|Zr mesh, (c) ATA|Zr coil, and (d) ATA|Zr mesh. The applied bias was 1.0 V. (e) Amount of NH_3 formed with the two-electrode system using Au-NPs/$SrTiO_3$|Zr coil (hollow squares), Au-NPs/$SrTiO_3$|Zr mesh (filled squares), ATA|Zr coil (hollow diamonds), and ATA|Zr mesh (filled diamonds). The applied bias was 1.0 V. The irradiation wavelength was 410–810 nm. (f) FEs corresponding to the panel (e) during the reaction. *Source*: Oshikiri et al. [75] © 2020, John Wiley & Sons.

in Figure 6.15c. This result indicates that the Zr mesh has sufficient surface area to consume the electrons supplied by the excitation of ATA. In addition, both a high absorption and a large reaction area are indispensable in the hybrid mode-assisted NH_3 photosynthesis system.

Next, NH_3 production was directly observed under the irradiation of visible light from a Xe lamp (410–800 nm) in the two-electrode systems with an applied bias of 1.0 V. Figure 6.15e shows the amount of NH_3 produced as a function of irradiation time. The reaction rates of the Au-NPs/$SrTiO_3$|Zr coil, Au-NPs/$SrTiO_3$|Zr mesh, and

ATA|Zr coil were comparable, as expected from the discussion of the IPCE action spectra. In contrast, the NH_3 production rate of the ATA|Zr mesh was approximately twice that of the other three systems. The FEs of the four types of two-electrode system were obtained as the ratio of the measured amount of produced NH_3 and the theoretical amount of produced NH_3 calculated from the photocurrent, as shown in Figure 6.15f. The high FEs in the cases of Au-NPs/$SrTiO_3$|Zr coil, Au-NPs/$SrTiO_3$|Zr mesh, and ATA|Zr coil indicate the selective formation of NH_3 under visible light irradiation. The FE of the ATA|Zr mesh was smaller, suggesting that the huge electron supply from ATA induces subreactions such as hydrogen production on the Zr mesh.

To improve the reaction efficiency of NH_3 photosynthesis via HET between a plasmonic metal and a semiconductor in the future, innovative development of the photoanode, such as modal strong coupling between the LSPR and other photonic modes to enhance hot-carrier utilization, is a practical approach [76–78]. To utilize modal strong coupling more efficiently, an understanding of the near-field decay and the generation and transportation of photocarriers is indispensable.

6.6 Outlook

As mentioned in this chapter, LSPR makes it possible to synthesize NH_3 by reducing N_2 using water as an electron source via irradiation with visible light. This means obtaining fuel and fertilizer from the air, water, and sun. Therefore, NH_3 photosynthesis using LSPR is one of the most promising ways to realize the upcoming low-carbon society. This N_2 fixation system is just an emerging methodology, and recently, the number of reports has been increasing. However, the energy conversion efficiency of these plasmon-induced NH_3 syntheses is still insufficient for practical use, and the development of an innovative methodology is required.

One of the factors of the present NH_3 photosynthesis system that requires further innovation is the extension of the response wavelength. To utilize the broad spectrum of sunlight, the use of LSPR is the strongest approach because the resonant wavelength can be tuned by the size and shape of plasmonic NPs. However, the conventional LSPR band is not yet sufficient and too weak to cover the broad solar spectrum, as introduced in this chapter. Modal strong coupling between the LSPR and other photonic modes, such as nanocavity modes, can solve this problem, as shown in the last part of Section 6.5.2.

Efficient water oxidation is also another challenging problem. To achieve 4-electron oxidation of water, hole accumulation mechanisms, such as co-catalyst and surface states, should be investigated. Study of the spatial relationship between hole sinks and plasmon hot sites is also required.

In addition, to enhance N_2 reduction, the development and understanding of active sites for the adsorption and activation of N_2 molecules are needed. Co-catalysts such as Ru and Zr and OVs are known as influential active sites. Furthermore, a new approach for increasing the storage performance to concentrate the reactant near the photocatalyst surface is required because the concentration of N_2 molecules near the catalyst is minimal.

Although three kinds of mechanisms for plasmon-induced NH_3 photosynthesis based on (i) N_2 fixation through NFE, (ii) DHEI into N_2 molecules, and (iii) HET from a plasmonic metal to a semiconductor were focused on, a *p*-type semiconductor loaded with metal NPs is a promising candidate as a new mechanism. Because the photogenerated hot holes in the metal are transferred into the VB of the *p*-type semiconductor and hot electrons are injected into the N_2 molecules, both efficient charge separation and modulation of the hot-electron energy can be achieved.

It is expected that N_2 fixation by LSPR, which contributes to practical use, will be realized by the progress in these innovative methodological developments.

References

1 Lan, R., Irvine, J.T.S., and Tao, S. (2012). *Int. J. Hydrogen Energy* 37: 1482–1494.
2 Fenwick, A.Q., Gregoire, J.M., and Luca, O.R. (2015). *J. Photochem. Photobiol. B* 152: 47–57.
3 Schlogl, R. (2003). *Angew. Chem. Int. Ed.* 42: 2004–2008.
4 Studt, F. and Tuczek, F. (2005). *Angew. Chem. Int. Ed.* 44: 5639–5642.
5 R. F. Service (2014). *Science* 345: 610.
6 Aika, K. (1986). *Angew. Chem. Int. Ed.* 25: 558–559.
7 Aika, K., Chiristiansen, L.J., Dybkjacr, I. et al. (1995). *Ammonia Springer*. Germany.
8 Kitano, M., Inoue, Y., Yamazaki, Y. et al. (2012). *Nat. Chem.* 4: 934–940.
9 Kitano, M., Kanbara, S., Inoue, Y. et al. (2015). *Nat. Commun.* 6: 6731.
10 Yandulov, D.V. and Schrock, R.R. (2003). *Science* 301: 76–78.
11 Arashiba, K., Miyake, Y., and Nishibayashi, Y. (2011). *Nat. Chem.* 3: 120–125.
12 Ashida, Y., Arashiba, K., Nakajima, K., and Nishibayashi, Y. (2019). *Nature* 568: 536–540.
13 Andersen, S.Z., Colic, V., Yang, S. et al. (2019). *Nature* 570: 504–508.
14 Yao, Y., Zhu, S., Wang, H. et al. (2018). *J. Am. Chem. Soc.* 140: 1496–1501.
15 Singh, A.R., Rohr, B.A., Schwalbe, J.A. et al. (2017). *ACS Cat.* 7: 706–709.
16 Schrauzer, G.N. and Guth, T.D. (1977). *J. Am. Chem. Soc.* 99: 7189–7193.
17 Schrauzer, G.N., Strampach, N., Hui, L.N. et al. (1983). *Proc. Natl. Acad. Sci. U. S. A.* 80: 3873–3876.
18 Fujishima, A. and Honda, K. (1971). *Bull. Chem. Soc. Jpn.* 44: 1148–1150.
19 Fujishima, A. and Honda, K. (1972). *Nature* 238: 37–38.
20 Miyama, H., Fujii, N., and Nagae, Y. (1980). *Chem. Phys. Lett.* 74: 523–524.
21 Yue, P.L., Khan, F., and Rizzuti, L. (1983). *Chem. Eng. Sci.* 38: 1893–1900.
22 Endoh, E., Leland, J.K., and Bard, A.J. (1986). *J. Phys. Chem.* 90: 6223–6226.
23 Boucher, D.L., Davies, J.A., Edwards, J.G. et al. (1995). *Photobiol. A Chem.* 88: 53–64.
24 Rusina, O., Eremenko, A., Frank, G. et al. (2001). *Angew. Chem. Int. Ed.* 40: 3993–3995.

25 Rusina, O., Linnik, O., Eremenko, A., and Kisch, H. (2003). *Chem. A Eur. J.* 9: 561–565.
26 Linnik, O.P. and Kisch, H. (2008). *Mendeleev Commun.* 18: 10–11.
27 Kisch, H. (2013). *Angew. Chem. Int. Ed.* 52: 812–847.
28 Hirakawa, H., Hashimoto, M., Shiraishi, Y., and Hirai, T. (2017). *J. Am. Chem. Soc.* 139: 10929–10936.
29 Link, S. and El-Sayed, M.A. (1999). *J. Phys. Chem. B* 103: 8410–8426.
30 Kelly, K.L., Coronado, E., Zhao, L.L., and Schatz, G.C. (2003). *J. Phys. Chem. B* 107: 668–677.
31 Ueno, K., Mizeikis, V., Juodkazis, S. et al. (2005). *Opt. Lett.* 30: 2158–2160.
32 Amendola, V., Saija, R., Marago, O.M., and Iati, M.A. (2015). *Nanoscale* 7: 8782–8792.
33 Sönnichsen, C., Franzl, T., Wilk, T. et al. (2002). *Phys. Rev. Lett.* 88: 077402.
34 Mubeen, S., Lee, J., Singh, N. et al. (2013). *Nat. Nanotechnol.* 8: 247–251.
35 Clavero, C. (2014). *Nat. Photonics* 8: 95–103.
36 Nakamura, K., Oshikiri, T., Ueno, K. et al. (2016). *J. Phys. Chem. Lett.* 7: 1004–1009.
37 Zhao, G., Kozuka, H., and Yoko, T. (1996). *Thin Solid Films* 277: 147–154.
38 Mukherjee, S., Libisch, F., Large, N. et al. (2013). *Nano Lett.* 13: 240–247.
39 Yu, S., Wilson, A.J., Heo, J., and Jain, P.K. (2018). *Nano Lett.* 18: 2189–2194.
40 Kim, Y., Smith, J.G., and Jain, P.K. (2018). *Nat. Chem.* 10: 763–769.
41 Mao, C., Yu, L., Li, J. et al. (2018). *Appl. Catal. Environ.* 224: 612–620.
42 Swearer, D.F., Knowles, N.R., Everitt, H.O., and Halas, N.J. (2019). *ACS Energy Lett.* 4: 1505–1512.
43 Xiao, C., Hu, H., Zhang, X., and MacFarlane, D.R. (2017). *ACS Sustainable. Chem. Eng.* 5: 10858–10863.
44 Oshikiri, T., Ueno, K., and Misawa, H. (2019). *Green Chem.* 21: 4443–4448.
45 Shipman, M.A. and Symes, M.D. (2017). *Catal. Today* 286: 57–68.
46 Hoffman, B.M., Dean, D.R., and Seefeldt, L.C. (2009). *Acc. Chem. Res.* 42: 609–619.
47 Rod, T.H., Logadottir, A., and Nørskov, J.K. (2000). *J. Chem. Phys.* 112: 5343–5347.
48 Nishijima, Y., Ueno, K., Kotake, Y. et al. (2012). *J. Phys. Chem. Lett.* 3: 1248–1252.
49 Zhao, J., Pinchuk, A.O., McMahon, J.M. et al. (2008). *Acc. Chem. Res.* 41: 1710–1720.
50 Imura, K., Nagahara, T., and Okamoto, H. (2004). *J. Phys. Chem. B* 108: 16344–16347.
51 Achermann, M., Shuford, K.L., Schatz, G.C. et al. (2007). *Opt. Lett.* 32: 2254–2256.
52 Chaturvedi, P., Hsu, K.H., Kumar, A. et al. (2009). *ACS Nano* 3: 2965–2974.
53 Kuttge, M., Garcia de Abajo, F.J., and Polman, A. (2010). *Nano Lett.* 10: 1537–1541.
54 Koh, A.L., Fernandez-Dominguez, A.I., McComb, D.W. et al. (2011). *Nano Lett.* 11: 1323–1330.

55 Sun, Q., Ueno, K., Yu, H. et al. (2013). *Light: Sci. Appl.* 2: e118.
56 Ali, M., Zhou, F., Chen, K. et al. (2016). *Nat. Commun.* 7: 11335.
57 Li, C., Wang, T., Zhao, Z.J. et al. (2018). *Angew. Chem. Int. Ed.* 57: 5278–5282.
58 Martirez, J.M.P. and Carter, E.A. (2017). *Sci. Adv.* 3: eaao4710.
59 Hu, C., Chen, X., Jin, J. et al. (2019). *J. Am. Chem. Soc.* 141: 7807–7814.
60 Zeng, H., Terazono, S., and Tanuma, T. (2015). *Cat. Com.* 59: 40–44.
61 Wu, H., Li, X., Cheng, Y. et al. (2020). *J. Mater. Chem. A* 8: 2827–2835.
62 Harutyunyan, H., Martinson, A.B., Rosenmann, D. et al. (2015). *Nat. Nanotechnol.* 10: 770–774.
63 Brown, A.M., Sundararaman, R., Narang, P. et al. (2016). *ACS Nano* 10: 957–966.
64 Liu, J.G., Zhang, H., Link, S., and Nordlander, P. (2017). *ACS Photonics* 5: 2584–2595.
65 Xing, P., Wu, S., Chen, Y. et al. (2019). *ACS Sustain. Chem. Eng.*
66 Liu, J., Chen, G., Li, Z., and Zhang, Z. (2007). *Int. J. Hydrogen Energy* 32: 2269–2272.
67 Yang, J., Guo, Y., Jiang, R. et al. (2018). *J. Am. Chem. Soc.* 140: 8497–8508.
68 Jia, H., Du, A., Zhang, H. et al. (2019). *J. Am. Chem. Soc.* 141: 5083–5086.
69 Oshikiri, T., Ueno, K., and Misawa, H. (2014). *Angew. Chem. Int. Ed.* 53: 9802–9805.
70 Siporin, S. (2004). *J. Catal.* 225: 359–368.
71 Rosowski, F., Hornung, A., Hinrichsen, O. et al. (1997). *Appl. Cat. A General* 151: 443–460.
72 Skulason, E., Bligaard, T., Gudmundsdottir, S. et al. (2012). *Phys. Chem. Chem. Phys.* 14: 1235–1245.
73 Abghoui, Y., Garden, A.L., Hlynsson, V.F. et al. (2015). *Phys. Chem. Chem. Phys.* 17: 4909–4918.
74 Oshikiri, T., Ueno, K., and Misawa, H. (2016). *Angew. Chem. Int. Ed.* 55: 3942–3946.
75 Oshikiri, T., Shi, X., and Misawa, H. (2020). *Eur. J. Inorg. Chem.* 2020: 1396–1401.
76 Shi, X., Ueno, K., Oshikiri, T. et al. (2018). *Nat. Nanotechnol.* 13: 953–958.
77 Yang, J., Sun, Q., Ueno, K. et al. (2018). *Nat. Commun.* 9: 4858.
78 Cao, Y., Oshikiri, T., Shi, X. et al. (2019). *ChemNanoMat* 5: 1008–1014.

7

Untangling Thermal and Nonthermal Effects in Plasmonic Photocatalysis

Xueqian Li[1,3,4], Jie Liu[1], and Henry O. Everitt[2,5]

[1] *Department of Chemistry, Duke University, Durham, NC, USA*
[2] *Department of Physics, Duke University, Durham, NC, USA*
[3] *Thomas J. Watson Laboratory of Applied Physics, California Institute of Technology, Pasadena, CA, USA*
[4] *Joint Center for Artificial Photosynthesis, California Institute of Technology, Pasadena, CA, USA*
[5] *U.S. Army CCDC Aviation & Missile Center, Redstone Arsenal, AL, USA*

7.1 Introduction

Plasmonic photocatalysis exploits the strong light–matter interactions in metal nanoparticles (NPs) to offer a sustainable route for the synthesis of societally important chemicals such as hydrocarbons (C_xH_y) and ammonia (NH_3). Illumination of tailored plasmonic photocatalysts or traditional thermal catalysts with intrinsic plasmonic properties leads to several photophysical effects, the most important of which are (i) generation of hot carriers, (ii) photothermal heating, and (iii) local enhancement of the electric field.

Of these three light-induced effects, photogenerated hot carriers and photothermal heating have been recognized as key enhancement factors in catalytic reactions. Local light-to-heat conversion and collective heating effects in metal NP solutions have been utilized for NP synthesis, where the plasmon-mediated heating can be orders of magnitude more energy efficient than conventional heating methods [1–7]. Considering reaction kinetics, collective heating effects can benefit both exothermic and endothermic reactions since higher temperatures lead to increased reaction rates [8–10].

Photothermal catalysis highlights the potential of synergy between photo and thermal effects in the utilization of the entire solar spectrum. Tunable localized surface plasmon resonances (LSPRs) enable enhanced light absorption which turn metal NPs into ideal concentrated sources of heat remotely controllable by light [11]. With gold (Au) as a model metal, theoretical and experimental investigations have demonstrated the local and collective plasmonic heating power of illuminated NP suspensions [12–16]. These light-activated nanoscopic heaters have found applications in photothermal cancer therapy [17–19], seawater desalination [20], and direct solar steam generation [21–23]. Heterogeneous catalysis mediated by plasmon heating presents a green alternative to traditional thermal catalysis

Plasmonic Catalysis: From Fundamentals to Applications, First Edition.
Edited by Pedro H.C. Camargo and Emiliano Cortés.

[2, 24–29]. However, commonly studied metals such as Au, silver (Ag), and copper (Cu), have resonant frequencies limited to the visible and near-infrared regions and poor catalytic properties [30]. For applications in catalytic reactions, tailored heterostructures in the form of binary NPs [31, 32], alloys [33–36], and core-shell geometries [37–39] have been shown to lower costs, offer higher light harvesting efficiency, and promote photocatalytic activity through synergistic effects. More complex core-shell structures with a plasmonic core and a shell of metal-organic-framework (MOF) [40–42], reduced graphene oxide (r-GO) [43–45], or semiconductors (Al_2O_3, SiO_2) [46–49] have also been synthesized. Tuning the composition and shell thickness in these core-shell heterostructures allows one to maximize the plasmonic-catalytic efficiency by adjusting the relative contributions of photothermal heating and hot-carrier generation/transfer [50].

Recently, the plasmonic research community has demonstrated that the transfer of hot carriers into antibonding orbitals of surface adsorbates at the rate-determining step (RDS) may select products and accelerate chemical reactions by weakening certain chemical bonds in reaction intermediates [51–59]. Nevertheless, absorbed light also heats these metal NPs, so, claims about the beneficial effects of hot carriers must first convincingly distinguish the effects of "nonthermal" hot carriers from "thermal" effects [60]. Unfortunately, it is quite difficult to measure the temperature of individual metal NPs, so this discrimination has primarily depended on calculations that estimate the degree of photothermal heating or on simple measurements of the catalyst's bulk temperature. Often, the bulk and surface temperature of the catalyst are different due to compromises in designing the reactor chamber to enable the use of spectroscopic techniques and illuminated conditions [61]. For traditional thermal catalytic reactions, comparison of Arrhenius plots enables a temperature correction. It has even been argued that claims regarding nonthermal reductions of activation barriers may instead be ascribed to unaccounted thermal effects using this relationship [62–66]. However, this simple correction is not possible for plasmon-enhanced catalytic reactions where both thermal and nonthermal effects can contribute to the overall reaction rate at the same time.

Thus, the intertwined "thermal" and "nonthermal" effects in plasmonic catalysis make it difficult to discriminate between the contributions from light-induced heating and injection of hot carriers from the metal NPs. The fervent debate about whether observed photocatalytic effects can be explained by simple plasmonic photothermal heating or more interesting nonthermal effects motivated this quest to ascertain the true role of light. To understand the underlying mechanism in these plasmon-driven processes, the intertwined thermal and nonthermal effects from light must be untangled, where nonthermal reactions are defined as contributions to the overall reaction that cannot be explained or achieved by traditional heating or photothermal heating. This chapter summarizes efforts in establishing theoretical and experimental techniques to distinguish between thermal and nonthermal contributions accurately in gas phase heterogeneous plasmonic photocatalysis. The intrinsic plasmonic and catalytic properties of supported rhodium (Rh) and ruthenium (Ru) catalysts are investigated in two model reactions of plasmonic photocatalysis: NH_3 synthesis and carbon dioxide (CO_2) hydrogenation.

Section 7.2 describes a multi-thermocouple experimental strategy to monitor the complex thermal profiles within the catalyst layer under reaction conditions. Using this method, an effective thermal contribution may be derived. This approach is then applied in Section 7.3 to investigate the photothermal effects of a traditional Ru-based catalyst for thermal NH_3 synthesis. It is found that light can be used to produce controlled thermal gradients within the catalyst layer to achieve a balance between thermodynamics and kinetics. Due to the contrast in thermal profiles produced under dark and illuminated conditions, residual effects related to photothermal heating may still be mistakenly ascribed as nonthermal effects. This problem may be solved by introducing a new experimental method that uses a thin layer of inactive photothermal material on top of the catalyst for "indirect" illumination. By using indirect illumination of the catalyst to account for both photothermal and thermal contributions, it is demonstrated that this system is dominated by photothermal effects.

In Section 7.4, these techniques will be applied to CO_2 hydrogenation on Rh-based photocatalysts to measure thermal gradients under illuminated conditions, to discriminate thermal and nonthermal effects, and to explore how photogenerated carriers enhance the nonthermal reaction rate. The observation of product selectivity by illuminated plasmonic photocatalysts cannot be explained by thermal effects, and the indirect illumination technique verifies the contribution of this nonthermal hot-carrier mediated process. As further confirmation, the extracted nonthermal methane (CH_4) production rate is found to have a linear dependence on the top surface temperature, distinctly different from an exponential dependence for thermal catalysis.

Section 7.5 summarizes the strategies introduced – discriminating thermal from nonthermal contributions and extracting the reaction rate and efficiency of hot-carrier driven reactions – as well as how they may be applied universally to other explorations of plasmonic photocatalysis. The outlook continues to be bright for plasmonic photocatalysis by providing new control over the catalytic behaviors of metal nanostructures whether the mechanism is thermal, nonthermal, or both.

7.2 Tools and Techniques for Product Analysis and Temperature Measurement

The heated debate in plasmonic photocatalysis regarding the relative importance of nonthermal contributions from hot carriers and thermal contributions from photothermal heating boils down to the accurate accounting of all thermal effects. It has been shown that illumination of a plasmonic photocatalyst can lower the activation barrier and accelerate the rate of a reaction [58], while others have argued that such apparent nonthermal rate enhancements can be more naturally explained as thermal effects incompletely accounted because of the difficulty in obtaining accurate measurements of catalyst temperature profiles [62–64, 66–69]. Both may be correct, so the challenge becomes quantifying how much the reaction rate is enhanced by each contribution.

Experimental efforts to improve quantification of thermal effects are therefore of paramount importance. The following seven procedures have been proposed as a trustworthy starting point to distinguish photothermal from hot-carrier processes in plasmonics [70]: varying illumination power, varying light beam diameter, combining thermocouple and infrared measurements, observing temporal evolution, calibrating with bubble formation (in liquid phase reactions), varying polarization, and varying wavelength. Regarding temporal evolution, the development of a plasmonic photoelectrode that can monitor photoelectrochemical processes on the millisecond timescale can be used to distinguish charge carrier and photothermal contributions [71]. Additional approaches include nanoscale thermometry techniques such as IR pyrometry [4, 32, 58, 72], molecular thermometers [73, 74], scanning probe thermometry [75], and most notably SERS via analysis of Stokes and anti-Stokes intensities [47, 76–78]. However, implementing these techniques while simultaneously monitoring reaction dynamics remains a challenge. For example, while light can cause the top surface temperature to rise due to photothermal heating effects, the release of heat from exothermic reactions can also influence the catalyst temperature.

When thin photocatalyst layers are subjected to high gas flow rates and illuminated by low duty cycle, short-pulsed lasers, photothermal heating is minimized and isothermal conditions may be assumed. Such conditions facilitate demonstrations of nonthermal plasmonic effects such as transfer of photogenerated hot carriers and/or enhanced local fields [55, 58, 79]. The effects of hot-carrier driven processes may be identified through certain signatures, such as a super-linear, sub-exponential dependence of photocatalytic reaction rates on light intensity [53, 80]. However, when thicker layers of catalysts are used, the vertical and lateral temperature profiles in the catalyst layers become more complicated. In this chapter, we discuss some of the tools and techniques used to measure temperature and the reaction dynamics accurately.

7.2.1 Gas Phase Reaction Chamber

All reactions described in this chapter were carried out in a gaseous reaction system that consists of a gas delivery system, a fixed-bed reactor equipped with a quartz window, LED/laser light sources, an online mass spectrometer, and a multi-thermocouple setup with programmable temperature controller (Figure 7.1a). For the experimental results discussed below, the photocatalytic reactor (Harrick, HVC-MRA-5), with a stainless-steel body, a copper catalyst cup, and a detachable 8 mm diameter quartz window (Figure 7.1b), was cooled by circulating chilled water. The catalyst cup (~6 mm diameter, ~3 mm tall) was illuminated through the quartz window by light-emitting diodes (LEDs) of varied intensity and wavelength (Prizmatix, UHP-F: 365, 385, 405, 455, 525 nm, and broad spectrum white light) or an NIR laser diode (Opto Engine LLC, MDL-H-800-200MW, 805 nm).

Powdered metal photocatalysts were loosely packed into the copper catalyst cup but thick enough to ensure complete absorption of light for the catalytic measurements. The light source was coupled into a Lumatic liquid light guide

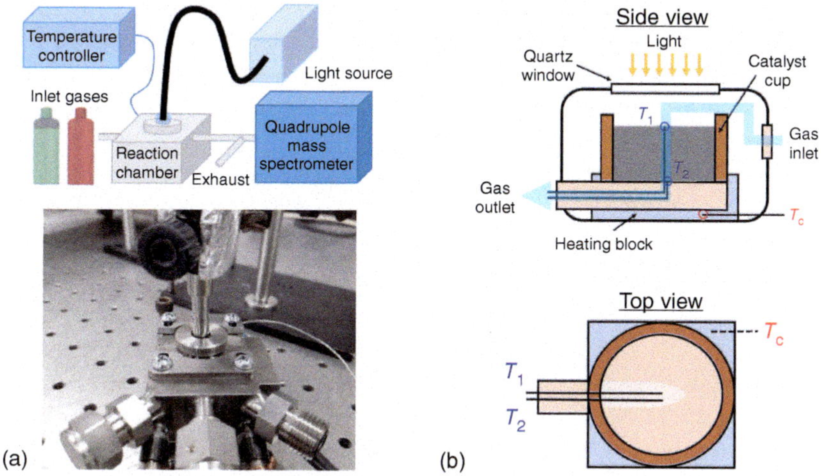

Figure 7.1 Typical gaseous reaction system for investigating plasmonic photocatalysis, including multiple thermocouples for *in situ* measurements of catalyst bed temperatures. (a) Schematic diagram and photograph of the gas delivery system. (b) Side and top views, illustrating how thermocouples inserted from the gas outlet port and through a flat mesh catalyst holder are positioned at the center of the top (T_1) and bottom (T_2) surface of the catalyst, respectively. T_c is the built-in thermocouple that measures the temperature of the heating block. Blue shading indicates the route taken by the flowing gases.

with a diameter of 5 mm and delivered to the top of the catalyst through the quartz window. The LEDs were controlled by 0–5 V analog signal, while the NIR laser was manually controlled and its output power was measured with a thermopile power meter (Thorlabs, PM310D). The temperature of the photocatalyst was controlled by a PID temperature controller kit (Harrick, ATK-024-3).

Reactant gas flow rates were individually controlled using mass flow controllers (GFC, 200 sccm, Aalborg). All reactions were operated under atmospheric pressure. The effluent gas was analyzed by a quadrupole mass spectrometer (Hidden, HPR-20) equipped with a Faraday cup and secondary electron multiplier detector. The responsivity for each reactant and product gas was calibrated with commercial calibration gas or house-made gas mixtures with known composition. For each heating and illumination condition, at least 30 minutes elapsed before ten sequential measurements were performed to ascertain the steady-state concentration of each gas and the associated reaction rates and uncertainties.

7.2.2 Temperature Measurement

The catalyst temperature can be difficult to measure due to evolving reaction conditions and limitations in the reactor design, and the addition of a light component further complicates this situation. In most commercial reactors, the internal reaction chamber thermocouple (T_c, 0.8 mm sheath diameter) is located within the heating block and is used to regulate the thermal contribution from the heating element and maintain a set temperature (Figure 7.1b). Although the sample cup is made of highly thermally conductive copper, the remote location of T_c from the catalyst bed, coupled with dynamic reaction conditions, suggests T_c may not adequately measure the catalyst temperature. Improved methods are needed to measure the catalyst temperature more accurately.

During illumination, photothermal heating causes the top surface temperature to rise. The light penetration depth (δ) is defined as the depth at which the intensity of the radiation inside the material falls to 1/e of its original value. The penetration depth for an electromagnetic wave at normal incidence is given by

$$\frac{1}{\delta} = \frac{4\pi}{\lambda}k, \tag{7.1}$$

where λ is the wavelength of the illuminating electromagnetic field, and k is the imaginary part of the refractive index of the material, which can be calculated from its effective dielectric function. To estimate k, the catalyst may be modeled as an oxide host with inclusions of dipolar metallic NPs using Maxwell-Garnett effective medium theory [81]. The effective dielectric function (ε_{eff}) is thereby given through

$$\frac{\varepsilon_{\text{eff}} - \varepsilon_{\text{h}}}{\varepsilon_{\text{eff}} + 2\varepsilon_{\text{h}}} = f\frac{\varepsilon_{\text{i}} - \varepsilon_{\text{h}}}{\varepsilon_{\text{i}} + 2\varepsilon_{\text{h}}}, \tag{7.2}$$

where ε_{h} and ε_{i} are the dielectric functions of the host oxide [82] and metal inclusions [83], respectively. f is the volume density of inclusions embedded in the host, derived for each experimental configuration from the weight percentage of metal (wt%) considering the metal NP volumes and the metal and oxide densities. From the value ε_{eff}, k is calculated through $k = Im\left(\sqrt{\varepsilon_{\text{eff}}}\right)$ to estimate the penetration depth.

For the example of Rh spherical NPs (Rh-s) on a continuous TiO_2 support with a 5:95 mass ratio of the respective constituents, the penetration depth into this catalyst is only ~0.1 μm for LEDs with wavelengths above and below the TiO_2 bandgap. This indicates that the absorption is dominated by the Rh NPs. Even when the porous structure of the TiO_2 support and light scattering at the gas–solid interface are considered, a Rh-s/TiO_2 catalyst still absorbs all illumination in much less than 1 μm depth. Obviously, nonthermal reactions requiring direct light–matter interactions, which involve (i) hot-electron driven reactions [53, 54], (ii) enhanced near-field effects [84], and (iii) photomodification of catalyst composition [85] can only occur in this very thin layer, and the temperature of that thin layer is best measured by T_1 (blue area in Figure 7.2).

Of course, (iv) thermal reactions can occur throughout the catalyst bed and only depend on the temperature of catalyst, regardless the source of heat (red area in Figure 7.2). This is even true for photothermal heating because the thermal conductivities of the catalyst and flowing gases transport surface heating throughout the catalyst layer. The local temperature within the catalyst layer generally increases with increasing illumination but decreases with increasing gas flow and may further increase or decrease depending on how exothermic or endothermic the reaction is. Consequently, the temperature profile in the catalyst evolves in a complex, dynamical manner as a function of many factors.

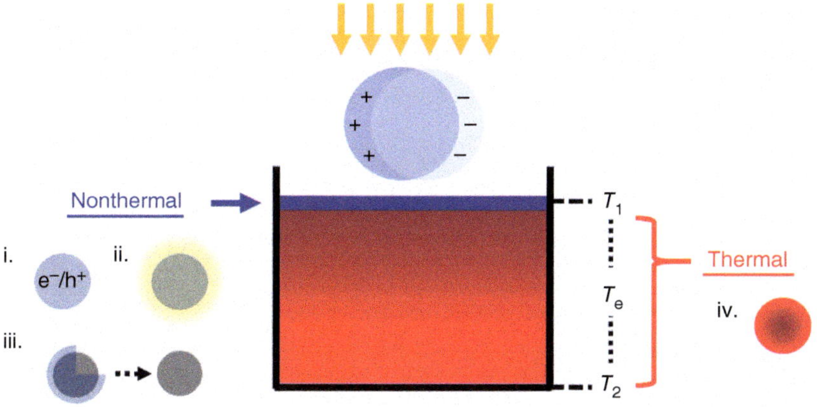

Figure 7.2 Temperature gradient model of the catalyst bed. Due to the short penetration depth of light, the nonthermal reaction due to (i) photogenerated hot carriers, (ii) enhanced near-field effects, (iii) and photomodification of the catalyst is restricted to the top surface, whereas thermal reaction from (iv) photothermal or dark thermal heating can occur throughout the catalyst. Using measured T_1 and T_2 temperatures, an equivalent temperature, T_e can be calculated. *Source*: Adapted with permission from Ref. [60]. Copyright 2018 American Chemical Society.

7.2.3 Thermal Gradients

To capture this dynamic thermal profile, consider a simple model describing a vertical temperature gradient within the catalyst bed. Under steady state conditions, the approximation of a linear vertical temperature gradient and a constant lateral temperature profile adequately captures the excellent thermal conduction throughout both the catalyst bed and the gaseous reaction mixture under uniform illumination. Since the relationship between the thermal reaction rate and the reaction temperature follows an Arrhenius equation under conditions of low reactant conversion (<10%), an equivalent temperature T_e may be defined through the relationship

$$e^{-\frac{E_a}{kT_e}} = \frac{1}{T_2 - T_1} \int_{T_1}^{T_2} e^{-\frac{E_a}{kT}}\, dT. \tag{7.3}$$

This approximation conveniently describes the overall reactivity of thermal reactions in a catalyst bed with temperature gradients in both dark and light conditions. Due to the interdependent relationship of T_e and the apparent activation energy E_a, the equation is solved by an iterative method using the measured values of T_1, T_2, and the dark thermal reaction rate.

To monitor the thermal profile of the catalyst, *K*-type thermocouple probes with a sheath diameter of 0.5 mm (Omega) may be inserted into the reaction chamber through the gas outlet port and bent upward to the desired top and bottom surface positions (Figure 7.3a). In the experiments described here, a piece of stainless steel filter cloth (635 Mesh T316, 20 micron opening, TWP Inc.) was cut into a 6 mm diameter circle and punctured with a 25 gauge needle through the middle. This created a ~0.5 mm hole for the inserted top surface thermocouple (Figure 7.3b). Subsequently, the catalyst was loaded over the mesh holder and just barely covered the tip of the T_1

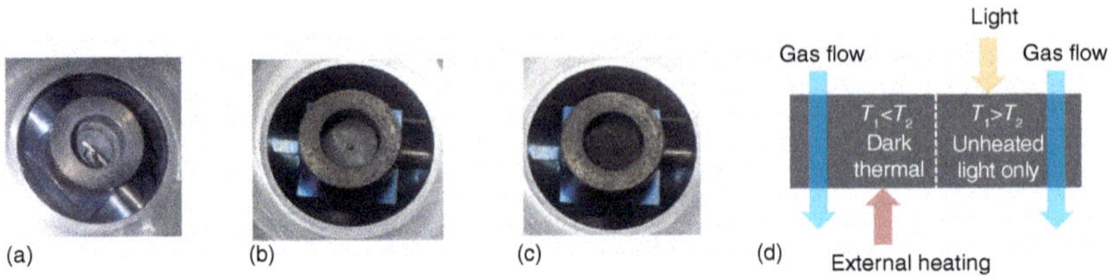

Figure 7.3 Photographs of a multi-thermocouple setup. (a) 0.5 mm sheath diameter thermocouples are inserted through the gas outlet port and bent to central positions to measure the top and bottom surface temperature. (b) Stainless steel mesh holder with hole for inserted T_1 thermocouple. (c) Catalyst is loaded over mesh holder and barely covering the tip of T_1 thermocouple. (d) Representation of thermal profile for dark thermal vs. unheated light-only conditions.

thermocouple (Figure 7.3c). Having a minimal amount of catalyst covering the tip ensured more accurate and reliable measurements of the top surface temperature of the catalyst under reaction conditions by preventing direct illumination from light sources and changes in the location of the thermocouple due to flowing gases. T_1 and T_2 represent the temperature measured at the center of the top and bottom surface of the catalyst layer, respectively.

What is the difference between reaction rates under dark thermal and light-only conditions, illustrated in Figure 7.3d? In all conditions, gas flows through the catalyst from the top to the bottom. Since the gas is not preheated, flowing room temperature gas cools the catalyst from the top while a heating block underneath the catalyst bed warms the catalyst. Consequently, an illumination-dependent thermal gradient ($\nabla T = T_2 - T_1$) may be created within the catalyst bed: in dark conditions the flowing gas cools the surface and produces a positive gradient (increasing temperature with depth), but upon illumination this evolves to a negative gradient as photoheating of the surface reduces the external heating required to maintain T_1. Thus, nonisothermal conditions are common in catalysts thicker than the light penetration depth, in contrast with early investigations of plasmonic photocatalysis that operated under isothermal conditions [86–90].

7.2.4 Thermocouple Diameter

It has been shown that decreasing the thickness of the sheath diameter increases the accuracy of the measured temperature [91]. To consider the effect of thermocouple thickness, consider "thin" and "thick" thermocouples with a sheath diameter of 0.25 and 0.5 mm, respectively. If these thermocouples are inserted in ~20 mg of a Ru-Cs/MgO catalyst, different T_1 temperatures are measured, even when efforts are made to ensure that the tip of the T_1 thermocouple is covered with a very small amount of catalyst to prevent direct illumination. Specifically, for each T_2 set under dark conditions, T_1 measured by the thin thermocouple is lower than that of the thick thermocouple. In NH_3 synthesis using a 3 mm thick version of this Ru-based photocatalyst, thin thermocouples indicate a thermal gradient of ~60 °C at $T_2 = 350$ °C, whereas thick thermocouples show a gradient of only ~30 °C.

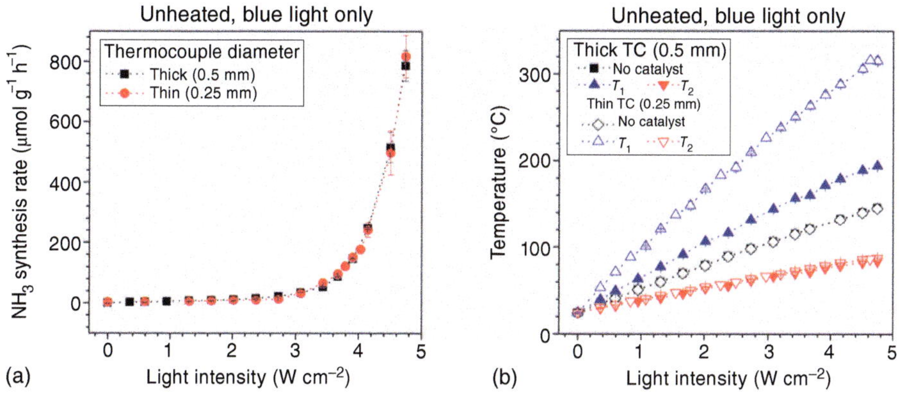

Figure 7.4 Comparison of thick and thin thermocouples under unheated, blue light-only conditions. Measured (a) NH_3 synthesis rate and (b) T_1, T_2 temperatures as a function of I_{blue} for thick and thin thermocouples.

Under unheated, blue light-only experiments, the influence of the thermocouple thickness is amplified. Consider Figure 7.4a, which confirms the reaction conditions and rates are nearly identical for both thermocouples for all light intensities. Although the corresponding T_2 temperatures are identical, Figure 7.4b reveals that thick and thin thermocouples measure a T_1 of ~190°C ($\nabla T \sim 120$°C) and ~310 °C ($\nabla T \sim 240$°C), respectively for 4.7 W cm^{-2}. Both exceed the T_1 measured by the bare thermocouple without catalyst (~130 °C), a difference caused by exothermic heating from the reaction.

The results show that changing from thick to thin thermocouples increases the accuracy of the measured top surface temperature. Consequently, the thinnest possible thermocouples should be used for properly accounting the contribution from thermal effects on the total reaction rate measured.

7.2.5 Additional Thermometry Methods in Plasmonic Photocatalysis

A multi-thermocouple setup such as this allows the thermal profiles of catalysts to be monitored, even set, under reaction conditions [60, 92]. While the temperature of solutions is well equilibrated in liquid reaction systems [25, 52, 93], the above discussion shows that thermal gradients can be quite severe in gaseous reaction systems [61]. Inside a porous catalyst bed, thermal gradients are inevitable due to a combination of illumination, external heating, gas flow, and reaction energetics.

To isolate and study nonthermal light effects, the thickness of the catalyst should be reduced as much as possible to minimize thermal gradients while ensuring reaction rates remain measurable. However, as will be shown next, photothermally induced gradients can actually improve the performance of a catalyst for certain reactions.

A challenge remains in obtaining a meaningful definition of temperature for the system. In addition to using traditional thermocouples, incorporation of improved nanoscale thermometry techniques requires modifications to the reaction chamber

[73, 75, 76, 78, 94–96]. Since the temperature measured by thermocouples may underestimate the surface temperature, temperature measurements from a noncontact infrared pyrometer may help to circumvent physical challenges associated with the use of thermocouples. To position an IR pyrometer perpendicular to the catalyst surface, a confocal geometry may be used, requiring a 45° dichroic beamsplitter filter that reflects >90% of the UV and visible light while transmitting >80% of the NIR and IR wavelengths. If the reaction allows, the quartz window over the reactor must be replaced with KBr or sapphire transparent from the near UV to IR ($\lambda = 0.25 - 25\,\mu m$).

Although IR temperature measurements can be easily obtained with IR pyrometers, the real challenge is in understanding the difference between apparent and actual IR temperatures. Emissivity is the ratio of how well an actual material radiates compared to a perfect radiator called a "black body." Emissivity varies by surface condition, viewing angle, temperature, and spectral wavelength. Therefore, temperatures deduced from pyrometric measurements of materials with unknown emissivity can have huge errors [64, 66]. The emissivity of catalyst samples may be determined experimentally through material heating. The catalyst of interest should be heated to a known uniform steady-state temperature in a furnace or hot plate. In addition, the temperature measured by the IR pyrometer can be calibrated with a contact temperature probe to obtain the correct emissivity for the material of interest.

7.3 Photothermal Catalysis

As discussed above, the effects from photothermal heating are subtle and often inadequately accounted in investigations of plasmonic photocatalysis. To explore this challenge, consider the production of NH_3 by illumination of a conventional Ru-based catalyst much thicker than the light penetration depth [92]. While the weak, broad plasmon resonance of Ru NPs limits hot-carrier generation [97], it facilitates plasmonic photothermal heating when illuminated by broadband sources such as natural sunlight. Plasmonic photocatalytic NH_3 synthesis and hot-carrier driven processes have already begun to be explored in both solution and gas phases [88, 98–100]. As will be described below, intense photothermal surface heating of this thick Ru NP catalyst layer actually produces a beneficial non-isothermal temperature gradient. The rate-limiting nitrogen scission step appears to be accelerated in the hot surface region, while the cooler lower region preserves the NH_3 products.

7.3.1 Ru-based Catalysts for NH_3 Synthesis

This investigation of NH_3 synthesis uses a commonly studied cesium-promoted, magnesium oxide-supported, ruthenium (Ru-Cs/MgO) catalyst (Figure 7.5a) [87, 101–104], for which the light penetration depth is less than 100 μm [32, 58, 60, 105]. It has been well established that basic supports (MgO) and alkali promoters (Cs) assist in NH_3 synthesis via electron transfer to Ru [86, 87, 89, 102, 106]. An electron-rich surface of the Ru metal is more active for N_2 dissociative chemisorption, which is the RDS in NH_3 synthesis. Let us now explore how these thick, non-isothermal plasmonic photocatalysts may

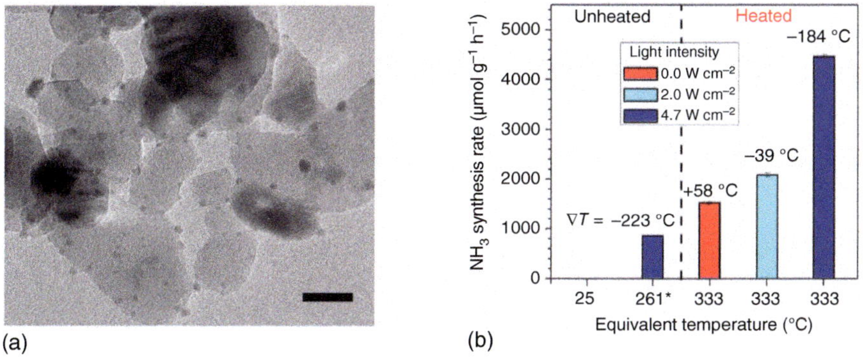

Figure 7.5 Morphology and size distribution of Ru-Cs/MgO. (a) TEM image, scale bar: 20 nm. (b) NH_3 synthesis rates under dark and illuminated conditions, with measured thermal gradients. *Blue LED illumination of 4.7 W cm^{-2} raises the "unheated" equivalent temperature of the catalyst bed from 25 °C to 261 °C. *Source*: Adapted with permission from Ref. [92]. Copyright 2019 American Chemical Society.

significantly enhance NH_3 reaction rates beyond what is possible under dark conditions (Figure 7.5b).

7.3.2 Thermal Gradients in Ru Catalysts

Using a catalyst only ~0.9 mm thick, values of $\nabla T = T_2 - T_1$ as large as −80 °C were observed for high intensity blue illumination (4.7 W cm^{-2}) and +30 °C under dark thermal conditions. Figure 7.6a illustrates the relationship between the thermal gradient and the equivalent temperature for dark conditions (positive gradient, red) and for illumination using a white LED that mimicked concentrated solar illumination of 27 sols (negative gradient, green). Notice that as external heating increases at the bottom of the catalyst, the smaller positive thermal gradient increases under dark conditions while the larger negative thermal gradient decreases under illumination.

To assess the potential of this catalyst for solar-driven NH_3 synthesis, a broadband white LED was used. The NH_3 production rate is consistently higher for combined light and heat conditions than for dark thermal conditions at the same T_e; indeed, it is almost 100 times greater for the most energy-efficient scenario when no external heat is applied (Figure 7.6b). Illumination clearly improves catalytic activity over dark conditions for the same equivalent temperature and the same absolute magnitude of the thermal gradient. In fact, the illuminated catalyst with no additional heating produces more than 100 μmol g^{-1} h^{-1} of ammonia, a rate not achieved for the unilluminated catalyst until heated to an equivalent temperature of almost 300 °C (Figure 7.6b, circled green data point).

This enhancement is surprising: the catalyst was optimized for dark thermal conditions so that for T_e = 333 °C and ∇T = +58 °C, the measured NH_3 synthesis rate was 1530 μmol g^{-1} h^{-1} (Figure 7.5b). However, at the same equivalent temperature, illumination from a blue LED (455 nm) at 2.0 and 4.7 W cm^{-2} produced a negative thermal gradient that increased the NH_3 synthesis rates by 36% (2088 μmol g^{-1} h^{-1}, ∇T = −39 °C) and 192% (4464 μmol g^{-1} h^{-1}, ∇T = −184 °C), respectively. Clearly

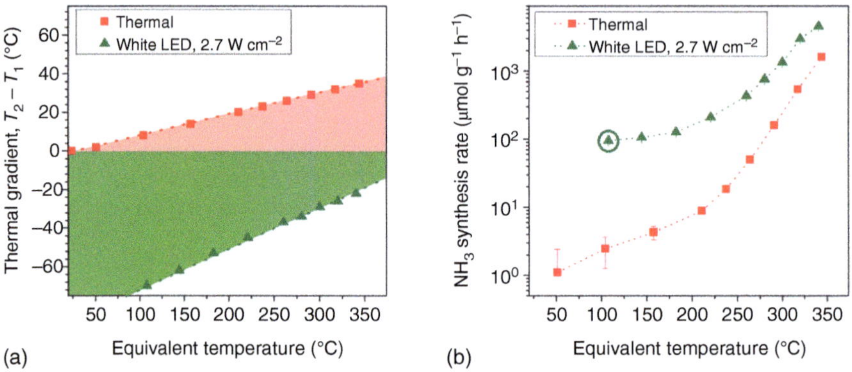

Figure 7.6 NH_3 synthesis under dark thermal and heated white light illumination. (a) Measured thermal gradients and (b) measured NH_3 synthesis rates as a function of equivalent temperature under dark thermal and heated white light illumination. The circled data point indicates the light-only condition with no external heating applied. *Source:* Adapted with permission from Ref. [92]. Copyright 2019 American Chemical Society.

the strong negative thermal gradient created by illumination significantly improved the catalytic activity of the Ru-Cs/MgO catalyst.

When a temperature gradient exists, thermophoretic forces tend to move molecules from high to low temperature regions [107]. Under a negative gradient, thermophoretic forces align with the flow of gases through the catalyst and improve yield as NH_3 products are drawn away from the hottest region to avoid the reverse decomposition reaction. A negative gradient produced different behaviors than a positive gradient, with a negative gradient being more effective at increasing reaction rates and product yields than a positive gradient of the same absolute value. Evidently, for $\nabla T < 0$ the rate-limiting nitrogen scission step is favored in the illuminated hot surface region while the NH_3 products are increasingly preserved in the cooler dark lower regions.

For a "room temperature" test of plasmonic photothermal heating efficacy within the Ru-Cs/MgO catalyst, light-only NH_3 synthesis experiments were conducted without applying any external heating. At these room temperature, atmospheric pressure, and dark conditions, NH_3 synthesis rates are unmeasurably slow. However, NH_3 is copiously produced with blue illumination as the sole energy source (Figure 7.7a). At 4.7 W cm^{-2} of blue illumination, T_1 and T_2 climb above room temperature to 311 and 88 °C, respectively, so T_e = 261 °C, ∇T = −223 °C, and the reaction rate reaches 858 μmol g^{-1} h^{-1} (Figure 7.7a,b).

Clearly, illumination-augmented gradients within the catalyst cause significant deviations from isothermal conditions, with total conversions that appear to exceed thermodynamic limits if the effect of a non-isothermal gradient is ignored. Simply stated, light-induced non-isothermal negative gradients resolve conflicting requirements for catalysis: the hotter T_1 temperature allows for increased catalytic activities while the cooler T_2 temperature maintains higher conversion yields.

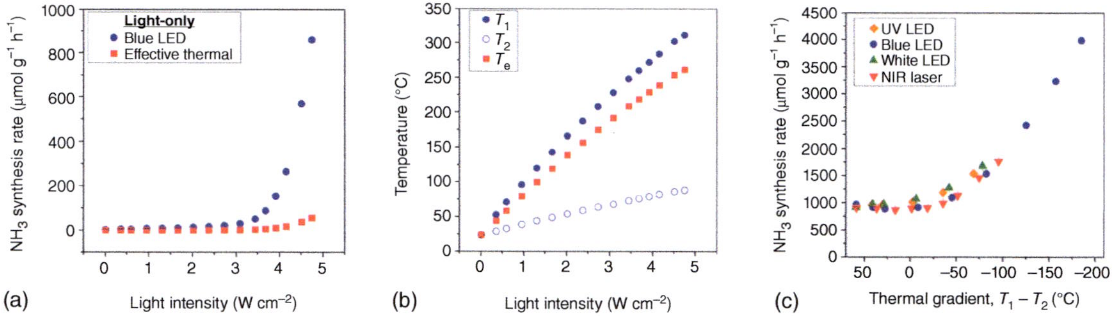

Figure 7.7 Light-only NH_3 synthesis rates and measured temperatures. (a) Measured unheated, light-only NH_3 synthesis rates and calculated effective thermal rates as a function of blue light intensity, I_{blue}. (b) Measured T_1, T_2, and calculated T_e temperatures as a function of I_{blue}. (c) NH_3 synthesis rates as a function of the thermal gradient for $T_e = 325\,°C$ for various light sources. *Source*: Adapted with permission from Ref. [92]. Copyright 2019 American Chemical Society.

7.3.3 Intensity- and Wavelength-Dependent Behavior

Consider next the dependence of thermal gradients on illumination wavelength and intensity. The optical properties of Ru-Cs/MgO were measured by diffuse-reflectance UV–vis spectroscopy as shown in Figure 7.8a. The pure MgO support did not show any absorption in the measured 270–800 nm wavelength range due to its large bandgap. The absorption of light by Ru-Cs/MgO in the UV–vis–NIR regions can be attributed to the properties of Ru NPs.

Examination of the intensity-dependent reaction rate for the four light sources suggests that the catalyst has a wavelength dependence (Figure 7.8c). At first glance this would appear to contradict the predicted behavior from the broad absorption of Ru NPs in the catalyst [88, 97]. But UV and visible LEDs of the same intensity produced similar heating regardless of wavelength. For the equivalent temperature of 325 °C, the dependence of T_1 and T_2 on illumination intensity for each LED is shown in Figure 7.8b. Only in the NIR were deviations observed, where the illuminated reaction was only half as efficient at similar light intensities. The lower absorption of NIR light by the catalyst allowed it to penetrate deeper [60]; consequently, the same intensity of light effectively heated a larger portion of the catalyst, and the overall thermal gradient shrank. Thus, the smaller ∇T produced by NIR illumination reduced the production rate.

So not only does T_e determine reaction rates, the size and sign of ∇T are critically important. The thermal gradient changed from positive to negative with increasing light intensity. The NH_3 synthesis rate is found to have no significant dependence on illumination wavelength for identical equivalent temperatures and thermal gradients, but it has a strong dependence on the sign of ∇T. However, as illumination intensity increases and thermal gradients evolve from positive to negative, the reaction rate first slightly decreases to a minimum and then significantly accelerates (Figure 7.7c), regardless of the wavelength. The minimum occurs for isothermal conditions where the thermal gradient is zero ($T_1 = T_2$). The same trend is seen at $T_e = 275$ and 300 °C, but the NH_3 synthesis rates are correspondingly lower. This confirms that the reaction rates are dependent on both the equivalent temperature and the overall thermal gradient.

7.3.4 Direct and Indirect Illumination

Given the wide range of thermal gradients that can produce a given value of T_e and the possibility of superheated metal NPs, it is imperative to compare light and dark conditions for catalysts of identical temperature profiles. It is also important to ascertain whether any of these photoenhancements might be caused by nonthermal effects such as hot electron transfer. This can be achieved by comparing NH_3 synthesis using the same Ru-Cs/MgO catalyst under both direct and indirect illumination conditions (Figure 7.9a).

In the direct illumination case, the photocatalyst was directly illuminated from the top by a UV LED (Figure 7.9a). Only the top submicron depth can contain any nonthermal (e.g. hot-electron driven) light effects, whereas the remaining portion of the catalyst beyond the penetration depth of light contains only the thermal contribution produced by illumination and external heating [60].

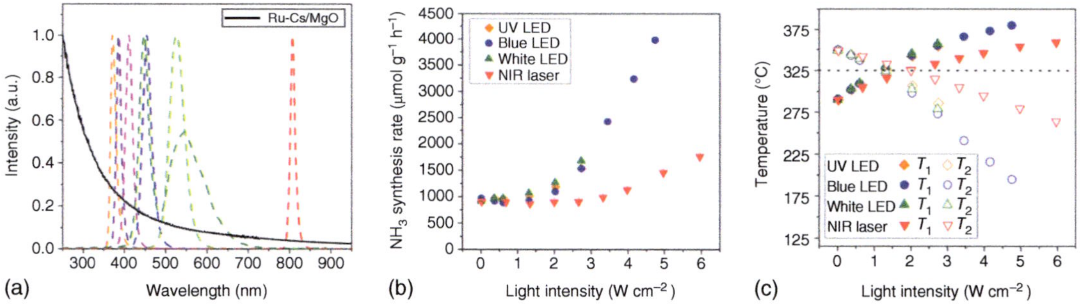

Figure 7.8 NH_3 synthesis and light-induced thermal gradients for $T_e = 325\,°C$. (a) Ru-Cs/MgO absorption spectrum and emission spectra of light sources. The normalized optical spectra of Ru-Cs/MgO (black line) is measured by diffuse reflectance in an integrating sphere. (b) Measured T_1 and T_2 temperatures of the catalyst using UV, blue, and white LEDs and the NIR laser. UV and white LEDs have a maximum intensity of 2.73 and 2.84 W cm^{-2}, respectively. (c) Measured NH_3 synthesis rates as a function of light intensity. Source: Adapted with permission from Ref. [92]. Copyright 2019 American Chemical Society.

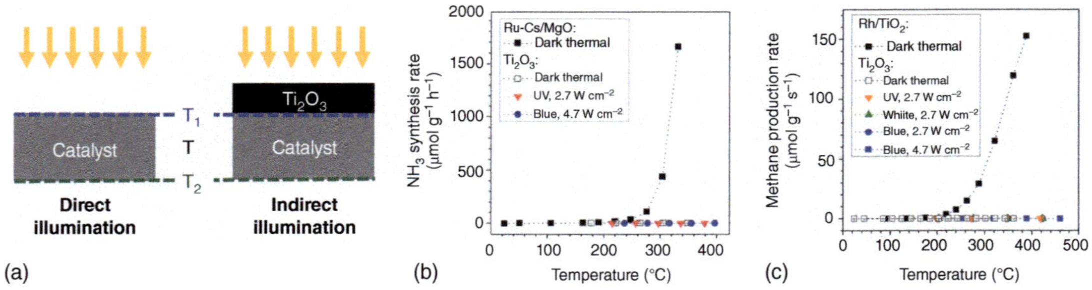

Figure 7.9 Direct and indirect illumination. (a) Schematic representation of photocatalytic reactors under direct and indirect illumination. The T_1 and T_2 are maintained at the same position within the catalyst to ensure identical thermal gradients can be created and compared. Control tests for Ti_2O_3 under reaction conditions for (b) NH_3 synthesis and (c) CO_2 methanation, compared with actual catalysts under dark thermal conditions at the same temperature. Commercial Ti_2O_3 does not produce any products under either dark or illuminated conditions.

For the indirect illumination case, a layer (~1 mm) of Ti_2O_3 was placed on top of the catalyst (Figures 7.9a, 7.10a). This prevents direct illumination of the active catalyst while keeping the measurement locations of the T_1 and T_2 thermocouples unchanged. Upon illumination, the Ti_2O_3 absorbs all the light and acts as a photothermal heater to achieve the same T_1, T_2, T_e, and temperature profile within the catalyst as would occur without the Ti_2O_3 overlayer [60, 92]. As a control test, a ~0.5 mm thick sample of Ti_2O_3 powder was placed on a quartz substrate. Due to the strong broad absorption of Ti_2O_3 [108], no light passed through the sample. Consequently, the ~1 mm Ti_2O_3 overlayer absorbed all light to facilitate indirect illumination of the active catalyst. The NPs in the underlying catalyst were not illuminated and therefore must be at the same temperature as the support.

Under all conditions for both NH_3 synthesis and CO_2 hydrogenation, no products are formed on a full catalyst cup of Ti_2O_3, confirming that it is inactive for both reactions (Figure 7.9b,c). As with the directly illuminated catalyst, photothermal heating of the light absorptive Ti_2O_3 overlayer also produced negative gradients within the catalyst (Figure 7.10b). More importantly, this technique allows one to achieve identical T_1 and T_2 temperatures and thermal gradients through both direct and indirect illumination, so that the thermal and nonthermal effects of light may be compared directly. For the same equivalent temperature of 325 °C and illumination by the blue LED over the range of 0–4.7 W cm^{-2} (Figure 7.10b), similar rate enhancements for NH_3 synthesis were observed with both direct and indirect light illumination of the Ru-Cs/MgO catalyst for equivalent thermal gradients (Figure 7.10c).

This observation confirms that any contributions from nonthermal effects are insignificant in this reaction [67]. Thus, the negative gradient produced through photothermal heating of the top surface must be the dominant factor responsible for the enhanced reaction rates and product yields in light-induced NH_3 synthesis. The results also confirm that the experimental technique of direct and indirect illumination successfully extracts the contribution from photothermal heating in the reaction. In the next section, this experimental technique will be applied on a different reaction system exhibiting both thermal and nonthermal contributions.

7.4 Discriminating Thermal and Nonthermal Effects

Although enhanced reaction rates may arise from thermal contributions, as demonstrated in the last section, plasmon-induced product selectivity remains a feature of plasmonic photocatalysis that cannot be simply explained using thermal effects. For example, the CO_2 hydrogenation reaction on Rh NPs with an Al_2O_3 support produces CH_4 and CO with nearly equal selectivity under purely thermal conditions. But under illumination from blue or UV LEDs, the photocatalytic reactions produced CH_4 with selectivity >98% over CO, with a reaction rate twice that of the thermal catalytic reaction rate at 350 °C [59].

In this section, the CO_2 hydrogenation reaction on Rh-based catalysts will be used to illustrate how the thermal and nonthermal contributions may be deduced accurately [60]. Since the technique presented here may be used for any catalyst layers thicker than the light penetration depth, this section presents a thorough

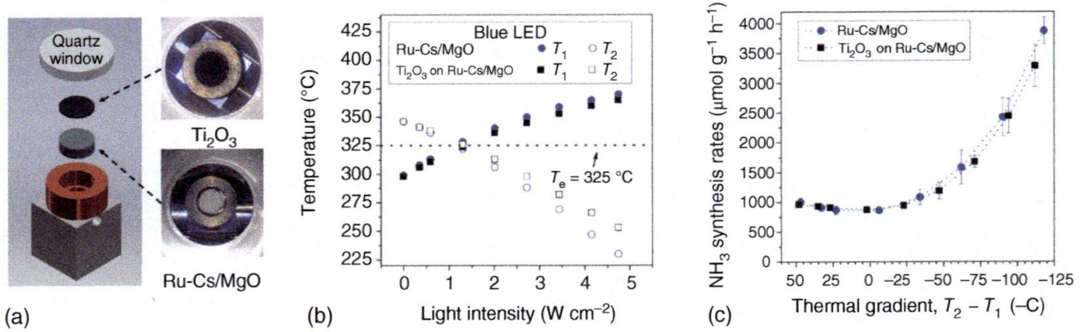

Figure 7.10 Direct and indirect illumination of Ru-Cs/MgO for NH_3 synthesis for $T_e = 325\,°C$. (a) Schematic for direct and indirect photothermal heating along with photographs of the Ru-Cs/MgO catalyst (gray) swithout and with a top layer of Ti_2O_3 (black). (b) Top- and bottom-surface temperatures of the Ru-Cs/MgO catalyst as a function of blue light intensity. (c) Measured NH_3 synthesis rates as a function of the thermal gradient for direct and indirect illumination of Ru-Cs/MgO by the blue LED. *Source*: Adapted with permission from Ref. [92]. Copyright 2019 American Chemical Society.

experimental exploration and discrimination of the thermal and nonthermal catalytic activities of this illustrative reaction and catalyst. To account for thermal gradients and distinguish between thermal and nonthermal catalytic effects experimentally, controlled thermal profiles are created and measured to reveal the effective thermal and nonthermal reaction rates under illumination with and without external heating. Measured dark thermal reaction rates using purely external heating can then be used to approximate the effects of photothermal heating caused by illumination [60, 62]. This section concludes by discussing how the direct and indirect illumination techniques introduced in Section 7.3 can be applied to reactions with both thermal and nonthermal attributes, thereby illustrating how the catalytic behaviors of metal NPs may be controlled by their plasmonic properties.

7.4.1 Rhodium Catalysts for CO_2 Hydrogenation

Supported Rh NPs and molecular Rh complexes are widely used as catalysts [109–114] in hydrogenation, hydroformylation, oxidative coupling, and reduction of nitrogen oxides (in three-way catalytic converters) [109, 112, 114–117]. CO_2 hydrogenation on transition metals at atmospheric pressure proceeds through two main competing pathways: CO_2 methanation ($CO_2 + 4H_2 \rightarrow CH_4 + 2H_2O$) and reverse water gas shift (RWGS, $CO_2 + H_2 \rightarrow CO + H_2O$) [118]. Prior experimental and theoretical investigations of the reaction mechanism on supported Rh and Ru catalysts suggest that CO_2 first dissociatively adsorbs on the catalyst surface as CO and O, and then CO is hydrogenated to CHO [113, 114, 118–122]. The dissociation of CH-O into CH and O is identified to be the RDS of CO_2 methanation, followed by rapid hydrogenation of CH to produce CH_4.

As a noble metal, Rh is extremely stable against oxidative and aqueous environments. This lack of an oxide layer on Rh allows for direct contact between the Rh nanostructures and adsorbates, a property that is essential for efficient charge transfer in plasmonic photocatalysis. The UV plasmonic properties of Rh NPs were only recently discovered, including demonstrations of plasmon-enhanced photocatalysis [59, 84, 97, 123]. The superior catalytic activities, plasmonic properties, and chemical stability of Rh nanostructures promise a compelling system in which to study plasmonic photocatalysis.

Both Rh nanosphere (Rh-s) and nanocube (Rh-c) geometries were considered here, recognizing that the most catalytically actives sites are at corners, edges, and tight curvature surfaces where high adsorbate coordination takes place. To compare the effect of the catalyst host, two metal oxide supports were used for these Rh NP catalysts: alumina (Al_2O_3) and titania (TiO_2). TEM images and XPS analysis of the Rh spheres (Rh-s, usually supported on TiO_2) indicate an average size of only ~6 nm with an Rh loading of ~6 wt% (Figure 7.11a), while the average size of the Rh nanocube catalysts was 37 nm (Rh-c, usually supported on Al_2O_3) with similar loading (Figure 7.11b).

Rh NPs strongly absorb light in the UV region [84, 97, 123, 124], and the LSP resonance peaks of these Rh spheres and Rh cubes in an ethanol suspension reside in the deep UV region (>6 eV) and near 3.71 eV, respectively. When supported on Al_2O_3

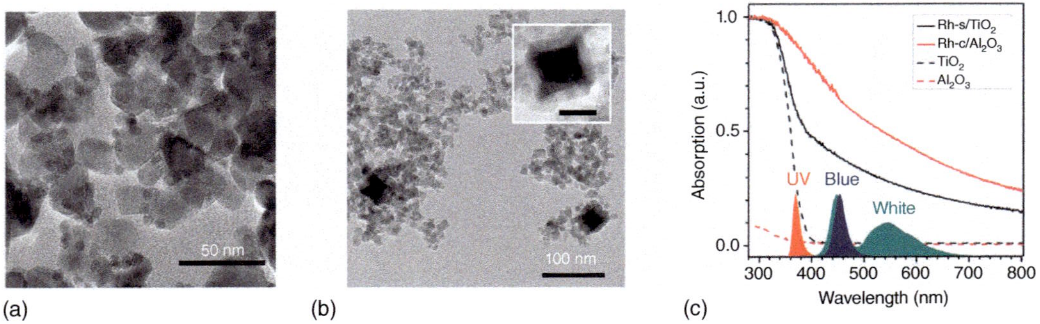

(a) (b) (c)

Figure 7.11 Characterization of Rh-based photocatalyst. TEM images of (a) Rh-s/TiO_2 and (b) Rh-c/Al_2O_3. (c) Spectroscopic characterization of these photocatalysts and LED emission. *Source*: Adapted with permission from Ref. [59, 60]. Copyright 2018 American Chemical Society.

and TiO_2, the peak broadens and blue shifts (Figure 7.11c). In both cases, the broad tail of their plasmon resonance extends well into the blue/violet spectral region and absorbs light from the UV and blue LEDs. The nonresonant excitation of Rh NPs by these LEDs is known to produce hot electrons with near free-electron behavior [59, 124].

The alumina support was transparent for all illumination wavelengths used, but the ~3 eV (413 nm) bandgap of the titania support strongly absorbs light from the UV LED (365 nm) while remaining transparent to the blue LED (455 nm) used. Nevertheless, the corresponding light penetration depth into both Rh-c/Al_2O_3 and Rh-s/TiO_2 photocatalysts is well below 1 μm throughout the UV and visible wavelength region. As with the Ru catalysts considered in the previous section, this means that nonthermal effects can only occur within the illuminated Rh NPs in the surface region, while photothermal heating of the surface can only affect the rest of the catalyst through subsequent heat conduction and diffusion mediated by the unilluminated catalyst and the flowing gases.

High surface-area catalyst supports not only play a passive role in improving the thermal stability of catalytic metal NPs, but they also actively affect the reaction rate and product selectivity. Consistent with previous reports, Rh supported on TiO_2 was found to be more active than that supported on Al_2O_3 for both light and dark conditions [113]. This occurs from modifications in the electronic structures of the metal NPs, strong interactions at the metal-support interface, and interactions with reaction intermediates which alter the energetics of reaction pathways. These effects are more pronounced on reducible metal oxide supports, such as TiO_2. Control experiments on pure TiO_2 and Al_2O_3 as well as isotopic labeling experiments with deuterium indicated that both were inactive for CO_2 hydrogenation under dark and light conditions [59, 60]. CH_4 and CO products were produced from photocatalytic reactions involving Rh rather than from contaminants or supports alone, confirming that the origin of the catalytic activities was from the Rh nanostructures and their interfaces with the support. No other carbon-containing product was detected in these experiments, and the reaction rates responded to light instantly and reversibly.

7.4.2 Plasmonic Photocatalytic Reduction of CO_2

The combined plasmonic and catalytic properties of Rh NPs simultaneously enhance reaction rates, lower activation energies, and produce strong product photoselectivity in the CO_2 hydrogenation reaction [59]. When Rh is fully reduced to its metallic state, CO_2 hydrogenation by Rh-c/Al_2O_3 photocatalysts under UV and blue LED illumination exhibited selectivity toward CH_4 of >98 and >86%, respectively, much higher than the selectivity of ~60% under dark thermal conditions (Figure 7.12a) [60, 125]. These results indicated that photons almost exclusively promoted the production of CH_4 with little perturbation in the production of CO. But is this selectivity due to photothermal or nonthermal effects?

Density functional theory (DFT) calculations suggest that the photoselectivity of Rh photocatalysts can be attributed to the alignment of the hot electron distribution with the antibonding orbital of the critical reaction intermediate, CHO, which activates the CO_2 methanation pathway [113, 114, 118–122, 126, 127]. Consider also the influence of excitation wavelength using varied light sources (Figure 7.12b). At

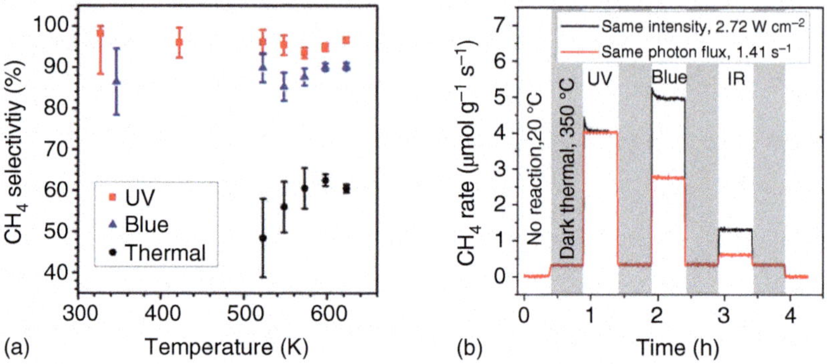

Figure 7.12 CH_4 selectivity and production rate under illuminated and dark conditions using the Rh-c/Al_2O_3 catalyst. (a) CH_4 selectivity is plotted as a function of measured chamber temperature, T_c, under dark, UV (365 nm), and blue (455 nm) illumination. (b) Overall CH_4 production rates under dark thermal conditions, compared to UV, blue, and NIR (805 nm) illumination with the same intensity (black curve) or the same photon flux (red curve). *Source*: Adapted with permission from Ref. [59].

350 °C and illumination intensity is held constant, photoenhanced CH_4 production is observed for all light sources used. Indeed, the rate enhancement was even more significant at lower reaction temperatures (not shown). However, when photon flux is kept constant, the production rate is enhanced to a lesser extent as the excitation wavelength increases. As the mismatch widens between LSP resonance and excitation wavelengths, lower energy photons are increasingly unable to generate hot carriers with the appropriate energy to drive reactions, so the photothermal effect becomes more dominant.

While these initial experimental results and DFT calculations support a hot-carrier mediated mechanism, care must be taken not to overestimate the nonthermal effect through imprecise temperature measurements. Consider Figure 7.13a, which illustrates the dependence of T_1 and T_2 on illumination intensity by a modest intensity UV LED (UV Intensity, $I_{uv} < 3\,W\,cm^{-2}$, 365 nm) with a ~3 mm thick Rh-s/TiO_2 catalyst thermostatically held at T_c = 300 °C and 350 °C using a flow of 50 sccm CO_2, 150 sccm H_2, and 50 sccm Ar gases. The flowing gas produces $T_2 < T_c$ for all light intensities while T_1 increases rapidly from below to well above T_c with increasing light intensity. Temperature gradients >100 °C across a 3 mm thick catalyst are observed at the highest light intensity.

On the illuminated Rh-s/TiO_2 catalyst, where CO_2 methanation is the dominant reaction pathway [119], the CH_4 reaction rates exhibit monotonic increases with increasing I_{uv} for a given T_c (Figure 7.13b) until plateauing (e.g. $I_{uv} \approx 0.6\,W\,cm^{-2}$ for T_c = 350 °C) because the reaction becomes diffusion limited (i.e. CO_2 conversion >10%). The observed photoenhanced reaction rate R_{tot} (Figure 7.13b) includes both "thermal" and "nonthermal" contributions, but how may these be parsed? An important clue comes from the methane/photon ratio which is commonly referred to as the quantum efficiency or quantum yield. This ratio is defined as the difference in the illuminated and dark reaction rates at a common chamber temperature (T_c)

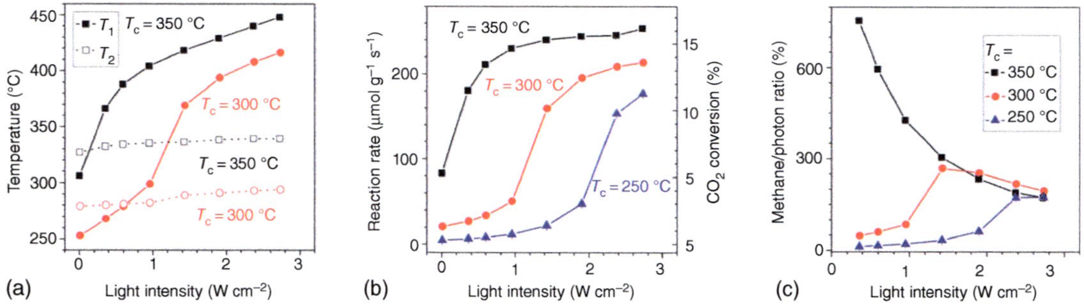

Figure 7.13 CO_2 methanation reactions, carried out with a ~3 mm thick Rh-s/TiO_2 catalyst and 50sccm CO_2, 150 sccm H_2, and 50 sccm Ar. (a) Measured T_1 and T_2 temperatures as a function of I_{uv} for $T_c = 300, 350\,°C$. Under dark thermal conditions, $T_1 < T_2$, whereas illumination increases the surface temperature so that $T_2 > T_1$. (b) Measured total reaction rates and CO_2 conversion for $T_c = 250, 300, 350\,°C$ as a function of I_{uv}. (c) Methane/photon ratios for $T_c = 250, 300, 350\,°C$ as a function of I_{uv}. The methane/photon ratio is calculated as difference between methane production rate under light and dark conditions at the same T_c, divided by photon flux. *Source*: Adapted with permission from Ref. [60]. Copyright 2018 American Chemical Society.

divided by the photon flux [53, 57]

$$\frac{\text{net methane produced}}{\text{photons used}} = \frac{R_{\text{illuminated}} - R_{\text{dark}}}{\text{photon flux}}. \tag{7.4}$$

However, inadequate accounting of the thermal contribution may misrepresent the relationship between nonthermal reaction and photon flux. Using this definition, one might incorrectly deduce a measured net methane/photon ratio as large as ~800% under low intensity UV illumination ($I_{uv} \approx 0.4\,\text{W cm}^{-2}$) with $T_c = 350\,°\text{C}$ (Figure 7.13c). This clearly seems unphysical, as if each photon creates a hot carrier that breaks bonds of as many as eight different intermediate adsorbates at the RDS [128–130]. Furthermore, this ratio decreases with increasing light intensity, as if adding more photons makes the reaction less efficient.

The only explanation is that somehow the subtraction of thermal effects was incomplete. Indeed, the critical assumption that illumination only generates nonthermal contributions is incorrect because it omits the effects of photothermal heating and the associated thermal gradient, effects clearly seen in Figure 7.13a. Thus, to ascertain the extent of any nonthermal contributions, the reaction rate caused by all thermal processes R_t, including photothermal heating, must also be characterized and subtracted, just as the reaction rate for unilluminated thermal heating was.

7.4.3 Unheated, Light-Only Photocatalysis

Consider the limit of unheated, light-only conditions, where the catalyst is only heated by photothermal effects. At low temperatures the thermal reaction rates are negligible, so one may suspect that the measured reaction rate for UV illumination is dominated by nonthermal effects. But without precise temperature profile measurements, the photothermal contributions to plasmon-driven reactions conducted at room temperature may be under- or overestimated. For a given T_e, UV light-only conditions produce more CH_4 than dark thermal conditions (Figure 7.14a); however, these comparisons are incomplete as vastly different thermal gradients exist. For example, when UV light is the sole energy input source, T_1 and T_2 rise to 282 and 94 °C, respectively, producing $T_e = 224\,°\text{C}$ and $\nabla T = -188\,°\text{C}$ for the maximum LED intensity of $I_{uv} = 2.73\,\text{W cm}^{-2}$ (Figure 7.14b). Under dark thermal conditions for the same T_e, the (opposite sign) thermal gradient is only $\nabla T = +27\,°\text{C}$.

When compared with the measured thermal rates ($R_{t,m}$) obtained through the indirect illumination technique for identical T_1 and T_2 temperatures, the calculated thermal rate $R_{t,c}$ for effective temperature T_e accurately reproduces $R_{t,m}$ at most light intensities (Figure 7.14c). Slight deviations occur at higher light intensities in which large light-induced negative gradients exist. In this regime, because $R_{t,c}$ only considers the T_e, it does not account for additional photothermal effects due to an elevated T_1. As a result, under non-isothermal conditions the indirect illumination that generates identical thermal profiles becomes a more reliable method to ascertain the total thermal contribution. It can be seen that for directly illuminated Rh/TiO_2 with $I_{uv} = 2.73\,\text{W cm}^{-2}$, $R_{t,m}$ accounts for only 27.6% of the total measured CH_4 production rate R_{tot}. This indicates that while a portion of the total CH_4 production under unheated UV light conditions can be attributed to thermal effects, the majority is due to nonthermal effects from light.

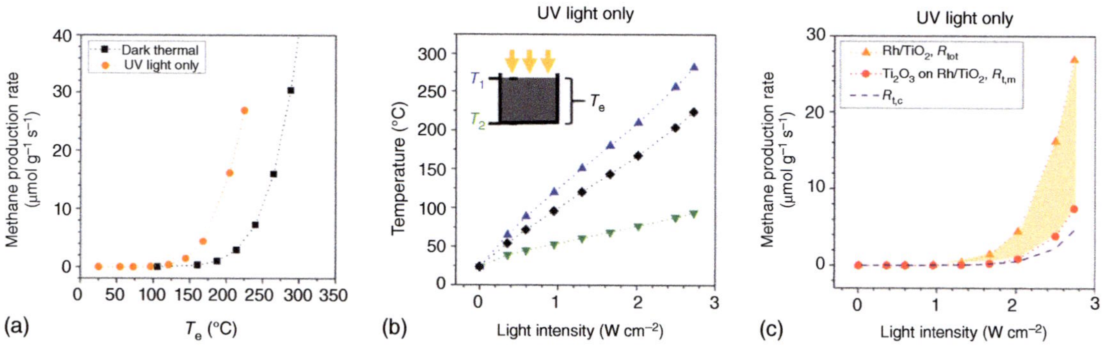

Figure 7.14 Comparison of heated (dark thermal) and unheated (light-only) CO_2 methanation. (a) Measured CH_4 production rates for unheated, UV light-only conditions (orange triangles), and dark thermal conditions (black squares) as a function of T_e on a bare Rh/TiO_2 catalyst. The flow is CO_2 20 sccm, H_2 60 sccm, and Ar 120 sccm for a total 200 sccm. (b) Measured top surface (T_1), bottom surface (T_2), and calculated equivalent (T_e) temperatures as a function of UV intensity light only. (c) Measured total CH_4 production rate (R_{tot}) for unheated, UV light only (orange triangles) is shown as a function of UV light intensity. Calculated ($R_{t,c}$) thermal CH_4 production rates are based on corresponding T_e (purple dashes). Measured ($R_{t,m}$) thermal CH_4 production rate from indirectly illuminated Rh/TiO_2 for identical T_1 and T_2 temperatures (red circles). The yellow shaded region represents the nonthermal contribution. *Source*: Adapted with permission from Ref. [60, 131]. Copyright 2018 American Chemical Society.

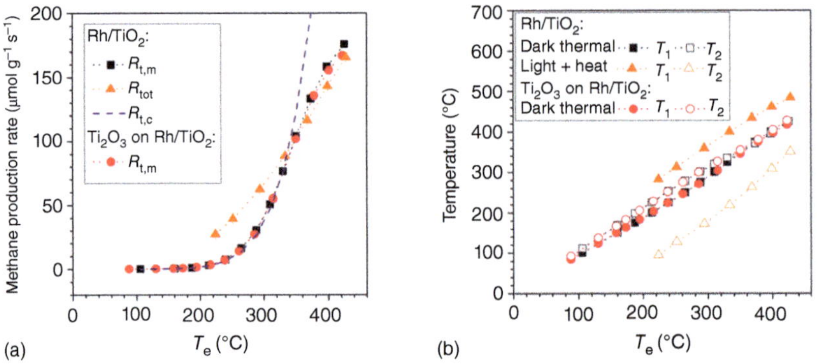

Figure 7.15 CO_2 methanation under dark thermal and illuminated conditions. (a) CH_4 production rates as a function of T_e, including the measured rate for the uncovered Rh/TiO_2 catalyst under dark thermal conditions ($R_{t,m}$, black squares) and under direct illumination by 2.73 W cm^{-2} of UV light with additional external heating (R_{tot}, orange triangles), the measured rate for the covered catalyst under dark thermal conditions ($R_{t,m}$, red circles), and the calculated thermal CH_4 production rate ($R_{t,c}$, purple dashed lines) as a function of T_e. (b) Corresponding measured T_1 and T_2 temperatures for dark thermal and illuminated conditions for covered and uncovered Rh/TiO_2. The flow is CO_2 20 sccm, H_2 60 sccm, and Ar 120 sccm for a total 200 sccm. *Source*: Adapted with permission from Ref. [131].

7.4.4 Light Intensity Dependence of Heated Photocatalysts

Upon illumination with external heating, the observed CH_4 production rate includes thermal, photothermal, and nonthermal contributions. The addition of external heating to illumination reveals a synergistic dependence on light intensity in plasmon-driven reactions. This synergy may be observed using direct and indirect illumination so that the measured dark thermal CH_4 production rate ($R_{t,m}$) of Rh/TiO_2 with identical thermal gradients for covered and uncovered catalysts may be compared over a wide range of T_e temperatures. Identical reaction rates for both (Figure 7.15a) further confirm that Ti_2O_3 is inactive for the reaction.

For a heated Rh/TiO_2 photocatalyst under direct light illumination, heat and light combine to enhance CH_4 production rates for $T_e < {\sim}350\,°C$ (Figure 7.15a). However, at higher temperatures, the measured total CH_4 production rate (R_{tot}) is lower than that of dark thermal conditions at the same T_e. For exothermic reversible reactions such as CO_2 methanation, the optimal temperature is a compromise between kinetic and thermodynamic factors. Although the CH_4 production rate initially grows exponentially with operating temperature, the reverse reaction of CH_4 reforming starts to become more favorable at higher temperatures [132–135]. The effects of this reverse reaction cause the deviation seen in Figure 7.15a for $T_e > {\sim}350\,°C$, and the addition of illumination reduces the rate even more. It is also interesting to note that when $T_e > {\sim}350\,°C$, it is possible for $T_1 > T_2$ even under dark conditions due to the exothermic nature of this reaction (Figure 7.15b). Self-heating of the catalyst renders the thermal profile to be nearly isothermal. As a result, the calculated thermal rate ($R_{t,c}$) becomes unreliable under high intensity illumination where non-isothermal conditions are amplified since $R_{t,c}$ does not consider the magnitude or direction of the thermal gradient and assumes near isothermal conditions.

Next, by maintaining the top surface at a constant temperature T_1 while varying the light intensity and external heating, the total CH_4 production rate and thermal profile may be monitored as a function of I_{uv} and T_2 (Figure 7.16). Under direct illumination, the measured total methane production rate of Rh/TiO_2 shows a characteristic "U" shape due to evolving light intensities and temperatures. At the lowest intensities of light, maximal external heating is required ($T_2 > T_1$), and the reaction is almost completely thermal in nature. The reaction rate decreases with increasing I_{uv} as less external heating is required and T_2 decreases toward T_1. As light intensities increase further and ∇T changes sign, the surface becomes the hottest portion of the catalyst, and R_{tot} increases as photothermal and nonthermal effects begin to dominate the reaction. Figure 7.16 reveals that the shape of the "U" depends on T_1 but that nonthermal effects become significant for $I_{uv} > 0.5\ W\ cm^{-2}$ for all T_1.

Using the methodology of indirect illumination to reproduce and characterize the light intensity-dependent photothermal gradient without superheating the NPs, the relative contributions of thermal and nonthermal effects may confidently be discerned. From the thermal profile, the T_e and corresponding $R_{t,c}$ can be calculated. Indirect illumination with various LEDs and light intensities can reproduce thermal profiles to reveal the thermal portion of the R_{tot} rates produced in the directly illuminated scenarios. Since Ti_2O_3 is inactive for CO_2 methanation and acts solely as a light absorber, the measured CH_4 production rates are independent of light wavelength. For most light intensities, the measured and calculated thermal portions are in close agreement (Figure 7.16). Although calculations of the thermal contribution become less accurate at high intensities where the gradient is increased, the nonthermal light contribution can be obtained by subtracting the measured thermal contribution using $R_{nt} = R_{tot} - R_{t,m}$.

One might argue that this extracted rate still contains a combination of effects from nonthermal hot carriers and superheated NPs. However, the temperature increase caused by plasmonic heating of metal NPs above its environment is negligible on supports with high thermal conductivity [5, 12, 52, 59]. Under direct illumination, any increase in the temperature of Rh NPs quickly equilibrates with the temperature of TiO_2 support. Therefore, the measured T_1 is a reasonable representation of the surface temperature. In addition, these investigations of a purely photothermal system showed that the measured total production rate and corresponding thermal contribution are identical under direct and indirect illumination [92]. Therefore, the experimental approach described here captures the majority of the thermal contribution from light, and the residual reaction rate is mainly due to nonthermal effects. The use of direct and indirect illumination techniques thereby experimentally distinguishes between thermal and nonthermal effects and demonstrates that light is not simply another heat source.

7.4.5 Nonthermal Photocatalytic Behaviors

Once the thermal contribution is appropriately accounted, the extracted nonthermal CH_4 production rate R_{nt} from the Rh-s/TiO_2 photocatalyst can be plotted as

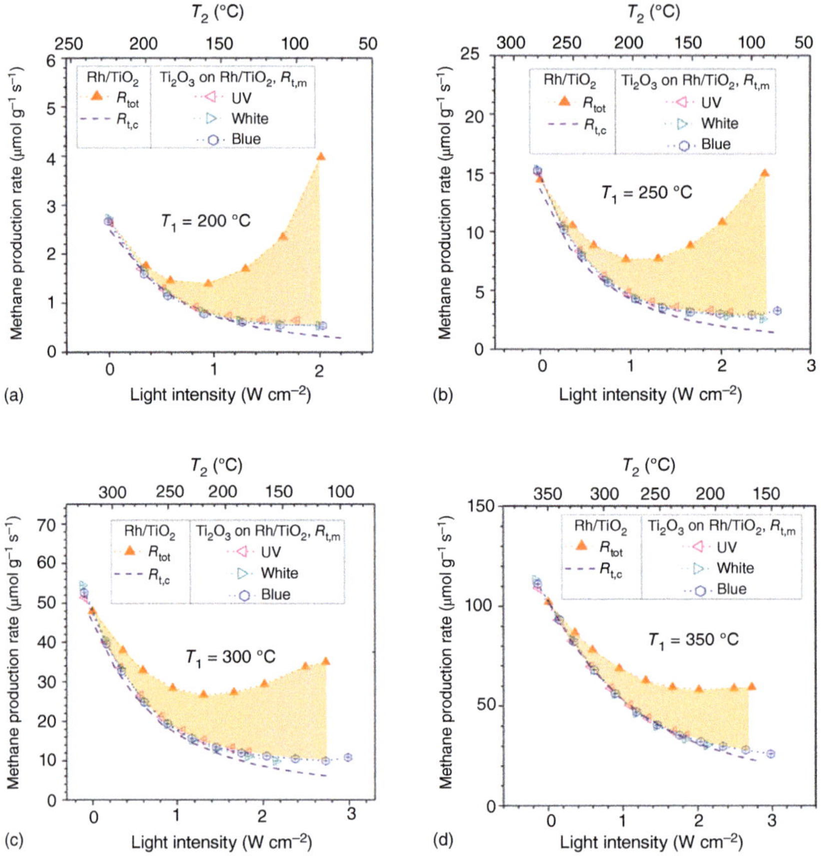

Figure 7.16 CO_2 methanation using direct vs. indirect illumination. For top surface temperature T_1 = (a) 200, (b) 250, (c) 300, and (d) 350 °C, the CH_4 production rates are plotted as a function of UV light intensity and corresponding bottom surface (T_2) temperature. Under direct illumination, measured total rate R_{tot} (orange triangles) and calculated $R_{t,c}$ (purple dashed lines) are shown. Under indirect illumination, measured CH_4 production rates, $R_{t,m}$, are plotted using UV (pink left triangles), white (green right triangles), and blue (navy hexagons) LEDs. The yellow shaded region represents the nonthermal contribution. *Source*: Adapted with permission from Ref. [131].

a function of I_{uv} (Figure 7.17a). For T_1 = 200 and 250 °C, R_{nt} has a superlinear dependence on I_{uv}, presumably due to a hot-carrier mediated process [53, 80, 136]. However, for T_1 = 300 °C, R_{nt} is linear, and for T_1 = 350 °C, R_{nt} becomes sublinear at the highest light intensities as CH_4 reactivity suffers from diffusion limitations and effects of the reverse reaction of CH_4 reforming. Since the nonthermal reaction is limited to the thin submicron surface region penetrable by light, the nonthermal CH_4 production rate per unit catalyst mass is indeed several orders of magnitude higher than the typical thermal CH_4 production rate at the same temperature [62].

As observed in previous investigations [60], the near linearity of R_{nt} with I_{uv} means the apparent quantum efficiency (AQE_{nt}) is fairly independent of I_{uv} (Figure 7.17b). AQE_{nt} is defined as the ratio of the deduced nonthermal rate R_{nt} to the photon flux

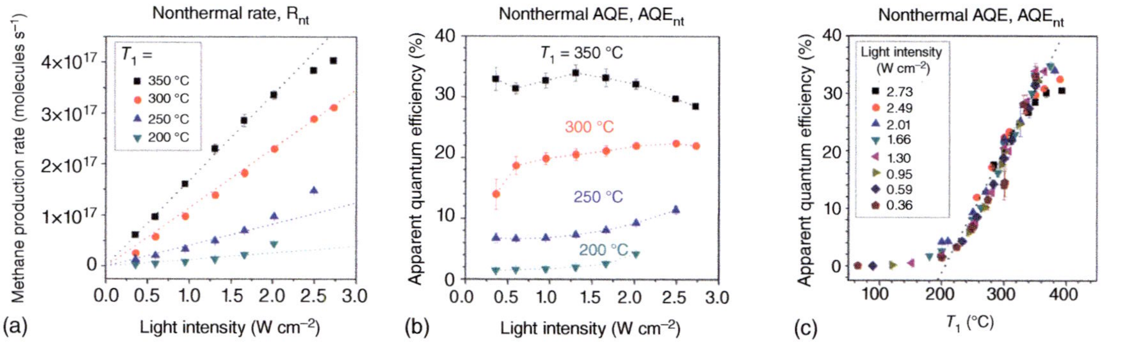

Figure 7.17 Extracted nonthermal reaction rate and apparent quantum efficiency (AQE_{nt}) for the Rh-s/TiO_2 photocatalyst. (a) Nonthermal CH_4 production rate and (b) nonthermal AQE_{nt} as a function of UV light intensity (I_{uv}) for varied top surface (T_1) temperatures. (c) Nonthermal apparent quantum efficiency as a function of T_1 for varied I_{uv}. *Source*: Adapted with permission from Ref. [131].

delivered to catalyst after all thermal effects (R_t) are subtracted out:

$$AQE_{nt} = \frac{R_{tot} - R_t}{\text{photon flux}} = \frac{R_{nt}}{\text{photon flux}}. \tag{7.5}$$

The AQE_{nt}, which may be used to evaluate the efficiency of plasmonic photocatalysis, does not involve, and should be independent of, the total weight (thickness) of catalyst.

Interestingly, the nonthermal rate and AQE_{nt} exhibit a strong dependence on top surface temperature (T_1), revealing the synergistic relationship between heat and light (Figure 7.17c). The $AQE_{nt} \approx 0\%$ at $T_1 \approx 200\,°C$, the same initiation temperature for dark thermal conditions (Figure 7.15a). It is no oxymoron that the nonthermal portion of the reaction depends on temperature, since the nonthermal part is not defined as an athermal contribution but as the part of the illuminated reaction rate that exceeds $R_{t,m}$.

Several light-activated mechanisms may exhibit a temperature dependence. For example, it has been shown that illumination may reduce the activation barrier of the RDS so that at elevated temperatures, the increased relative population of adsorbate excited vibrational states requires less energy gain to overcome the activation barrier [58–60]. In addition, the probability of gaining a specific number of vibrational quanta increases when the molecule is initially in an excited vibrational state [53, 136]. Of course, this linear relationship between R_{nt} and T_1 cannot continue indefinitely, lest AQE_{nt} unphysically exceed 100%. Indeed, as T_1 increases toward $T_e \approx 350\,°C$, AQE_{nt} begins to deviate from the linear trend as the reverse reaction also becomes enhanced by the addition of light to heat.

7.5 Outlook

This chapter has introduced a methodology to distinguish thermal and nonthermal contributions from an illuminated, plasmon-enhanced catalyst. Although nonthermal effects in light-driven reactions are deservedly drawing much attention in the literature, photothermal effects may prove to be even more beneficial because of the way illumination can tailor thermal profiles within a catalyst. The technique presented here extracts the effective thermal and nonthermal reaction rates under illumination by simultaneously measuring the total reaction rate and the top- and bottom-surface temperatures of the catalyst bed. Using these measured temperatures, a simplified model of the catalyst thermal profile and effective thermal reaction rates were deduced for the illuminated catalyst for two reactions. Specifically, monometallic photocatalysts with intrinsic plasmonic and catalytic properties for NH_3 synthesis and CO_2 hydrogenation were used to illustrate this technique and examine thermal, photothermal, and nonthermal reaction rates as well as their effect on product selectivity. Through innovative experimental techniques, the thermal and nonthermal contributions may be systematically evaluated to understand the underlying synergistic mechanisms in plasmonic photocatalysis and extract the nonthermal contribution from the total measured reaction rate.

For the purely thermal reaction presented in Section 7.3 (NH_3 synthesis by Ru-Cs/MgO catalysts), it was shown that thermal gradients can be favorably tuned by varying the light intensity, even without external heating, to enhance both the yield and the reaction rate in the ammonia synthesis reaction. The short penetration depth of light, previously considered a major weakness in photocatalysis, is ideal for creating the desired non-isothermal conditions. The presence of thermal gradients within a catalyst acts as a thermodynamic pump to shift the global equilibrium to improve catalytic activities and product yield simultaneously. This non-isothermal environment enhanced both the reaction rate and yield by balancing the conflicting requirements of kinetics and thermodynamics, heralding the use of optically controlled thermal gradients as a universal, scalable strategy for the catalysis of many exothermic chemical reactions.

Next, it was shown that the thermal contribution of a directly illuminated catalyst can be captured through measurements of the reaction rate produced by indirectly illuminated catalysts with identical thermal profiles. Comparison of light and dark conditions with identical thermal gradients under direct and indirect illumination confirmed that surface photothermal heating is the dominant factor for enhanced reaction rates in light-driven NH_3 synthesis.

Third, for a reaction that has both thermal and nonthermal contributions (CO_2 hydrogenation by Rh-c/Al_2O_3 or Rh-s/TiO_2 catalysts), direct and indirect illumination was used to demonstrate the nonthermal production of CH_4 and CO from CO_2 and H_2. This strategy of direct and indirect light illumination can be universally applied to untangle intertwined thermal and nonthermal effects and to enhance yield and reaction rates for other exothermic reactions under mild ambient conditions, thus simplifying reactor designs and reducing energy costs. After properly accounting for the thermal contribution, photoenhanced CH_4 selectivity was confirmed in the nonthermal reaction, and the apparent activation energy was reduced by a hot-carrier mediated process.

Significant enhancement of the reaction rate and reduced activation energies indicate that light and heat can work synergistically in plasmonic photocatalysis. The overall thermal and photothermal contribution of a directly illuminated catalyst was captured through measurements of the reaction rate produced by indirectly illuminated catalysts with identical thermal profiles. Interestingly, both the nonthermal rate and AQE_{nt} show a striking dependence on the top surface temperature. At low to moderate temperatures, light and heat work synergistically to accelerate CH_4 production. However, past a threshold temperature of $\sim$350 °C, heat begins to affect the light-driven reaction negatively as the reverse reaction of CH_4 reforming is also enhanced.

As the field of plasmonic photocatalysis continues to grow, it becomes increasingly important to develop a fundamental understanding of the underlying mechanisms to avoid misinterpretations of experimental results. Under illumination, the thermal contribution to the overall reaction includes traditional heating and photothermal heating effects, while the nonthermal contribution remains as an umbrella term representing plasmonic effects that cannot be achieved under thermal conditions. By carefully quantifying the thermal effects, this approach provides first step for understanding the nonthermal effects in plasmonic photocatalysis. Moving forward, *in situ* techniques such as infrared and ultrafast spectroscopic surface analysis are required to capture discrete adsorbates/products formed dynamically on the surface. These insights can aid the development of microkinetic models to understand the true meaning of changes in the apparent activation energy in plasmonic photocatalysis. In addition, light-coupled environmental TEM can be used to explore the influence of particle structure on photocatalytic behavior at the sub-NP level *in situ* and in real time.

The promise of plasmonic photocatalysis rests in its capability for precise control over product selectivity beyond what is possible in traditional thermal catalysis. Fulfilling this promise requires both experimental and theoretical efforts to identify desired structures and guide the rational design of plasmonic photocatalysts. One day, one may envision that the entire solar spectrum will be used to its full photocatalytic potential, with a portion of the spectrum providing photothermal heating while another promoting tailored nonthermal enhancements to the overall reaction system.

References

1 Sutter, P., Li, Y., Argyropoulos, C., and Sutter, E. (2017). In situ electron microscopy of plasmon-mediated nanocrystal synthesis. *J. Am. Chem. Soc.* 139: 6771–6776.

2 Cao, L., Barsic, D.N., Guichard, A.R., and Brongersma, M.L. (2007). Plasmon-assisted local temperature control to pattern individual semiconductor nanowires and carbon nanotubes. *Nano Lett.* 7 (11): 3523–3527.

3 Robert, H.M.L., Kundrat, F., Bermúdez-Ureña, E. et al. (2016). Light-assisted solvothermal chemistry using plasmonic nanoparticles. *ACS Omega* 1 (1): 2–8.

4 Kamarudheen, R., Castellanos, G.W., Kamp, L.P.J. et al. (2018). Quantifying photothermal and hot charge carrier effects in plasmon-driven nanoparticle syntheses. *ACS Nano*.

5 Govorov, A.O. and Richardson, H.H. (2007). Generating heat with metal nanoparticles. *Nano Today* 2 (1): 30–38.

6 Ahmad, M., Anguita, J.V., Stolojan, V. et al. (2013). Efficient coupling of optical energy for rapid catalyzed nanomaterial growth: high-quality carbon nanotube synthesis at low substrate temperatures. *ACS Appl. Mater. Interfaces* 5 (9): 3861–3866.

7 Qiu, J. and Wei, W.D. (2014). Surface plasmon-mediated photothermal chemistry. *J. Phys. Chem. C* 118 (36): 20735–20749.

8 Wang, F., Huang, Y., Chai, Z. et al. (2016). Photothermal-enhanced catalysis in core-shell plasmonic hierarchical Cu_7S_4 microsphere@zeolitic imidazole framework-8. *Chem. Sci.* 7 (12): 6887–6893.

9 Wang, S.-S., Hu, W.-C., Liu, F.-F. et al. (2019). Insights into direct plasmon-activated electrocatalysis on gold nanostar via efficient photothermal effect and reduced activation energy. *Electrochim. Acta* 301: 359–365.

10 Zhou, Y., Doronkin, D.E., Zhao, Z. et al. (2018). Photothermal Catalysis over Nonplasmonic Pt/TiO_2 Studied by Operando HERFD-XANES, Resonant XES, and DRIFTS. *ACS Catal.*: 11398–11406.

11 Baffou, G. and Quidant, R. (2013). Thermo-plasmonics: using metallic nanostructures as nano-sources of heat. *Laser Photon. Rev.* 7 (2): 171–187.

12 Baffou, G., Quidant, R., and García de Abajo, F.J. (2010). Nanoscale control of optical heating in complex plasmonic systems. *ACS Nano* 4: 709–716.

13 Chen, X., Chen, Y., Yan, M., and Qiu, M. (2012). Nanosecond photothermal effects in plasmonic nanostructures. *ACS Nano* 6 (3): 2550–2557.

14 Roper, K.D., Ahn, W., and Hoepfner, M. (2007). Microscale heat transfer transduced by surface plasmon resonant gold nanoparticles. *J. Phys. Chem. C* 111: 3636–3641.

15 Chen, H., Shao, L., Ming, T. et al. (2010). Understanding the photothermal conversion efficiency of gold nanocrystals. *Small* 6 (20): 2272–2280.

16 Cheemalapati, S., Ladanov, M., Pang, B. et al. (2016). Dynamic visualization of photothermal heating by gold nanocages using thermoresponsive elastin like polypeptides. *Nanoscale* 8 (45): 18912–18920.

17 Jain, P.K., Huang, X., El-Sayed, I.H., and El-Sayed, M.A. (2008). Noble metals on the nanoscale: optical and photothermal properties and some applications in imaging, sensing, biology, and medicine. *Acc. Chem. Res.* 41 (12): 1578–1586.

18 Huang, X., Jain, P.K., El-Sayed, I.H., and El-Sayed, M.A. (2008). Plasmonic photothermal therapy (PPTT) using gold nanoparticles. *Lasers Med. Sci.* 23 (3): 217–228.

19 Bakhtiari, A.B., Hsiao, D., Jin, G. et al. (2009). An efficient method based on the photothermal effect for the release of molecules from metal nanoparticle surfaces. *Angew. Chem. Int. Ed. Engl.* 48 (23): 4166–4169.

20 Gao, M., Connor, P.K.N., and Ho, G.W. (2016). Plasmonic photothermic directed broadband sunlight harnessing for seawater catalysis and desalination. *Energ. Environ. Sci.* 9 (10): 3151–3160.

21 Zhou, L., Tan, Y., Ji, D. et al. (2016). Self-assembly of highly efficient, broadband plasmonic absorbers for solar steam generation. *Sci. Adv.* 2 (4): e1501227.

22 Wang, X., He, Y., Liu, X. et al. (2017). Investigation of photothermal heating enabled by plasmonic nanofluids for direct solar steam generation. *Solar Energy.* 157: 35–46.

23 Guo, A., Fu, Y., Wang, G., and Wang, X. (2017). Diameter effect of gold nanoparticles on photothermal conversion for solar steam generation. *RSC Adv.* 7 (8): 4815–4824.

24 Boyd, D.A., Greengard, L., Brongersma, M.L. et al. (2006). Plasmon-assisted chemical vapor deposition. *Nano Lett.* 6 (11): 2592–2597.

25 Adleman, J.R., Boyd, D.A., Goodwin, D.G., and Psaltis, D. (2009). Heterogenous catalysis mediated by plasmon heating. *Nano Lett.* 9 (12): 4417–4423.

26 Wang, F., Li, C., Chen, H. et al. (2013). Plasmonic harvesting of light energy for Suzuki coupling reactions. *J. Am. Chem. Soc.* 135 (15): 5588–5601.

27 Fasciani, C., Alejo, C.J.B., Grenier, M. et al. (2011). High-temperature organic reactions at room temperature using plasmon excitation: decomposition of dicumyl peroxide. *Org. Lett.* 13 (2): 204–207.

28 Qiu, J., Wu, Y.C., Wang, Y.C. et al. (2013). Surface plasmon mediated chemical solution deposition of gold nanoparticles on a nanostructured silver surface at room temperature. *J. Am. Chem. Soc.* 135 (1): 38–41.

29 Meng, X., Wang, T., Liu, L. et al. (2014). Photothermal conversion of CO(2) into CH(4) with H(2) over Group VIII nanocatalysts: an alternative approach for solar fuel production. *Angew. Chem. Int. Ed. Engl.* 53 (43): 11478–11482.

30 Linic, S., Christopher, P., and Ingram, D.B. (2011). Plasmonic-metal nanostructures for efficient conversion of solar to chemical energy. *Nat. Mater.* 10 (12): 911–921.

31 Kim, C., Suh, B.L., Yun, H. et al. (2017). Suface plasmon aided ethanol dehydrogenation using Ag-Ni binary nanoparticles. *ACS Catal.* 7: 2294–2302.

32 Robatjazi, H., Zhao, H., Swearer, D.F. et al. (2017). Plasmon-induced selective carbon dioxide conversion on earth-abundant aluminum-cuprous oxide antenna-reactor nanoparticles. *Nat. Commun.* 8 (1): 27.

33 Tada, S., Watanabe, F., Kiyota, K. et al. (2017). Ag addition to CuO-ZrO_2 catalysts promotes methanol synthesis via CO_2 hydrogenation. *J. Catal.* 351: 107–118.

34 Rani, S., Bao, N., and Roy, S.C. (2014). Solar spectrum photocatalytic conversion of CO_2 and water vapor into hydrocarbons using TiO_2 nanoparticle membranes. *Appl. Surf. Sci.* 289: 203–208.

35 Xu, Z., Kibria, M.G., AlOtaibi, B. et al. (2018). Towards enhancing photocatalytic hydrogen generation: which is more important, alloy synergistic effect or plasmonic effect? *Appl Catal B* 221: 77–85.

36 Stanley, J.N.G., Garcia-Garcia, I., Perfrement, T. et al. (2019). Plasmonic effects on CO_2 reduction over bimetallic Ni-Au catalysts. *Chem. Eng. Sci.* 194: 94–104.

37 Cui, J., Li, Y., Liu, L. et al. (2015). Near-infrared plasmonic-enhanced solar energy harvest for highly efficient photocatalytic reactions. *Nano Lett.* 15 (10): 6295–6301.

38 Guo, J., Zhang, Y., Shi, L. et al. (2017). Boosting hot electrons in hetero-superstructures for plasmon-enhanced catalysis. *J. Am. Chem. Soc.* 139: 17964–17972.

39 Salmon-Gamboa, J.U., Romero-Gomez, M., Roth, D.J. et al. (2019). Optimizing hot carrier effects in Pt-decorated plasmonic heterostructures. *Faraday Discuss.*

40 Zhang, W., Wang, L., Wang, K. et al. (2017). Integration of photothermal effect and heat insulation to efficiently reduce reaction temperature of CO_2 hydrogenation. *Small* 13 (7).

41 Yang, Q., Xu, Q., Yu, S.-H., and Jiang, H.-L. (2016). Pd nanocubes@ZIF-8: integration of plasmon-driven photothermal conversion with a metal-organic framework for efficient and selective catalysis. *Angew. Chem.* 128: 3749–3753.

42 Robatjazi, H., Weinberg, D., Swearer, D.F. et al. (2019). Metal-organic frameworks tailor the properties of aluminum nanocrystals. *Sci. Adv.* 5: eaav5340.

43 Lee, J.-E., Mota, F.M., Choi, C.H. et al. (2019). Plasmon-enhanced electrocatalytic properties of rationally designed hybrid nanostructures at a catalytic interface. *Adv. Mater.* 6: 1801144.

44 Lim, D.K., Barhoumi, A., Wylie, R.G. et al. (2013). Enhanced photothermal effect of plasmonic nanoparticles coated with reduced graphene oxide. *Nano Lett.* 13 (9): 4075–4079.

45 Kumar, D., Lee, A., Lee, T. et al. (2016). Ultrafast and efficient transport of hot plasmonic electrons by graphene for Pt free, highly efficient visible-light responsive photocatalyst. *Nano Lett.* 16 (3): 1760–1767.

46 Zhao, H., Zheng, X., Feng, X., and Li, Y. (2018). CO_2 reduction by plasmonic Au nanoparticle-decorated TiO_2 photocatalyst with an ultrathin Al_2O_3 interlayer. *J. Phys. Chem. C* 122: 18949–18956.

47 Keller, E.L., Kang, H., Haynes, C.L., and Frontiera, R.R. (2018). Effect of silica supports on plasmonic heating of molecular adsorbates as measured by ultrafast surface-enhanced raman thermometry. *ACS Appl. Mater. Interfaces* 10: 40577–40584.

48 Gutierrez, Y., Ortiz, D., Sanz, J.M. et al. (2016). How an oxide shell affects the ultraviolet plasmonic behavior of Ga, Mg, and Al nanostructures. *Opt. Express* 24 (18): 20621–20631.

49 Mankidy, B.D., Joseph, B., and Gupta, V.K. (2013). Photo-conversion of CO_2 using titanium dioxide: enhancements by plasmonic and co-catalytic nanoparticles. *Nanotechnology* 24: 405402.

50 Li, Q., Zhang, W., Zhao, D., and Qiu, M. (2014). photothermal enhancement in core-shell structured plasmonic nanoparticles. *Plasmonics* 9 (3): 623–630.

51 Watanabe, K., Menzel, D., Nilius, N., and Freund, H.-J. (2006). Photochemistry on metal nanoparticles. *Chem. Rev.* 106 (10): 4301–4320.

52 Christopher, P., Xin, H., and Linic, S. (2011). Visible-light-enhanced catalytic oxidation reactions on plasmonic silver nanostructures. *Nat. Chem.* 3 (6): 467–472.

53 Christopher, P., Xin, H., Marimuthu, A., and Linic, S. (2012). Singular characteristics and unique chemical bond activation mechanisms of photocatalytic reactions on plasmonic nanostructures. *Nat. Mater.* 11 (12): 1044–1050.

54 Mukherjee, S., Libisch, F., Large, N. et al. (2013). Hot electrons do the impossible: plasmon-induced dissociation of H_2 on Au. *Nano Lett.* 13 (1): 240–247.

55 Linic, S., Aslam, U., Boerigter, C., and Morabito, M. (2015). Photochemical transformations on plasmonic metal nanoparticles. *Nat. Mater.* 14 (6): 567–576.

56 Wu, B., Lee, J., Mubeen, S. et al. (2016). Plasmon-mediated photocatalytic decomposition of formic acid on palladium nanostructures. *Adv Opt Mater.* 4 (7): 1041–1046.

57 Mukherjee, S., Zhou, L., Goodman, A.M. et al. (2014). Hot-electron-induced dissociation of H_2 on gold nanoparticles supported on SiO_2. *J. Am. Chem. Soc.* 136 (1): 64–67.

58 Zhou, L., Swearer, D.F., Zhang, C. et al. (2018). Quantifying hot carrier and thermal contributions in plasmonic photocatalysis. *Science* 362 (6410): 69–72.

59 Zhang, X., Li, X., Zhang, D. et al. (2017). Product selectivity in plasmonic photocatalysis for carbon dioxide hydrogenation. *Nat. Commun.* 8: 14542.

60 Zhang, X., Li, X., Reish, M.E. et al. (2018). Plasmon-enhanced catalysis: distinguishing thermal and nonthermal effects. *Nano Lett.* 18 (3): 1714–1723.

61 Li, H., Rivallan, M., Thibault-Starzyk, F. et al. (2013). Effective bulk and surface temperatures of the catalyst bed of FT-IR cells used for in situ and operando studies. *Phys. Chem. Chem. Phys.* 15 (19): 7321–7327.

62 Sivan, Y., Un, I.W., and Dubi, Y. (2018). Assistance of metal nanoparticles to photo-catalysis - nothing more than a classical heat source. *Faraday Discuss.*

63 Dubi, Y. and Sivan, Y. (2019). "Hot" electrons in metallic nanostructures-non-thermal carriers or heating? *Light Sci. Appl.* 8: 89.

64 Sivan, Y., Baraban, J.H., and Dubi, Y. (2020). Experimental practices required to isolate thermal effects in plasmonic photo-catalysis: lessons from recent experiments. *OSA Continuum.* 3 (3).

65 Jain, P.K. (2019). Taking the heat off of plasmonic chemistry. *J. Phys. Chem. C* 123 (40): 24347–24351.

66 Sivan, Y., Baraban, J., Un, I.W., and Dubi, Y. (2019). Comment on "quantifying hot carrier and thermal contributions in plasmonic photocatalysis". *Science.*

67 Sivan, Y. and Chu, S.-W. (2017). Nonlinear plasmonics at high temperatures. *Nanophotonics* 6 (1): 317–328.

68 Boltersdorf, J., Forcherio, G.T., McClure, J.P. et al. (2018). Visible light-promoted plasmon resonance to induce "hot" hole transfer and photothermal conversion for catalytic oxidation. *J. Phys. Chem. C* 122 (50): 28934–28948.

69 Dubi, Y., Un, I.W., and Sivan, Y. (2020). Thermal effects – an alternative mechanism for plasmon-assisted photocatalysis. *Chem. Sci.* 11 (19): 5017–5027.

70 Baffou, G., Bordacchini, I., Baldi, A., and Quidant, R. (2020). Simple experimental procedures to distinguish photothermal from hot-carrier processes in plasmonics. *Light Sci. Appl.* 9: 108.

71 Zhan, C., Liu, B.W., Huang, Y.F. et al. (2019). Disentangling charge carrier from photothermal effects in plasmonic metal nanostructures. *Nat. Commun.* 10 (1): 2671.

72 Swearer, D.F., Zhao, H., Zhou, L. et al. (2016). Heterometallic antenna-reactor complexes for photocatalysis. *Proc. Natl. Acad. Sci. U. S. A.* 113 (32): 8916–8920.

73 Brites, C.D., Lima, P.P., Silva, N.J. et al. (2012). Thermometry at the nanoscale. *Nanoscale* 4 (16): 4799–4829.

74 Brites, C.D., Lima, P.P., Silva, N.J. et al. (2010). A luminescent molecular thermometer for long-term absolute temperature measurements at the nanoscale. *Adv. Mater.* 22 (40): 4499–4504.

75 Menges, F., Mensch, P., Schmid, H. et al. (2016). Temperatrue mapping of operating nanoscale devices by scanning probe thermometry. *Nat. Commun.* 7: 10874.

76 Pozzi, E.A., Zrimsek, A.B., Lethiec, C.M. et al. (2015). Evaluating single-molecule Stokes and anti-Stokes SERS for nanoscale thermometry. *J. Phys. Chem. C* 119: 21116–21124.

77 Zeng, Z.-C., Wang, H., Johns, P. et al. (2017). Photothermal microscopy of coupled nanostructures and the impact of nanoscale heating in surface-enhanced Raman spectroscopy. *J. Phys. Chem. C* 121: 11623–11631.

78 Xie, X. and Cahill, D.G. (2016). Thermometry of plasmonic nanostructures by anti-Stokes electronic Raman scattering. *Appl. Phys. Lett.* 109: 183104.

79 Kale, M.J., Avanesian, T., and Christopher, P. (2014). Direct photocatalysis by Plasmonic nanostructures. *ACS Catal.* 4 (1): 116–128.

80 Olsen, T. and Schiøtz, J. (2009). Origin of power laws for reactions at metal surfaces mediated by hot electrons. *Phys. Rev. Lett.* 103 (23): 238301.

81 Garnett, J.C.M. (1906). Colours in metal glasses, in metallic films, and in metallic solutions. *II. Philos. Trans. R Soc. A Math. Phys. Eng. Sci.* 205 (387-401): 237–288.

82 DeVore, J.R. (1951). Refractive indices of rutile and sphalerite. *J. Opt. Soc. Am.* 41 (6): 416–419.

83 Palik, E. and Bennett, J.M. (1995). *Handbook of Optical Constants of Solids I and II*. New York: Academic Press.

84 Watson, A.M., Zhang, X., Alcaraz de la Osa, R. et al. (2015). Rhodium nanoparticles for ultraviolet plasmonics. *Nano Lett.* 15 (2): 1095–1100.

85 Marimuthu, A., Zhang, J., and Linic, S. (2013). Tuning selectivity in propylene epoxidation by plasmon mediated photo-switching of Cu oxidation state. *Science* 339 (6127): 1590–1593.

86 Kitano, M., Inoue, Y., Yamazaki, Y. et al. (2012). Ammonia synthesis using a stable electride as an electron donor and reversible hydrogen store. *Nat. Chem.* 4 (11): 934–940.

87 Wu, S., Chen, J., Zheng, X. et al. (2003). Novel preparation of nanocrystalline magnesia-supported caesium-promoted ruthenium catalyst with high activity for ammonia synthesis. *Chem. Commun.* (19): 2488–2489.

88 Mao, C., Yu, L., Li, J. et al. (2018). Energy-confined solar thermal ammonia synthesis with $K/Ru/TiO_{2-x}H_x$. *Appl Catal B* 224: 612–620.

89 Kitano, M., Kanbara, S., Inoue, Y. et al. (2015). Electride support boosts nitrogen dissociation over ruthenium catalyst and shifts the bottleneck in ammonia synthesis. *Nat. Commun.* 6: 6731.

90 Lu, Y., Li, J., Tada, T. et al. (2016). Water durable electride Y_5Si_3: electronic structure and catalytic activity for ammonia synthesis. *J. Am. Chem. Soc.* 138 (12): 3970–3973.

91 Hindasageri, V., Vedula, R.P., and Prabhu, S.V. (2013). Thermocouple error correction for measuring the flame temperature with determination of emissivity and heat transfer coefficient. *Rev. Sci. Instrum.* 84 (2): 024902.

92 Li, X., Zhang, X., Everitt, H.O., and Liu, J. (2019). Light-induced thermal gradients in ruthenium catalysts significantly enhance ammonia production. *Nano Lett.* 19 (3): 1706–1711.

93 Yu, S., Wilson, A.J., Kumari, G. et al. (2017). Opportunities and challenges of solar-energy-driven carbon dioxide to fuel conversion with plasmonic catalysts. *ACS Energy Lett.* 2 (9): 2058–2070.

94 Lundt, N., Kelly, S.T., Rodel, T. et al. (2013). High spatial resolution Raman thermometry analysis of TiO_2 microparticles. *Rev. Sci. Instrum.* 84 (10): 104906.
95 Sadat, S., Meyhofer, E., and Reddy, P. (2012). High resolution resistive thermometry for micro/nanoscale measurements. *Rev. Sci. Instrum.* 83 (8): 084902.
96 Yue, Y. and Wang, X. (2012). Nanoscale thermal probing. *Nanotechnol. Rev.* 3.
97 Sanz, J.M., Ortiz, D., Alcaraz de la Osa, R. et al. (2013). UV plasmonic behavior of various metal nanoparticles in the near- and far-field regimes: geometry and substrate effects. *J. Phys. Chem. C* 117 (38): 19606–19615.
98 Li, C., Wang, T., Zhao, Z.J. et al. (2018). Promoted fixation of molecular nitrogen with surface oxygen vacancies on plasmon-enhanced TiO_2 photoelectrodes. *Angew. Chem. Int. Ed. Engl.* 57 (19): 5278–5282.
99 Yang, J., Guo, Y., Jiang, R. et al. (2018). High-efficiency "working-in-tandem" nitrogen photofixation achieved by assembling plasmonic gold nanocrystals on ultrathin titania nanosheets. *J. Am. Chem. Soc.* 140 (27): 8497–8508.
100 Swearer, D.F., Knowles, N.R., Everitt, H.O., and Halas, N.J. (2019). Light-driven chemical looping for ammonia synthesis. *ACS Energy Lett.* 4 (7): 1505–1512.
101 Aika, K. and Tamara, K. (1995). Ammonia synthesis over non-iron catalysts and related phenomena. In: *Ammonia Catalysis and Manufacture* (ed. A. Nielsen). Berlin, Heidelberg: Springer.
102 Jennings, J.R. (1991). *Catalytic Ammonia Synthesis: Fundamentals and Practice.* New York: Plenum Press.
103 Larichev, Y.V., Moroz, B.L., Zaikovskii, V.I. et al. (2007). XPS and TEM studies on the role of the support and alkali promoter in Ru/MgO and Ru-Cs+/MgO catalysts for ammonia synthesis. *J. Phys. Chem. C* 111: 9427–9436.
104 Larichev, Y.V. (2011). Effect of Cs+ promoter in Ru/MgO catalysts. *J. Phys. Chem. C* 115: 631–635.
105 Hogan, N.J., Urban, A.S., Ayala-Orozco, C. et al. (2014). Nanoparticles heat through light localization. *Nano Lett.* 14 (8): 4640–4645.
106 Szmigiel, D., Bielawa, H., Kurtz, M. et al. (2002). The kinetics of ammonia synthesis over ruthenium-based catalysts: the role of barium and cesium. *J. Catal.* 205 (1): 205–212.
107 Duhr, S. and Braun, D. (2006). Why molecules move along a temperature gradient. *Proc. Natl. Acad. Sci. U. S. A.* 103 (52): 19678–19682.
108 Wang, J., Li, Y., Deng, L. et al. (2017). High-performance photothermal conversion of narrow-bandgap Ti_2O_3 nanoparticles. *Adv. Mater.* 29 (3): 1603730.
109 Xie, S., Liu, X.Y., and Xia, Y. (2015). Shape-controlled syntheses of rhodium nanocrystals for the enhancement of their catalytic properties. *Nano Res.* 8 (1): 82–96.
110 Chambers, M.B., Wang, X., Elgrishi, N. et al. (2015). Photocatalytic carbon dioxide reduction with rhodium-based catalysts in solution and heterogenized within metal–organic frameworks. *ChemSusChem* 8 (4): 603–608.
111 Hou, C., Zhao, G., Ji, Y. et al. (2014). Hydroformylation of alkenes over rhodium supported on the metal-organic framework ZIF-8. *Nano Res.* 7 (9): 1364–1369.

112 Zhang, Y., Grass, M.E., Kuhn, J.N. et al. (2008). Highly selective synthesis of catalytically active monodisperse rhodium nanocubes. *J. Am. Chem. Soc.* 130 (18): 5868–5869.

113 Solymosi, F., Erdöhelyi, A., and Bánsági, T. (1981). Methanation of CO_2 on supported rhodium catalyst. *J. Catal.* 68 (2): 371–382.

114 Sexton, B.A. and Somorjai, G.A. (1977). The hydrogenation of CO and CO_2 over polycrystalline rhodium: correlation of surface composition, kinetics and product distributions. *J. Catal.* 46 (2): 167–189.

115 Satoh, T. and Miura, M. (2010). Oxidative coupling of aromatic substrates with alkynes and alkenes under rhodium catalysis. *Chem. A Eur. J.* 16 (37): 11212–11222.

116 Shen, X. and Buchwald, S.L. (2010). Rhodium-catalyzed asymmetric intramolecular hydroamination of unactivated alkenes. *Angew. Chem. Int. Ed.* 49 (3): 564–567.

117 Knowles, W.S. and Sabacky, M.J. (1968). Catalytic asymmetric hydrogenation employing a soluble, optically active, rhodium complex. *Chem. Commun. (Camb)* (22): 1445–1446.

118 Avanesian, T., Gusmão, G.S., and Christopher, P. (2016). Mechanism of CO_2 reduction by H_2 on Ru(0001) and general selectivity descriptors for late-transition metal catalysts. *J. Catal.*

119 Karelovic, A. and Ruiz, P. (2013). Mechanistic study of low temperature CO_2 methanation over Rh/TiO_2 catalysts. *J. Catal.* 301: 141–153.

120 Jacquemin, M., Beuls, A., and Ruiz, P. (2010). Catalytic production of methane from CO_2 and H_2 at low temperature: insight on the reaction mechanism. *Catal Today* 157 (1–4): 462–466.

121 Williams, K.J., Boffa, A.B., Salmeron, M. et al. (1991). The kinetics of CO_2 hydrogenation on a Rh foil promoted by titania overlayers. *Catal. Lett.* 9 (5-6): 415–426.

122 Henderson, M.A. and Worley, S.D. (1985). An infrared study of the hydrogenation of carbon dioxide on supported rhodium catalysts. *J. Phys. Chem.* 89: 1417–1423.

123 Zhang, X., Li, P., Barreda, Á. et al. (2016). Size-tunable rhodium nanostructures for wavelength-tunable ultraviolet plasmonics. *Nanoscale Horiz.* 1 (1): 75–80.

124 Christensen, N.E. (1973). The band structure of rhodium and its relation to photoemission experiments. *Phys. Stat. Solidi (b)* 55 (1): 117–127.

125 Kohno, Y., Hayashi, H., Takenaka, S. et al. (1999). Photo-enhanced reduction of carbon dioxide with hydrogen over Rh/TiO_2. *J. Photochem. Photobiol. A.* 126: 117–123.

126 Matsubu, J.C., Yang, V.N., and Christopher, P. (2015). Isolated metal active site concentration and stability control catalytic CO_2 reduction selectivity. *J. Am. Chem. Soc.* 137 (8): 3076–3084.

127 Goodman, D.W., Peebles, D.E., and White, J.M. (1984). CO_2 dissociation on rhodium: measurement of the specific rates on Rh(111). *Surf. Sci.* 140 (1): L239–L243.

128 Narang, P., Sundararaman, R., and Atwater, H.A. (2016). Plasmonic hot carrier dynamics in solid-state and chemical systems for energy conversion. *Nanophotonics* 5 (1): 96.

129 Demichel, O., Petit, M., Viarbitskaya, S. et al. (2016). Dynamics, efficiency, and energy distribution of nonlinear plasmon-assisted generation of hot carriers. *ACS Photonics* 3 (5): 791–795.

130 Brown, A.M., Sundararaman, R., Narang, P. et al. (2016). Nonradiative plasmon decay and hot carrier dynamics: effects of phonons, surfaces, and geometry. *ACS Nano* 10 (1): 957–966.

131 Li, X., Everitt, H.O., and Liu, J. (2019). Confirming nonthermal plasmonic effects enhance CO_2 methanation on Rh/TiO_2 catalysts. *Nano Res.* 12 (8): 1906–1911.

132 Schaaf, T., Grünig, J., Schuster, M.R. et al. (2014). Methanation of CO_2 - storage of renewable energy in a gas distribution system. *Energy Sustain. Soc.* 4 (1).

133 Halabi, M.H., de Croon, M.H.J.M., van der Schaaf, J. et al. (2010). Low temperature catalytic methane steam reforming over ceria–zirconia supported rhodium. *Appl. Catal. Gen.* 389 (1-2): 68–79.

134 Mark, M.F. and Maier, W.F. (1996). CO_2-reforming of methane on supported Rh and Ir catalysts. *J. Catal.* 164: 122–130.

135 Nakamura, J., Aikawa, K., Sato, K., and Uchijima, T. (1994). Role of support in reforming of CH_4 with CO_2 over Rh catalysts. *Catal. Lett.* 25: 265–270.

136 Brongersma, M.L., Halas, N.J., and Nordlander, P. (2015). Plasmon-induced hot carrier science and technology. *Nat. Nanotechnol.* 10 (1): 25–34.

8

Earth-Abundant Plasmonic Catalysts

Hefeng Cheng[1,2,*], *Yasutaka Kuwahara*[1,3,4] *and Hiromi Yamashita*[1,3,*]

[1] *Division of Materials and Manufacturing Science, Graduate School of Engineering, Osaka University, Osaka, Japan*
[2] *State Key Laboratory of Crystal Materials, Shandong University, Jinan, China*
[3] *Elements Strategy Initiative for Catalysts Batteries (ESICB), Kyoto University, Kyoto, Japan*
[4] *JST, PRESTO, Kawaguchi, Japan*

8.1 Introduction

The noble metal nanoparticles (NPs) exhibit a unique localized surface plasmon resonance (LSPR) feature [1–5], and this could date back to as early as the Lycurgus cup created by the Romans in 400 CE, where the plasmonic metal particles are embedded in the glass. When the wavelength of the incident light is equal to or longer than the sizes of metal NPs, the free electron cloud of the noble metal NPs polarizes and produces the accumulated negatively charged centers against the positive nuclei, forming an electric dipole (Figure 8.1) [6]. As a result, the curved surface of metal NPs exerts a restoring force on the driven electrons, and thus the resonance takes place and causes the enhancement in the vicinity of the near-field area. Generally, the electromagnetic field enhancement is as large as several magnitudes of the incident light. This process enables metal NPs to harvest and amplify the energy of incident light near their surface, and then convert light energy into the energy of excited charge carriers [4]. Such near-field enhancement has been employed to enable or enhance a wide range of applications involving surface-enhanced Raman spectroscopy (SERS), ultrasensitive detection, as well as heterogeneous catalysis [2, 4, 5].

Conventional plasmonic materials are mainly noble metals, such as Au and Ag. Because of their relatively high free electron concentration (c. 5.9×10^{22} cm^{-3} for Ag and Au), their NPs show very intense and narrow optical absorption in the lower wavelength region of visible light range [5]. Given the abundance of the chemical elements in Earth's upper continental crust, noble metals like Au, Ag, and Pt suffer from the earth-rarity and high-cost (Figure 8.1b) [7] it would be more appealing and promising to develop alternative plasmonic candidates with earth-abundant transition metal elements like Mo, W, Cu, Zn, Sn, and Si. Intriguingly, the degenerately doped semiconductors have recently emerged as another class of plasmonic materials [8–14]. In semiconductors, in order to sustain the LSPR, high density of free

Plasmonic Catalysis: From Fundamentals to Applications, First Edition.
Edited by Pedro H.C. Camargo and Emiliano Cortés.

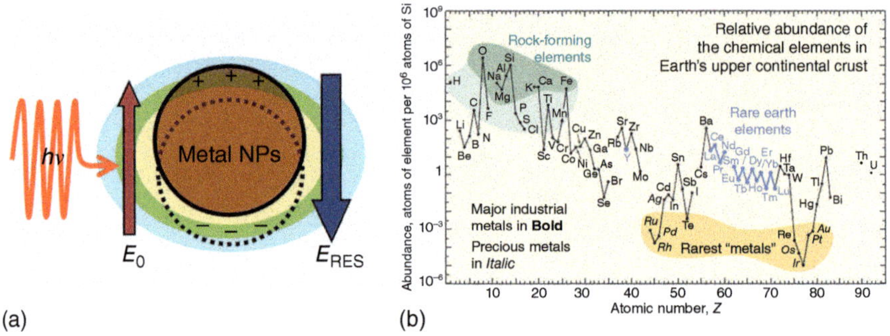

Figure 8.1 (a) The schematic illustration of plasmonic excitation on metallic nanoparticles. *Source*: Cheng et al. [6]. (b) Relative abundance (atom fraction) of the chemical elements in Earth's upper continental crust as a function of atomic number. *Source*: Rare Earth Elements—Critical Resources for High Technology Retrived from: https://pubs.usgs.gov/fs/2002/fs087-02/.

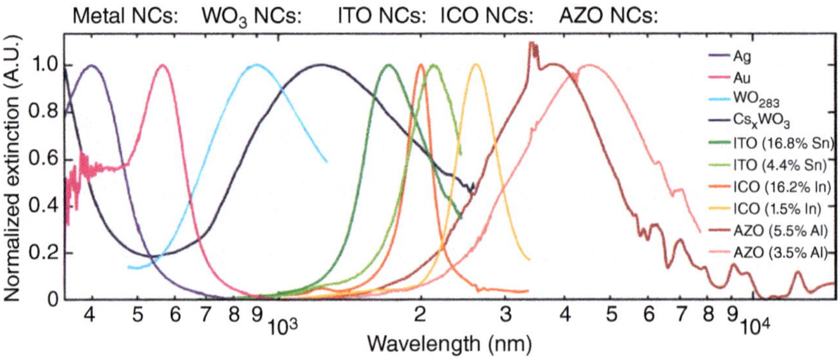

Figure 8.2 The normalized optical extinction due to LSPR in solutions and films of metal and metal oxide nanocrystals. *Source*: Lounis et al. [12]. © 2014, American Chemical Society.

carriers are introduced by controlled doping, and their free carrier concentration is usually lower than 10^{22} cm^{-3}. As a consequence, degenerately doped semiconductors generally have lower free carrier concentration than noble metals, allowing the observation of plasmonic resonance across the visible, near-infrared (NIR), and mid-infrared regime (Figure 8.2). For instance, the conducting oxides such as RuO_2, aluminum-doped zinc oxide (AZO), and indium tin oxide (ITO) display broader absorptions than noble metals (Figure 8.2b), mainly because of their higher optical losses [15]. In noble metals, the plasmon resonance is tunable upon size, shape, and dielectric surrounding environments [4, 16]. Besides these conventional parameters, in the degenerately doped semiconductors, the plasmonic resonance could also be intriguingly tailored by changing the stoichiometric compositions, dopant concentrations, or phase transitions [10, 11]. The emergence of doped semiconductor plasmonic materials not only expands the scope of plasmonic materials, but also extends the utilization of the infrared region of the solar spectrum and finds applications such as photothermal cancer therapy as well as THz communication.

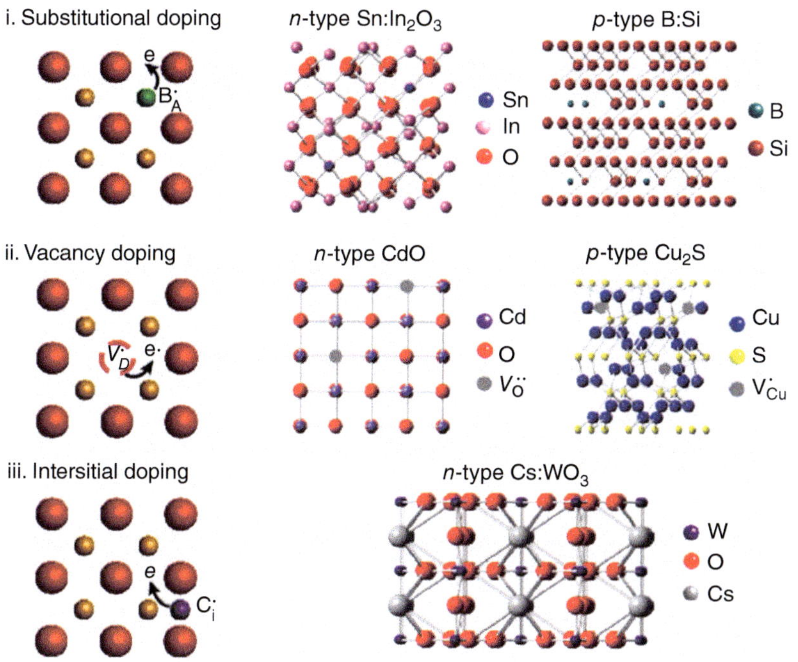

Figure 8.3 Several doping mechanisms with host cations (orange spheres) and anions (red spheres) and corresponding examples of doped semiconductors [14]. Three common donor doping types are (i) extrinsic aliovalent substitutional doping, (ii) intrinsic lattice vacancies formation, and (iii) extrinsic interstitial doping. *Source*: Agrawal et al. [14]. © 2018, American Chemical Society.

Typically, in plasmonic semiconductor materials, free carriers at high concentration are required to sustain the LSPR effect. The introduction of free carriers (electrons or holes) in semiconductors is realized through controlled doping and is generally classified into three categories with the corresponding doping mechanisms (Figure 8.3) [10, 14]:

(i) Extrinsic substitutional doping by aliovalent atoms with higher or lower valence contributes free electrons or holes to the semiconductor host materials. This category accounts for a majority of doped semiconductors, and has been extensively used in semiconductor electronics and solar cells. Typical examples are transparent conductive oxides (TCO) of Sn:In_2O_3 (ITO), Al:ZnO (AZO), and B-doped Si (p-Si), P-doped Si (n-Si), and so on [17–20]. In these materials, the extra free carriers are provided by the dopant atoms.

(ii) Intrinsic doping by lattice (anion or cation) vacancies with sufficient free carrier concentration could support LSPR effect. These materials include n-type oxygen-deficient metal oxides (e.g. MoO_{3-x}, WO_{3-x}) [21–28] and p-type copper-deficient chalcogenides ($Cu_{2-x}E$, E = S, Se, Te) [8, 9, 29–33]. In this case, free carriers arise from the variation in the oxidation state of the nonstoichiometric phases. To compensate the sub-stoichiometry, it results in extra

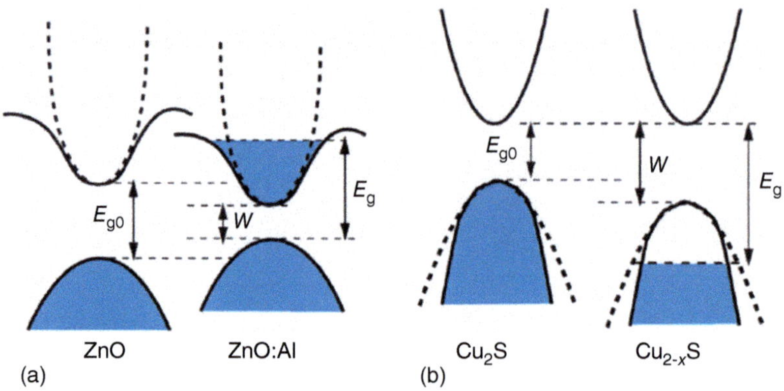

Figure 8.4 Two typical different band structures of plasmonic semiconductors: (a) AZO for *n*-doped semiconductor and (b) $Cu_{2-x}S$ for *p*-doped semiconductor. *Source*: Comin et al. [10]. © 2014, Royal Society of Chemistry.

electrons in the conduction band (CB) or holes in the valence band (VB) of the semiconductor materials.

(iii) Extrinsic interstitial doping, though less common, could also be able to introduce extra electrons into the CB of the semiconductors to sustain plasmonic effect. Typical examples of this category are molybdenum or tungsten oxide bronzes (A_xMO_3, A = H, Li, Na, K, etc., and M = Mo, W) [34–40]. The relative open crystal structure of MoO_3 and WO_3 host materials allows the interaction of ions with small radius, and thus the delocalized free electrons are accumulated in the CB.

In addition, apart from the common doping mechanisms mentioned above, there is another class of nonmetal plasmonic materials, and that is intrinsic conductive metallic oxides. This includes conductive RuO_2, ReO_3, VO_2, TiN, and so on [41–44]. In these materials, the plasmonic resonance is due to their intrinsic electronic structure with partially filled *d*-band to show conductive behavior [45].

In the light of free carriers (electrons or holes), the degenerately doped plasmonic semiconductors are divided into two different classes of *n*-type and *p*-type semiconductors. Figure 8.4 illustrates the band structures for typical *n*-type AZO and *p*-type $Cu_{2-x}S$ doped semiconductors [10]. The bandgap is defined as the photon energy needed to excite an electron transition from the top of the VB to the bottom of the CB. Therefore, for an intrinsic semiconductor, the bandgap is defined as the difference between the bottom of CB and the top of VB. Due to the Burstein–Moss effect, the optical bandgaps of the degenerately doped semiconductor are dependent on their free carrier concentrations [10]. In highly *n*-doped semiconductor AZO, the lowest energy levels of CB are occupied by the extra electrons from the dopants. Consequently, the intrinsic bandgap of AZO is smaller than that of undoped ZnO, but the optical bandgap is augmented due to the state filling effect. With increasing carrier concentration, the curvature of the CB and the effective mass increases. Analogously, in *p*-doped semiconductor $Cu_{2-x}S$, the highest energy levels of VB are populated by the extra holes. In this case, both the intrinsic and the optical bandgaps

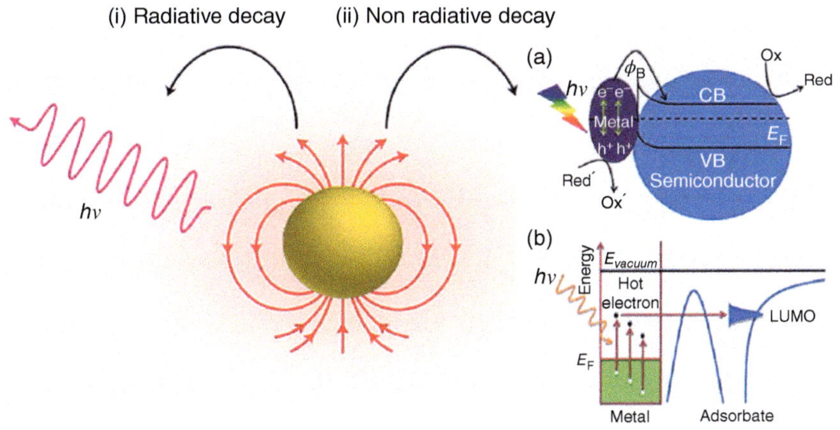

Figure 8.5 A schematic illustration of the surface plasmon decay process via (i) radiative and (ii) nonradiative routes [6]. (a) The transfer of the hot electrons to the nearby *n*-type semiconductor. (b) The direct injection of the hot electrons to the LUMO of the adsorbates. *Source*: Cheng et al. [6].

become larger when the doping level is increased. The curvature of VB increases as the Fermi energy level decreases, and hence the effective mass decreases as the number of vacancies increases.

Upon plasmonic excitation, the delocalized free carriers near the Fermi level of materials oscillate collectively, thus resulting in the enhanced electromagnetic field in the vicinity of NPs surface and the subsequent generation of "hot" electron and hole carriers [5, 6]. As illustrated in Figure 8.5, after light absorption and plasmonic excitation, the electromagnetic decay occurs on a timescale of femtosecond via the radiative re-emission of photons or nonradiative energy transfer to hot electrons. In the nonradiative process, the hot electrons could overcome the interfacial Schottky barrier to migrate to the adjacent appropriate semiconductors (Figure 8.5a), or otherwise inject into the lowest unoccupied molecular orbital (LUMO) of the adsorbate (Figure 8.5b). Due to electronic depletion, the plasmonic nanostructures are left positively charged after injection of hot electrons. Both the transferred hot electrons and holes left could activate and stimulate the redox catalytic reactions, leading to the plasmon-induced catalysis.

Plasmon-induced energy transfer facilitates the separation of hot electrons and holes, and the resulting enhanced catalytic reactions have been well documented in the plasmonic noble metal nanostructures [4, 46–48]. For instance, plasmonic excitation of Au NPs has been shown to activate CO oxidation, hydrogenation of carbonyls, dissociation of H_2, reduction of nitroaromatics, and among others [46]. However, suffering from earth rarity and high cost, the noble metals are prevented from large-scale practical applications. On the other hand, nonprecious metals (e.g. Cu, Al, Mg) have also showed interesting plasmonic properties toward hydrogenation, epoxidation, H_2 sensing, and so on [49–53]. Taking Mg for example, the reversible transition between metal Mg and dielectric hydride (MgH_2) offers plasmonic chirality switching, and dynamic color displays at visible frequencies.

However, they are prone to oxidation and fail to be used stable plasmonic catalysts at ambient conditions. Therefore, compared to plasmonic metals, it is more promising to use the degenerately doped semiconductor counterparts with earth-abundant elements. Nonetheless, the plasmonic-doped semiconductors are usually subject to the following drawbacks: (i) longer wavelength of solar spectrum (NIR to mid-IR region) light absorption with relatively lower carrier concentration, and (ii) lower surface feasibility by surface capping agents during conventional colloidal synthesis.

To utilize solar light effectively and achieve efficient catalysis, it is of paramount significance to develop visible light-driven earth-abundant plasmonic catalysts without surfactant capping agents. Intriguingly, recently a variety of earth-abundant plasmonic catalysts, such as MoO_{3-x} nanosheets [24], Pd/MoO_{3-x} hybrid [25], mesoporous $WO_{2.83}$ [28], Pd/H_xMoO_3 [38], Pt/H_xMoO_{3-y} [39], $Rb_{0.33}WO_3$ [40], Cu_7S_4@Pd heteronanostructures [32], and Pd/$Cu_{2-x}S$ nanowires hybrid [54], have been rationally prepared and shown efficient or enhanced catalytic activities toward NH_3BH_3 dehydrogenation, Suzuki coupling reaction, hydrogen evolution reaction, *p*-nitrophenol reduction, sulfoxide deoxygenation, CO_2 reduction, selective oxidation of benzyl alcohol, nitrobenzene hydrogenation, and so on. Hereafter, we will mainly summarize the recent findings of plasmon-enhanced catalysis based on earth-abundant degenerately doped semiconductors in this burgeoning field. The following parts could be mainly classified into three categories: (i) MoO_{3-x} and WO_{3-x}-based plasmonic catalysts, (ii) molybdenum- and tungsten bronzes-based plasmonic catalysts, and (iii) $Cu_{2-x}E$ (E = S, Se, Te)-based plasmonic catalysts.

8.2 MoO_{3-x}- and WO_{3-x}-Based Plasmonic Catalysts

As important *n*-type transition metal oxide semiconductors, MoO_3 (bandgap of 3.2 eV) and WO_3 (bandgap of 2.8 eV), have found applications ranging from photochromic and electrochromic materials to photoelectrochemical water splitting, and gas sensors [55–57]. Interestingly, once the oxygen vacancy concentration exceeds the threshold, the unique character of delocalized outer-d valence electrons in MoO_{3-x} and WO_{3-x} enables them to exhibit LSPR effect [22, 24]. The abundant defects enable self-doped semiconductors of MoO_{3-x} and WO_{3-x} to have more active sites for reactions than the pristine counterparts, and could be promising for heterogeneous catalysis.

MoO_{3-x} nanosheets have been the first example to show plasmon-enhanced photocatalysis over earth-abundant plasmonic catalysts [24]. We prepared the MoO_{3-x} nanosheets by oxidizing metal Mo powder with H_2O_2 in ethanol solution and the subsequent solvothermal treatment. The as-prepared MoO_{3-x} nanosheets exhibit intense blue color (Figure 8.6a). With an LSPR peak around 680 nm, the oxygen-deficient MoO_{3-x} nanosheets display strong visible light absorption (Figure 8.6b). In contrast, commercial MoO_3 sample exhibits only UV-light response, with the absorption edge at about 400 nm corresponding to its wide bandgap of c. 3.1 eV. The MoO_{3-x} nanosheets are 200 nm to 1 μm in length and c. 20–30 nm in thickness (Figure 8.6c,d). It is found that experimental parameters

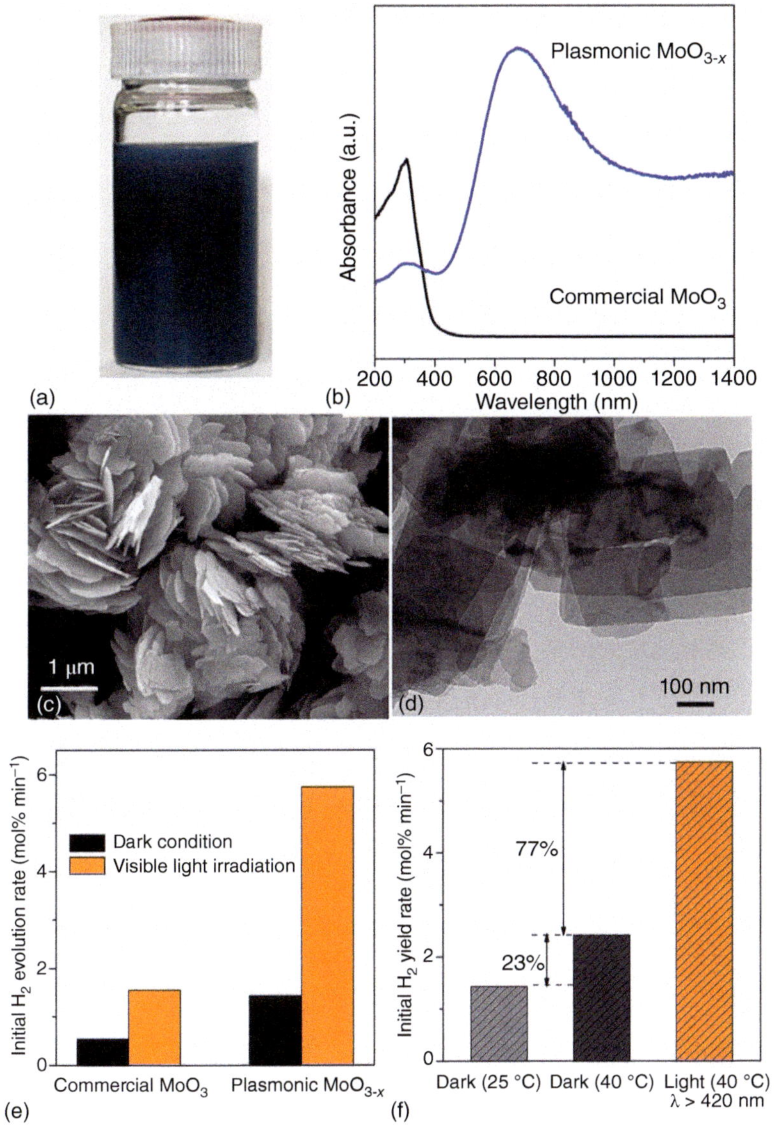

Figure 8.6 (a) Photograph of the MoO_{3-x} nanosheets dispersed in ethanol solution. (b) The optical absorption of plasmonic MoO_{3-x} nanosheets and commercial pristine MoO_3 sample. (c) Typical FE-SEM and (d) TEM images of the plasmonic MoO_{3-x} nanosheets. (e) Comparison of initial H_2 yield rate over plasmonic MoO_{3-x} nanosheets and commercial MoO_3 samples with and without light irradiation in dehydrogenation of ammonia borane (NH_3BH_3). (f) Thermal effect on the catalytic performance of plasmonic MoO_{3-x} nanosheets in dark condition and under visible light irradiation. *Source*: Cheng et al. [24] © 2014, John Wiley & Sons.

(i.e. solvothermal temperatures, H_2O_2 volumes added, and solvents) play great roles in determining the phase structures and plasmonic resonances of the MoO_{3-x} products. The LSPR peak is varied from 680 to 950 nm, thus offering the tunable light harvesting in a wide visible–NIR light range. The presence of oxygen vacancies is further verified by the X-ray photoelectron spectroscopy (XPS), where Mo^{5+} cations account for 77.3% of total Mo species, providing abundant delocalized d-electrons to sustain its visible light plasmon absorption.

It was then found that the plasmonic MoO_{3-x} nanosheets could be used as a highly efficient catalyst to enhance ammonia borane (NH_3BH_3) dehydrogenation. NH_3BH_3 contains as high as 19.6 wt% of hydrogen, which exceeds that of gasoline; therefore, it is an attractive candidate for the next-generation hydrogen storage materials [58]. With a suitable catalyst, the dehydrogenation of NH_3BH_3 would take place and H_2 is thus obtained by following (8.1):

$$NH_3BH_3 + 2H_2O \overset{\text{catalyst}}{\rightarrow} NH_4^+ + BO_2^- + 3H_2 \tag{8.1}$$

Conventional efficient catalysts are confined to noble metal nanostructures (e.g. Ru, Rh, Pt) [58]. In terms of practical applications, development of earth-abundant catalysts with low cost and high efficiency is crucially important. The as-obtained plasmonic MoO_{3-x} nanosheets outperformed commercial pristine MoO_3 both in dark and under visible light irradiation (Figure 8.6e). Notably, visible light dramatically improved the catalytic activity by four times on the plasmonic MoO_{3-x} nanosheets, even exceeding the plasmonic noble metals of Ag/SBA-15 [59]. The control experiments showed that the plasmonic effect accounted for c. 77% of the catalytic enhancement (Figure 8.6f). This work demonstrates that the plasmonic-doped semiconductors can also be employed for efficient visible light-driven photocatalysis, and it offers the potential utilization of the earth-abundant elements instead of the precious noble metals to a certain extent. Following this, high-surface-area MoO_{3-x} nanosheets [60] and MoO_{3-x} NPs [61] were also prepared by different aqueous routes, which showed intense visible light plasmonic absorption with LSPR peaks around 623 and 682 nm, respectively. Because of the oxygen vacancies introduced, both of them displayed enhanced H_2 yield in the catalytic performance toward NH_3BH_3 dehydrogenation.

The plasmonic hybrids that integrate different components with diverse properties usually exhibit synergistic multilevel coupling effect, outperforming the single counterparts. To this end, we prepared plasmonic Pd/MoO_{3-x} hybrid by coupling Pd NPs and a plasmonic degenerately doped MoO_{3-x}, which was endowed with improved solar harvesting and enhanced catalytic performances [25]. The plasmonic Pd/MoO_{3-x} hybrid was obtained through a solution-processed impregnation-reduction method, having Pd NPs of about 10 nm anchored on the MoO_{3-x} microsized plates (Figure 8.7a). While pristine MoO_3 responds to UV light only, the as-obtained Pd/MoO_{3-x} hybrid shows strong visible light response, with a prominent symmetrical LSPR peak at around 640 nm (Figure 8.7b). It is noted that the Pd/MoO_{3-x} hybrid exhibits reversible tunability in its plasmon resonance upon O_2 oxidation or $NaBH_4$ reduction (Figure 8.7c,d). Gradual O_2 oxidation decreases the delocalized electron concentration of the Pd/MoO_{3-x} hybrid, thus leading to its red shift of the LSPR peak; while the oxidized product could be recovered by reduction with $NaBH_4$ in solution, resulting in the blue shift of the LSPR peak. Such plasmonic Pd/MoO_{3-x} hybrid exhibited dramatic enhancement in NH_3BH_3 dehydrogenation, where the wavelength-dependent activity enhancement was observed with different LED (i.e. blue light 470 nm, green light 530 nm and red light 650 nm) irradiation (Figure 8.7e,f). In addition, the plasmonic Pd/MoO_{3-x} hybrid also showed enhanced catalytic activity toward Suzuki–Miyaura coupling reactions under visible light, illustrating the synergetic effect between plasmonic-doped semiconductor support and catalytic metal NPs contribute together to efficient catalysis.

Analogously, the $Mo_xW_{1-x}O_{3-y}$ hybrid [62] displayed strong LSPR in the visible light region, and the plasmonic $Mo_xW_{1-x}O_{3-y}$ hybrid at optimized condition could be used as a highly efficient catalyst that dramatically enhanced the dehydrogenation activity from NH_3BH_3 under visible light irradiation, outperforming any single components of MoO_{3-x} or WO_{3-x}. Moreover, further coupling of plasmonic $Mo_xW_{1-x}O_{3-y}$ hybrid with Pd NPs could greatly improve the catalytic activity toward NH_3BH_3 dehydrogenation upon exposure to visible light [63]. These works verify that the rational coupling between a plasmonic degenerately doped semiconductor and a catalytic metal could lead to highly efficient heterogeneous catalysis.

Different from MoO_{3-x}, oxygen vacancy enables WO_{3-x} $(0 < x \leq 0.28)$ with the presence of a number of defined sub-stoichiometric oxides [56, 65–67], including $WO_{2.9}$ $(W_{20}O_{58})$, $WO_{2.83}$ $(W_{24}O_{68})$, and $WO_{2.72}$ $(W_{18}O_{49})$. The selective synthesis of their sub-stoichiometric oxides with a defined composition is rather challenging. Through a controlled oxygen vacancy formation in mesoporous WO_3 host, well-defined sub-stoichiometric $WO_{2.83}$ with mesoporous structure was obtained [28]. Different from bulk WO_3, the use of a mesoporous WO_3 (Meso-WO_3) not only greatly decreases the H_2 reduction temperature, but also causes reduction to $WO_{2.83}$ with high selectivity. Oxygen vacancies enable exceptional plasmon resonance in mesoporous $WO_{2.83}$ (Meso-$WO_{2.83}$), with LSPR peak at approximately 650 nm (Figure 8.8a). The LSPR absorption of Meso-$WO_{2.83}$ could be tailored in a wide range through redox chemistry. Furthermore, mesoporous $WO_{2.83}$ exhibits superior hydrogen evolution reaction (HER) activity compared to pristine Meso-WO_3 (Figure 8.8b), and it can be shown that the HER activity is crucially dependent on the amount of oxygen vacancies.

The plasmonic WO_{3-x} has also been employed to efficiently catalyze ethanol reforming to produce value-added products [64]. Through facile solvothermal treatment, several WO_{3-x} nanostructures were prepared with different oxygen vacancies. The plasmonic WO_{3-x} nanowire bundles are demonstrated to possess the maximum amount of oxygen vacancies and exhibit an optimal activity for ethanol dehydration. The prepared WO_{3-x} nanowire bundles show strong optical absorption in the NIR region (Figure 8.8c) that arises from the oxygen vacancies. The oxygen vacancies facilitate utilization for full-spectrum solar energy and also serve as active sites for improved ethanol adsorption, greatly promoting photocatalysis. Under full-spectrum light irradiation, the WO_{3-x}-5 nanowires exhibit a remarkable ethylene generation rate of 16.9 mmol g^{-1} h^{-1} (selectivity of 94.9%) within 180 minutes, much higher than 1.0 mmol g^{-1} h^{-1} (selectivity of 90.5%) of

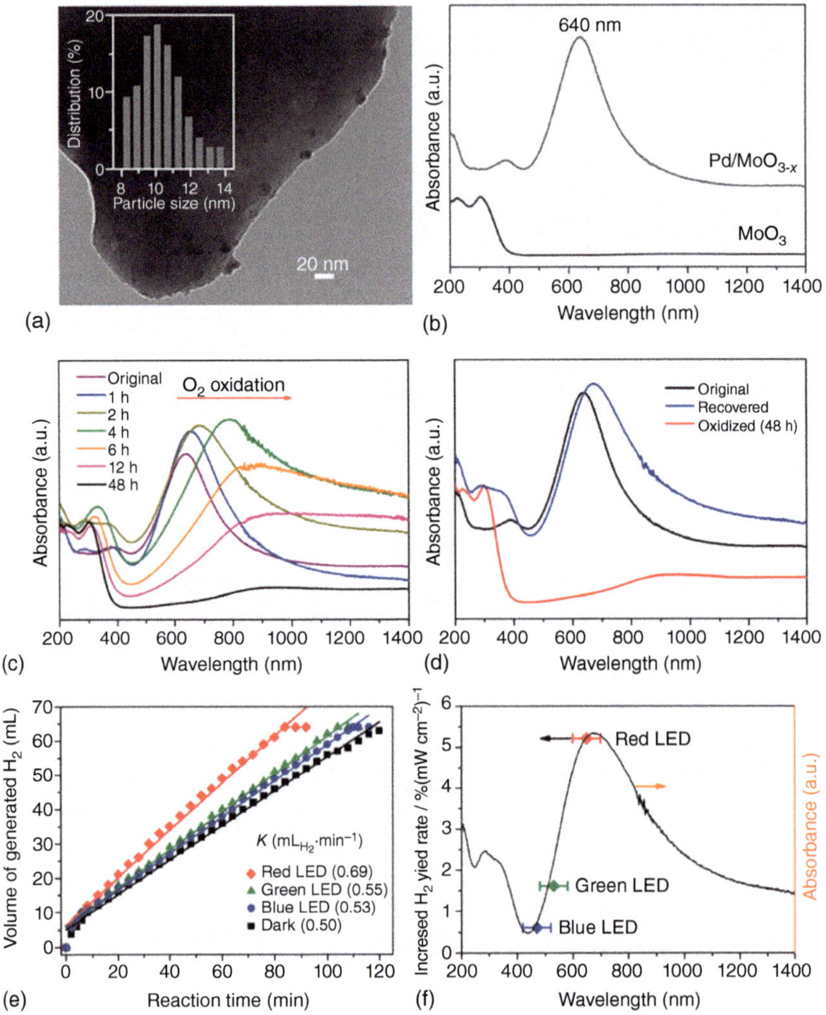

Figure 8.7 (a) TEM image of the Pd/MoO_{3-x} hybrid and (inset) particle size distribution of Pd NPs. (b) UV/vis–NIR diffuse reflectance spectra of the Pd/MoO_{3-x} hybrid and pristine MoO_3. The tunable plasmonic absorption of the Pd/MoO_{3-x} hybrid upon (c) O_2 oxidation and (d) reduction by $NaBH_4$. (e) Time course of H_2 generation over the Pd/MoO_{3-x} catalyst under blue, green, and red LED illumination together with dark condition in dehydrogenation of ammonia borane (NH_3BH_3). (f) The increased H_2 yield rate as a function of the wavelength by LED irradiation over the Pd/MoO_{3-x} catalyst. *Source*: Cheng et al. [25]. © 2015, John Wiley & Sons.

WO_{3-x}-20 nanoplates with low oxygen vacancies atmosphere (Figure 8.8d). The excited wavelength on the photocatalytic ethanol reforming has been investigated, and oxygen-free atmosphere and NIR irradiation are favorable for highly selective ethylene generation. The results also reveal that the plasmonic hot electrons and photothermal effect synergistically contribute to the superior photocatalytic performance of WO_{3-x} nanowires in ethanol dehydration.

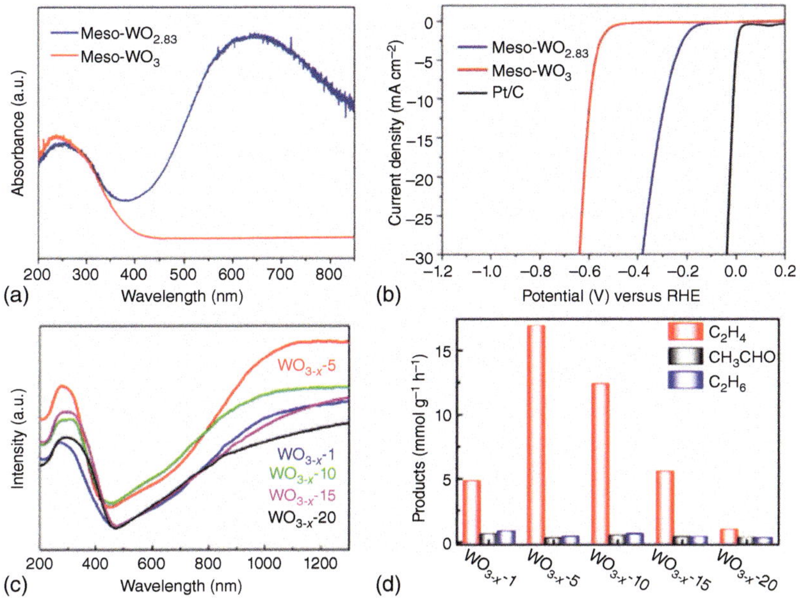

Figure 8.8 (a) The UV/vis diffuse reflectance spectra of Meso-$WO_{2.83}$ and Meso-WO_3, (b) Polarization curves for hydrogen evolution reaction activities over Meso-$WO_{2.83}$, along with Meso-WO_3 and commercial Pt/C. *Source*: Cheng et al. [28]. (c) The UV/vis–NIR absorption spectra of the five WO_{3-x} samples. (d) The main product selectivity of ethanol reforming over different plasmonic WO_{3-x} samples under UV–vis–NIR light irradiation. *Source*: Li et al. [64]. © 2020, Elsevier.

In addition, plasmonic WO_{3-x} also promotes photocatalytic water oxidation [26]. The WO_{3-x} nanosheets were synthesized by exfoliation of layered tungstic acid and subsequent introduction of oxygen vacancies via vacuum treatment (WO_{3-x}-VT) or hydrogen treatment (WO_{3-x}-HT). The introduction of oxygen vacancies to pristine WO_3 yields LSPR absorption in NIR region. The as-prepared WO_{3-x} nanosheets showed dramatically enhanced performance in both photocurrent responses and photocatalytic water oxidation. It was stated that the presence of oxygen vacancies promotes the light harvesting both in the NIR region (LSPR) and in the UV–vis range.

8.3 Molybdenum and Tungsten Bronzes-Based Plasmonic Catalysts

Among two types of extrinsic doping that contribute to the plasmonic degenerately doped semiconductors, substitutional doping is the replacement of the crystal lattice atom with the dopant atom possessing higher or lower valence. To sustain the plasmon resonance, high concentration of aliovalent atoms are required to provide the abundant free electrons or holes. However, these excessive substitutional aliovalent atoms tend to become impurity or recombination centers for photoexcited electrons and holes in semiconductor photocatalysis [68]. Therefore, plasmonic degenerately doped semiconductors are usually not appropriate for heterogeneous catalysis. Alternatively, interstitial doping allows a new atom to be incorporated into the crystal structure of host material, providing extra electrons to semiconductor to support plasmon resonance. One common class of interstitially doped semiconductors is known as molybdenum or tungsten bronzes having the formula A_xMO_3 (M = Mo, W), where A is H, Li, Na, K, Rb, and Cs [35–37]. The crystal structures of MoO_3 and WO_3 are composed of distorted MO_6 (M = Mo, W) octahedra with corner- or edge-sharing configuration to create one-dimensional tunnel structure, allowing the incorporation of hydrogen or alkali metal atoms. The ionization of introduced atoms provide the free electrons to the CB of MoO_3 and WO_3, and gives rise to plasmon resonance once exceeding the threshold [38–40, 69].

Being the smallest atom in nature, hydrogen atom is able to migrate and insert into abundant interstitial sites of transition metal oxides, without inducing the large structural variation. Hydrogen molybdenum bronzes H_xMoO_3 ($0 < x \leq 2$) and hydrogen tungsten bronzes H_xWO_3 ($0 < x < 0.6$) have been found as relatively new classes, in which hydrogen is topotactically inserted into the MoO_3 and WO_3 matrix to yield several unique phases [34–36]. In the case of H_xMoO_3, four different phases have been well documented with different hydrogen contents: blue orthorhombic ($0.23 < x < 0.4$), blue monoclinic ($0.85 < x < 1.04$), red monoclinic ($1.55 < x < 1.72$), and green monoclinic ($x = 2$). For H_xWO_3, three distinct phases have been characterized: blue tetragonal ($0.09 < x < 0.16$), blue tetragonal ($0.31 < x < 0.5$), and red cubic ($0.5 < x < 0.6$). The structure of each phase is closely related to the host matrix of MoO_3 and WO_3, while the property is dependent on the hydrogen content [35].

In a recent study, we have reported hydrogen-doped MoO_3 and WO_3 to yield H_xMoO_3 and H_xWO_3, having exceptional and tunable LSPR in the visible light region [38]. To circumvent the conventional hydrogen doping in metal oxides that requires high temperature and high pressure, a H-spillover approach is adopted. As illustrated in Figure 8.9a, in this process, the chemisorbed hydrogen molecules dissociate at metal sites (e.g. Pd) to produce highly reactive hydrogen atoms, which would migrate to the surface of MoO_3 supports, and further diffuse into the bulk, thereby leading to the reduction of MoO_3 matrix. Upon H_2 reduction at room temperature, the as-prepared product (denoted as Pd/MoO_3 H_2-RT) exhibits strong optical absorption in the visible regime, with an LSPR peak pinning at approximately 565 nm. At the same time, orthorhombic MoO_3 is converted into monoclinic $H_{1.68}MoO_3$. Through variation of their stoichiometric compositions, tunable plasmon resonances could be observed in a wide range (Figure 8.9b–e), which hinge upon the reduction temperatures, metal species, the nature and the size of metal oxide supports in the synthetic H_2 reduction process, as well as oxidation treatment in the postsynthetic process.

To reveal the origin of surface plasmon resonance in the hydrogen bronzes and understand the drastic changes induced by hydrogen-doping process, first-principle DFT calculations was conducted (Figure 8.10a–d). With a semiconductor-like band structure, pristine MoO_3 exhibits no defect state; however, heavily doped hydrogen molybdenum bronze ($H_{1.68}MoO_3$) displays strong metallic feature with continuous VB across the Fermi level into the CB. The drastic changes in electronic structures

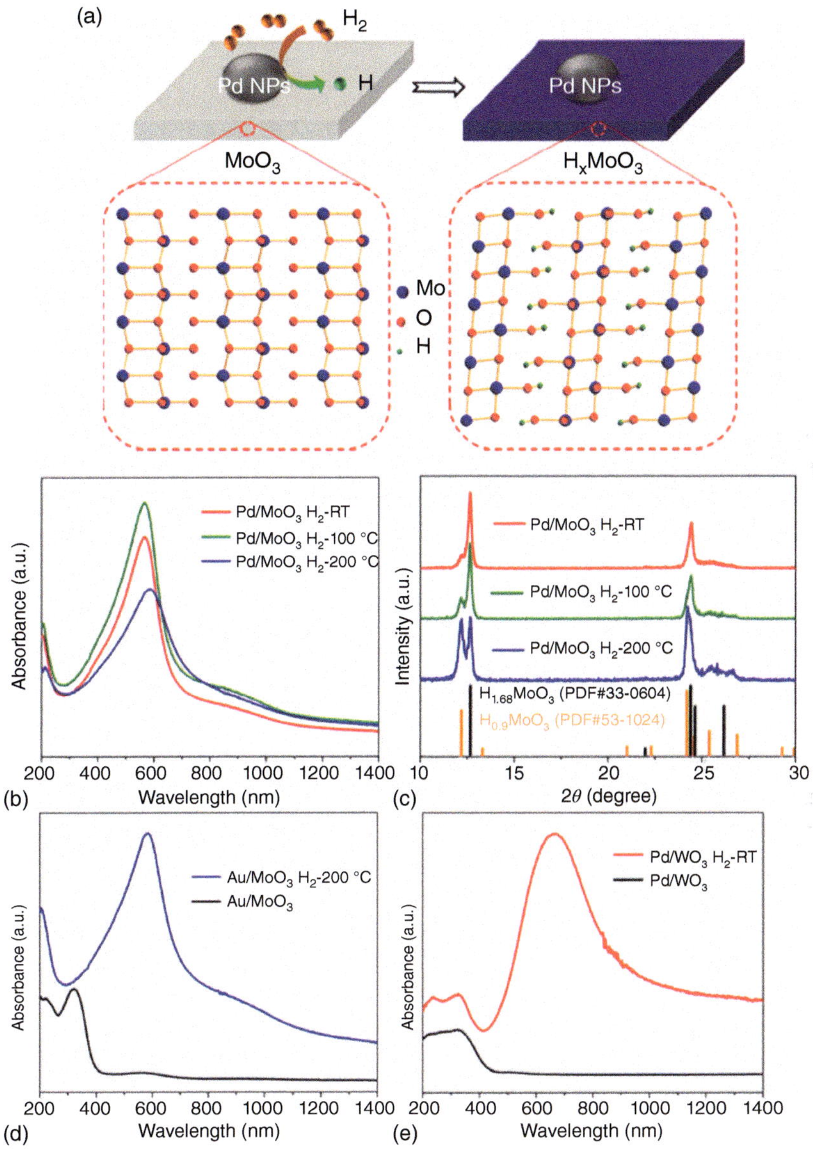

Figure 8.9 (a) A schematic illustration to prepare hydrogen molybdenum bronzes via H-spillover route. (b) UV/vis–NIR diffuse reflectance spectra and (c) XRD patterns of the Pd/MoO_3 products upon H_2 reduction at different temperatures. (d) UV/vis–NIR diffuse reflectance spectra of Au/MoO_3 samples before and after H_2 reduction at 200 °C. (e) UV/vis–NIR diffuse reflectance spectra of Pd/WO_3 samples before and after H_2 reduction at room temperature. *Source*: Cheng et al. [38]. © 2016, American Chemical Society.

of hydrogen molybdenum bronzes stem from the intercalation of hydrogen atoms into the MoO_3 matrix and subsequent charge transfer, leading to the distortion of MoO_6 octahedra in the resultant $H_{1.68}MoO_3$ framework. According to the three-dimensional (3D) visualization of electronic charge density distribution for $H_{1.68}MoO_3$, the introduced electronic charge is distributed uniformly on Mo atoms, indicative of the delocalization nature of Mo 4d electrons. This results in the shift of Fermi level within the CB. With high degree of delocalization, free electrons donated by hydrogen atoms are expected to, when exceeding the critical concentration, support the observed LSPR in the visible light spectrum. The as-prepared Pd/H_xMoO_3 has been used as plasmonic catalyst for *p*-nitrophenol reduction to *p*-aminophenol using ammonia borane as a reductant (Figure 8.10e,f). In contrast to dark condition, the plasmonic H_xMoO_3 support boosts Pd-catalyzed reduction of *p*-nitrophenol upon exposure to visible light. These findings provide direct evidence for achieving plasmon resonances in hydrogen-doped metal oxide semiconductors, and demonstrate that their coupling with ultrafine metal NPs will find more efficient heterogeneous catalysis.

Subsequently, the plasmonic Ru/H_xMoO_{3-y} hybrid was prepared and showed superior catalytic activity to photocatalytic reduction of *p*-nitrophenol in contrast to dark conditions, in which H_xMoO_{3-y} acts as electron-donor centers and Ru NPs act as active centers [70]. Furthermore, with the synergistic effect of bimetallic NPs, the RuPd NPs anchored on the plasmonic H_xMoO_{3-y} support were successfully prepared, and exhibited greatly enhanced activity toward *p*-nitrophenol reduction under visible light [71]. The as-synthesized plasmonic bimetallic $RuPd/H_xMoO_{3-y}$ hybrid could finish the *p*-nitrophenol reduction within five minutes, superior to the Ru/H_xMoO_{3-y} and Pd/H_xMoO_{3-y} hybrids with single metallic NPs. These results demonstrate the promising application of the plasmonic molybdenum oxide hybrid as an efficient and stable deoxygenation catalyst for green chemistry.

In addition to *p*-nitrophenol reduction, the plasmonic Pt/H_xMoO_{3-y} hybrid showed highly enhanced catalytic activity toward deoxygenation of sulfoxides with 1 atm of H_2 at room temperature under visible light irradiation relative to dark condition [39]. The Pt/H_xMoO_{3-y} hybrid was prepared by hydrogen-doped MoO_3 coupled with Pt NPs through a H-spillover process. After H_2 reduction, the Pt/H_xMoO_{3-y} hybrid exhibits strong absorption in the visible to NIR region with an intense LSPR peak pinning at around 556 nm, while the unreduced Pt/MoO_3 only shows absorption below 400 nm. As the H_2 reduction temperature increases, a blue shift of the plasmonic wavelength from 564 to 556 nm is observed (Figure 8.11a). Coupling the plasmonic H_xMoO_{3-y} with the catalytic Pt NPs leads to highly efficient plasmonic visible light-driven catalyst in the deoxygenation of sulfoxides with 1 atm of H_2 at room temperature (Figure 8.11b–d). Over various catalysts, the as-prepared Pt/H_xMoO_{3-y} hybrid exhibited the highest catalytic activity, affording diphenyl sulfide product in >99% yield in seven hours of reaction (Figure 8.11b).

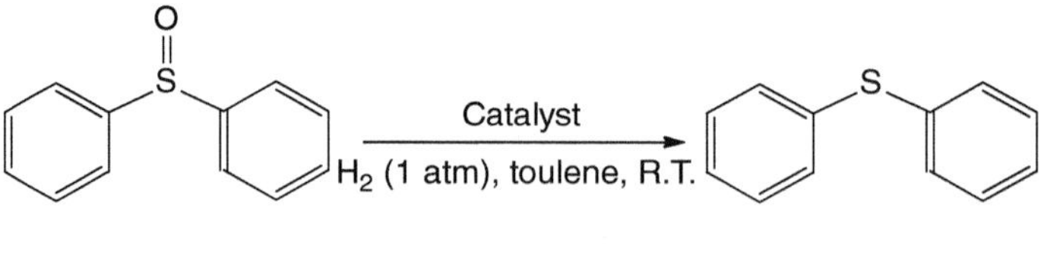

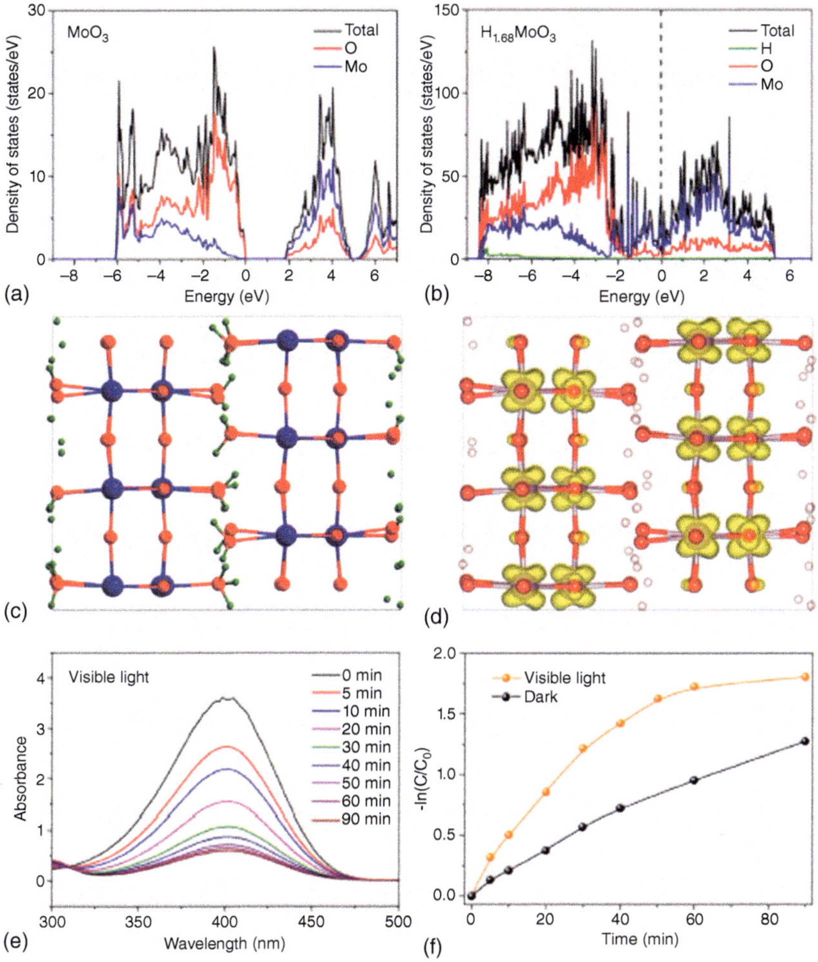

Figure 8.10 TDOS and PDOS of (a) pristine MoO_3 and (b) heavily hydrogen-doped $H_{1.68}MoO_3$. (c) Structure illustration for $H_{1.68}MoO_3$ in (010) projection. Blue balls represent Mo atoms, red for O, and green for H atoms. (d) Three-dimensional visualization of electronic charge density distribution around the Fermi level for $H_{1.68}MoO_3$. (e) Time course evolution of UV/vis absorption spectra of reaction solutions in catalytic reduction of *p*-nitrophenol with ammonia borane over Pd/MoO_3 H_2-RT catalyst under visible light irradiation. (f) Plots of $-\ln(C/C_0)$ against the initial point at 400 nm peak for *p*-nitrophenol as a function of time. *Source*: Cheng et al. [38]. © 2016, American Chemical Society.

Intriguingly, a significant improvement of catalytic efficiency under visible light was observed over Pt/H_xMoO_{3-y} hybrid, giving a twofold faster reaction rate relative to dark condition. This plasmonic enhancement in the activity was not found on nonplasmonic catalysts of Pt/SiO_2, Pt/WO_3, and Pt/MoO_2 under visible light. Given the wavelength-dependent catalytic performance, the activity enhancement is driven by the plasmonic effect of hydrogen molybdenum bronze. The reaction pathway for the deoxygenation of sulfoxide over Pt/H_xMoO_{3-y} hybrid was

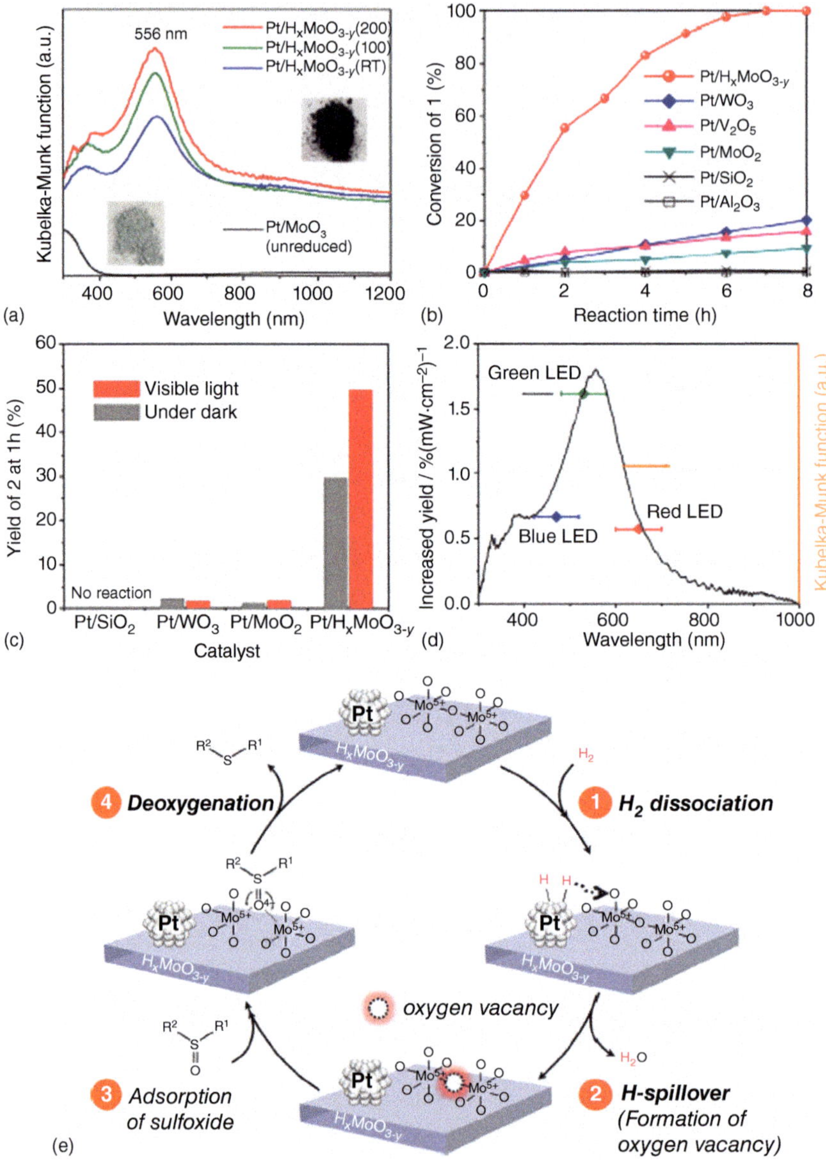

Figure 8.11 (a) UV–vis–NIR diffuse reflectance spectra for the Pt/H_xMoO_{3-y} hybrids reduced at varied temperatures and for the unreduced Pt/MoO_3. (b) Time course in the deoxygenation of sulfoxide over various Pt-loaded oxide catalysts. (c) Comparison of catalytic activities over various catalysts in the dark or under visible light irradiation conditions. (d) The increased yield of diphenyl sulfide as a function of the wavelength by LED irradiation over Pt/H_xMoO_{3-y}(200) catalyst (left axis) and its Kubelka–Munk function (right axis). (e) Possible reaction pathway of the deoxygenation of sulfoxide over Pt/H_xMoO_{3-y} hybrid catalyst using molecular H_2 as a reductant. *Source*: Kuwahara et al. [39]. © 2018, American Chemical Society.

proposed through four steps of H_2 dissociation, H-spillover to form oxygen vacancy, adsorption of sulfoxide and the final deoxygenation to complete the reaction cycle (Figure 8.11e). During the reaction, the massive oxygen vacancies and the reversible redox property of Mo atoms in H_xMoO_{3-y} provide an effective catalytic system under mild conditions. The extraordinary performance of the plasmonic Pt/H_xMoO_{3-y} hybrid presents a great opportunity for utilizing degenerately doped plasmonic molybdenum oxide as a light-harvesting platform in heterogeneous catalysis, which can boost diverse selective organic transformations under solar light.

In addition to hydrogen, the alkali metals-containing molybdenum and tungsten bronzes A_xMO_3 (A = Li, Na, K, Rb, and Cs; M = Mo, W) are an important class that possess intense color and metallic luster with resistance upon corrosion. Depending on the crystallographic structures and content of alkali metals, either metallic or semiconducting properties are characterized in these bronzes [35]. The tungsten bronzes $M_{0.33}WO_3$ (M = K, Rb, Cs) have shown efficient full spectrum (UV, visible, and NIR lights)-induced photocatalytic CO_2 reduction performance directly from the air at ambient pressure [40]. In the tungsten bronze crystal, the corner-sharing WO_6 octahedron build up the tungsten–oxygen framework to form uniformly dispersed trigonal and hexagonal tunnels, and large amounts of alkali metal ions could be regularly introduced in the hexagonal tunnel (Figure 8.12a). The prepared $M_{0.33}WO_3$ (M = K, Rb, Cs) showed LSPR absorption in the visible light region (Figure 8.12b). The calculated dielectric functions (Figure 8.12c) and the energy loss function (Figure 8.12d) of the sample $Rb_{0.33}WO_3$ further confirmed the presence of plasmon energy around 1.4 eV, corresponding to LSPR peak approximately at 885 nm. In the photocatalytic CO_2 conversion to methanol, the as-prepared $M_{0.33}WO_3$ (M = K, Rb, Cs) showed superior activity both under full spectrum light irradiation and NIR light irradiation compared to WO_3 and $WO_{18}O_{49}$. Particularly, after four hours of NIR light irradiation, c. 4.32% CO_2 in the air could be converted into CH_3OH with 98.35% selectivity for $Rb_{0.33}WO_3$. Both the experiments and theoretical calculations unveil that the introduced alkali metal atom occupy the tunnel of hexagonal structure and donate more free electrons to reconstruct the electronic structure of $M_{0.33}WO_3$, which can enhance the polaron transition, modify the energy band structure, selectively adsorb CO_2 rather than O_2 from the air, decrease the activation energy of CO_2 reaction, and finally make the effective CO_2 reduction in the air a reality.

In addition to CO_2 reduction, the tungsten bronzes have also been used for photocatalytic water splitting. Tang et al. prepared the WO_2-Na_xWO_3 hybrid and found that it showed efficient IR-driven photocatalytic water splitting ability [72]. The WO_2-Na_xWO_3 hybrid was prepared by high-temperature reduction of semiconductor Na_xWO_3 (x < 0.25) nanowire bundles (Figure 8.13a). With increasing the reduction temperature, the color of the products changes from yellowish-green to gray, blue, deep-violet, and dark-red (Figure 8.13b) in the range of 180–1000 °C. The as-prepared WO_2-Na_xWO_3 hybrid exhibits broad spectral absorption in the whole UV–vis–NIR range, and an LSPR peak is clearly observed in the visible light region (Figure 8.13c). At low value of x (x < 0.25), sodium tungsten bronzes (Na_xWO_3) are semiconducting; when x value is greater than approximately 0.25, it undergoes

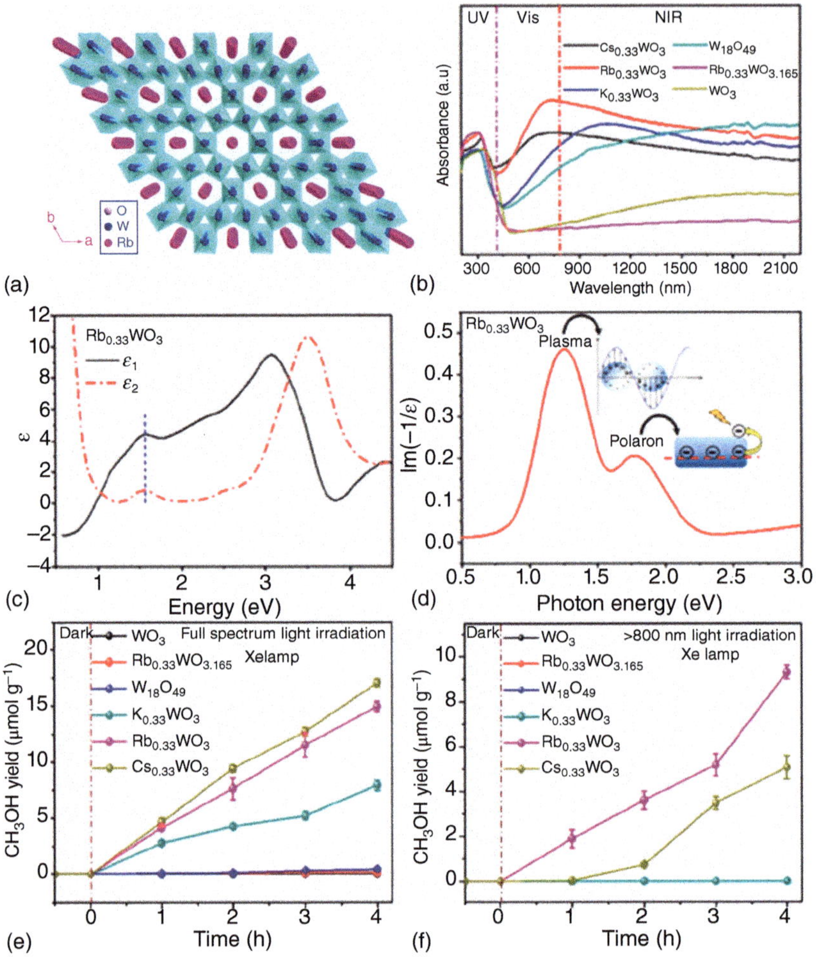

Figure 8.12 (a) The crystal structure of $Rb_{0.33}WO_3$. (b) UV–vis–NIR diffuse reflectance spectra of the samples. (c) Calculated dielectric functions of the $Rb_{0.33}WO_3$ sample. (d) Energy loss function of $Rb_{0.33}WO_3$ derived from (c). The CO_2 reduction activities of the samples under (e) full spectrum light irradiation and (f) NIR light irradiation. *Source*: Wu et al. [40]. © 2019, American Chemical Society.

semiconducting-metallic transition to become a conductor [73]. Interestingly, among the WO_2-Na_xWO_3 hybrids, the product prepared at reduction temperature of 900 °C (NaHTB-C900) exhibited the highest H_2 evolution ability for water splitting, of which consists of 37.5% WO_2 and 62.5% Na_xWO_3 ($x \approx 0.54$), even under IR light irradiation (Figure 8.13d). The efficient IR-driven photocatalytic water splitting ability of the WO_2-Na_xWO_3 hybrid was considered to be associated with its special ladder-type energy band and the barrier-free interface carrier transport due to the electronic conductivity of WO_2 and Na_xWO_3.

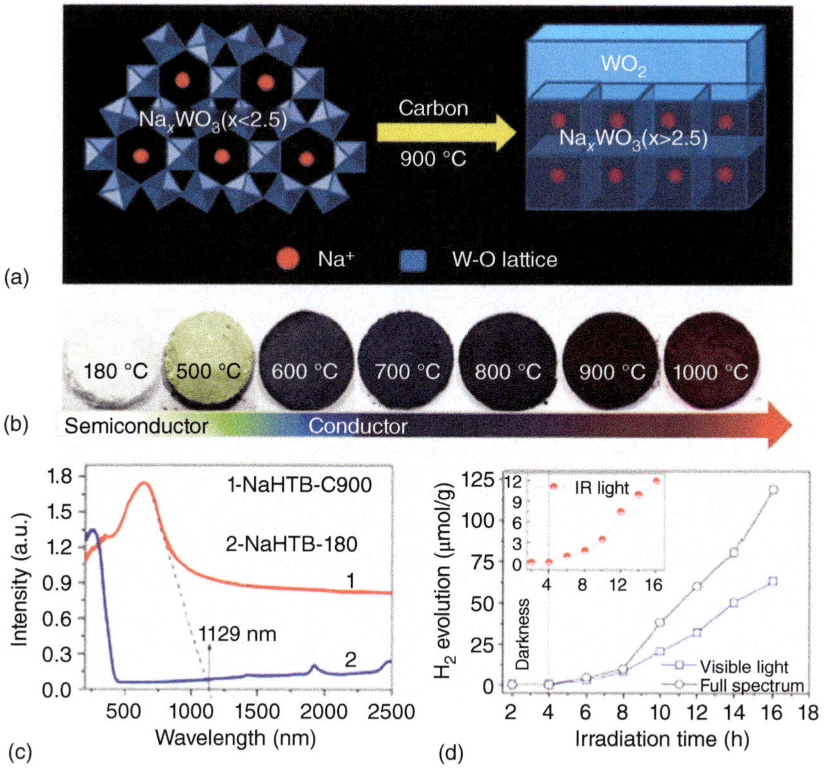

Figure 8.13 (a) Illustration of the process to prepare WO_2-Na_xWO_3 hybrid. (b) The color change for the different samples obtained under increased calcination temperatures. (c) UV–vis–NIR diffuse reflectance spectra of the products. (d) Photocatalytic H_2 evolution rates for the WO_2-Na_xWO_3 hybrid product under different light wavelength region. *Source*: Adapted with permission from [72]. Copyright 2015 American Chemical Society.

8.4 $Cu_{2-x}E$ (E = S, Se, Te)-Based Plasmonic Catalysts

Since the discovery of LSPR in $Cu_{2-x}S$ nanocrystals at NIR region, copper-deficient copper chalcogenides ($Cu_{2-x}E$, E = S, Se, Te) nanostructures as plasmonic candidates have been extensively studied [8, 9, 29–33]. The $Cu_{2-x}E$ (E = S, Se, Te) are typical *p*-type semiconductors. In copper chalcogenides, the top of the VB has a strong contribution from the chalcogen *p* orbitals. The bottom of the CB has mainly contributions from Cu 4s and 4p orbitals [10]. For a fully stoichiometric Cu_2E compound, the VB is completely filled and the material would behave as an intrinsic semiconductor. When a Cu atom is removed from the lattice, a hole in created in the top of the VB. With sufficient free holes in the VB, $Cu_{2-x}E$ (E = S, Se, Te) could behave surface plasmon resonance in the NIR region. Recently, plasmonic $Cu_{2-x}E$ (E = S, Se, Te) nanostructures have been used as potential catalysts under NIR light irradiation [32, 54, 74, 75].

The integration of plasmonic $Cu_{2-x}S$ and catalytic Pd NPs leads to an efficient Cu_7S_4@Pd hybrid photocatalyst for Suzuki coupling reaction, hydrogenation of nitrobenzene, and selective oxidation of benzyl alcohol [32]. Through a combined hot-injection method, the Cu_7S_4@Pd hybrid was synthesized, having Cu_7S_4 with the average size of 14 nm and Pd NPs of c. 4.3 nm grown onside. The Cu_7S_4@Pd hybrid exhibits strong NIR light absorption with an LSPR peak around 2000 nm (Figure 8.14a), which is red shifted with respect to pure Cu_7S_4. The catalytic activity of the Cu_7S_4@Pd hybrid was assessed by Suzuki-coupling reaction between iodobenzene and phenylboronic acid (Figure 8.14b). Intriguingly, the Cu_7S_4@Pd hybrid showed wavelength-dependent activity, and the highest activity was obtained under irradiation at 1500 nm, approaching its LSPR absorption peak. In catalysis toward selective oxidation of benzyl alcohol and hydrogenation of nitrobenzene, the Cu_7S_4@Pd hybrid also showed efficient and similar wavelength-dependence (Figure 8.14c,d). Upon 1500 nm irritation and LSPR absorption, hot holes can be generated on Cu_7S_4, which can inject into Pd domain to render hole-rich Pd surface. This hole-rich Pd surface may serve as effective catalytic sites for Suzuki coupling, oxidation and hydrogenation, illustrating the potential use of near-infrared plasmons to harvest solar energy and enhance the photocatalytic reactions.

In addition, plasmonic $Cu_{2-x}S$ nanowires coupled with Pd NPs have also displayed efficient catalytic activity toward NH_3BH_3 dehydrogenation [54]. Through cation exchange between CdS nanowires and Cu ions (Cu^+ and Cu^{2+}), the $Cu_{2-x}S$ nanowires was obtained. With x ranging from 0 to 1, the tuned $Cu_{2-x}S$ nanowires exhibit plasmonic absorption from 740 to 930 nm (Figure 8.15a,b). After decorated with Pd NPs, the resulting Pd/$Cu_{2-x}S$ hybrid possesses well-dispersed Pd NPs on Cu_2S nanowires (Figure 8.15c). As shown in Figure 8.15d,e, the great enhancement in the H_2 production was observed with $Cu_{2-x}S$ nanowires decorated with Pd NPs relative to undecorated pristine ones. The results not only shed light on the innovative synthesis method to plasmonic $Cu_{2-x}S$ nanowires, but may lead to an effective strategy for the design and development of LSPR materials for photocatalytic applications.

The plasmon-assisted energy conversion to drive chemical reactions on $Cu_{2-x}E$ ($E = S$, Se, Te)-based nanostructures has also been employed in SERS photoredox chemistry and photocatalytic H_2 evolution. Recently, Millstone and coworkers reported the plasmon-driven dimerization of 4-nitrobenzenethiol (NBT) to 4,4′-dimercaptoazobenzene (DMAB) on $Cu_{2-x}Se$ surfaces with yields comparable to those of noble metal NPs [74]. Synthesized by the hot-injection method, the oleylamine-capped $Cu_{2-x}Se$ NPs have the diameter of c. 16.4 nm and exhibit an LSPR at approximately 1100 nm (Figure 8.16a). The SERS property was studied by introducing NBT as a reporter molecule for the SERS measurement. Within five minutes of irradiation at 1035 nm, a solution containing NBT adsorbed to the $Cu_{2-x}Se$ surface exhibited DMAB product peaks in the Raman spectra (Figure 8.16b). Interestingly, the product peaks are shifted by ~10–20 cm^{-1}, comparable to those observed on noble metal plasmonic substrate. In another study, Teranishi et al. reported IR-responsive plasmonic energy conversion system for photocatalytic H_2 evolution reaction on plasmonic *p–n* junction, that is, CdS/Cu_7S_4

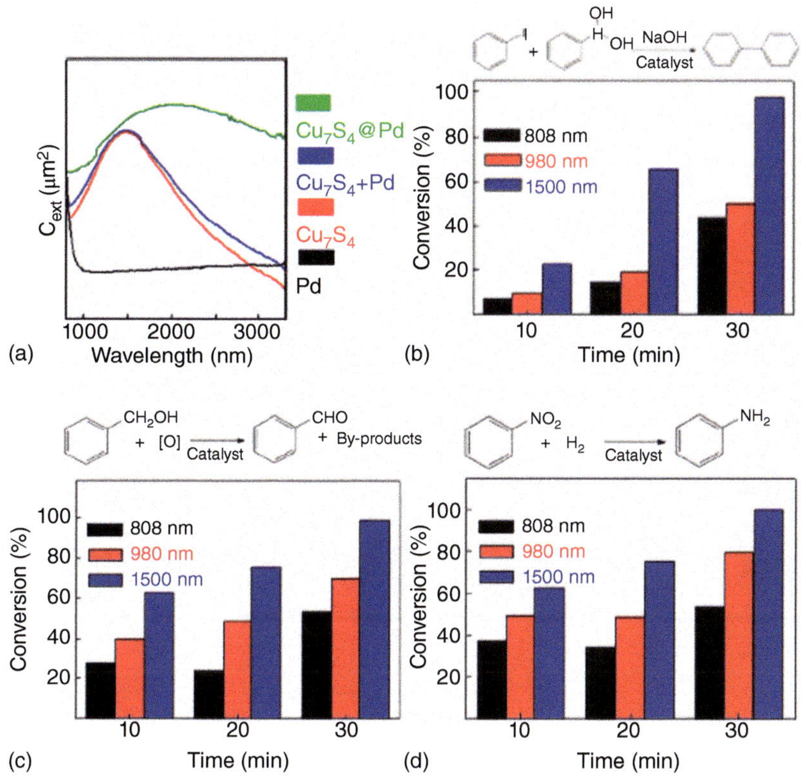

Figure 8.14 (a) Optical absorption spectra of the obtained different nanostructures. (b–d) Photocatalytic activity of Cu_7S_4@Pd nanostructure for Suzuki coupling reactions, benzyl alcohol oxidation, and nitrobenzene hydrogenation under different light wavelength irradiation. *Source*: Cui et al. [32]. © 2015, American Chemical Society.

heterostructured nanocrystals (HNCs) [75]. Compared to single Cu_7S_4 nanocrystals (NCs) with LSPR peak at 1585 nm, the as-synthesized CdS/Cu_7S_4 HNCs display a blue-shifted LSPR peak at around 1115 nm (Figure 8.16c). The *p*-type Cu_7S_4 and *n*-type CdS form the plasmonic *p*–*n* junction at the interface. The Cu_7S_4 NCs exhibited no H_2 evolution activity, whereas the CdS/Cu_7S_4 HNCs showed high activity. The apparent quantum yield (AQYs) obtained at monochromic light with different wavelengths were consistent with the LSPR spectrum, confirming the LSPR excitation-induced photocatalytic reaction (Figure 8.16d). These results pave the way for developing novel efficient systems based on IR solar energy, and might allow the implementation of plasmonic sensors and detectors responsive to IR light energy.

8.5 Outlook

As a burgeoning research field, degenerately doped semiconductors have recently emerged as an important class of plasmonic candidates and have received extensive

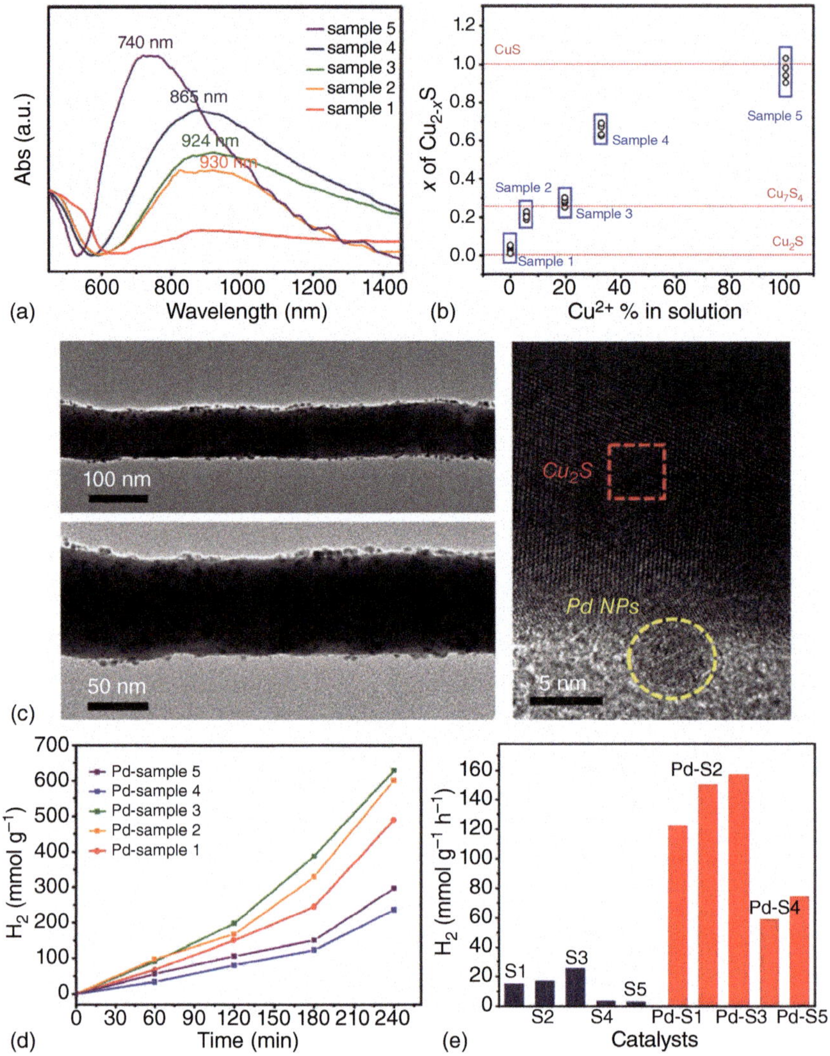

Figure 8.15 (a) The vis-NIR spectra of different $Cu_{2-x}S$ nanowires. (b) The plot of x values of $Cu_{2-x}S$ nanowires produced with various synthesis conditions. (c) TEM images of Pd NPs-decorated sample-1 structure. (d) The H_2 generation versus time during the catalytic hydrolysis of NH_3BH_3 in Pd-Sample 1 to Pd-Sample 5. (e) The H_2 generation rates achieved by bare $Cu_{2-x}S$ samples (in blue) and Pd-decorated materials (in red). *Source*: Adapted with permission from [54]. Copyright 2017 Elsevier.

attention. LSPR in degenerately doped semiconductor could be tuned by free carrier concentration, which is dependent on the size, shape, surrounding dielectric environment, dopant species, as well as the postsynthetic oxidation/reduction treatment, allowing one to tailor the spectral response from visible light to NIR region in a controlled manner. Upon plasmonic excitation, the hot electrons or holes in degenerately doped semiconductors could separate effectively and participate in

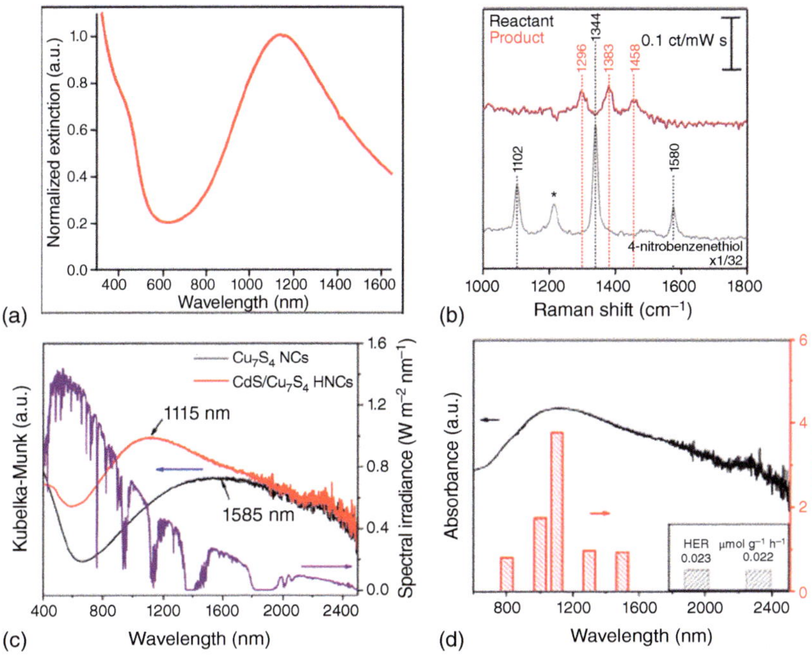

Figure 8.16 (a) The extinction spectrum of the NPs dispersed in $CHCl_3$ oleylamine-capped $Cu_{2-x}Se$ NPs. (b) The SER difference spectrum of oleylamine-capped $Cu_{2-x}Se$ NPs (maroon) for dimerization of the reactant 4-nitrobenzenethiol (NBT) (black dashed lines) to the product 4,4′-dimercaptoazobenzene (DMAB). *Source*: Gan et al. [74]. © 2019, American Chemical Society. (c) The diffuse reflectance spectra of Cu_7S_4 NCs and CdS/Cu_7S_4 HNCs together with AM1.5 solar spectrum. (d) The absorption spectrum and AQYs for hydrogen evolution of the CdS/Cu_7S_4 HNCs under monochromic light. *Source*: Lian et al. [75]. © 2019, American Chemical Society.

the redox reactions efficiently, paving the way to harvest and utilize the solar energy to drive chemical reactions. The plasmonic degenerately doped semiconductors have shown efficient or enhanced catalytic activities toward NH_3BH_3 dehydrogenation, Suzuki coupling reaction, *p*-nitrophenol reduction, hydrogen evolution reaction, CO_2 reduction, selective oxidation of benzyl alcohol, nitrobenzene hydrogenation, sulfoxide deoxygenation, 4-nitrobenzenethiol dimerization, and water splitting. In contrast to noble metals, degenerately doped semiconductors with earth-abundant elements would be more appealing for large-scale practical applications.

Albeit with much progress, plasmonic degenerately doped semiconductors are still at the infant stage, many issues need to be addressed and many new discoveries need to be explored:

(i) To gain insight into the origin of plasmon resonance in the degenerately doped semiconductors, the temporal and spatial carriers distribution profile is a requisite. The far- and near-field spectral response are also important. Therefore, first-principle theoretical calculations and finite difference time domain (FDTD) simulations are useful tools to build the structure–property relationship in the plasmonic degenerately doped semiconductors.

(ii) The conventional degenerately doped semiconductors are usually prepared by hot-injection route with the capping agents. However, the use of capping agents lowers the surface feasibility and may lead to unpredictable cytotoxic effects. The surfactant-free, facile, and green routes to plasmonic doped semiconductors are highly significant for both experimental studies and industrial implementation.

(iii) The plasmon resonance in semiconductors is realized through heavy doping. The dopants are prone to be balanced or compensated, and they are susceptible to surrounding oxidation/reduction condition, which makes the plasmonic-doped semiconductors changeable or not stable at long time exposure as compared to their noble metal counterparts. How to maintain the original plasmonic feature of degenerately doped semiconductors remains a challenging task.

(iv) The plasmon resonance of doped semiconductors is closely related to their local crystallographic structure as well as dopant distribution. The collective characterization tools like synchrotron radiation light source, and the spatial resolution techniques like aberration-corrected high-resolution electron microscopy are useful means to unravel these connections.

The plasmonic degenerately doped semiconductors have immensely expanded the library of materials displaying unique LSPR feature. Depending on the free carrier concentration induced by dopants, the degenerately doped semiconductors undergo semiconducting-metallic transition, which covers the gap between metals and semiconductors. Apart from harvesting and converting solar energy to drive chemical reactions, the plasmonic degenerately doped semiconductors could find more other applications in photonic, optoelectronic, sensing, and spectroscopy. With a rapid development of this research field and better understanding of the origin mechanism, local crystallographic structure, as well as dopant distribution, which are assisted by the advanced characterizations and theoretical calculations, it is expected that highly efficient plasmonic doped semiconductors catalysts with stronger light response and better stability can be rationally designed and prepared in the following future.

References

1 Johnson, P.B. and Christy, R.-W. (1972). Optical constants of the noble metals. *Phys. Rev. B* 6: 4370.

2 Murray, W.A. and Barnes, W.L. (2007). Plasmonic materials. *Adv. Mater.* 19: 3771–3782.

3 Lu, X., Rycenga, M., Skrabalak, S.E. et al. (2009). Chemical synthesis of novel plasmonic nanoparticles. *Annu. Rev. Phys. Chem.* 60: 167–192.

4 Linic, S., Christopher, P., and Ingram, D.B. (2011). Plasmonic-metal nanostructures for efficient conversion of solar to chemical energy. *Nat. Mater.* 10: 911–921.
5 Clavero, C. (2014). Plasmon-induced hot-electron generation at nanoparticle/metal-oxide interfaces for photovoltaic and photocatalytic devices. *Nat. Photonics* 8: 95–103.
6 Cheng, H., Fuku, K., Kuwahara, Y. et al. (2015). Harnessing single-active plasmonic nanostructures for enhanced photocatalysis under visible light. *J. Mater. Chem. A* 3: 5244–5258.
7 Haxel, G. (2002). *Rare Earth Elements: Critical Resources for High Technology*, vol. 87. US Department of the Interior, US Geological Survey.
8 Zhao, Y., Pan, H., Lou, Y. et al. (2009). Plasmonic $Cu_{2-x}S$ nanocrystals: optical and structural properties of copper-deficient copper (I) sulfides. *J. Am. Chem. Soc.* 131: 4253–4261.
9 Luther, J.M., Jain, P.K., Ewers, T., and Alivisatos, A.P. (2011). Localized surface plasmon resonances arising from free carriers in doped quantum dots. *Nat. Mater.* 10: 361–366.
10 Comin, A. and Manna, L. (2014). New materials for tunable plasmonic colloidal nanocrystals. *Chem. Soc. Rev.* 43: 3957–3975.
11 Liu, X. and Swihart, M.T. (2014). Heavily-doped colloidal semiconductor and metal oxide nanocrystals: an emerging new class of plasmonic nanomaterials. *Chem. Soc. Rev.* 43: 3908–3920.
12 Lounis, S.D., Runnerstrom, E.L., Llordés, A., and Milliron, D.J. (2014). Defect chemistry and plasmon physics of colloidal metal oxide nanocrystals. *J. Phys. Chem. Lett.* 5: 1564–1574.
13 Kriegel, I., Scotognella, F., and Manna, L. (2017). Plasmonic doped semiconductor nanocrystals: properties, fabrication, applications and perspectives. *Phys. Rep.* 674: 1–52.
14 Agrawal, A., Cho, S.H., Zandi, O. et al. (2018). Localized surface plasmon resonance in semiconductor nanocrystals. *Chem. Rev.* 118: 3121–3207.
15 Hövel, H., Fritz, S., Hilger, A. et al. (1993). Width of cluster plasmon resonances: bulk dielectric functions and chemical interface damping. *Phys. Rev. B* 48: 18178.
16 Kelly, K.L., Coronado, E., Zhao, L.L., and Schatz, G.C. (2003). The optical properties of metal nanoparticles: the influence of size, shape, and dielectric environment. *J. Phys. Chem. B* 107: 668–677.
17 Kanehara, M., Koike, H., Yoshinaga, T., and Teranishi, T. (2009). Indium tin oxide nanoparticles with compositionally tunable surface plasmon resonance frequencies in the near-IR region. *J. Am. Chem. Soc.* 131: 17736–17737.
18 Buonsanti, R., Llordes, A., Aloni, S. et al. (2011). Tunable infrared absorption and visible transparency of colloidal aluminum-doped zinc oxide nanocrystals. *Nano Lett.* 11: 4706–4710.
19 Chou, L.-W. and Filler, M.A. (2013). Engineering multimodal localized surface plasmon resonances in silicon nanowires. *Angew. Chem. Int. Ed.* 52: 8079–8083.

20 Kramer, N.J., Schramke, K.S., and Kortshagen, U.R. (2015). Plasmonic properties of silicon nanocrystals doped with boron and phosphorus. *Nano Lett.* 15: 5597–5603.
21 Huang, Q., Hu, S., Zhuang, J., and Wang, X. (2012). MoO_{3-x}-based hybrids with tunable localized surface plasmon resonances: chemical oxidation driving transformation from ultrathin nanosheets to nanotubes. *Chem. A Eur. J.* 18: 15283–15287.
22 Manthiram, K. and Alivisatos, A.P. (2012). Tunable localized surface plasmon resonances in tungsten oxide nanocrystals. *J. Am. Chem. Soc.* 134: 3995–3998.
23 Alsaif, M.M.Y.A., Latham, K., Field, M.R. et al. (2014). Tunable plasmon resonances in two-dimensional molybdenum oxide nanoflakes. *Adv. Mater.* 26: 3931–3937.
24 Cheng, H., Kamegawa, T., Mori, K., and Yamashita, H. (2014). Surfactant-free nonaqueous synthesis of plasmonic molybdenum oxide nanosheets with enhanced catalytic activity for hydrogen generation from ammonia borane under visible light. *Angew. Chem. Int. Ed.* 53: 2910–2914.
25 Cheng, H.F., Qian, X.F., Kuwahara, Y. et al. (2015). A plasmonic molybdenum oxide hybrid with reversible tunability for visible-light-enhanced catalytic reactions. *Adv. Mater.* 27: 4616–4621.
26 Yan, J., Wang, T., Wu, G. et al. (2015). Tungsten oxide single crystal nanosheets for enhanced multichannel solar light harvesting. *Adv. Mater.* 27: 1580–1586.
27 Liu, W., Xu, Q., Cui, W. et al. (2017). CO_2-assisted fabrication of two-dimensional amorphous molybdenum oxide nanosheets for enhanced plasmon resonances. *Angew. Chem. Int. Ed.* 56: 1600–1604.
28 Cheng, H.F., Klapproth, M., Sagaltchik, A. et al. (2018). Ordered mesoporous $WO_{2.83}$: selective reduction synthesis, exceptional localized surface plasmon resonance and enhanced hydrogen evolution reaction activity. *J. Mater. Chem. A* 6: 2249–2256.
29 Dorfs, D., Härtling, T., Miszta, K. et al. (2011). Reversible tunability of the near-infrared valence band plasmon resonance in $Cu_{2-x}Se$ nanocrystals. *J. Am. Chem. Soc.* 133: 11175–11180.
30 Kriegel, I., Jiang, C., Rodríguez-Fernández, J. et al. (2012). Tuning the excitonic and plasmonic properties of copper chalcogenide nanocrystals. *J. Am. Chem. Soc.* 134: 1583–1590.
31 Li, W., Zamani, R., Rivera Gil, P. et al. (2013). CuTe nanocrystals: shape and size control, plasmonic properties, and use as SERS probes and photothermal agents. *J. Am. Chem. Soc.* 135: 7098–7101.
32 Cui, J., Li, Y., Liu, L. et al. (2015). Near-infrared plasmonic-enhanced solar energy harvest for highly efficient photocatalytic reactions. *Nano Lett.* 15: 6295–6301.
33 Wang, F., Li, Q., Lin, L. et al. (2015). Monodisperse copper chalcogenide nanocrystals: controllable synthesis and the pinning of plasmonic resonance absorption. *J. Am. Chem. Soc.* 137: 12006–12012.
34 Dickens, P.G., Hibble, S.J., and Jarman, R.H. (1981). Hydrogen insertion compounds of transition metal oxides. *J. Electron. Mater.* 10: 999–1009.

35 Greenblatt, M. (1988). Molybdenum oxide bronzes with quasi-low-dimensional properties. *Chem. Rev.* 88: 31–53.

36 Mehandru, S.P. and Anderson, A.B. (1988). H_xMoO_3 bronzes: structures, stabilities, and electronic properties. *J. Am. Chem. Soc.* 110: 2061–2065.

37 Haldolaarachchige, N., Gibson, Q., Krizan, J., and Cava, R.J. (2014). Superconducting properties of the K_xWO_3 tetragonal tungsten bronze and the superconducting phase diagram of the tungsten bronze family. *Phys. Rev. B* 89: 104520.

38 Cheng, H., Wen, M., Ma, X. et al. (2016). Hydrogen doped metal oxide semiconductors with exceptional and tunable localized surface plasmon resonances. *J. Am. Chem. Soc.* 138: 9316–9324.

39 Kuwahara, Y., Yoshimura, Y., Haematsu, K., and Yamashita, H. (2018). Mild deoxygenation of sulfoxides over plasmonic molybdenum oxide hybrid with dramatic activity enhancement under visible light. *J. Am. Chem. Soc.* 140: 9203–9210.

40 Wu, X., Li, Y., Zhang, G. et al. (2019). Photocatalytic CO_2 conversion of $M_{0.33}WO_3$ directly from the air with high selectivity: insight into full spectrum-induced reaction mechanism. *J. Am. Chem. Soc.* 141: 5267–5274.

41 Lopez, R., Haynes, T.E., Boatner, L.A. et al. (2002). Temperature-controlled surface plasmon resonance in VO_2 nanorods. *Opt. Lett.* 27: 1327–1329.

42 Biswas, K. and Rao, C.N.R. (2006). Metallic ReO_3 nanoparticles. *J. Phys. Chem. B* 110: 842–845.

43 Wang, L., Clavero, C., Yang, K. et al. (2012). Bulk and surface plasmon polariton excitation in RuO_2 for low-loss plasmonic applications in NIR. *Opt. Express* 20: 8618–8628.

44 Rej, S., Mascaretti, L., Santiago, E.Y. et al. (2020). Determining plasmonic hot electrons and photothermal effects during H_2 evolution with TiN–Pt nanohybrids. *ACS Catal.* 10: 5261–5271.

45 Greiner, M.T. and Lu, Z.-H. (2013). Thin-film metal oxides in organic semiconductor devices: their electronic structures, work functions and interfaces. *NPG Asia Mater.* 5: e55.

46 Chen, H., Shao, L., Li, Q., and Wang, J. (2013). Gold nanorods and their plasmonic properties. *Chem. Soc. Rev.* 42: 2679–2724.

47 Aslam, U., Rao, V.G., Chavez, S., and Linic, S. (2018). Catalytic conversion of solar to chemical energy on plasmonic metal nanostructures. *Nat. Catal.* 1: 656–665.

48 Mascaretti, L., Dutta, A., Kment, Š. et al. (2019). Plasmon-enhanced photoelectrochemical water splitting for efficient renewable energy storage. *Adv. Mater.* 31: 1805513.

49 Lignier, P., Bellabarba, R., and Tooze, R.P. (2012). Scalable strategies for the synthesis of well-defined copper metal and oxide nanocrystals. *Chem. Soc. Rev.* 41: 1708–1720.

50 Marimuthu, A., Zhang, J., and Linic, S. (2013). Tuning selectivity in propylene epoxidation by plasmon mediated photo-switching of cu oxidation state. *Science* 339: 1590–1593.

51 Knight, M.W., King, N.S., Liu, L. et al. (2014). Aluminum for plasmonics. *ACS Nano* 8: 834–840.

52 Duan, X. and Liu, N. (2019). Magnesium for dynamic nanoplasmonics. *Acc. Chem. Res.* 52: 1979–1989.

53 Robatjazi, H., Bao, J.L., Zhang, M. et al. (2020). Plasmon-driven carbon–fluorine (C(sp3)–F) bond activation with mechanistic insights into hot-carrier-mediated pathways. *Nat. Catal.* 3: 564–573.

54 Liu, P.-H., Wen, M., Tan, C.-S. et al. (2017). Surface plasmon resonance enhancement of production of H_2 from ammonia borane solution with tunable $Cu_{2-x}S$ nanowires decorated by Pd nanoparticles. *Nano Energy* 31: 57–63.

55 Huang, Z.-F., Song, J., Pan, L. et al. (2015). Tungsten oxides for photocatalysis, electrochemistry, and phototherapy. *Adv. Mater.* 27: 5309–5327.

56 Cong, S., Geng, F., and Zhao, Z. (2016). Tungsten oxide materials for optoelectronic applications. *Adv. Mater.* 28: 10518–10528.

57 de Castro, I.A., Datta, R.S., Ou, J.Z. et al. (2017). Molybdenum oxides – from fundamentals to functionality. *Adv. Mater.* 29: 1701619.

58 Zhu, Q.-L. and Xu, Q. (2015). Liquid organic and inorganic chemical hydrides for high-capacity hydrogen storage. *Energ. Environ. Sci.* 8: 478–512.

59 Fuku, K., Hayashi, R., Takakura, S. et al. (2013). The synthesis of size- and color-controlled silver nanoparticles by using microwave heating and their enhanced catalytic activity by localized surface plasmon resonance. *Angew. Chem. Int. Ed.* 52: 7446–7450.

60 Yin, H., Kuwahara, Y., Mori, K. et al. (2017). High-surface-area plasmonic MoO_{3-x}: rational synthesis and enhanced ammonia borane dehydrogenation activity. *J. Mater. Chem. A* 5: 8946–8953.

61 Shi, J., Kuwahara, Y., Wen, M. et al. (2016). Room-temperature and aqueous-phase synthesis of plasmonic molybdenum oxide nanoparticles for visible-light-enhanced hydrogen generation. *Chem. Asian J.* 11: 2377–2381.

62 Yin, H., Kuwahara, Y., Mori, K. et al. (2017). Localized surface plasmon resonances in plasmonic molybdenum tungsten oxide hybrid for visible-light-enhanced catalytic reaction. *J. Phys. Chem. C* 121: 23531–23540.

63 Yin, H., Kuwahara, Y., Mori, K., and Yamashita, H. (2018). Plasmonic metal/$Mo_xW_{1-x}O_{3-y}$ for visible-light-enhanced H_2 production from ammonia borane. *J. Mater. Chem. A* 6: 10932–10938.

64 Li, J., Chen, G., Yan, J. et al. (2020). Solar-driven plasmonic tungsten oxides as catalyst enhancing ethanol dehydration for highly selective ethylene production. *Appl Catal B* 264: 118517.

65 Lee, K., Seo, W.S., and Park, J.T. (2003). Synthesis and optical properties of colloidal tungsten oxide nanorods. *J. Am. Chem. Soc.* 125: 3408–3409.

66 Polleux, J., Gurlo, A., Barsan, N. et al. (2006). Template-free synthesis and assembly of single-crystalline tungsten oxide nanowires and their gas-sensing properties. *Angew. Chem. Int. Ed.* 45: 261–265.

67 Li, Y.H., Liu, P.F., Pan, L.F. et al. (2015). Local atomic structure modulations activate metal oxide as electrocatalyst for hydrogen evolution in acidic water. *Nat. Commun.* 6: 8064.

68 Linsebigler, A.L., Lu, G., and Yates, J.T. (1995). Photocatalysis on TiO_2 surfaces: principles, mechanisms, and selected results. *Chem. Rev.* 95: 735–758.

69 Mattox, T.M., Bergerud, A., Agrawal, A., and Milliron, D.J. (2014). Influence of shape on the surface plasmon resonance of tungsten bronze nanocrystals. *Chem. Mater.* 26: 1779–1784.

70 Yin, H., Kuwahara, Y., Mori, K. et al. (2019). Plasmonic Ru/hydrogen molybdenum bronzes with tunable oxygen vacancies for light-driven reduction of *p*-nitrophenol. *J. Mater. Chem. A* 7: 3783–3789.

71 Yin, H., Kuwahara, Y., Mori, K., and Yamashita, H. (2019). RuPd alloy nanoparticles supported on plasmonic H_xMoO_{3-y} for efficient photocatalytic reduction of *p*-nitrophenol. *Eur. J. Inorg. Chem.* 2019: 3745–3752.

72 Cui, G., Wang, W., Ma, M. et al. (2015). IR-driven photocatalytic water splitting with WO_2–Na_xWO_3 hybrid conductor material. *Nano Lett.* 15: 7199–7203.

73 Chaitanya Lekshmi, I. and Hegde, M.S. (2005). Synthesis and electrical properties of cubic Na_xWO_3 thin films across the metal–insulator transition. *Mater. Res. Bull.* 40: 1443–1450.

74 Gan, X.Y., Keller, E.L., Warkentin, C.L. et al. (2019). Plasmon-enhanced chemical conversion using copper selenide nanoparticles. *Nano Lett.* 19: 2384–2388.

75 Lian, Z., Sakamoto, M., Vequizo, J.J.M. et al. (2019). Plasmonic p–n junction for infrared light to chemical energy conversion. *J. Am. Chem. Soc.* 141: 2446–2450.

9

Plasmon-Enhanced Electrocatalysis

Subin Yu[1,], Nur Aqlili Riana Che Mohamad[1,*], Minju Kim[1], Yoonseo Nah[2], Filipe Marques Mota[1] and Dong Ha Kim[1,2]*

[1] *Department of Chemistry and Nano Science, Division of Molecular and Life Sciences, College of Natural Sciences, Ewha Womans University, Seoul, Republic of Korea*
[2] *Department of Chemical Engineering and Materials Science, College of Engineering, Ewha Womans University, Seoul, Republic of Korea*

9.1 Introduction

In recent years, driven by a systematic growth of novel energy technologies, renewables have surpassed coal to become the primary source of electricity. This steep progress is largely ascribed to the wide implementation of wind-derived energy and the steady development of solar cells in the photovoltaic industry with uprising efficiency and cost-effectiveness [1]. As a result, the decrease in electricity prices has enhanced the attractiveness of non-carbon-emitting energy sources as alternatives to fossil fuels. Utilizing low-cost environmental-friendly renewable electricity sources that can be operated at remote locations and under mild conditions is a promising answer today to accommodate ever-rising energy consumptions and decreasing oil supplies. Exploiting the redox chemistry to convert earth-abundant renewable feedstocks such as water and CO_2 into value-added fuels may, on the other hand, assist a necessary output leveling of wind and solar intermittent energy sources, strongly dependent on season and daylight [2]. Feasible application of electrocatalytic processes depends, nonetheless, on local feedstock/petrochemical markets/technologies, the expansion of carbon-capture platforms, and both social and political incentives for the decarbonization of the society. Also, the direct coupling of electrocatalytic platforms with photovoltaic technologies is necessary to avoid additional electricity storage costs and to enhance the attractiveness of these systems against conventional counterparts.

The design and envisioned implementation of electrocatalytic systems and applied materials have been extensively surveyed in the literature. Against photocatalytic technologies possessing superior promise toward commercialization and facile reactor design, electrocatalytic systems have been distinctly distinguished for their

* These authors contributed equally to this work.

Plasmonic Catalysis: From Fundamentals to Applications, First Edition.
Edited by Pedro H.C. Camargo and Emiliano Cortés.

notably superior energy efficiencies. Splitting the water molecule to produce H_2 and reducing CO_2 to valuable feedstocks are expected to be crucial in the development of a sustainable society. However, high-energy barriers and large overpotentials in corresponding electrocatalytic reactions emerge as impeding drawbacks toward an anticipated operation of said technologies on a global-scale demand. Among viable solutions to enhance the energy efficiency of benchmark electrocatalytic materials, the incorporation of light-responsive materials toward solar energy harvesting in the extended UV–Vis–NIR range (occupying ~5, ~49, and ~46% of the solar light range, respectively) is of great promise. The combination of materials with distinct optoelectronic fingerprints in photoelectrochemical (PEC) systems has received rising notice since Honda and Fujishima paired the semiconductor TiO_2 and a Pt counter-electrode to split the water molecule in their seminal works [3].

As an alternative, enhancing the performance of electrocatalytic systems has been achieved through the incorporation of plasmonic metal nanoparticles (e.g. Au, Ag, Cu, and Pd) with plasmonic resonances deriving from the collective oscillation of abundant *quasi*-free conduction electrons. The exploitation of this class of materials has been suitably extended to a wide range of research fields including optical sensing, light emission, photodetectors, waveguiding, data storage, and theragnosis [4–7]. In the catalysis realm, the tunable surface plasmon resonance (SPR) properties directly correlated with the nanoparticle geometry, size, and shape of distinct metals offer an extended versatility in the construction of plasmon-enhanced electrocatalytic systems. Whereas the reported catalytic activity, selectivity, and stability of the resulting hybrid materials have stressed the promise of this approach, to date, understanding the concepts and mechanisms enhancing the inherent performance of state-of-the-art electrocatalysts still requires a crucial focus. In agreement, in the sections below we unveil the promise of plasmon-enhanced electrocatalytic systems while rationalizing the properties of plasmonic nanoparticles.

9.2 Principles and Mechanism

9.2.1 Introducing Plasmonic Nanostructures in Electrocatalytic Systems

To date, SPR excitation is primarily exploited through indirect photocatalysis via a plasmon-mediated electron-transfer path to the surface of adjacently coupled semiconductors or to other adjacent metals in photo(electro)catalytic systems [8, 9], which facilitates a separation of generated charge carriers. As opposed to the conventional incorporation of plasmonic nanostructures on semiconductors, plasmon-enhanced electrocatalysis exploits the strong interaction between the incident light and surface plasmons in nanoparticles either (i) incorporated into benchmark electrocatalysts or (ii) directly serving as catalytically active surfaces. In both approaches, the performance and mechanism of plasmon-enhanced catalytic systems are strongly correlated with the uniqueness of the hybrid architecture. Increasingly sophisticated metal nanoparticles offer today a wide degree of morphological features, and optical, physical, and electrochemical properties of these materials, including light-manipulating capabilities and polarization control [10]. On the other hand, methodologies to incorporate plasmonic metals on benchmark electrocatalysts offer wide versatility in the construction of hybrid structures, following the know-how established in PEC systems [11]. Careful analysis is thus required to understand the interaction and the resulting interface between the electrocatalytic surface and the introduced plasmonic nanostructures.

However, the direct exploitation of plasmonic materials with the direct conversion of reactants adsorbed on the metal surface remains relatively unexplored due to the intrinsically poor activity of plasmonic nanostructures and the swift recombination of charge carriers detrimental to the desired process efficiency [12]. Herein, comprehending the effect of the nanoparticle geometry/shape/size on the efficiency, selectivity, and catalytic stability in addition to its plasmonic properties, SPR wavelength, and primary plasmon decay mechanism is critical. Rational selection of crystal facets offers distinct surface energies, reaction kinetics, and resulting catalytic activities [13]. Conversely, shape dependence has been reflected through model reactions, with Au nanostars and nanoplates showing superior enhancement against nanospheres and nanorods for the representative electro-oxidation of ascorbic acid [14].

9.2.2 Disentangling Mechanism Pathways

Conventionally, coupling light with oscillating surface plasmons induces near-field generation surrounding the nanoparticles through a concentration of far-field radiation. The plasmonic energy eventually dephases and relaxes through distinct radiative or nonradiative decay channels. As detailed in previous chapters, energy redistribution occurs in the excited carriers through electron–electron interaction within a 100 fs time frame (Figure 9.1a) [12]. Electron–photon cooling can alternatively induce a Fermi–Dirac distribution of carriers. Subsequent relaxation of hot carriers takes place through electron–phonon scattering, with the nanoparticle lattice being heated and eventually establishing a thermal equilibrium with the surrounding medium [18]. Among rationally exploited plasmonic effects, e.g. produced energetic hot carriers by Landau damping, nonradiative near-field enhancement, plasmon resonance energy transfer, and photothermal, the former two are commonly attributed to superior electrocatalytic performances (Figure 9.1b). Detailed discussion in the context of electrocatalytic systems is introduced below.

Hot carriers. Because the electrocatalytic systems described in this chapter serve primarily redox reactions requiring an efficient electron transfer between reactants and active sites, hot carriers are primarily italicized in reported works and remain the focus of mechanistic studies in this field. Photoinduced hot electrons, either directly utilized at the surface of plasmonic nanoparticles or transferred to neighboring electrocatalytic centers, offer a great promise to enhance the activity of electrochemical reactions. In the same manner, remaining hot holes close to the metal surface offer the promise of accelerating the oxidation of reactant species [19, 20].

The exploitation of hot carriers in catalysis is challenging as carrier recombination of metals is quick, and hot electrons can swiftly relax to lower energy levels in

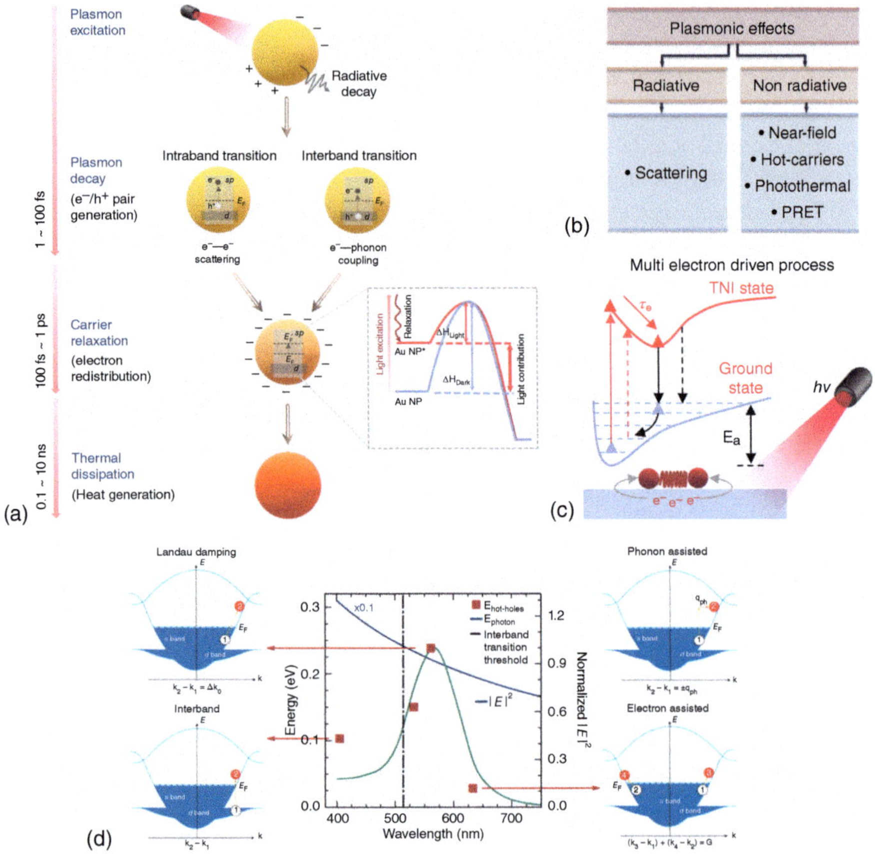

Figure 9.1 (a) Dynamics of the plasmon excitation/relaxation process in a metal nanoparticle. (b) Plasmonic effects upon light irradiation of plasmonic nanostructures. (c) Multielectron-driven mechanism in superlinear regime, during which the adsorbate species can undergo multiple electron excitations before overcoming the activation barrier. (d) Hot holes energy versus wavelength and energy of the incident photons versus wavelength on Au nanospheres. Red arrows correlate selected wavelengths and presumably predominant absorption mechanisms. *Source*: Gargiulo et al. [15] (based on the works of Pensa et al. [16] and Khurgin [17]). © 2019, American Chemical Society.

the noble metal nanostructure. Charge separation, which should occur within the sub-picosecond timescale to prevent undesirable recombination, is a clear drawback when the metal surface serves as the active site for electrocatalytic redox reactions. This difficulty drastically contrasts with PEC systems in which the quick electron injection to adjacent semiconductors favors the exploitation of hot carriers. As illustrated by the works of Halas [21–23] and Jain [12, 20], this efficiency is here primarily determined by the correlation between the activation energy of plasmon-driven chemical reactions and the incident laser power and excitation wavelength [24, 25]. To overcome ultrafast nonradiative decay dynamics, research has focused on distinct approaches, e.g. the incorporation of scavenging species [19, 20, 26, 27]. The design of hybrids based on an anisotropic distribution of hot electrons can be also exploited for the availability of concentrated hot holes on the surface of plasmonic nanoparticles [28]. When plasmonic nanostructures are incorporated on benchmark electrocatalytic materials, efficient charge separation can also be promoted by the rational construction of the interface established between both nanocomponents (e.g. band alignment). However, molecular linkers or insulating layers, allowing a direct attachment of nanostructured noble metals to neighboring nanocomponents, dramatically hinder, low-energy hot charge collection [29–31].

Electrocatalytic reduction reactions benefitting from accelerated (direct) hot electron injection can be rationalized based on earlier photocatalytic studies. Generated hot carriers impact the rate of reaction by strengthening the adsorption of active species, enhancing the surface charge heterogeneity, and charge density [12, 15]. Light irradiation can excite the electrons and overall Fermi level (E_f), minimizing the electron transfer barrier and increasing the hot electron injection rate (Figure 9.1b), with chemically absorbed reactants further enhancing the carrier capture abilities [20]. Perhaps similar to conventional photocatalytic systems, an electronic hybridization between metal and adsorbate favors fast electron transportation without any barrier of the tunneling effect. Electron backscattering into the metal is, nonetheless, proposed to occur, even if catalytic activation during the initial electron transfer to the adsorbate is observed. This chemical interface damping (backscattered process), such as electron–surface or electron–adsorbate interaction, occurs within the sub-picosecond time range for effective vibrational activation of adsorbates before thermalization via energy dissipation [32, 33]. Adsorbate species can also go through multiple electronic transitions if electronic excitation events occur sufficiently fast compared with the rate of molecular relaxation (Figure 9.1c) [34]. Multiple electronic transitions can be supported by the nonlinear trend of the photon flux versus electron transfer rate [12, 34–36]. However, it is important to understand that these single-electron excitations are events with the limited occurrence and relatively long in-between periods, both notably dependent on the light source and intensity (Section 9.3.3) [37].

Distinct underlying mechanisms leading to the generation of hot carriers can be identified based on the excitation wavelength and the resulting plasmonic energy decay. As a result, light source selection plays a key role to selectively manipulate the reaction mechanism and resulting reaction products (Figure 9.1d) [16, 17]. If the excitation wavelength is well-matched with the LSPR of selected plasmonic nanoparticles, maximum effective energy reflected in the electric field ($|E|$ [2]) results can be attained [15]. Landau damping process primarily dominates the conversion of photon energy to single intraband transition, inducing nonthermal electrons and holes [38]. Low-energy photon absorption induces the excitation of electrons close to the Fermi edge, whereas hot electrons far away from the Fermi edge can be excited by higher photon irradiation [39–41]. Generated nonthermal hot electrons, on the surface of the plasmonic nanostructure or transferred to neighboring benchmark electrocatalysts, play a key role in enhanced reaction rates. Analogously, Landau damping-induced holes close to the metal surface are proposed to serve as reactive carriers [15, 42].

In hot charge carrier-enhanced mechanisms, it is important to simultaneously assess the contribution of thermal hot carriers and occurring interband transitions.

Following LSPR excitation, excited carriers can undergo redistribution within the electronic cloud within 100 fs, leading to thermal hot carriers. This phenomenon also induces a temperature increase of the electron cloud, which follows a linear relationship with the heat capacity of the irradiated material and depends on the extent of optical energy [38, 43, 44]. Thermal hot carriers primarily contribute to the reaction rate [42], with an efficiency that follows a linear relationship with optical power but does not directly depend on the incident light wavelength. Also, interband transitions, directly correlated with the wavelength and power of the light source, can play a key role in electrocatalytic reactions. In the presence of incorporated Au nanoparticles under <450 nm light irradiation, for instance, 5*d* to 6*sp* transitions have been shown to suitably direct the overall water splitting (OWS) reaction. Conversely, even though interband transitions from the *d*-band above the Fermi level can be directed by higher photon energy, *d* holes with high energy have small mobility, and, therefore, limited utilization [44].

Near-field enhancement. Local electric fields can be enhanced when light excitation and the LSPR overlap. Normalized electric fields with absorption depend on the excitation wavelength. Reaction rate enhancement by local field if molecular (anti)bonding orbitals of reactants are promoted via local field enhancement. In detail, a strong polarized electrostatic field can be induced by the enhanced local electric field due to the facilitated correlation between catalytic sites and reactant. Comparatively, near-field enhancement can play a stronger role in electrocatalytic systems in which reactants are adsorbed to the metal surface (against conventional semiconductor-based PEC platforms). The normalized intensity of near-field enhancement to the absorbed optical power is increased toward longer wavelengths. Thus, the activity of the reactants becomes strong or proton tunneling is increased [45]. However, with complex electronic configurations, a necessary focus to comprehend the relationship between the reaction rate enhancement of adsorbed molecules and incident power is needed [42].

Thermal effect (local heating). Nonradiative decay of surface plasmons with induced heat generation is of critical importance if the chemical reaction rate is related in an Arrhenius type with the temperature. The relationship between reaction rate and incident optical power reveals, in this case, an exponential profile [46]. Phonons can be efficiently excited by nonthermal and thermalized carriers in the host lattice, a phenomenon well demonstrated in small structures with ultrashort laser pulses irradiation [47]. These acoustic phonons from nonequilibrium carriers are near the center of the Brillouin zone, leading to secondary nonequilibrium carriers *per* second. Thus, the phonon energy distribution significantly differs from the specific Bose–Einstein distribution at a certain temperature, leading to further relaxation of energy by phonon–phonon interaction [48]. Finally, generated heated particles resulting from thermal distribution facilitate thermal catalysis by reducing the activation energy of adsorbates and the Gibbs free energy of an electrocatalytic reaction, and by enhancing mass transport and the solubility of reactants toward hindered solution resistance [15, 49, 50]. Further investigation of the correlation between resulting catalytic performances and selected light sources/intensity and temperature ranges is thus expected to unveil details about the mechanisms.

Plasmon resonance energy transfer. If the plasmonic nanostructures show significantly enhanced electric field under LSPR excitation, this energy can be dissipated to surrounding nonplasmonic metal shells or small molecules such as Cytochrome *c* [51], cy 5 [52], through a so-called plasmon (coupled) resonance energy transfer [53, 54]. In detail, PRET can induce the activation of the reactant via vibration or electron spin state. PRET plays a key role in PEC catalytic systems in which photoresponsive nanocomponents (primarily semiconductor supports) are coupled with plasmonic metals. The phenomenon can be compared with the process occurring over photosensitizers, in which collected photon energy is transferred to surrounding oxygen molecules, altering the spin state of the oxygen and forming singlet oxygen [55]. In electrocatalytic systems, this strategy may be applied to plasmon-enhanced platforms to enhance the overall reactivity of adsorbates and facilitate the activation of these species at the catalytic surface.

9.2.3 Defining Approaches

The complexity of numerous involved mechanisms pinpoints, in a way, the plasmonic enhancement in electrocatalysis as a controversial approach. The necessity to disentangle single contributions in the overall catalytic performance demands, therefore, comprehensive identification of contributing modes, while providing guidelines toward superior light-matter exploitation and enhanced photocatalytic quantum efficiencies. A poor understanding of the underlying mechanism pathways is, in addition, critical for the rational construction of optimized hybrid systems. Because photo(electro)chemical systems coupling plasmonics and semiconductors have paved the way to understand operating SPR modes in catalysis, detailed studies in electrocatalytic platforms have followed at seminal stages similar methodologies for understanding the operating mechanisms. In agreement, mechanistic studies are fundamentally focused on near-field enhancement and the transfer of hot charge carriers as a key approach in redox reactions. Works investigating reduction reactions (e.g. H_2 evolution, O_2 reduction, and CO_2 reduction) predominantly investigate the injection of hot electrons and the possibility to hinder charge recombination at the metal surface. Conversely, a greater effort has been devoted to understand the complexity toward oxidation reactions with hot electrons being efficiently scavenged or extracted to the external circuit. As extensively demonstrated for the representative O_2 evolution reaction, the remaining hot holes can therefore induce an electronic reconstruction of benchmark electrocatalysts, altering their oxidation state, which can favorably accelerate the oxidation reaction (Section 9.3.1).

The efforts here proposed are expected to provide better clarity of the promise and practicality to lower reaction kinetic barriers and to enhance the rates and efficiency of electrocatalytic systems. When the reaction takes place in the surface of adjacent nanocomponents, the established interface with the plasmonic metal and charge transfer pathways also requires an effective recognition. The limited number of reports in the literature further emphasizes the need to establish unifying guidelines and a consistent comparison to effectively determine the enhancement factor in presence of plasmonic nanostructures. To unscramble involved mechanistic pathways, *in situ* characterization techniques are expected to play a key role in the future, in a manner consistent with leading approaches in current electrocatalytic studies. *In situ* XRD, XPS, and XANES can be crucial to determine the phases and oxidation state of catalytic centers electronically reconstructed in the presence of incorporated plasmonic nanoparticles. Direct imaging of hot carriers at the locations where these species are generated and transferred is here representatively underlined, and a presumed valuable step notwithstanding the challenge associated with the ultrafast dynamics of hot carriers [56]. Finally, another valuable approach is to consider theoretical calculations to establish property–performance correlating descriptors, which are believed to be helpful to rationalize previous claims in distinct catalytic systems.

9.3 Plasmon-Enhanced Electrocatalytic Systems

Using redox chemistry to transform earth-abundant renewable feedstocks such as water (H_2O), carbon dioxide (CO_2), and molecular nitrogen (N_2) into value-added fuels as a response to the ever-rising global energy demand has been reflected in the plethora of emerging research works (Figure 9.2) [57]. When additionally relying on renewable electricity and direct solar energy, this approach represents a promising pathway to redefine the promise of a decarbonized global society. As so, proposed electrocatalytic technologies and the nanomaterials exploited for the reutilization

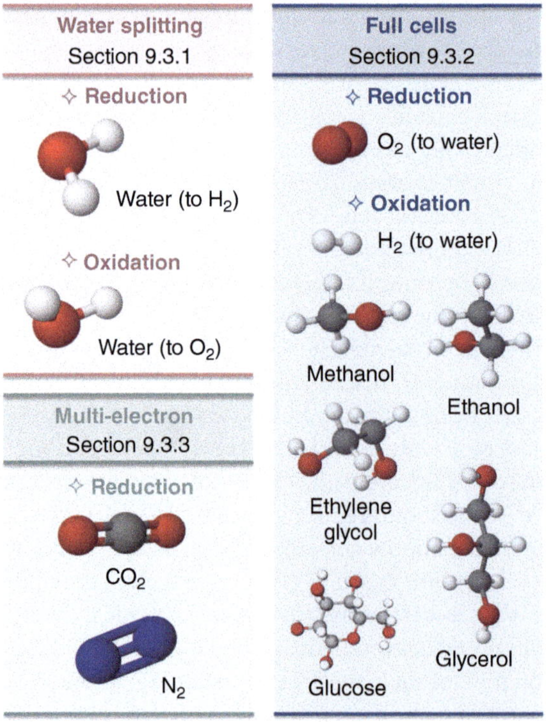

Figure 9.2 Representative reactions under focus in this chapter, reflecting the emerging interest of plasmon-enhanced electrocatalysis toward sustainability [57].

of these small molecules have unveiled a striking performance improvement in the past years. Commercially appropriate cost-effective electrocatalysts operate at low overpotentials, and assure selective and stable production of intended products at high current densities. Complex multielectron-based mechanistic pathways involving multiple reaction intermediates often reflect low faradaic efficiencies (FE, as the ratio of transferred charges to a product and the total number of charges passed through the circuit) making the process undesirable at a commercial scale. Albeit the increasing attention ascribed to their distinctive light harnessing capabilities and SPR fingerprints, the number of reported works assessing the promise of boosting the electrocatalytic performance of the catalytic systems above listed through a utilization of plasmonic nanoparticles has still been limited to date. In other cases, exploratory claims have demonstrated that plasmonic metals can act not only as light absorbers but also as catalytic sites for specific reactions, which requires careful dissection of the underlying roles of the plasmonic material. Interpretative reviews establishing a summary of drawn conclusions while stimulating progress in the community are increasingly necessary [49, 58]. In the following sections, benchmark electrocatalysts are first introduced for each reaction under focus. The plasmonic nanomaterials, as well as the associated plasmonic operating modes and underlying mechanisms under light irradiation against dark conditions, are discussed.

9.3.1 Water Splitting: Hydrogen and Oxygen Evolution

With a gravimetric heating value (141.9 MJ kg^{-1}) superior to conventional fuels, molecular hydrogen (H_2) has been hypothesized as a key pathway toward a decarbonized and sustainable society in the past decades [59]. With H_2 being primarily derived from carbon-based processes including methane steam reforming (48%), reforming of heavy oil/naphtha (30%), and coal gasification (18%), a necessary shift is compulsory to validate the promise of H_2 as a renewable and clean energy source [60, 61]. Carbon-free high-purity H_2 production through an efficient split of the water molecule and utilizing solar, geothermal, wave, and wind powers have, in this sense, gathered necessary focus. The water splitting process included the hydrogen evolution reaction (HER) and the oxygen evolution reaction (OER), which occur on the cathode and anode of a water splitting electrolyzer, respectively. The thermodynamically uphill process requires a Gibbs free energy (ΔG) of ~237.2 kJ mol^{-1}, corresponding to a standard potential (ΔE) of 1.23 V versus reversible hydrogen electrode (V_{RHE}) [62]. The sluggish kinetic pathways of the two-electron reduction of protons into H_2 and, particularly, of the four-electron–proton-coupled OER reaction induce large-energy barriers and severely limit the electrocatalytic process [63]. Optimized electrocatalysts are expected to reduce associated large overpotentials of each reaction and to improve the energy conversion efficiency of electricity to H_2. At present, benchmark Pt- and Ir/Ru-based HER and OER electrocatalysts simultaneously suffer from high cost and scarcity, which severely hinders their utilization in commercial electrolyzers (Figure 9.3a) [64, 65, 67, 68]. Honda and Fujishima first assessed the exploitation of solar energy to effectively split the water molecule through the incorporation of a semiconductor as a photoresponsive anode

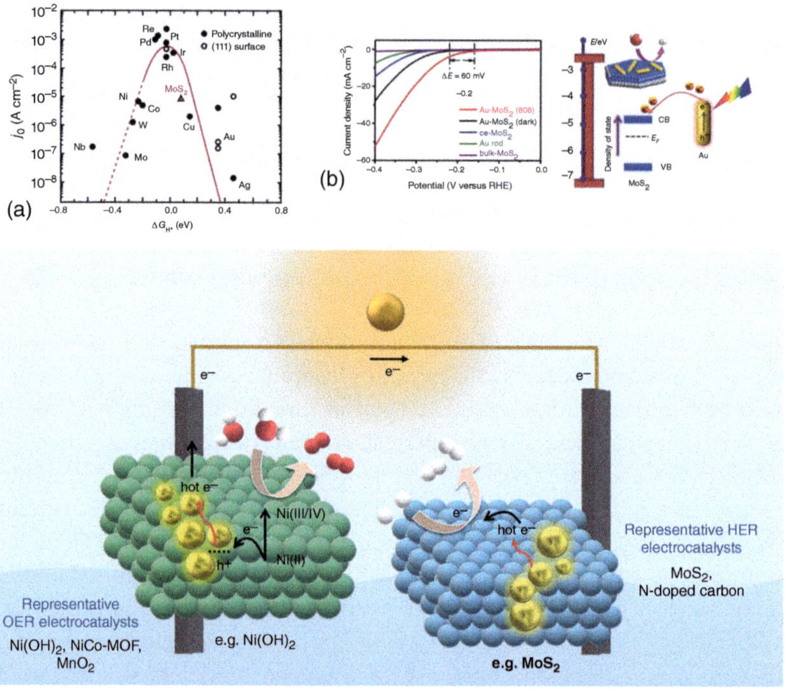

Figure 9.3 (a) HER volcano plot for metals and benchmark MoS_2 and Ni_2P electrocatalysts. The solid line reflects the simple kinetic model proposed by Nørskov [64]. *Source*: Morales-Guio et al. [65]. © 2014, The Royal Society of Chemistry. (b) HER performance of the Au–MoS_2 hybrid, and schematic representation with energy level diagram depicting hot electron injection and change of the Fermi level of MoS_2. *Source*: Shi et al. [66]. © 2015, American Chemical Society. (c) Schematization of the OWS electrocatalytic system for both hydrogen and OERs. Representative strategy for water oxidation with energetic holes driving a continuous extraction of electrons from active transition metal sites, producing oxidized centers suitable for O_2 evolution, as claimed in $Ni(OH)_2$, NiCo-MOF, and MnO_2-based systems.

[3]. Since then, PEC systems have aspired to unveil the promise and durability of novel materials as either photoanodes or photocathodes [69]. Alternatively, utilizing solar energy through the integration of plasmonic nanostructures in light-assisted electrocatalytic water splitting has been demonstrated in several reports.

Hydrogen evolution reaction. In 2015, seminal works of Xia and colleagues assessed the incorporation of plasmonic metal Au nanorods on molybdenum disulfide (MoS_2) nanosheets, one of the most promising Pt-alternative electrocatalysts with outstanding HER activity and scalable cost [66]. Fast-forwarding past decades during which MoS_2 was believed inactive toward HER, reported DFT calculations on the Mo(1010) edge of MoS_2 at 50% H_2 coverage unveiled a hydrogen adsorption free energy (ΔG_H) of 0.08 eV, near the 0 eV optimal value [70]. In agreement with the previously reported poor HER activity of bulk MoS_2, a value of 1.92 eV was calculated for the basal planes [71]. In the works of Xia, a synergistic effect was established between both nanocomponents and reflected by a dramatic threefold

HER activity improvement upon the excitation of Au LSPR (808 nm). Low loading assured hindered aggregation and plasmonic coupling of the loaded nanorods. A significant increment in the carrier density on MoS_2 and an improvement of the electron–hole separation resulted in a decrease of the HER overpotential to ~120 mV (against ~320 mV under dark conditions at 10 mA cm^{-2}) (Figure 9.3b). The pioneer works by Xia reflected for the first time the promise of harnessing the energy from hot electrons for electricity-driven water splitting to produce H_2. Relatively unexplored thermal effects deduced for both Au and MoS_2 and further mechanistic analysis have since catapulted several investigations considering the proposed system as a model.

With an ascribed rationality still under discussion nonetheless, enhanced activity of 1T-MoS_2 has been denoted over similarly prepared 2H-MoS_2. The 1T-phase MoS_2 synthesized via conventional methods such as chemical exfoliation is metastable and tends to devolve back into the thermally preferred 2H-phase due to its relatively higher energy. Stabilization of the 1T-phase has been in this sense actively surveyed, with a clear focus on the possible valuable role of neighboring plasmonic centers. Relevant strategies have recently surveyed showcasing the effect of a local injection of electrons to demonstrate a multiphase transition in this material. The incorporation of 2D electrode crystals on MoS_2 was, for instance, recently reported to induce a gradual evolution of lattice symmetry and band-structure normalization in bulk MoS_2, with multiphases, e.g. heavily doped 2H, metallic 1T phases, and a possible combination of both phases being observed by Raman shift [72]. In the context of this chapter, Fang successfully demonstrated a reversible 2H–1T phase transition in an Au nanoparticles-deposited MoS_2 monolayer upon resonant plasmon excitation. The 1T phase MoS_2 monolayer facilitated the utilization of the induced plasmonic hot electrons for HER. Taken together, the combination of local stress, charge transfer effects, the electrocatalytic properties of the 1T phase of MoS_2 monolayer, and the plasmon resonance further enhanced HER [73, 74]. A proposed synthesis method exploiting bimetallic assembly of Pd nanosheets and single-side nucleated Au nanoislands on pristine 2H-MoS_2 uncovered a novel mechanism of lattice-mismatch-induced strain to produce ultrastable 1T-MoS_2 with excellent HER activity under the light source [75]. First-principles calculations revealed that the lattice mismatch and covalent bonds between the Pd and S atoms prompted mechanical and compressive strains, triggering phase transition from 2H to the highly stable 1T phase.

Emerging reports have underlined to date the need to assess the established interface between the HER sites on MoS_2 nanosheets and introduced (dispersed) Au centers, with distinct plasmonic nanostructures being equally surveyed [76]. Recent studies investigated the effect of Mo-terminated defects of MoS_2 to guide multisite attachment of ultrasmall Au nanocrystals [77]. Most importantly, against plane-attached Au-decorated hybrids, defect-attached Au nanocrystals were proposed to offer a more valid plasmonic-excited hot-electron transfer pathway to neighboring catalytic sites, based on the principle of proximity. The presence of defects on layered semiconductors has been today pinpointed to direct electric and catalytic unique properties, which can be manipulated toward superior performances. The reduced Schottky barrier of HER was ascribed to the synergistic dual function of Au nanocrystals on MoS_2 which accelerated the transport rates of conducting electrons between conducting electrodes and catalysts while serving as a source of plasmonic-induced hot electrons. To enhance the contact between both nanocomponents and increase the efficiency of the charge transfer, in 2019, Chen reported a 3D bicontinuous nanoporous heterostructure by growing MoS_2 on the internal surfaces of 3D nanoporous Au [78]. A red-shift of the LSPR band (from 544 to 556 nm) of the hybrid structures was attributed to the increased dielectric constant of MoS_2 with an increasing number of layers [79]. Low onset potentials (−70 mV) and the determined Tafel slope (38 mV dec^{-1}) under visible light illumination were attributed to the superior specific surface area and high electric conductivity. The effect of a rational structural arrangement has been further demonstrated in the photocatalytic production of H_2, which is believed to serve as a valuable guideline for electrocatalytic platforms. Induced strong near-field coupling established at the interparticle gap of Au nanoparticles was, for instance, proposed to yield an absorption boost, which subsequently leads to an effective exciton generation and dissociation in the Au-MoS_2 junctions [80]. In the report by Wei et al., improved H_2 production by 240% was also seen on a rational structural arrangement of Au multimer composed of three to six nanoparticles with 5–10 nm interparticle distance on MoS_2.

Following a similar strategy, superior HER activity increment was claimed on Au nanoparticles embedded in metal–organic framework (MOF)-derived abundant N-doped carbon [81]. The architecture showcased above fourfold current boost and a manifest overpotential decrease (99 mV against 196 mV at 10 mA cm^{-2} with an Au-free counterpart). The superior performance was deduced from the efficient charge transfer on the N-doped carbon support, which could further prompt a swift injection of hot electrons from the plasmonic Au. Complementary results evidenced a charge transfer resistance decreased by 50% upon light irradiation. In a distinct study, significant HER activity enhancement under Vis and NIR light was achieved on plasmonic Pd-tipped Au nanorods bimetallic structures showcasing a more than twofold lower overpotential (40 mV against 108 mV at 10 mA cm^{-2} under dark conditions) [82]. The superior HER performance of Pd-tipped against fully covered counterparts (including higher exchange current density and good stability) was correlated with the photothermal effect of Au nanoparticles, which manifested an efficient transfer of the highly active free electrons on the Au surface to the Pd sites.

Oxygen evolution reaction. Sharing parallelism with the oxygen reduction reaction described in the following section, the sluggish four-electron OER involves multiple intermediates (OOH*, OH*, and O*) and imposes a clear bottleneck toward OWS. The promise of precious metals, e.g. Pt, Ru, and Ir, is unsuitably balanced by their high cost and scarcity. As opposed, nonprecious nickel, cobalt, and manganese oxides, and multication perovskites reflect valuable alternatives, although their stability is limited to alkaline conditions and much higher overpotentials are demanded [83–86]. An extensive assessment of oxide candidates has been reported on experimental and theoretical studies, in which volcano plots using $\Delta G_O - \Delta G_{OH}$ as the catalytic descriptor has served as a valuable guideline for upcoming exploratory works [87]. As a representative effort to enhance the OER activity utilizing plasmonic nanostructures, in 2016, Ye et al. reported a fourfold activity boost with a remarkable improvement in overpotential over an Au-doped $Ni(OH)_2$ nanosheets (270 mV against 338 mV without Au incorporation at 10 mA cm^{-2}) upon SPR excitation with a 532 nm laser (Figure 9.3c) [88]. The determined turnover frequency (TOF, based on the BET surface area), 20×10^{-3} s^{-1}, was remarkably superior to the $Ni(OH)_2$ reference material (4.2×10^{-3} s^{-1}). Hot electrons on the Au nanoparticles promptly transferred to the glassy carbon electrode leave hot holes on the surface of the metallic nanostructures. As a result, the energetic holes drive a continuous extraction of the electron from the Ni^{II} sites, producing Ni^{III} centers, detected through electron spin resonance spectroscopy measurements, suitable for water oxidation. The generated hot electrons could therefore be correlated with the enhanced generation of high valence states of the Ni metal active species and improved charge transfer. The reported performance surpassed those of IrO_2 and RuO_2 benchmark materials and representative Ni-based OER catalysts. The plasmon-enhanced OER performance was further denoted over Au-doped CoO and FeOOH electrocatalysts, suggesting a promising extension of the claimed methodology.

In light of the findings above, Xia and collaborators recently exploited the formation of hot electrons on Au nanorods to increase the generation of high valence Ni* active species in a 2D ultrathin NiCo metal–organic framework (MOF) with suitably matched energy levels and hindered OER activation energy [89]. Reported MOF systems in electrocatalytic platforms exhibit poor conductivity, low mass permeability, and metal centers with limited accessibility due to present organic ligands [90]. Incorporation of unsaturated transition metals (Fe, Ni, Cu, Co, Cr, and Mn) with variable valences represents a promising strategy toward superior performances. In this work, the Au rods were decorated onto the surface of the NiCo-MOF. The superior OER performance of the NiCo-MOF under Au SPR excitation was reflected in the decreased overpotential of 240 mV (against 283 mV at 10 mA cm^{-2}) and an abridged Tafel slope from 83 to 69 mV dec^{-1}. Chen followed a similar approach with the introduction of plasmonic Au nanoparticles (9.9 nm in radius) on MnO_2 nanosheets to drive a facile OER performance under low laser power (100 and 200 mW) [91]. Analogously, the confinement of the outer electrons of Mn species by generated hot holes was proposed to induce the formation of Mn^{n+} centers, which could actively extract electrons from OH^-, facilitating the formation and subsequent deprotonation of OOH intermediates (Figure 9.3c). The activity enhancement was found to be dependent on the LSPR excitation, with an overpotential decrease from 320 mV (against 380 mV at 10 mA cm^{-2}) when tuning the laser intensity from 100 to 200 mW. The performance was comparable to benchmark IrO_2 and RuO_2 with the pristine MnO_2 nanosheet and gold nanospheres showed weak OER performance regardless of the light irradiation.

However, with the increasing exploitation of Au in OER electrocatalytic platforms, the evaluation of Au-included electrocatalysts requires an additional differentiation of the properties of Au as a catalytic center and as a plasmonic material [92]. The presence of an Au substrate has, for instance, been reported to play a role in favorable binding energies of the OER intermediates [93]. An optimized electronic reconstruction established between Au nanoclusters and $CoSe_2$ was also proposed to favor the formation of reaction intermediates as well as the reduction of the undesirable catalyst poisoning caused by the exposure to molecular O_2 [94].

Overall water splitting. In line with the discussion above, it is clear that the design of bifunctional electrocatalysts for OWS has remained particularly challenging in the community. Recently our group reported the rational design of Pd-doped reduced graphene oxide (GO)-wrapped Au nanoparticles (AuNP@rGO@Pd) [95]. The efficiency of this novel nanohybrid structure was investigated as a HER and OER bifunctional catalyst based on the activity of the incorporated Pd nanoparticles, and the optimized composite was compared against a commercial Pd-dispersed Vulcan XC 72. The thickness of the rGO shell determining the proximity of the Pd and plasmonic Au nanoparticles, resulted in a careful optimization of the near-field plasmonic enhancement in the hybrid structures. A single layer of the rGO shell (rGO(1)) was further shown to prevent the Pd leaching and aggregation, while a thicker shell induced clear obstruction of the direct charge transfer between metals and hindered light exposure on Au surface. Under the light, AuNP@rGO(1)@Pd revealed an overpotential of 291 mV (against 342 mV at 10 mA cm^{-2} in dark), reflecting a twofold HER increment, and showcased superior stability up to 10 000 seconds against Pd/C. The superior performance was ascribed to an efficient electron transfer between Au metal core and the highly conductive rGO, as well as the demonstrated near electromagnetic field in the vicinity of the Au nanoparticles. The composite material also showed an increment of ~1.1-fold in OER activity, reflecting a positive potential shift to 542 mV (against 553 mV at 10 mA cm^{-2}). Our effort in disentangling the plasmonic effects operative in the enhancement of the electrocatalytic properties are also expected to be representative of future works in the community. Representatively, integrated plasmonic Au nanoparticles and Ni and/or Co hydroxide semiconductors were recently selected to assess the role of the interfacial charge transfer [96]. Performance-wise, the Au/NiCo layered double hydroxides (LDH) disclosed an overpotential of 119 mV (against 200 mV without Au at 10 mA cm^{-2}) for HER and an overpotential of 203 mV (against 234 mV without Au) for OER. Both Tafel slopes (57.5 and 42.3 mV dec^{-1} for HER and OER, respectively) were significantly lower against the individually evaluated NiCo LDH. By using wavelength-dependent experiments, the enhanced HER and OER activity was attributed to the significant LSPR effect of the hybrid. Most interestingly, however, by using surface-enhanced Raman spectroscopy (SERS), a strong spontaneous interfacial charge transfer between Au and NiCo LDH was found to significantly facilitate the water splitting reactions. Both water splitting reactions showed unidirectional electron transfer in the plasmonic metal–semiconductor. Through selective thiol-coating on the Au surface, the authors successfully identified the LDH semiconductor and the Au surface as the real catalytic sites for plasmon-promoted OER and HER, respectively.

9.3.2 Fuel Cells: Oxygen Reduction and Alcohols Electro-oxidation

Closing the cycle of carbon-free technologies initiated by the evolution of H_2 and O_2 from water, conventional fuel cells drive the oxidation of hydrogen (HOR) and the reduction of oxygen (ORR) (Figure 9.4a) [97, 102]. The sluggish transfer of four electrons for the ORR to water with unpractical overpotentials has required greater attention. The unfeasible exploitation of scarce and cost-ineffective Pt electrocatalysts (Figure 9.4b) [98, 103, 104] emphasizes the need to develop novel systems or promising alternatives to reduce the utilization of Pt. Developing Pt-included binary and ternary alloys [105] and core–shell nanostructures, or prompting an electronic reconstruction of the Pt overlayer toward optimized oxygen absorption energies are common approaches [106–111]. Representatively, research has disclosed the effects of tuning compressive surface strains inducing a downshift in the *d*-band center [112], or the result of manipulating neighboring atoms with a distinct electronic environment. Exploiting plasmonic nanoarchitectures is theoretically promising for fuel cells toward a complete four-electrons transfer pathway [113], guaranteed that the intrinsic surface properties of incorporated metal nanoparticles is not propitious for H_2O_2 formation (two-electrons transfer) [100]. Also, an additional need to deploy critical engineering efforts to alter the conventional fuel cell design toward suitable light harvesting shall be considered.

In 2017, Yang and colleagues investigated the promise of plasmonic Pd nanorods, with a highly tunable LSPR (651–945 nm range) with increasing nanorod length, as an ORR electrocatalyst [114]. The ORR linear sweep voltammetry curves revealed that current densities upon light excitation could be increased linearly with an increase of the incident resonant excitation power. Under an 850 nm laser, a 4.4-fold ORR current enhancement was claimed. Koutecky–Levich plots suggested that regardless of illumination the rate-determining step of the ORR remained unchanged, in the confirmed four-electron process. Arrhenius plots obtained by extrapolation further revealed an increase of the apparent activation energy (one-fourth higher) and the kinetic pre-exponential factor (two-orders of magnitude higher), which were ascribed to a change in the Pd morphology and to an SPR-induced local electric field enhancement, respectively. In particular, an increasingly pronounced quantum-mechanical proton tunneling effect leading to superior pre-exponential factor. The measured kinetic H/D isotope effect ($KIE = j_k(E)H_2O/j_k(E)D_2O$) shed light on these observations. The KIE value increased with increasing ORR overpotential. Abnormally large KIE values offer evidence for a quantum-mechanical proton tunneling contribution in the pre-exponential factor. At the highest overpotential (0.729 V), the KIE value reached 98.25 under 0.6 W cm^{-2} light irradiation against 15.98 under dark and initial temperature of 22.5 °C.

Chen exploited the SPR properties of edgeless plasmonic Ag-Pt bimetallic nanocages with a hollow interior to provide empirical evidence of a clear suppression of H_2O_2 (side product, two-electron process) through plasmonic effects (Figure 9.4c) [99]. The synthesis of Ag-Pt nanostructures was carried out by galvanic replacement reaction utilizing sacrificial Ag nanocubes under increasing $PtCl_6^{2-}$ concentration. The Pt content could be tuned from 4.4 to 23.2%, inducing a continuous red-shift from 526 to 645 nm, respectively. To investigate the LSPR effect, x-ray absorption near edge structure (XANES) of the Pt L_3-edge (involving the transition from 2*p* to 5*d* states) was conducted. As a result of plasmon-induced hot electron

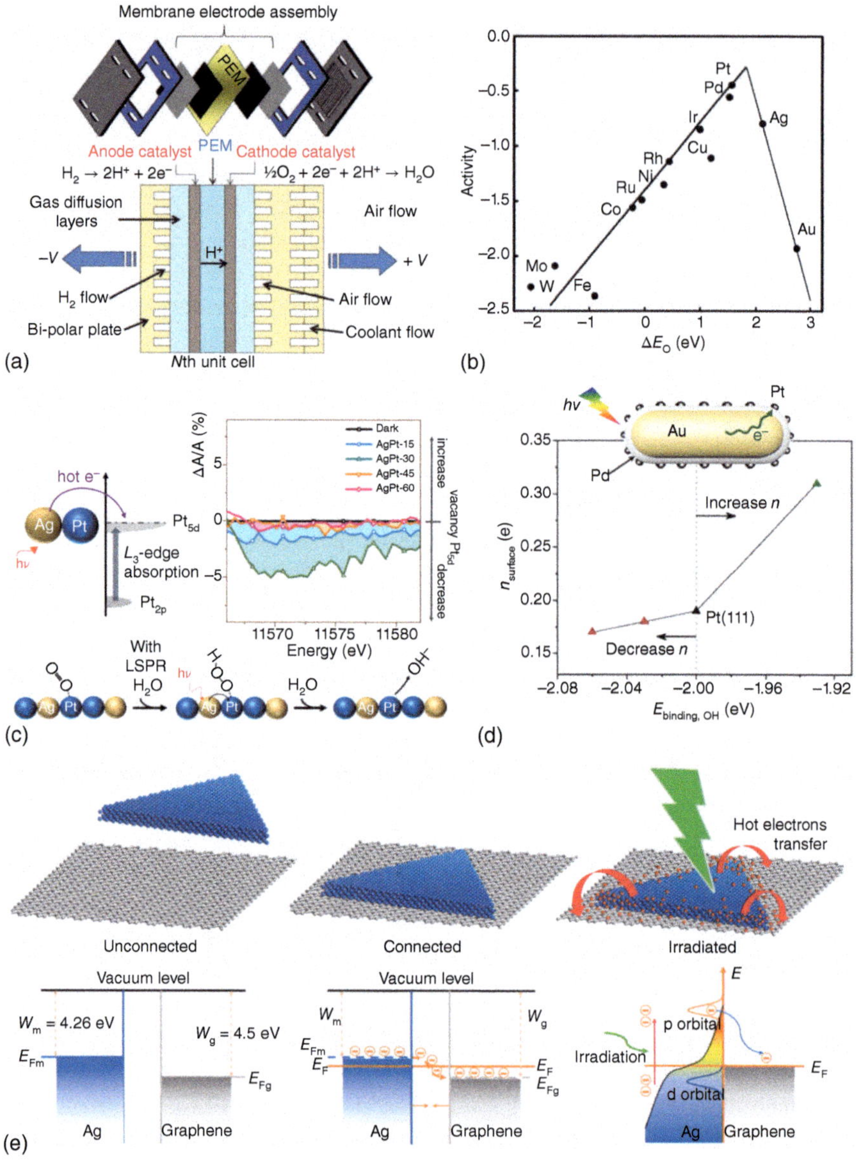

Figure 9.4 (a) Unit cell cross-section showing the components of an expanded MEA. *Source*: Debe [97]. © 2012, Springer Nature. (b) ORR volcano plot for metals. *Source*: Nørskov et al. [98]. © 2004, American Chemical Society. (c) Representative schematic representation with hot electron transfer and relative vacancies ($\Delta A/A$) of Pt L_3-edge XANES under light. The bottom scheme represents the reaction mechanism with hindered H_2O_2 formation via hot electron transfer. *Source*: Lin et al. [99]. © 2017, American Chemical Society. (d) Number of injected hot electrons versus calculated adsorbed OH binding energy. *Source*: Zheng et al. [100]. © 2017, Elsevier. (e) Plasmon-enhanced ORR mechanism through hot electron transfer over unconnected, connected, and irradiated Ag nanotriangles on a graphene support, and corresponding energy band diagrams. *Source*: Shi et al. [101]. © 2019, American Chemical Society.

transfer to Pt active centers, less 5*d* vacancies could be denoted through a decrease in the corresponding XANES intensity. Plasmon-induced effects in ORR were analyzed by using a customized real-time monitor of the change of peroxide yield. Under dark and with an increase of Pt loading, the onset potentials and corresponding half-wave potential ($E_{1/2}$) values were shifted to the higher potential region. When the light was shed, an optimum Pt loading of 6.22% exhibited the highest performance in limited current enhancement and peroxide yield suppression. This was due to the LSPR effects and electromagnetic field increment in Ag atoms with the adjacent Pt atoms that facilitated the effective reactant adsorption/desorption. The swift transfer of hot electrons was proposed to enhance the electron population in the antibonding orbital of O_2, weakening the O-O bond, and facilitating H_2O formation via a four-electron transfer. An excess of Pt loading covered most of the Ag-surrounded Pt atoms and lead to a deficient transfer of hot electrons to Pt.

Following a similar approach, Amal assessed Pt electrocatalysts with plasmonic nanocores (i.e. Au-Pd-Pt nanorods), for ORR [100]. HAADF-STEM and corresponding EDS mapping images of single nanorods confirmed the morphology of these architectures, whereas the collected extinction spectrum revealed a strong and broadening longitudinal plasmon absorption. Performance-wise, these Au-Pd-Pt nanorods on carbon unveiled 25.6- and 27.5-fold superior mass and specific activities upon visible-light illumination against a commercial Pt/C at 0.9 V_{RHE} with inherent weak light-absorption-ability and absent SPR absorption. The results further italicized a 2.2-fold enhancement against dark conditions. Against the LSPR single-particle photoluminescence (PL) intensity of the pristine Au, Au-Pt, and Au-Pd-Pt architectures revealed a quenching efficiency of ~89% and nearly 100%, respectively, showcasing an efficient hot-electron sequential-transfer from Au to Pd and then to Pt by Landau damping or chemical interface damping. DFT calculations further revealed that the increasing number of electrons on the Pt surface monolayer on the Pd substrate could weaken the OH* binding energy, enhancing the ORR activity (Figure 9.4d). In addition to this charge carrier-driven and photothermal mode induced by the local relaxation of generated hot electrons, 3D FDTD simulations showcased the spatial distribution of near-field enhancement and its contribution to the ORR. The results further inferred that incorporated Pd and Pt layers did not entirely suppress the SPR effect of the Au nanorods. Analogous tri-metallic Au@PtPd mesoporous nanorods (MNR) employing Pt and Pd precursors to Au nanorod substrates were later introduced by Gu and colleagues, showcasing the contribution of electronic interaction between nanocomponents [115]. Whereas in this study the SPR effect is not induced in the Au@PtPd mesoporous structures, the interaction between the *d* orbitals of Pt and Pd was shown to play a key role to induce the dissociation of adsorbed oxygen, toward improved ORR efficiency and lower onset potential compared to Au-Pd-Pt nanorod counterparts. This work emphasizes the need to disentangle optophysical, electrochemical, and electronic properties of each nanocomponent, the resulting interaction between active phases, as well as the impact of the active surface area, which was confirmed to play a key role in the activity of these Au@PtPd mesoporous structures. Performance-wise, Au@PtPd MNR revealed a much more positive onset (1.01 V) and $E_{1/2}$ (0.92 V) potentials, with superior limiting current density (5.4 mA cm^{-2}) than a PtPd (0.96 V, 0.89 V, and 4.6 mA cm^{-2}, respectively) and Au@Pt MNR counterparts (0.98 V, 0.90 V, and 5.0 mA cm^{-2}, respectively), and a performance comparable to Pt/C (0.95 V, 0.88 V, and 5.5 mA cm^{-2}, respectively).

The effect of a local temperature increase has also been investigated in the enhancement of the ORR activity. Recently, Wu employed graphene as a support material for Ag nanotriangles, which against graphite and CNT is proposed to minimize the thermal contribution of the support on the plasmon-enhanced ORR performance (Figure 9.4e) [101]. Ag nanowires, nanospheres, and nanotriangles under laser irradiation (532 nm) revealed $E_{1/2}$ shifts positively increase from 18.2 to 21.1 mV and 22.6 mV, respectively. It was found that the Ag nanotriangles possessed stronger electric field distribution confined at the architecture corners and the largest absorption efficiency, with superior mass activity at 0.8 V among the surveyed Ag nanocrystals. Localized resonant collective oscillation of free electrons in the *d* orbitals of Ag in contact with graphene takes place upon illumination, with a subsequent decay leading to hot electrons into *p* orbitals, which are then spontaneously transferred to the neighboring graphene with lower energy level. The continuous process induces electron redistribution and a desired ORR performance enhancement, as corroborated by x-ray photoelectron spectroscopy measurements and electronic band structure calculations based on first-principles DFT. However, detailed studies focusing on the interface established between metals and graphene, and on the intrinsic properties of graphene, remain insufficient in the literature [116, 117].

Alcohol oxidation: Methanol and ethanol electro-oxidation. Flexible exploitation of renewable fuels as feedstocks, e.g. methanol and ethanol through corresponding anodic methanol and ethanol oxidation reaction (MOR and EOR, respectively) coupled with the cathodic ORR toward electricity generation have received increasing attention in the development of direct alcohol fuel cells (Figure 9.5a) [97]. Sluggish alcohol electro-oxidation reactions requiring challenging C–C bond splitting, high cost of membranes, and suitable catalysts have, however, limited the development of these devices [122, 123]. Complete MOR requires the transfer of six electrons, with Pt unveiling superior catalytic activity, and an emerging interest being reported over Pd-incorporated oxides [124]. The complete EOR, a promising candidate for fuel cell with superior energy density and lower toxicity compared to methanol, requires the transfer of 12 electrons. Unlike MOR, EOR is best achieved when employing Pd-based catalysts [125, 126]. Analogously, the hot electrons in plasmonic nanostructures can be conveniently extracted leaving behind hot holes capable of enhancing the anodic electro-oxidation of the selected alcohols. The extraction of hot electrons can be driven to the external circuit, electron scavengers, or neighboring nanocomponents acting as electron sinks. Alternatively, electrons can be extracted to the conduction band of semiconductors as supports for the plasmonic metal nanoparticle in conventional photoelectrocatalytic platforms. Representatively, Cui and colleagues claimed a 1.82-fold enhancement of net peak currents ascribed to the plasmon-induced charge separation on the generated heterointerface of a highly ordered periodic Au nanoparticle-decorated bilayer TiO_2 nanotube architecture for EOR exploiting Vis light [127].

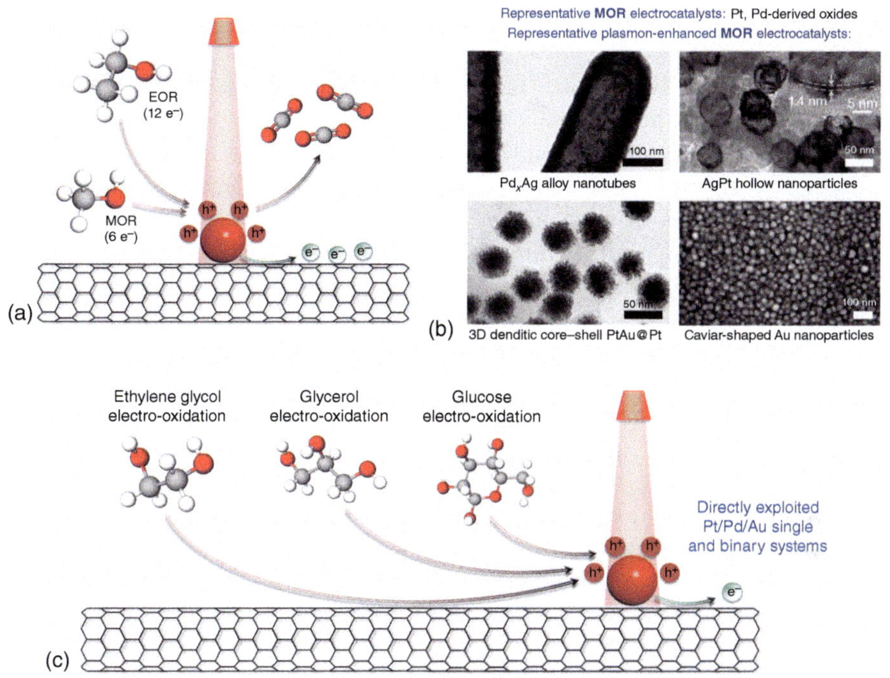

Figure 9.5 (a) Schematic representation of the complete methanol (MOR) and ethanol oxidation reactions (EOR) following the transfer of 6 and 12 electrons, respectively. A direct exploitation of the hot holes on the surface of plasmonic structures is depicted herein. (b) TEM and SEM photographs of representative plasmon-enhanced electrocatalytic systems reported for MOR, including Pd$_x$Ag alloy nanotubes. (*Source*: Huang et al. [118]. © 2019, John Wiley and Sons), AgPt hollow nanoparticles (*Source*: Bi et al. [119]. © 2019, The Royal Society of Chemistry), 3D dendritic core–shell PtAu@Pt (*Source*: Xu et al. [120]. © 2018, American Chemical Society), and caviar-shaped Au nanoparticles (*Source*: Liu et al. [121]. © 2018, The Royal Society of Chemistry). (c) Representative electro-oxidation of EG, glycerol, and glucose on Pt/Pd/Au-based single and binary electrocatalytic systems.

Plasmon-enhanced electrocatalytic systems have similarly emphasized the promise of this strategy. Recently, Jain and collaborators investigated Pd$_x$Ag alloy nanotubes for MOR, combining the Vis-responsive properties of Ag and the electrocatalytic activity of the Pd metal (Figure 9.5b) [118]. The enhanced activity under LSPR excitation was reflected by superior current densities and positively shifted potentials. Conversely, negligible photo-enhancement was witnessed over Pd/C and Ag nanowires/CNTs. The improvement degree was directly correlated with the Pd to Ag molar ratio (x) of the prepared architectures, ranging from $x = 0.65$, 0.52, 0.41, and 0.33, as determined by inductively coupled plasma-atomic emission spectroscopy. With photoexcited electrons transferred to the external circuit or adsorbates operating as electron sinks, the resulting hot holes remained on the Ag surface. Notwithstanding the clear role played by the photothermal heating of the electrochemical interface to the enhanced MOR activity, the generated hot holes were found to be critical to facilitate the electro-oxidation reaction. The open-circuit potential (OCP) of the prepared Pd$_{0.52}$Ag/CNT was measured in an aqueous solution of 1.0 M CH_3OH and 0.5 M NaOH as a function of time under the illumination of various intensities using a 530 nm laser. A light intensity increase was shown to induce an OCP decrease, ascribed to a larger downshift of the quasi-Fermi level due to an increase in the population of hot holes. Interestingly, however, at sufficiently high light intensities, electron–hole recombination was considered predominant, with the hot hole population and corresponding OCP reaching saturation.

Similar electrocatalytic systems have been reported over recent years [128]. Yang recently assessed the promise of bimetallic AgPt hollow nanoparticles with an ultra-thin shell for MOR (Figure 9.5b) [119]. The photothermal effect was claimed to play a key role in the activity enhancement. More importantly, upon light source (500 nm) irradiation, the periodic oscillation of carriers was dominant near the Pt side. This heterogeneity in surface charge was proposed to decrease the *d*-band center character of Pt sites and to hinder CO absorption as the main intermediate in MOR [99, 129]. As a result, AgPt architectures exhibited 2.0 and 4.79-fold improvement in specific activities against dark conditions and a commercial 20 wt% Pt/C, respectively. Exploratory works by Du and collaborators focused on PtAu electrocatalysts for MOR. Dendritic structures were first reported to exhibit negatively shifted onset potentials, smaller resistance, and higher electrical conductivity under Vis light against dark conditions [130]. In addition, the mass activity of dendritic PtAu catalysts was 2261 mA mg^{-1}, which reflected 1.8- and 9.7-fold improvements against dark conditions and a commercial Pt/C. Following a similar approach, Du et al. later introduced a 3D dendritic core–shell PtAu@Pt with a further enhancement of the electrocatalytic activity toward MOR and EOR under Vis light (Figure 9.5b) [120].

Direct MOR has analogously been carried out directly on the surface of Au nanostructures. In 2018, Chen reported enhanced MOR performance by employing a caviar-shaped assembly composed of Au nanoparticles, with controlled particle size by adjusting the ion sputtering time (Figure 9.5b) [121]. Under the solar light source, the caviar-like assembly exhibited an anodic peak current density of 249 μA μg^{-1}, reflecting a 2.15-fold increase against dark conditions. Comparatively, the MOR catalytic activity of free-standing nanoporous Au was recently investigated, with a current density enhancement up to nearly fourfold when the light was shed over these nanoporous structures with an average ligament/nanopore size of 18.3 nm (Figure 9.5b). Most strikingly, facet-rich rationally designed nanostructures revealed up to a 531 μA cm^{-2}, claimed to be the highest value reported for Au-based catalysts at the time the work was published [131]. Hindered recombination of electrons and holes in these Schottky barriers-free 3D bicontinuous structures prepared from dealloying $Ag_{65}Au_{35}$ leaves in HNO_3 was shown to result in a high energetic hole yield (0.486%, fourfold against discrete AuAg nanoparticles). A maximization of LSPR-induced hot carriers was tentatively correlated with dominant {111} facets compared with {100} and {110} facets. Photothermal effects were deduced from control experiments carried out at high temperatures in the dark. Drawn conclusions were extended to the electro-oxidation of ethanol, isopropanol, and ethylene glycol (EG) (cf. below).

Ethylene glycol electro-oxidation. Utilization of EG with biological renewability, low toxicity, facile storability, high solubility in aqueous solutions, and relatively high reactivity in direct EG fuel cells has gained increasing focus based on a promising high power density (7.56 kW h dm^{-3}) and large theoretical capacity (4.8 A h ml^{-1}) [132–134]. A comparison with previous sections places Pt as the suitable anode electrocatalyst with inherent cost-ineffectiveness, inappropriate abundance, driven sluggish kinetics, and susceptible to surface poisoning by carbonaceous intermediates (e.g. carbon monoxide) [135]. As an alternative, palladium has gathered modest recognition primarily ascribed to its resistance to CO poisoning [136]. Improved catalytic and antipoisoning activity was for instance reported with Pt–Pd bimetallic structures on rGO as effectual active carbon support [134]. However, cases exploiting the plasmonic properties of Pd remain scarce in this field. Recently, Gao and colleagues decorated a snowflake-like narrow bandgap cuprous sulfide (Cu_2S) semiconductor with Pd nanoparticles to achieve broad light absorption but drawn mechanistic conclusions pinpoint occurring light-harvesting steps to the semiconductor material [137]. The direct exploitation of plasmonic nanomaterials as the catalytic active surface was further demonstrated in 2018 with a bowl-like AuAg nanocatalyst with rough open surfaces [138]. In the electro-oxidation of EG, Au nanostructures emerge as promising candidates to boost the intermediate oxidation and to accelerate the reaction process with long-term durability and capacity [133]. In the works of Du et al., the Au_1Ag_1 modified electrode displayed the highest current density of 5653.1 mA mg^{-1} under visible light irradiation, reflecting a 2.3-, 5.7-, and 9.4-fold boost against Au_1Ag_1 under dark, bare Au under light irradiation, and bare Au under ambient reaction, respectively. The plasmon-enhanced electrocatalytic activity, superior to those of Pt/C and Pd/C commercial references, was ascribed to the synergistic and electronic effects established between Au and Ag, the more exposed surface active sites, and the exploited SPR effect of the metal components (Figure 9.5c).

Glycerol electro-oxidation. As a major byproduct of the steady development of biodiesel, increasing demand for the utilization of low-cost glycerol is denoted [139]. Glycerol can be selectively transformed into valuable chemicals including dihydroxyacetone, glyceraldehyde, tartronic acid, and glyoxylic acid. Full glycerol oxidation in direct alcohol fuel cells is also of critical interest as it a nonflammable and environmental-friendly reactant, easily purified, and possessing high theoretical energy density. Pt, Pd, and Au are also benchmark materials for the glycerol electro-oxidation reaction [140, 141], with sluggish kinetics involving numerous intermediates and reaction pathways, CO-poisoning, and swift deactivation reflecting, however, to date key bottlenecks [142]. Glycerol electro-oxidation has also witnessed a survey of bimetallic structures and rationally designed core–shell architectures, which uncover the relevance of comprehending the electronic interaction and the synergism established between selected metals [143, 144]. Exploiting plasmonic nanostructures has been investigated in PEC platforms using semiconductor substrates [145], and in electrocatalytic systems where the plasmonic surface provides active sites. Surface plasmon-generated reactive oxygen species have also been proposed to facilitate the C–C bond splitting, with the resulting intermediate species being subsequently oxidized [146]. Notwithstanding the limited numbers of the plasmon-enhanced electrocatalytic oxidation of glycerol, conclusions can be drawn from analogous photocatalytic platforms similarly exploiting the plasmonic fingerprints of benchmark electrocatalysts in this field. In 2015, AuCu nanoparticles supported on mesoporous silica showed accelerated kinetics ascribed to the plasmonic properties, with a 2.5-fold boost in glycerol conversion against a monometallic Au catalyst under Vis light [147]. In 2019, Minteer et al. showcased the enhancement of the electrocatalytic oxidation of glycerol over binary Ag–Au plasmonic nanoparticles immobilized onto carbon electrodes, under Vis light [146]. Due to the SPR effect of the conventional plasmonic materials, the fuel cell had a 100% increase in both maximum current and power output under Vis light ($\lambda = 400$–425 nm), which agreed well with the peak plasmonic wavelength of the nanomaterial. Enhanced fuel cell power was primarily ascribed to hot electron transfer, with SPR-induced heating being confirmed minimal by CV collection. As a reference, Krejci et al. demonstrated that a temperature increase at the electrode surface induces mass transfer in the Nernst layer, with a sigmoidal CV being witnessed in a stagnant solution [148]. Recently, by coupling localized surface plasmons and surface plasmon polaritons (SPP), our group demonstrated an enhanced hot carrier generation with superior thin film-based photon-to-carrier conversion efficiency [149]. In detail, introduced AgPt-decorated Au nanobipyramids on Au film served as a model electrocatalytic system to assess glycerol electro-oxidation under the irradiation of a laser in Kretschmann configuration. A 2.5-fold photocurrent evolution enhancement was witnessed under the illumination of p-polarized light compared with the neat Au film. With SPPs excitation being attained solely under *p*-polarized light, our findings served as direct evidence for SPP-induced hot carrier generation. As a result, the photon-to-hot carrier conversion efficiency achieved 2.4×10^{-3} %.

Glucose electro-oxidation. Glucose has gathered focus in biofuel cells offering an alternative energy source for biologically implanted devices [150, 151]. To accelerate the intrinsically slow kinetics of glucose electro-oxidation at ambient conditions, efficient charge transfer between plasmonic nanostructures and supports is promoted, with hot holes generated on the plasmonic structure facilitating the reaction and hot electrons being swiftly transferred to the external electric circuit [152]. Directly exploited Au nanoparticles, as a Schottky-junction-free electrocatalytic system, on glassy carbon electrode, and under LSPR excitation was proposed as an efficient approach to harness hot charge carriers [153]. The energetic charge carriers are generated on the surface of the nanoparticles and injected into the external circuit, with the resulting hot holes being depleted by the oxidation of glucose due to their matched energy levels. The peak current could be increased up to 2.05 μA under illumination with a 532 nm laser corresponding to the plasmonic excitation peak (against 1.01 μA under dark conditions). By surveying wavelength- and solution-pH-dependent electrochemical oxidation rates and carrying *in situ* dark-field scattering spectroscopy measurements of a single nanoparticle under potential control, a direct plasmon-accelerated electrochemical reaction mechanism could be confirmed. Following prior studies investigating the promise of Au nanoparticles on graphene oxide sheets for glucose oxidation [154], Chen and colleagues considered an analogous electrocatalytic system to assess the potential of additionally tackling the Au plasmonic fingerprints [152]. The reduced graphene oxide with excellent conductivity and charge transport properties was proposed to receive the hot electrons from the introduced Au centers for subsequent glucose activation on the carbon-based surface. As a result, the hybrid architecture showed superior anodic current density against Au nanoparticles tested separately upon light irradiation and a threefold improvement against dark conditions. The enhanced performance above ascribed to the LSPR effect of Au and the excellent electron transfer capacity of rGO was corroborated by a wavelength-dependent current response, electrochemical impedance spectroscopy, open-circuit photovoltage, dark-field spectroscopy, and Raman spectra. In an attempt to further establish a distinction between thermal and nonthermal effects [46, 155], glucose electro-oxidation was recently investigated under hydroxide adsorption on Au nanoparticles attached to an ITO electrode in an alkaline electrolyte as a model reaction [42]. A strong dependence of the generated photocurrent and reaction rate normalized with the absorbed power on the irradiated light wavelength was witnessed, with a pronounced maximum around the plasmon resonance ($\lambda = 590\,\text{nm}$). These findings strongly emphasize that nonthermal carriers seem to predominantly be responsible for the reported plasmon-enhanced conversion enhancement. With an increasing focus in exploring the morphology of plasmonic structures to widen the library of optophysical and catalytic properties, Xu and colleagues recently reported the one-pot ecofriendly synthesis of penta-twinned icosahedral Au nanoparticles [156]. Correlated with their ample active sites, surface strain, good stability/chemical inertness, and easy functionalization, the reported nanoparticles showed superior catalytic activity with peak current densities of 1.06 and 6.87 mA cm^{-2} during measured linear sweep voltammetry, which reflect a 3.7- and 14-fold boost against a Turkevich Au reference. The SERS signals of rhodamine 6G on icosahedral Au was much stronger than that on other Au counterparts, demonstrating that abundant sharp corners and edges in the icosahedral structure act as potential hot spots for localized near-field enhancement effects.

9.3.3 The Multielectron CO_2 Reduction to Valuable Products

Assessing the promise of multielectron reactions. The design of plasmon-driven systems represents a promising strategy to enhance the abundance of charge carriers in reactions requiring multiple e^-/H^+ transfer and involving multiple intermediate and reaction products. However, the promise of this approach may directly contrast with the limited number of single-electron excitation events particularly at low light intensities, the short lifetimes of hot carriers, and the long period between excitation events [37]. In addition to the electro-oxidation of alcohols (Section 9.3.2), CO_2 reduction and N_2 fixation with increasing interest toward sustainability, serve as model multielectron reactions with foreseen analytical intricacy of related works. The increasing focus of a renewable N_2 fixation, the conversion of N_2 to NH_3, is ascribed to the utilization of NH_3 as an energy carrier, a fuel, and a fertilizer [157]. NH_3 is conventionally produced from H_2 and N_2 at high pressure (>200 bar) and temperature (>673 K) conditions through the Haber–Bosch process [158]. The electrocatalytic system suffers from a difficult activation of the N_2 molecule on the catalytic surface [159], the cleavage of the triple bond as the rate-determining step, and the need to transfer a total of eight electrons toward NH_3. Whereas the investigation of N_2 fixation over semiconductor-based photocatalysts incorporating plasmonic nanostructures has revealed promising seminal steps (Chapter 8) [160], plasmon-enhanced electrocatalytic systems remain unreported to date.

CO_2 reduction reaction. With global anthropogenic CO_2 emissions of ~35.5 Gt year^{-1} and concentration levels above 400 ppm in the atmosphere [161], a sustainable reutilization of CO_2 is one of the most groundbreaking research topics under assessment [162–164]. Despite an interest sparked nearly 200 years ago for the chemical transformation of CO_2 into chemicals, plastics, and fuels [165–167], commercialized technologies reutilizing CO_2 are still insufficient today. Critical focus is today attributed to the promise of "green" synthesis of products of higher value such as hydrocarbons and alcohols in already established infrastructures. The conventional thermocatalytic CO_2 reduction reaction (CO_2RR) over Ni and Co-based, and Fe-based catalysts has been exhaustively explored for the synthesis of methane and long carbon chains, respectively [168]. Opposing this promising break of the commercial barrier, electro/photochemical platforms remain at their infancy stage [169]. The electrochemical CO_2RR occurs at the positive electrode with water acting as a proton source, while O_2 evolves at the anode. Distinctly from photoresponsive benchmark (e.g. TiO_2) and emerging metal oxides CO_2RR photo(electro)catalysts (Chapter 7) [168], electrocatalytic systems primarily require the presence of metals to activate the inert CO_2 and drive the catalytic reaction with practical efficiency [170, 171]. To date, however, high overpotentials and energy input requirements have been plagued by high CO_2 activation energies, multiple electron transfer, ohmic losses, and mass transport limitations. Following a difficult direct electron injection to form $CO_2^{\bullet-}$ attributed to a negative electron affinity [172, 173], the binding energy of the available active centers with intermediate species primarily governs the CO_2RR selectivity (Figure 9.6a) [174]. Metals including Sn, Hg, Pb, and Cd, easily desorb the formed $HCOO^-$ ($n = 1$, with n as the number of transferred electrons), whereas the latter can be strongly bound over Au, Ag, and Zn to allow the breaking of the C=O bond. In these metal surfaces, however, the resulting CO is weakly adsorbed, swiftly evolving as a reaction product [176]. Conversely, metals, e.g. Pt, Ti, Ni, and Fe strongly bind CO (and subsequent intermediate species for $n > 2$), suffering from a resulting surface carbon-poisoning, and primarily evolving H_2 [174]. First pinpointed by the works of Hori and Suzuki first disclosing the formation of CH_4 and C_2H_4 [177, 178], Cu remains the benchmark material to date for the synthesis of hydrocarbons ($n \geq 8$) and alcohols ($n \geq 6$), notwithstanding the slow kinetics, highest overpotentials, and lowest FE in this research field [170, 179].

It is denoted that the discussed plasmonic materials overlap herein with state-of-the-art CO_2RR electrocatalysts for CO (Ag and Au) and hydrocarbons/alcohols (Cu). However, disentangling the underlying role of the plasmonic metal, and the intrinsic surface properties of the metal is deemed fundamental in this field of application. Fortunately, as electrocatalysts, these metals have been widely surveyed in the literature. The size, shape, and aspect ratio of CO-selective Au and Ag nanoparticles, for instance, have been correlated with the electrochemical

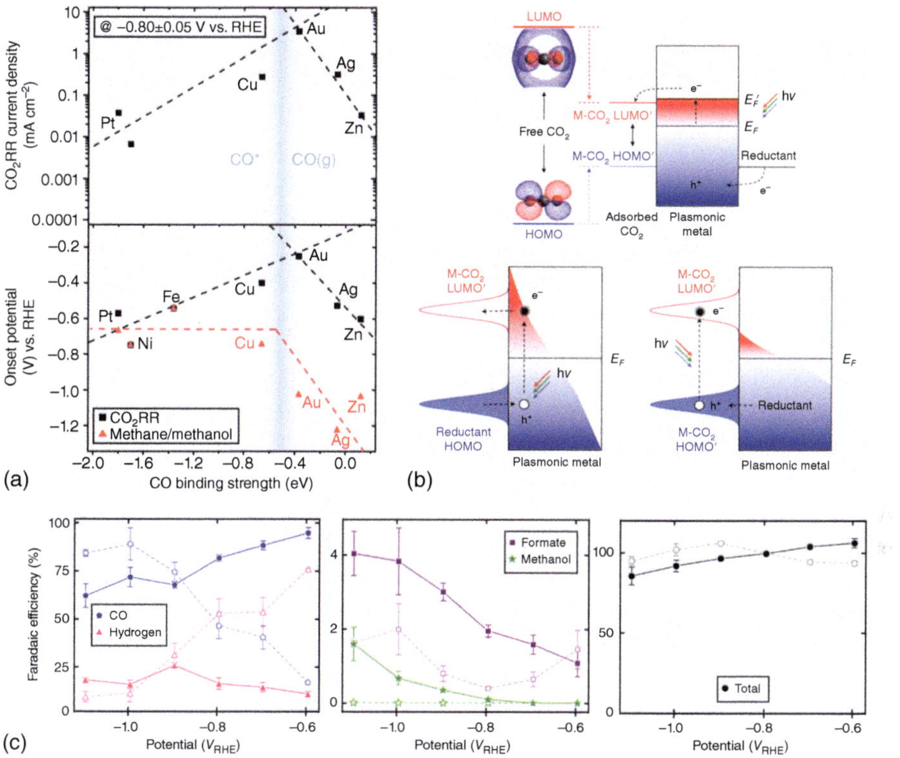

Figure 9.6 (a) Volcano plot of CO_2RR partial current density at −0.80 V against the CO binding strength and onset potentials plotted against the CO binding strength for the overall CO_2RR, and for both methane and methanol. *Source*: Kuhl et al. [174]. © 2014, American Chemical Society. (b) Mechanistic scenarios for CO_2RR on Au nanoparticles. The upper panel depicts electron injection/hole removal cathodically charging the nanoparticle and elevating the Fermi level. Energetic electrons occupying states above the quasi-Fermi level are swiftly injected into the LUMO of adsorbed CO_2 (LUMO′). Transient hot electron generation above the Fermi level is depicted in the bottom-left panel, with a fraction being scattered into unoccupied LUMO′ states, prompting a vibrational CO_2 activation. The bottom-right panel showcases a direct HOMO′–LUMO′ (with reduced gap magnitude) photoexcitation of the adsorbed CO_2. *Source*: Yu et al. [12]. © 2017, American Chemical Society. (c) Faradaic efficiency over illuminated (365 nm LED at 170 mW cm^{-2}, filled symbols) and dark silver cathodes (open symbols). *Source*: Creel et al. [175]. © 2019, American Chemical Society.

fingerprints of involved active centers, the electronic structure of adsorbate species, and the resulting efficiency, selectivity, and stability [180–183]. At more negative potentials, the metal surface oxophilicity of these metals has also been shown to play a role in the selectivity toward either methanol (with the least oxophilic Au) or both methane and methanol (Ag and Zn) [174]. Based on the design of the electrocatalytic system, a possible electronic reconstruction between plasmonic nanostructures and neighboring electrocatalysts or supports is of additional concern to understand the performance of the resulting systems [184]. On the other hand, if the plasmonic material simply serves as the catalytic active metal surface, we are now required to uncover the correlation between the photo-responsive properties of these metal nanoparticles, the electronic structure of adsorbate species, and their CO_2RR activity, selectivity, and stability. This is believed crucial to optimize systems with several intermediates and mechanistic pathways [35]. The exploitation of the photoresponsive properties of noble metal nanoparticles shall further address light absorption-inducing interband transitions (Section 9.2). The photon energy shed over the Au metal surface, for instance, is high enough to excite *d* band transitions, positioned above representative CO_2/C_2H_6, CO_2/CH_3OH, and CO_2/HCHO redox potentials [185].

Whereas works investigating a plasmon-enhanced electrocatalytic CO_2RR remain scarcely tackled, they are expected to benefit, both mechanistic-wise and in the design of valuable materials library, from prior know-how in the photocatalytic and PEC CO_2RR counterparts, in which CO_2 inertness and the sluggish kinetics of hot electron transfer process are key drawbacks [186, 187]. Jain et al. have assessed distinct mechanistic pathways driven by generated hot-electrons over the surface of Au nanoparticles toward CH_4 ($n = 8$) and C_2H_6 ($n = 12$) [12]. During the activation of free CO_2 on the metal surface, induced hybridization of its electronic states with those of Au was proposed, resulting in a new reduced HOMO–LUMO gap, which can be excited under Vis light (Figure 9.6b). Based on the authors' claims, the resulting electron injection (and hole removal in the presence of isopropyl alcohol electron donor) charges the nanoparticle and elevates the Fermi level to a quasi-Fermi level state. The energetic electrons occupying states above the quasi-Fermi level are facilely injected into the LUMO′ level, inducing the formation of a $CO_2^{\bullet-}$ radical. Besides, light selection (intensity and flux) could enhance the electron-transfer rate allowing the concentration of adsorbed $CO_2^{\bullet-}$ species on the surface (with inherently short residence time) to be sufficiently high to allow dimerization reactions with C–C coupling to occur. In agreement, C_2H_6 could be evolved when light intensity ranges surpassed 300 mW cm^{-2}. The importance of these findings is ascribed to a necessary swift reduction of involved intermediate species through rapid electron and proton-transfer to assure practical CO_2RR product selectivity.

With the knowledge above, the exploitation and effective interpretation of plasmon-enhanced electrocatalytic systems appear promising in the upcoming future. A plasmonic silver thin film electrode under applied potentials was recently reported to selectively generate CO, with hindered H_2 evolution (Figure 9.6c) [175]. Toward more negative potentials, the electrocatalyst was witnessed to promote the formation of both formate and methanol, whereas the latter would be only generated under illumination. The challenging formation of methanol ($n = 6$) at 0.8 V_{RHE} reflects a 550 mV overpotential decrease against a previously reported polycrystalline silver [188].

In the context of this section, it is also important to underline expected upcoming trends in exploratory works, which may find a direct correlation with plasmon-enhanced systems. Alloys and intermetallic materials, for example, offering tunable compositions and chemical states, and precise control of the distribution of active centers for effective binding energy manipulation of intermediate species and proton–catalyst interaction have been recently explored in

CO_2RR [179, 189–191]. This approach (when incorporating Au, Ag, Pd, and Cu) gathers particular interest in light of interpretative studies of plasmonic properties of alloy materials [192]. Examples include the synthesis of Cu-including alloys as an alternative to pricey Au and Ag metals for CO production, or toward a facile CO protonation and stabilization of CHO adsorbed intermediates for multielectron reaction ($n > 2$) [189, 193–195]. With CO as a key intermediate, other reports have assessed alloys between metals that bind CO strongly (e.g. Pd) and metals that bind CO weakly (e.g. Au), as an alternative to Cu [196]. Designing architectures rationally neighboring distinct active centers for sequential catalysis with superior activity and selectivity CO_2RR rates as found rationality in an increase of the local CO concentration, which is directly correlated with adsorbed CO coverage, and a favored C-C dimerization pathway-based product selectivity [197]. Herein too, neighboring Au/Ag-based CO-selective active centers to Cu sites may engender synergetic effects exploiting distinctly offered activity and selectivity fingerprints and the diffusional transport of CO [198–200]. Similarly, SPR-fingerprints of these materials can be individually tuned to manipulate the discrete reaction pathways and specifically tune the activity of each operating active site. Finally, nonprecious earth-abundant Al should also be here emphasized for promising exploitation in industrial-scale platforms. As introduced by Halas and colleagues, Al on neighboring Cu_2O unveiled improved surface reactivity, resulting hot-carrier generation to form $CO_2^{\bullet-}$ on the metal oxide shell, and enhanced CO production [201]. These hot electrons are proposed to transiently populate adsorbed CO_2 antibonding orbitals, before relaxation into an excited vibrational state in the ground state by releasing the electron to the Cu_2O support. The transferred energy into the vibrational mode of C–O induces vibrational excitation and C–O bond elongation, with subsequent dissociation.

9.4 Outlook

Notwithstanding the steady progress in the efficiency and overall performance of electrocatalytic technologies to reutilize small molecules such as H_2O, N_2, and CO_2, considerable economic barriers still predominate toward a break of the commercial barrier against the complex and well-established petrochemical industry. The development of plasmon-enhanced electrocatalytic systems, with a promising insertion within the current chemical supply chain, is a valuable tool to convert the energy from light while utilizing benchmark electrocatalysts and is expected to inspire a continuous interest growth in the scientific community. In this Chapter, fundamental notions and an assessment of the promise of plasmonics applied in electrocatalytic systems of marked importance are provided. To date, however, the number of comprehensive studies remains scarce and the promise of this strategy requiring the incorporation of cost-ineffective, scarce, and with the unsuitable thermal stability of plasmonic noble metals is notably dependent on future optimization of the exploitation of the SPR. The upsurge of refractory plasmonic materials with low-cost, chemical stability, and mechanical durability may reflect in this sense promising candidates to be considered in upcoming years [202]. A key drawback in this field is correlated with the poor understanding of the underlying mechanism pathways. For each reaction under analysis, the reported systems follow similar strategies and exploit conventional benchmark electrocatalysts. Only recent examples attempt to uncover the individual plasmonic effects introduced in Section 9.2.2. We believe additional efforts may serve as a valuable first step in a direction toward superior know-how and process efficiency.

Acknowledgements

This work was supported by National Research Foundation of Korea grant funded by the Korean Government (NRF-2020R 1A 2C 3003958) and by Creative Materials Discovery Program through the National Research Foundation of Korea (NRF) funded by Ministry of Science and ICT (2018M 3D 1A 1058536). F.M.M. acknowledges the support by the Brain Pool Program through the National Research Foundation of Korea (NRF) funded by the Ministry of Science and ICT (2017H1D3A1A02054206).

References

1 Haegel, N.M., Margolis, R., Buonassisi, T. et al. (2017). *Science* 356: 141.
2 Whipple, D.T. and Kenis, P.J.A. (2010). *J. Phys. Chem. Lett.* 1: 3451.
3 Fujishima, A. and Honda, K. (1972). *Nature* 238: 37.
4 Zeng, S., Baillargeat, D., Ho, H.-P., and Yong, K.-T. (2014). *Chem. Soc. Rev.* 43: 3426.
5 Knight, M.W., Sobhani, H., Nordlander, P., and Halas, N.J. (2011). *Science* 332: 702.
6 Warren, S.C. and Thimsen, E. (2012). *Energ. Environ. Sci.* 5: 5133.
7 Jang, Y.H., Jang, Y.J., Kim, S. et al. (2016). *Chem. Rev.* 116: 14982.
8 Boppella, R., Thomas Kochuveedu, S., Kim, H. et al. (2017). *ACS Appl. Mater. Interfaces* 9: 7075.
9 Boppella, R., Marques Mota, F., Lim, J.W. et al. (2019). *ACS Appl. Energy Mater.* 2: 3780.
10 Hentschel, M., Schäferling, M., Duan, X. et al. (2017). *Sci. Adv.* 3: e1602735.
11 Kochuveedu, S.T., Jang, Y.H., and Kim, D.H. (2013). *Chem. Soc. Rev.* 42: 8467.
12 Yu, S., Wilson, A.J., Kumari, G. et al. (2017). *ACS Energy Lett.* 2: 2058.
13 Kuo, T.-R., Lee, Y.-C., Chou, H.-L. et al. (2019). *Chem. Mater.* 31: 3722.
14 Wang, C., Zhao, X.-P., Xu, Q.-Y. et al. (2018). *ACS Appl. Nano Mater.* 1: 5805.
15 Gargiulo, J., Berté, R., Li, Y. et al. (2019). *Acc. Chem. Res.* 52: 2525.
16 Pensa, E., Gargiulo, J., Lauri, A. et al. (2019). *Nano Lett.* 19: 1867.
17 Khurgin, J.B. (2015). *Nat. Nanotechnol.* 10: 2.
18 Clavero, C. (2014). *Nat. Photonics* 8: 95.
19 Behl, M. and Jain, P.K. (2015). *Angew. Chem. Int. Ed.* 54: 992.
20 Kim, Y., Torres, D.D., and Jain, P.K. (2016). *Nano Lett.* 16: 3399.
21 Mukherjee, S., Libisch, F., Large, N. et al. (2013). *Nano Lett.* 13: 240.

22 Mukherjee, S., Zhou, L., Goodman, A.M. et al. (2014). *J. Am. Chem. Soc.* 136: 64.

23 Zhou, L., Swearer, D.F., Zhang, C. et al. (2018). *Science* 362: 69.

24 Furube, A. and Hashimoto, S. (2017). *NPG Asia Mater.* 9: e454.

25 Gao, M., Zhu, L., Peh, C.K., and Ho, G.W. (2019). *Energ. Environ. Sci.* 12: 841.

26 Di Valentin, C. and Fittipaldi, D. (2013). *J. Phys. Chem. Lett.* 4: 1901.

27 Wu, K., Chen, J., McBride, J.R., and Lian, T. (2015). *Science* 349: 632.

28 Forcherio, G.T., Baker, D.R., Boltersdorf, J. et al. (2018). *J. Phys. Chem. C* 122: 28901.

29 Liu, N., Prall, B.S., and Klimov, V.I. (2006). *J. Am. Chem. Soc.* 128: 15362.

30 Li, M., Cushing, S.K., Wang, Q. et al. (2011). *J. Phys. Chem. Lett.* 2: 2125.

31 Wu, K., Rodríguez-Córdoba, W.E., Yang, Y., and Lian, T. (2013). *Nano Lett.* 13: 5255.

32 Linnert, T., Mulvaney, P., and Henglein, A. (1993). *J. Phys. Chem.* 97: 679.

33 Bauer, C., Abid, J.-P., Fermin, D., and Girault, H.H. (2004). *J. Chem. Phys.* 120: 9302.

34 Christopher, P., Xin, H., Marimuthu, A., and Linic, S. (2012). *Nat. Mater.* 11: 1044.

35 Kale, M.J., Avanesian, T., and Christopher, P. (2014). *ACS Catal.* 4: 116.

36 Busch, D.G. and Ho, W. (1996). *Phys. Rev. Lett.* 77: 1338.

37 Aslam, U., Rao, V.G., Chavez, S., and Linic, S. (2018). *Nat. Catal.* 1: 656.

38 Khurgin, J.B. (2019). *Faraday Discuss.* 214: 35.

39 Bernardi, M., Mustafa, J., Neaton, J.B., and Louie, S.G. (2015). *Nat. Commun.* 6: 7044.

40 Brown, A.M., Sundararaman, R., Narang, P. et al. (2016). *ACS Nano* 10: 957.

41 Zhao, J., Nguyen, S.C., Ye, R. et al. (2017). *ACS Cent. Sci.* 3: 482.

42 Rodio, M., Graf, M., Schulz, F. et al. (2020). *ACS Catal.* 10: 2345.

43 Brown, A.M., Sundararaman, R., Narang, P. et al. (2016). *Phys. Rev. B* 94: 75120.

44 Boriskina, S.V., Cooper, T.A., Zeng, L. et al. (2017). *Adv. Opt. Photonics* 9: 775.

45 Etchegoin, P.G., Le Ru, E.C., and Meyer, M. (2006). *J. Chem. Phys.* 125: 164705.

46 Minutella, E., Schulz, F., and Lange, H. (2017). *J. Phys. Chem. Lett.* 8: 4925.

47 Crut, A., Maioli, P., Del Fatti, N., and Vallée, F. (2015). *Ultrasonics* 56: 98.

48 Ashcroft, N.W. and Mermin, N.D. (1977). *Phys. Today* 30: 61.

49 Choi, C.H., Chung, K., Nguyen, T.T.H., and Kim, D.H. (2018). *ACS Energy Lett.* 3: 1415.

50 Nurlaela, E., Shinagawa, T., Qureshi, M. et al. (2016). *ACS Catal.* 6: 1713.

51 Choi, Y., Kang, T., and Lee, L.P. (2009). *Nano Lett.* 9: 85.

52 Cao, Y., Xie, T., Qian, R.-C., and Long, Y.-T. (2017). *Small* 3: 1601955.

53 Chavez, S., Aslam, U., and Linic, S. (2018). *ACS Energy Lett.* 3: 1590.

54 Hsu, L.-Y., Ding, W., and Schatz, G.C. (2017). *J. Phys. Chem. Lett.* 8: 2357.

55 DeRosa, M.C. and Crutchley, R.J. (2002). *Coord. Chem. Rev.* 233–234: 351.

56 Jang, Y.J., Chung, K., Lee, J.S. et al. (2018). *ACS Photonics* 5: 4711.

57 Seh, Z.W., Kibsgaard, J., Dickens, C.F. et al. (2017). *Science* 355: eaad4998.

58 Wang, C., Shi, Y., Yang, D.-R., and Xia, X.-H. (2018). *Curr. Opin. Electrochem.* 7: 95.

59 Nikolaidis, P. and Poullikkas, A. (2017). *Renew. Sustain. Energy Rev.* 67: 597.
60 Levin, D.B. and Chahine, R. (2010). *Int. J. Hydrogen Energy* 35: 4962.
61 Saeidi, S., Fazlollahi, F., Najari, S. et al. (2017). *J. Ind. Eng. Chem.* 49: 1.
62 Wang, Z., Li, C., and Domen, K. (2019). *Chem. Soc. Rev.* 48: 2109.
63 Hurst, J.K. (2010). *Science* 328: 315.
64 Nørskov, J.K., Bligaard, T., Logadottir, A. et al. (2005). *J. Electrochem. Soc.* 152: J23.
65 Morales-Guio, C.G., Stern, L.-A., and Hu, X. (2014). *Chem. Soc. Rev.* 43: 6555.
66 Shi, Y., Wang, J., Wang, C. et al. (2015). *J. Am. Chem. Soc.* 137: 7365.
67 Walter, M.G., Warren, E.L., McKone, J.R. et al. (2010). *Chem. Rev.* 110: 6446.
68 Wang, J., Cui, W., Liu, Q. et al. (2016). *Adv. Mater.* 28: 215.
69 Che Mohamad, N.A.R., Marques Mota, F., and Kim, D.H. (2020). *Solar-to-Chemical Conversion: Photocatalytic and Photoelectrochemical Processes* (ed. H. Sun). Wiley-VCH.
70 Hinnemann, B., Moses, P.G., Bonde, J. et al. (2005). *J. Am. Chem. Soc.* 127: 5308.
71 Tsai, C., Chan, K., Nørskov, J.K., and Abild-Pedersen, F. (2015). *Surf. Sci.* 640: 133.
72 Dhakal, K.P., Ghimire, G., Chung, K. et al. (2019). *ACS Nano* 13: 14437.
73 Kang, Y., Najmaei, S., Liu, Z. et al. (2014). *Adv. Mater.* 26: 6467.
74 Kang, Y., Gong, Y., Hu, Z. et al. (2015). *Nanoscale* 7: 4482.
75 Shang, B., Cui, X., Jiao, L. et al. (2019). *Nano Lett.* 19: 2758.
76 Ali, A., Mangrio, F.A., Chen, X. et al. (2019). *Nanoscale* 11: 7813.
77 Li, B.L., Zou, H.L., Tian, J.K. et al. (2019). *Nano Energy* 60: 689.
78 Zhang, Y., Du, J., Luo, R. et al. (2019). *Nano Energy* 58: 552.
79 Lang, X., Qian, L., Guan, P. et al. (2011). *Appl. Phys. Lett.* 98: 93701.
80 Li, X., Guo, S., Kan, C. et al. (2016). *Nano Energy* 30: 549.
81 Zhang, H.-X., Li, Y., Li, M.-Y. et al. (2018). *Nanoscale* 10: 2236.
82 Wei, Y., Zhao, Z., and Yang, P. (2018). *ChemElectroChem* 5: 778.
83 Zhang, B., Zheng, X., Voznyy, O. et al. (2016). *Science* 352: 333.
84 McCrory, C.C.L., Jung, S., Peters, J.C., and Jaramillo, T.F. (2013). *J. Am. Chem. Soc.* 135: 16977.
85 Hong, W.T., Risch, M., Stoerzinger, K.A. et al. (2015). *Energ. Environ. Sci.* 8: 1404.
86 Dionigi, F. and Strasser, P. (2016). *Adv. Energy Mater.* 6: 1600621.
87 Man, I.C., Su, H., Calle-Vallejo, F. et al. (2011). *ChemCatChem* 3: 1159.
88 Liu, G., Li, P., Zhao, G. et al. (2016). *J. Am. Chem. Soc.* 138: 9128.
89 Hu, W.-C., Shi, Y., Zhou, Y. et al. (2019). *J. Mater. Chem. A* 7: 10601.
90 Aubrey, M.L., Wiers, B.M., Andrews, S.C. et al. (2018). *Nat. Mater.* 17: 625.
91 Xu, J., Gu, P., Birch, D.J.S., and Chen, Y. (2018). *Adv. Funct. Mater.* 28: 1801573.
92 Sayeed, M.A. and O'Mullane, A.P. (2017). *J. Mater. Chem. A* 5: 23776.
93 Ng, J.W.D., García-Melchor, M., Bajdich, M. et al. (2016). *Nat. Energy* 1: 16053.
94 Zhao, S., Jin, R., Abroshan, H. et al. (2017). *J. Am. Chem. Soc.* 139: 1077.
95 Lee, J.E., Marques Mota, F., Choi, C.H. et al. (2019). *Adv. Mater. Interfaces* 6: 1801144.

96 Du, L., Shi, G., Zhao, Y. et al. (2019). *Chem. Sci.* 10: 9605.
97 Debe, M.K. (2012). *Nature* 486: 43.
98 Nørskov, J.K., Rossmeisl, J., Logadottir, A. et al. (2004). *J. Phys. Chem. B* 108: 17886.
99 Lin, S.-C., Hsu, C.-S., Chiu, S.-Y. et al. (2017). *J. Am. Chem. Soc.* 139: 2224.
100 Zheng, Z., Xie, W., Li, M. et al. (2017). *Nano Energy* 41: 233.
101 Shi, F., He, J., Zhang, B. et al. (2019). *Nano Lett.* 19: 1371.
102 Wagner, F.T., Lakshmanan, B., and Mathias, M.F. (2010). *J. Phys. Chem. Lett.* 1: 2204.
103 Gasteiger, H.A., Kocha, S.S., Sompalli, B., and Wagne, F.T. (2005). *Appl. Catal. Environ.* 56: 9.
104 Marković, N.M., Schmidt, T.J., Stamenković, V., and Ross, P.N. (2001). *Fuel Cells* 1: 105.
105 Greeley, J., Stephens, I.E.L., Bondarenko, A.S. et al. (2009). *Nat. Chem.* 1: 552.
106 Huang, X., Zhao, Z., Cao, L. et al. (2015). *Science* 348: 1230.
107 Holewinski, A., Idrobo, J.-C., and Linic, S. (2014). *Nat. Chem.* 6: 828.
108 Sun, X., Li, D., Ding, Y. et al. (2014). *J. Am. Chem. Soc.* 136: 5745.
109 Chen, C., Kang, Y., Huo, Z. et al. (2014). *Science* 343: 1339.
110 Zhou, W.-P., Yang, X., Vukmirovic, M.B. et al. (2009). *J. Am. Chem. Soc.* 131: 12755.
111 Nørskov, J.K., Bligaard, T., Rossmeisl, J., and Christensen, C.H. (2009). *Nat. Chem.* 1: 39.
112 Bu, L., Zhang, N., Guo, S. et al. (2016). *Science* 354: 1410.
113 Zhang, B., Wang, S., Fan, W. et al. (2016). *Angew. Chem. Int. Ed.* 55: 14748.
114 Xiong, Y., Ren, M., Li, D. et al. (2017). *J. Catal.* 354: 160.
115 Deng, Y., Xue, H., Lu, S. et al. (2018). *ACS Appl. Energy Mater.* 1: 4891.
116 Chung, K., Lee, J.S., Kim, E. et al. (2018). *Adv. Mater. Interfaces* 5: 1800433.
117 Chung, K., Rani, A., Lee, J.-E. et al. (2015). *ACS Appl. Mater. Interfaces* 7: 144.
118 Huang, L., Zou, J., Ye, J. et al. (2019). *Angew. Chem. Int. Ed.* 58: 8794.
119 Bi, J., Cai, H., Wang, B. et al. (2019). *Chem. Commun.* 55: 3943.
120 Xu, H., Song, P., Wang, J. et al. (2018). *ACS Sustain. Chem. Eng.* 6: 7159.
121 Liu, Y., Chen, F., Wang, Q. et al. (2018). *J. Mater. Chem. A* 6: 10515.
122 Liu, H., Li, J., Wang, L. et al. (2017). *Nano Res.* 10: 3324.
123 Han, S., Liu, H., Chen, P. et al. (2018). *Adv. Energy Mater.* 8: 1801326.
124 Zhang, K.-F., Guo, D.-J., Liu, X. et al. (2006). *J. Power Sources* 162: 1077.
125 Xu, C., Shen, P.K., and Liu, Y. (2007). *J. Power Sources* 164: 527.
126 Nguyen, S.T., Ling Tan, D.S., Lee, J.-M. et al. (2011). *Int. J. Hydrogen Energy* 36: 9645.
127 Jin, Z., Wang, Q., Zheng, W., and Cui, X. (2016). *ACS Appl. Mater. Interfaces* 8: 5273.
128 Lee, J.-E., Chung, K., Jang, Y.H. et al. (2012). *Anal. Chem.* 84: 6494.
129 Zheng, Z., Tachikawa, T., and Majima, T. (2014). *J. Am. Chem. Soc.* 136: 6870.
130 Xu, H., Song, P., Wang, J. et al. (2018). *ChemElectroChem* 5: 1191.
131 Wang, Z., Du, J., Zhang, Y. et al. (2019). *Nano Energy* 56: 286.
132 An, L. and Chen, R. (2016). *J. Power Sources* 329: 484.

133 Xu, H., Yan, B., Zhang, K. et al. (2017). *ACS Sustain. Chem. Eng.* 5: 10490.

134 Raj kumar, T., Yoo, D.J., Kim, A.R. et al. (2018). *Chem* 42: 14386.

135 Selvaraj, V., Vinoba, M., and Alagar, M. (2008). *J. Colloid Interface Sci.* 322: 537.

136 Hong, W., Shang, C., Wang, J., and Wang, E. (2015). *Energ. Environ. Sci.* 8: 2910.

137 Gao, H., Zhai, C., Yuan, C. et al. (2020). *Electrochim. Acta* 330: 135214.

138 Xu, H., Song, P., Yan, B. et al. (2018). *ACS Sustain. Chem. Eng.* 6: 4138.

139 Huber, G.W., Iborra, S., and Corma, A. (2006). *Chem. Rev.* 106: 4044.

140 Dodekatos, G., Schünemann, S., and Tüysüz, H. (2018). *ACS Catal.* 8: 6301.

141 Zhang, Z., Xin, L., and Li, W. (2012). *Appl. Catal. Environ.* 119–120: 40.

142 Kim, Y., Kim, H.W., Lee, S. et al. (2017). *ChemCatChem* 9: 1683.

143 Shen, Y., Zhang, S., Li, H. et al. (2010). *Chem. A Eur. J.* 16: 7368.

144 Zhou, Y., Shen, Y., and Luo, X. (2020). *J. Catal.* 381: 130.

145 Guo, L., Sun, Q., Marcus, K. et al. (2018). *J. Mater. Chem. A* 44: 22005.

146 Rasmussen, M., Serov, A., Artyushkova, K. et al. (2019). *ChemElectroChem* 6: 241.

147 Schünemann, S., Dodekatos, G., and Tüysüz, H. (2015). *Chem. Mater.* 27: 7743.

148 Krejci, J., Sajdlova, Z., Krejci, J. Jr., and Marvanek, T. (2010). *Sensors* 10: 6821.

149 Chung, K., Zhu, X., Zhuo, X. et al. (2019). *Nanoscale* 11: 23234.

150 Zhao, C., Gai, P., Song, R. et al. (2017). *Chem. Soc. Rev.* 46: 1545.

151 Cosnier, S., Le Goff, A., and Holzinger, M. (2014). *Electrochem. Commun.* 38: 19.

152 Liao, X., Wang, S., Xu, G., and Wang, C. (2019). *ChemElectroChem* 6: 1419.

153 Wang, C., Nie, X.-G., Shi, Y. et al. (2017). *ACS Nano* 11: 5897.

154 Wojnicki, M., Luty-Błocho, M., Dobosz, I. et al. (2013). *Chem. Mater. Sci.* 4: 162.

155 Zhang, X., Li, X., Reish, M.E. et al. (2018). *Nano Lett.* 18: 1714.

156 Xu, M., Zhang, L., and Zhao, F. (2020). *ACS Appl. Mater. Interfaces* 12: 12186.

157 Lan, R., Irvine, J.T.S., and Tao, S. (2012). *Int. J. Hydrogen Energy* 37: 1482.

158 Haber, F. and Le Rossignol, R. (1909). Production of ammonia (BASF SE). US 1, 202, 995 A, 1–4.

159 Hirakawa, H., Hashimoto, M., Shiraishi, Y., and Hirai, T. (2017). *J. Am. Chem. Soc.* 139: 10929.

160 Yang, J., Guo, Y., Lu, W. et al. (2018). *Adv. Mater.* 30: 1802227.

161 Chu, S., Cui, Y., and Liu, N. (2017). *Nat. Mater.* 16: 16.

162 Karl, T.R. and Trenberth, K.E. (2003). *Science* 302: 1719.

163 Roemmich, D., Church, J., Gilson, J. et al. (2015). *Nat. Clim. Chang.* 5: 240.

164 Bernstein, A., Sargent, E.H., Aspuru-Guzik, A. et al. (2016). *Nature* 538: 30.

165 Kolbe, H. and Lautemann, E. (1869). *Justus Liebigs Ann. Chem.* 113: 125.

166 Solvay, E. (1882). Absorption of ammonia in soda ash production. US Patent 263, 981 A.

167 Bosch, C. and Meiser, W. (1922). Process of manufacturing urea (BASF SE). US Patent 1, 429, 483 A.

168 Marques Mota, F. and Kim, D.H. (2019). *Chem. Soc. Rev.* 48: 205.

169 De Luna, P., Hahn, C., Higgins, D. et al. (2019). *Science* 364: eaav3506.

170 Larrazábal, G.O., Martín, A.J., and Pérez-Ramírez, J. (2017). *J. Phys. Chem. Lett.* 8: 3933.
171 Watanabe, M., Shibata, M., Katoh, A. et al. (1991). *Denki Kagaku* 59: 508.
172 Gutsev, G.L., Bartlett, R.J., and Compton, R.N. (1998). *J. Chem. Phys.* 108: 6756.
173 Chang, X., Wang, T., and Gong, J. (2016). *Energ. Environ. Sci.* 9: 2177.
174 Kuhl, K.P., Hatsukade, T., Cave, E.R. et al. (2014). *J. Am. Chem. Soc.* 136: 14107.
175 Creel, E.B., Corson, E.R., Eichhorn, J. et al. (2019). *ACS Energy Lett.* 4: 1098.
176 Hori, Y., Wakebe, H., Tsukamoto, T., and Koga, O. (1994). *Electrochim. Acta* 39: 1833.
177 Hori, Y., Kikuchi, K., and Suzuki, S. (1985). *Chem. Lett.* 14: 1695.
178 Hori, Y., Kikuchi, K., Murata, A., and Suzuki, S. (1986). *Chem. Lett.* 15: 897.
179 Back, S., Kim, H., and Jung, Y. (2015). *ACS Catal.* 5: 965.
180 Hori, Y., Murata, A., Kikuchi, K., and Suzuki, S. (1987). *J. Chem. Soc. Chem. Commun.* 10: 728.
181 Zhu, W., Michalsky, R., Metin, Ö. et al. (2013). *J. Am. Chem. Soc.* 135: 16833.
182 Liu, M., Pang, Y., Zhang, B. et al. (2016). *Nature* 527: 382.
183 Saberi Safaei, T., Mepham, A., Zheng, X. et al. (2016). *Nano Lett.* 16: 7224.
184 Marques Mota, F., Nguyen, D.L.T., Lee, J.-E. et al. (2018). *ACS Catal.* 8: 4364.
185 Hou, W., Hung, W.H., Pavaskar, P. et al. (2011). *ACS Catal.* 1: 929.
186 Song, H., Meng, X., Dao, T.D. et al. (2018). *ACS Appl. Mater. Interfaces* 10: 408.
187 Collado, L., Reynal, A., Fresno, F. et al. (2018). *Nat. Commun.* 9: 4986.
188 Hatsukade, T., Kuhl, K.P., Cave, E.R. et al. (2014). *Phys. Chem. Chem. Phys.* 16: 13814.
189 Peterson, A.A. and Nørskov, J.K. (2012). *J. Phys. Chem. Lett.* 3: 251.
190 Kim, D., Resasco, J., Yu, Y. et al. (2014). *Nat. Commun.* 5: 4948.
191 He, J., Dettelbach, K.E., Salvatore, D.A. et al. (2017). *Angew. Chemie* 129: 6164.
192 Gong, C. and Leite, M.S. (2016). *ACS Photonics* 3: 507.
193 Yin, Z., Gao, D., Yao, S. et al. (2016). *Nano Energy* 27: 35.
194 Hirunsit, P., Soodsawang, W., and Limtrakul, J. (2015). *J. Phys. Chem. C* 119: 8238.
195 Zhao, W., Yang, L., Yin, Y., and Jin, M. (2014). *J. Mater. Chem. A* 2: 902.
196 Kortlever, R., Peters, I., Balemans, C. et al. (2016). *Chem. Commun.* 52: 10229.
197 Wang, X., Varela, A.S., Bergmann, A. et al. (2017). *ChemSusChem* 10: 4642.
198 Sandberg, R.B., Montoya, J.H., Chan, K., and Nørskov, J.K. (2016). *Surf. Sci.* 654: 56.
199 Lum, Y. and Ager, J.W. (2018). *Energ. Environ. Sci.* 11: 2935.
200 Morales-Guio, C.G., Cave, E.R., Nitopi, S.A. et al. (2018). *Nat. Catal.* 1: 764.
201 Robatjazi, H., Zhao, H., Swearer, D.F. et al. (2017). *Nat. Commun.* 8: 27.
202 Zeng, X., Choi, S.M., Bai, Y. et al. (2020). *ACS Appl. Energy Mater.* 3: 146.

10

Plasmonic Metal/Semiconductor Heterostructures

Wenxiao Guo, Jiawei Huang and Wei David Wei

Department of Chemistry and Center for Catalysis, University of Florida, Gainesville, FL, USA

10.1 Introduction

Plasmonic metal nanoparticles represent a new type of solar photocatalysts due to their strong absorption of visible light via surface plasmon resonance (SPR) and their capability to generate hot electron–hole pairs for driving photochemical processes [1–4]. However, using plasmonic metal nanoparticles alone for photocatalysis is significantly limited as lifetimes of those plasmon-generated hot carriers are much shorter than the timescale of chemical reactions [1, 4–7]. Constructing heterostructures of plasmonic metal nanoparticles and semiconductors provides an ideal solution to overcome this limitation. When noble metal nanoparticles are brought into contact with semiconductors, an energy barrier known as the Schottky barrier forms at the interface and facilitates the separation of hot electrons and hot holes via the plasmon-mediated hot-carrier transfer, prolonging lifetimes of hot carriers to match with the timescale of chemical reactions [1, 5–10]. Meanwhile, surface defects on semiconductors (e.g. oxygen vacancies) serve as active sites for target reactions [11–13], leading to an enhanced photocatalytic activity when compared to that of using metal nanoparticles alone.

In this chapter, we begin with a description of basic working principles of plasmonic metal/semiconductor heterostructures in photocatalysis. The subsequent section summarizes main methods for the fabrication of plasmonic heterostructures. Strategies for optimizing the performance of plasmonic heterostructures by material design based on their working principles are then given, followed by a summary of commonly encountered photochemical reactions conducted on plasmonic heterostructures.

10.2 Working Principles

Plasmonic metal/semiconductor heterostructures facilitate photocatalytic reactions mainly by achieving an efficient charge separation with the Schottky barrier

Plasmonic Catalysis: From Fundamentals to Applications, First Edition.
Edited by Pedro H.C. Camargo and Emiliano Cortés.

formed at the metal/semiconductor interface. For heterostructures consisting of metal nanoparticles and n-type semiconductors, the barrier allows the transfer of plasmon-generated hot electrons from the metal nanoparticle to the conduction band (CB) of semiconductor, with hot holes remaining on the metal nanoparticle. By contrast, when metal nanoparticles and p-type semiconductors are combined, hot holes are transferred to the semiconductor, but hot electrons are left on the metal nanoparticle. Such charge separations reduce the possibility of charge recombination, thus prolonging the lifetime of hot carriers and enabling their participation in chemical reactions.

10.2.1 Formation of the Schottky Barrier at the Metal/Semiconductor Interface

As shown in Figure 10.1a, when a metal nanoparticle is brought into contact with an n-type semiconductor (e.g. n-TiO_2), the work function (ϕ) difference between the semiconductor and the metal drives electrons to flow from the semiconductor to the metal until their Fermi levels are aligned. This electron flow depletes the semiconductor's electron density near the interface, forming a space charge region (or depletion layer for metal/n-type semiconductor heterostructures) that carries excessive positive charges compared to the bulk [8, 9]. Meanwhile, the electron flow causes the metal nanoparticle surface to be negatively charged, which combines with those positive charges on the semiconductor surface to establish a built-in electric field pointing to the metal (i.e. exerting an electrical force pointing to the bulk of semiconductor on electrons). This is reflected as an upward energy band bending in the energy diagram [8]. The energy band bending creates a barrier called the Schottky barrier at the metal/semiconductor interface. In convention, the height of the Schottky barrier Φ_{SB} is defined as:

$$\Phi_{SB} = \phi_m - \chi_s,$$

where ϕ_m stands for the work function of the metal and χ_s is the electron affinity of the semiconductor (i.e. the energy difference between the CB edge and the vacuum level). Recent studies indicated that in real cases, the value of Φ_{SB} may deviate from this ideal value due to the formation of an additional interface dipole (i.e. eD_{int}) that stems from the hybridization between isolated orbitals on the surface of metal and semiconductor, and the equation for calculating Φ_{SB} is modified to [8, 9]

$$\Phi_{SB} = \phi_m - \chi_s + eD_{int}$$

Other interfacial effects, such as the Fermi level pinning, further complicate the situation [9]. Nonetheless, using ϕ_m and χ_s to predict Φ_{SB} has been adapted for most metal/semiconductor heterostructures. Experimentally, Φ_{SB} can be determined by fitting the current–voltage (I–V) curve of a metal/semiconductor photoelectrode using the thermionic emission equation [9, 14, 15]. More sophisticated methods, such as measuring surface photovoltage and using modified scanning probe microscopy, have also been exploited to determine Φ_{SB} [8].

A similar situation is held for a heterostructure consisting of a metal and a p-type semiconductor (Figure 10.1b). In this case, electrons flow from the metal to the semiconductor to align their Fermi levels, generating a space charge region in the

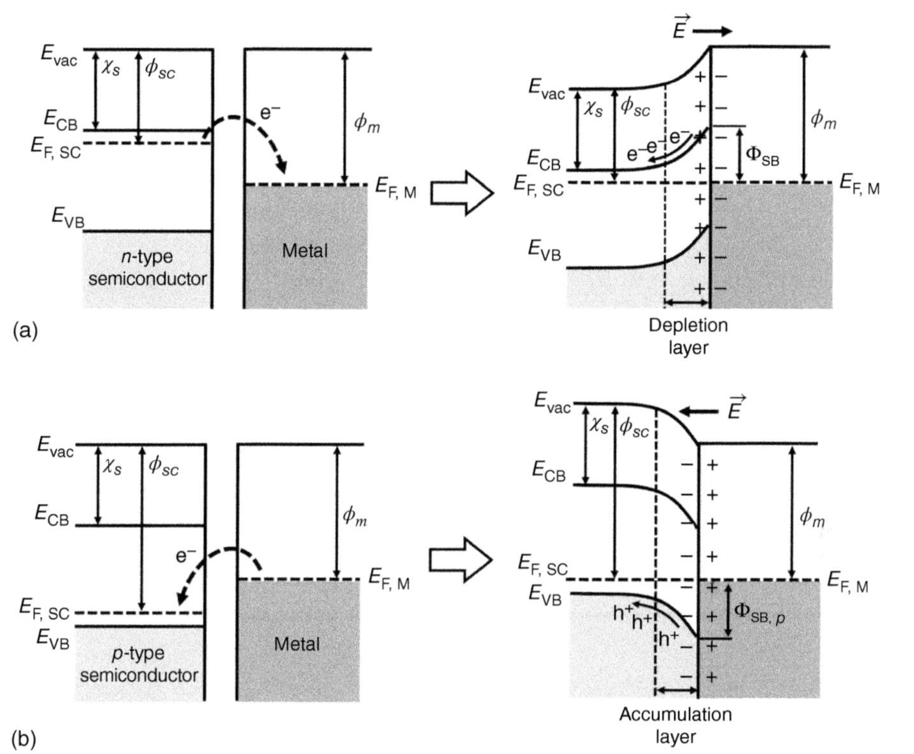

Figure 10.1 The formation of the Schottky barrier at a metal/semiconductor interface. (a) For hot electrons: at the interface between a metal and an *n*-type semiconductor, electrons are transferred from the semiconductor to the metal to align their Fermi levels. A depletion layer forms in the semiconductor near the interface, and a built-in electric field ($\vec{E}$) pointing to the metal is established, leading to an upward energy band bending and the formation of the Schottky barrier Φ_{SB}. (b) For hot holes: at the interface between a metal and a *p*-type semiconductor, electrons are transferred from the metal to the semiconductor to align their Fermi levels. An accumulation layer forms in the semiconductor, and a built-in electric field ($\vec{E}$) pointing to the bulk of semiconductor is established, leading to a downward energy band bending and the formation of the Schottky barrier $\Phi_{SB,\,p}$. In both (a) and (b), E_{vac} stands for the vacuum energy level, $E_{F,SC}$ and $E_{F,M}$ stand for Fermi levels of the semiconductor and the metal, respectively, and E_{CB} and E_{VB} represent energy levels of the CB and the valence band (VB) of semiconductor, respectively. *Source*: Figures adapted from ref. [8]. Copyright © 2012, American Chemical Society.

semiconductor that accumulates excessive negative charges (or accumulation layer). Hence, a built-in electric field pointing to the bulk of semiconductor and a downward energy band bending in the semiconductor form, giving rise to an energy barrier for the hot-hole transfer ($\Phi_{SB,p}$) at the metal/semiconductor interface.

10.2.2 Electron Transfer Across the Schottky Barrier

For a metal/*n*-type semiconductor heterostructure, plasmon-generated electrons in the metal with energy higher than the Schottky barrier can be transferred into the CB of semiconductor via the over-barrier transfer within tens to hundreds of femtoseconds [16, 17]. In a simplified case, the maximum internal quantum efficiency of electron transfer η_{IQE} from metal to semiconductor can be estimated with the classic Fowler equation [18–20]:

$$\eta_{\text{IQE}} = C_F \frac{(h\nu - q\Phi_{\text{SB}})^2}{h\upsilon},$$

where C_F is a structure-specific Fowler emission coefficient, q is the charge of a single electron, and $h\nu$ represents the energy of an incident photon. The Fowler equation predicts η_{IQE} to monotonically increase when increasing the incident photon energy (i.e. shorter incident wavelength) [20]. In the Fowler equation, assumptions are made that electrons are only generated via the intraband (sp → sp band) excitation, excited electrons have an isotropic momentum distribution in the metal, and only electrons with enough momentum along the z-direction (i.e. $\frac{p_z^2}{2m_e} > q\Phi_{\text{SB}}$) can be transferred across the Schottky barrier [18, 20, 21]. Consequently, the deviation from the monotonic trend predicted by the Fowler equation occurs in plasmonic metal/semiconductor heterostructures behaving in a way that contradicts these assumptions. For instance, both intraband and interband (d → sp band) transitions occur in Au nanoparticles under the SPR excitation due to the relatively low interband threshold of Au (ca. 2 eV, 620 nm), and the contribution of interband transition in the hot-carrier generation becomes more significant in the shorter-wavelength region [22–28]. Interband transitions generate highly energetic d-band holes and near-Fermi level electrons that cannot surpass the Schottky barrier, causing heterostructures incorporating Au nanoparticles to deviate from the Fowler theory by exhibiting a low η_{IQE} in the shorter-wavelength region. Deviations from the Fowler theory also exist for plasmonic metal/semiconductor heterostructures with imperfect metal/semiconductor interfaces or metal nanoparticles with anisotropic morphologies (i.e. anisotropic momentum distribution of hot electrons). Therefore, the case-dependent modification of Fowler theory is necessary in order to accurately predict η_{IQE} [19, 20, 29].

Electrons transferred to the CB of semiconductors are swept away from the interface by the built-in electric field located in the depletion layer (Figure 10.1a). Under the steady-state plasmonic excitation, hot electrons are accumulated in the CB of semiconductor to drive reduction reactions, while hot holes left on the metal are removed by hole scavengers. It is noted that in addition to the over-barrier transfer, quantum tunneling was suggested to also contribute to the transfer of electrons with energy below the Schottky barrier [21, 30–33]. The efficiency of tunneling, however, is less important in view of the width of depletion layers (ca. 1 nm for Au/TiO_2) [32], and is generally not considered unless additional strategies for improvement were taken.

The external quantum efficiency (EQE) of electron transfer for a metal/semiconductor heterostructure depends on the combination of the IQE and the absorption efficiency of light by the metal nanoparticle. Hence, the wavelength dependence of electron transfer in metal/semiconductor heterostructures usually exhibits the same trend of absorption spectra of metal nanoparticles (Figure 10.2) [16, 28, 29, 34]. Generally, the highest efficiency of electron transfer is observed

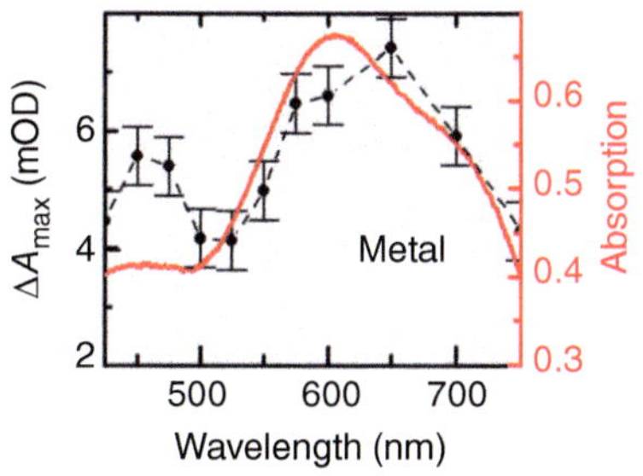

Figure 10.2 EQE of electron transfer in an Au/TiO_2 heterostructure. The red curve was the absorption spectrum of Au nanoparticles. Black data points obtained using the transient absorption spectroscopy represented the number of hot electrons transferred to and accumulated at the CB of TiO_2 when Au was excited with different wavelengths. The efficiency of electron transfer followed the absorption spectra of Au nanoparticles. *Source*: Figure reprinted with permission from ref. [29]. Copyright © 2017, American Chemical Society.

when the incident wavelength coincides with the SPR band due to the more efficient absorption of light.

10.3 Fabrication of Metal/Semiconductor Heterostructures

Plasmonic metal/semiconductor heterostructures were prepared either by assembling presynthesized metal nanoparticles and semiconductors supports via the colloidal deposition, or by directly depositing metal nanoparticles on semiconductors from metal precursors through deposition-precipitation or photodeposition methods. Schematics of those fabrication methods are summarized in Figure 10.3.

10.3.1 Colloidal Deposition Method

Colloidal deposition, as indicated by its name, involves the deposition of preformed colloidal metal nanoparticles on the surface of semiconductor supports (Figure 10.3a) [35, 36]. This is achieved by refluxing a solution containing metal nanoparticles capped with weakly binding surfactants (e.g. citrate) and semiconductors for a certain period, followed by solvent removal via centrifugation or evaporation [37, 38]. Colloidal deposition allows for an independent control on the size and morphology of metal nanoparticles, and is suitable for preparing heterostructures that require the incorporation of anisotropic metal nanostructures [38, 39]. However, heterostructures prepared from the colloidal deposition were reported to have a nonideal contact at the interface between metal nanoparticles and semiconductors, and the calcination at a high temperature (ca. 673 K) is necessary to optimize their catalytic activity [36, 40].

Recently, the self-assembly method has been developed for preparing metal/semiconductor heterostructures as a modification of the conventional colloidal deposition [41–44]. In this method, surfaces of metal nanoparticles and

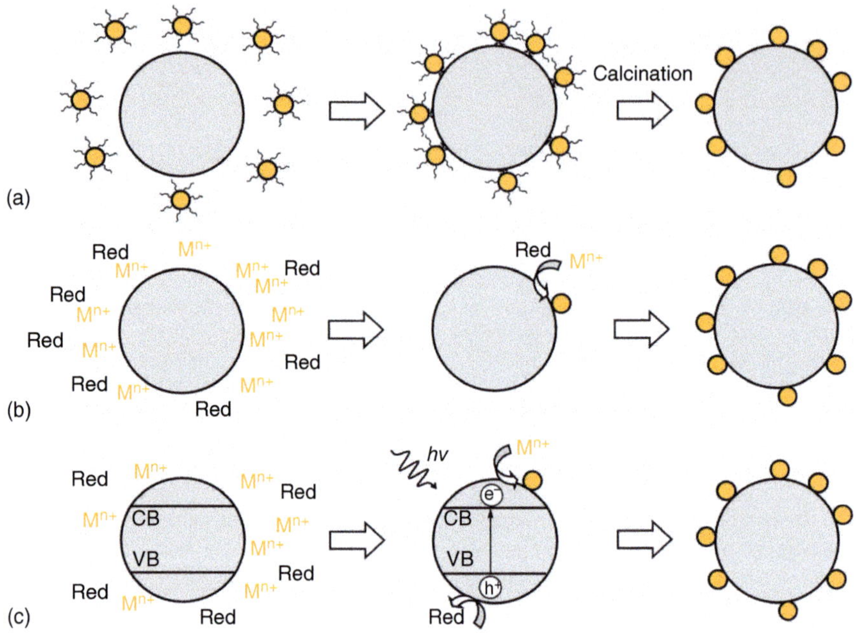

Figure 10.3 Schematics showing different strategies for fabricating metal/semiconductor heterostructures. (a) Colloidal deposition method: semiconductors and presynthesized metal nanoparticles capped with weak-binding surfactants are assembled together, followed by the calcination to remove residual surfactants. (b) Deposition-precipitation method: metal nanoparticles are directly deposited on the surface of semiconductors by certain weak reductants. (c) Photodeposition method: semiconductors are photoexcited, and electrons in the CB drive the reduction of metal precursors, while holes in the VB are removed by hole scavengers. In all figures, the large grey sphere represents a semiconductor, the small yellow spheres stand for metal nanoparticles, "M^{n+}" is the metal precursor for targeting metal nanoparticles, and "Red" stands for either reductants (in the deposition-precipitation method) or hole scavengers (in the photodeposition method).

semiconductors are functionalized with surfactants that exhibit strong interactions, facilitating the attachment of metal nanoparticles to semiconductors. The electrostatic attraction between negatively charged surfactants and positively charged ones is commonly utilized to achieve the self-assembly. For instance, functionalizing Au nanoparticles with positively charged 4-dimethylaminopyridine (DMAP) allowed for a rapid assembly of Au nanoparticles on a negatively charged WO_3 surface [43]. Similar to the conventional colloidal deposition, the calcination of heterostructures prepared from self-assembly is necessary to remove those organic surfactants and optimize the interface.

10.3.2 Deposition-Precipitation Method

In the deposition-precipitation method, metal nanoparticles are loaded onto semiconductors by directly reducing metal precursors on the surface of semiconductors with a mild reductant (Figure 10.3b) [11, 45–51]. For instance, for heterostructures

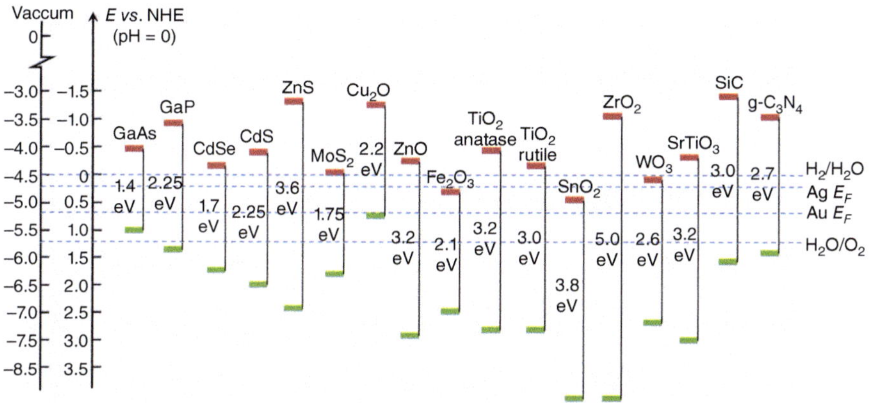

Figure 10.4 Schematics of energy levels of semiconductors that are potentially used to construct plasmonic metal/semiconductor heterostructures. *n*-Type semiconductors, such as TiO_2, possess Fermi levels very close to their CB edges, and the energy difference between the CB edge and the Fermi level of metal nanoparticles is used to estimate the Schottky barrier height. The position of Fermi levels (E_F) of Au (−5.2 eV vs. vacuum) and Ag (−4.7 eV vs. vacuum) are shown for reference. *Source*: Data were adapted from refs. [58, 61–64].

with Au nanoparticles, the commonly used method is to use urea to reduce $HAuCl_4$ on the surface of targeting semiconductors. Deposition-precipitation has been reported to produce an intimate interface between metal nanoparticles and semiconductors [52, 53], which is beneficial for interfacial electron transfer during photocatalytic processes. Following this method, Au/TiO_2 and $Au/SrTiO_3$ heterostructures have been successfully prepared [11, 45–49, 51].

Postsynthesis calcination of heterostructures prepared from deposition-precipitation can be used to further tune the size and dispersion of metal nanoparticles [11, 40, 53]. It has been reported that calcinating Au/TiO_2 heterostructures prepared from deposition-precipitation at 300 °C drove the migration and fusion of Au nanoparticles, increasing the diameter of Au nanoparticles from less than 3.5 nm to around 4 nm, and calcination at an even higher temperature (≥400 °C) further increased the size of Au nanoparticles to larger than 5 nm [11, 47].

10.3.3 Photodeposition Method

Semiconductors can efficiently absorb UV light to generate hot electrons capable of reducing precursors to form metal nanoparticles on the surface of semiconductors (Figure 10.3c) [52, 54–57], due to the fact that CB edges of commonly used semiconductors (Figure 10.4) have energy levels well above the reduction potential of plasmonic metal precursors ($AuCl_4^-$: 1.002 V vs. NHE; Ag^+: 0.7996 vs. NHE) [58]. During the photodeposition, photo-generated holes in the semiconductor are removed by scavengers such as alcohol molecules (e.g. methanol, ethanol, and isopropyl alcohol).

Photodeposition is advantageous for its ease of operation. However, studies have shown that the photoexcitation of semiconductors leads to the formation of defect states [53, 59, 60]. For instance, oxygen vacancies were found to form on TiO_2 during the photodeposition of Au nanoparticles [59, 60]. Hence, it is usually necessary to carefully judge the photocatalytic performance of heterostructures prepared from the photodeposition and consider the influence of aforementioned defects.

10.4 Design of Metal/Semiconductor Heterostructures

Designing plasmonic metal/semiconductor heterostructures as efficient photocatalysts requires to consider properties of both semiconductors and plasmonic metals. Attention should also be paid to the interface between the semiconductor and the metal to enhance the catalytic activity. In this section, we limit our discussion to metal/n-type semiconductor heterostructures for simplification. However, it is noted that the discussed design rules are also applicable to metal/p-type semiconductor heterostructures.

10.4.1 Design of Semiconductor Materials

Semiconductors in plasmonic heterostructures play a dominant role in determining the efficiency of charge separation and photocatalytic activity, as their properties govern the Schottky barrier height (Φ_{SB}) and the transport of electrons after being injected. Moreover, since semiconductors provide active sites for electron-driven chemical reactions, the inherent catalytic activity of semiconductors must also be considered when designing plasmonic heterostructures.

10.4.1.1 Optimization of the Schottky Barrier Height

As described in Section 10.2, Φ_{SB} in a heterostructure is determined by the work function difference between the semiconductor and the metal. Therefore, one of the most important criteria in designing metal/semiconductor photocatalysts is the selection of metal/semiconductor combinations that generate Φ_{SB} with appropriate values. A low Φ_{SB} is inefficient in preventing the back-transfer of electrons from semiconductors and the subsequent charge recombination. However, an extremely high Φ_{SB} is also unfavorable, since most of plasmon-generated hot electrons in the metal nanoparticle would have insufficient energy to surpass the barrier. Due to the limited choice of plasmonic metals (i.e., Ag and Au), the optimization of Φ_{SB} is usually executed by selecting the right semiconductor materials. For instance, for Au nanoparticles with Fermi level at ca. −5.2 eV vs. vacuum, TiO_2 is one of the commonly used semiconductors to build heterostructures because the Au/TiO_2 interface results in a suitable Φ_{SB} of ca. 1 eV [16]. Electron transfer efficiency in an Au/TiO_2 heterostructure was estimated as ~45% when Au nanoparticles were excited at 550 nm (Au SPR) [29]. ZnO has also been reported to form heterostructures with Au, as an Au/ZnO interface generates a Φ_{SB} of ca. 0.8 V [41]. By contrast, zero efficiency of the electron transfer was observed in an Au/Al_2O_3 heterostructure with Φ_{SB} of ca. 4 eV [29]. Figure 10.4 summarizes energy levels of semiconductors that can potentially be used to construct plasmonic metal/semiconductor heterostructures [58, 61–64].

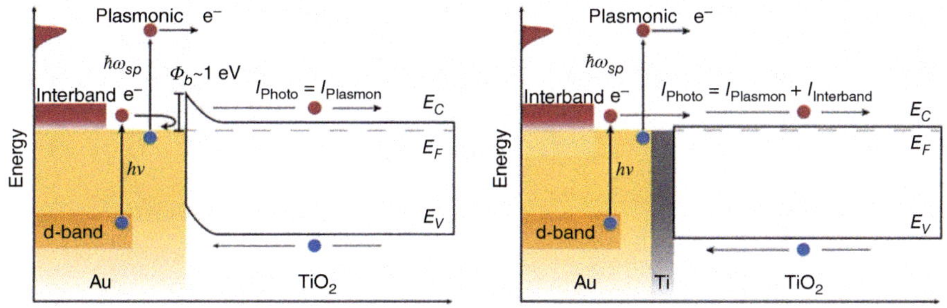

Figure 10.5 Controlling the electron transfer efficiency by tuning Φ_{SB}. Inserting a thin layer of Ti between Au and TiO_2 significantly decreased the value of Φ_{SB}, which allowed for a more efficient electron collection by TiO_2. *Source*: Figure reprinted from ref. [14]. Copyright © 2015, Springer Nature.

The facet-dependent work function of crystalline semiconductors offers one strategy to manipulate Φ_{SB}. For instance, BiOCl nanosheets were known to expose either (010) or (100) facets and the work function of BiOCl-(010) is 0.17 eV larger than that of BiOCl-(001), which led to a lower Φ_{SB} and a 3.5 times more efficient electron transfer in Au/BiOCl-(010) heterostructures [65].

Introducing additional layers of materials at the interface of heterostructures was also reported to alter Φ_{SB}. A recent study demonstrated that Φ_{SB} in an Au/TiO_2 heterostructure can be reduced to near-zero by introducing a thin layer of Ti between Au and TiO_2 (Figure 10.5) [14]. Ti has a Fermi level that naturally aligns with that of TiO_2, and the contact between Ti and TiO_2 only gave rise to a negligible Φ_{SB}. Consequently, in this Au/Ti/TiO_2 heterostructure, even electrons with energy near the Fermi level of Au were collected to the CB of TiO_2. Nonetheless, as previously mentioned, the rate of charge recombination would be high for such a small Φ_{SB}, and a balance between the efficiency of electron transfer and charge recombination should be achieved to optimize the performance of plasmonic heterostructures.

10.4.1.2 Optimization of Charge Transport in Semiconductors

Faster electron transport in semiconductors is beneficial for the charge separation, as electrons transferred to semiconductors can rapidly leave the interface, decreasing the possibility of electron–hole recombination. The facet-dependent electron affinity of crystalline semiconductors offers opportunities to control the electron transport by depositing metal nanoparticles onto semiconductors with site selectivity [39, 66–68]. In one study, Au nanoparticles were deposited on the top surface of TiO_2 nanosheets, and the surface heterojunction between top-(001) surface and edge-(101) surface of TiO_2 nanosheets drove a faster electron transport within TiO_2, leading to a better charge separation and a more efficient activation of O_2 by electrons on the surface of TiO_2 (Figure 10.6a) [68]. Similarly, in an Au/$La_2Ti_2O_7$ heterostructure with Au nanorods deposited at steps of $La_2Ti_2O_7$ (Figure 10.6b), the surface heterojunction between $La_2Ti_2O_7$-(010) and $La_2Ti_2O_7$-(012) facets at steps (0.2 eV driving force) drove a faster electron transport in $La_2Ti_2O_7$ and thus more efficient H_2 evolution [66].

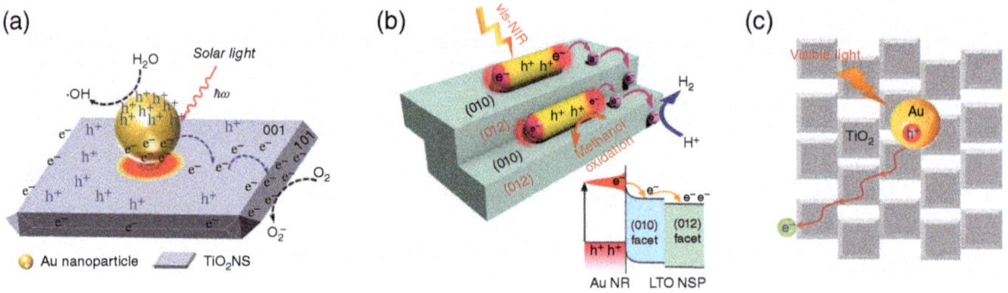

Figure 10.6 Optimizing the electron transport in semiconductors. (a) Faster electron transport in an Au/TiO_2 heterostructure facilitated by a surface heterojunction between TiO_2-(001) and TiO_2-(101). (b) Faster electron transport in an Au/$La_2Ti_2O_7$ heterostructure facilitated by a surface heterojunction between $La_2Ti_2O_7$-(010) and $La_2Ti_2O_7$-(012). (c) Faster electron transport in a heterostructure with Au nanoparticles deposited on the basal plane of a superstructure of plate-like mesocrystal anatase TiO_2. *Source*: Panel (a) reprinted with permission from ref. [68]. Copyright © 2017, American Chemical Society. Panel (b) reprinted with permission from ref. [66]. Copyright © 2018, American Chemical Society. Panel (c) reprinted with permission from ref. [39]. Copyright © 2014, American Chemical Society.

The control of electron transport can be further improved by fine-tuning the assembling pattern of crystalline semiconductors. In one study, Au nanoparticles were deposited on the basal plane of superstructures of plate-like mesocrystal anatase TiO_2 (MesoTiO_2, Figure 10.6c) [39]. While each TiO_2 plate possessed a surface heterojunction that facilitated the electron transport from the top surface to the edge, the superstructure further amplified the overall efficiency of electron transport along the basal plane of MesoTiO_2, eventually leading to a three times higher rate of photocatalytic H_2 evolution on Au/MesoTiO_2 compared with that on Au/TiO_2(P25).

Recently, two-dimensional (2D) metal-free semiconductors such as reduced graphene oxide (rGO) and graphitic carbon nitride (g-C_3N_4) were found to be excellent materials for electron transport. The mobility of electrons in these 2D materials is higher than that in TiO_2 in orders of magnitude [69–72]. Plasmonic photocatalysts based on Au/rGO and Au/g-C_3N_4 heterostructures were developed for various reactions, including H_2 evolution and the degradation of organic pollutants [54, 73, 74].

10.4.1.3 Catalytic Activity of Semiconductors

In metal/semiconductor heterogeneous photocatalysts, active sites on semiconductors are essential for hot-electron-driven reactions. Thus, how to generate and maintain active sites on semiconductors is an important consideration when designing heterostructures for the optimized photocatalytic performance. While inherent surface defects of commonly used oxide semiconductors (e.g. oxygen vacancies) can be utilized to adsorb and activate target reactants, the defect density on the surface of semiconductors is limited [59, 60]. A higher reaction rate is expected by adopting novel semiconductor materials exhibiting a higher density of active sites. For instance, heterostructures with MoS_2 as semiconductors have been developed to catalyze a plasmon-driven H_2 evolution due to the high affinity of MoS_2 itself toward H_2

on surface unsaturated sulfur atoms [42, 75]. Recently, plasmonic heterostructures built with metal–organic framework (MOF) materials have also been reported. The porous structure of MOFs not only exposes more active sites but also functions to trap and stabilize reactants. Different MOFs that incorporate Au nanoparticles were recently developed to promote H_2 evolution, and CO_2 reduction on Ag/UiO-67 MOF heterostructures was also reported [51, 76–78].

10.4.2 Design of Metal Nanoparticles

Plasmonic metal nanoparticles not only act to generate hot carriers to initiate photocatalytic reactions, but also provide surface active sites for hot-hole-driven reactions. Therefore, the design of both their morphology and composition is critical for optimizing their optical properties and overall photocatalytic activities.

10.4.2.1 Morphology of Metal Nanoparticles

Spherical metal nanoparticles are commonly adopted in plasmonic metal/semiconductor heterostructures due to the convenience of synthesis. Nonetheless, using anisotropic metal nanostructures offers opportunities to improve the photocatalytic activity of plasmonic heterostructures as a result of the morphology-dependent optical absorption of plasmonic metals (tunable from the visible to near-infrared region) [42, 74, 75, 79, 80]. In one study, Au nanocubes (SPR band at ca. 550 nm) and Au cubic nanocages (SPR band at ca. 760 nm) were simultaneously deposited on a TiO_2 nanosheet, forming a photocatalyst for H_2 evolution with a broadband light absorption [80].

Moreover, using anisotropic metal nanostructures allows for a larger contact area between metals and semiconductor supports, which is beneficial for the interfacial electron transfer, especially for heterostructures containing semiconductors with 2D structure. Heterostructures consisting of Au nanotriangles and 2D semiconductors with flat surfaces, such as reduced graphene oxide (r-GO) and molybdenum disulfide (MoS_2), have been exploited for photocatalytic H_2 evolution reaction [42, 74, 75].

Furthermore, anisotropic metal nanostructures are known to generate a larger population of hot carriers at sites with a higher curvature (i.e. hot spots). Selectively depositing semiconductors at those hot spots allows for a higher efficiency of hot-electron harvesting. For instance, TiO_2 was selectively deposited on tips of Au nanorods (Au NRs) via the plasmon-driven deposition to construct a photocatalyst for H_2 evolution (Figure 10.7a) [81]. Exciting the longitudinal SPR mode of Au NRs generated hot electrons that were more concentrated at tips, thus improving the efficiency of hot electron collection by TiO_2 and the rate of H_2 evolution on the TiO_2 surface. Similarly, heterostructures of Au hexoctahedral (HOH) and Cu_2O exhibited the highest activity toward the plasmon-driven H_2 evolution when Cu_2O was selectively grown on corners of Au HOH (Figure 10.7b) [82].

10.4.2.2 Materials of Metal Nanoparticles

While Au and Ag nanoparticles are most used to construct plasmonic photocatalysts, their high price and inherent chemical inertness limit their applications to

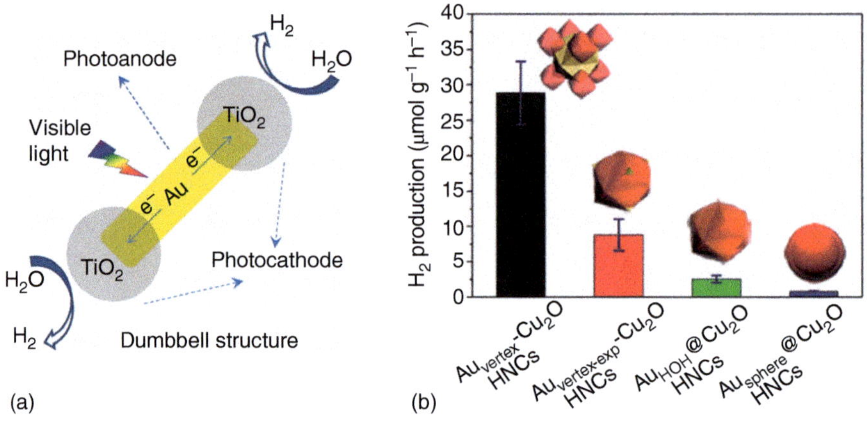

Figure 10.7 Metal/semiconductor heterostructures with anisotropic metal nanoparticles. (a) A Au/TiO_2 heterostructure with TiO_2 deposited on two tips (hot spots) of an Au NR. An efficient plasmon-driven HER was achieved due to the generation of more hot electrons at tips of the Au NR. (b) Au/Cu_2O with Cu_2O deposited on corners (hot spots) of Au HOH exhibiting the better performance in the plasmon-driven HER. *Source*: Panel A reprinted with permission from ref. [81]. Copyright © 2016, American Chemical Society. Panel B reprinted with permission from ref. [82]. Copyright © 2016, American Chemical Society.

a wider range of chemical reactions. By contrast, Cu is an earth-abundant metal with plasmonic absorption in the visible region and high catalytic activities in various reactions [83–86], making Cu/semiconductor heterostructures favorable potential candidates as plasmonic photocatalysts. However, only very few studies have reported the use of Cu/semiconductor heterostructures (e.g. Cu/TiO_2, Cu/g-C_3N_4) in photocatalytic H_2 evolution and CO_2 reduction [86–89], likely due to the fact that Cu surfaces are easily oxidized in the air, especially in the presence of reaction intermediates that strongly interact with Cu [84, 90, 91]. Therefore, more reliable surface-protection methods are needed to further explore the use of Cu/semiconductor heterostructures. As an alternative, multimetallic alloy nanoparticles have been used in plasmonic heterostructures to balance the low activity of noble metals and the poor stability of Cu [84, 92]. For instance, AuCu/TiO_2 heterostructures have been used to drive the photocatalytic reduction of CO_2 to CH_4 (Figure 10.8) [84]. In addition to being transferred to TiO_2, hot electrons generated in Au also kept Cu in its metal form (i.e. Cu^0) in order to serve as active sites for CO_2 reduction. In addition to Cu, plasmonic Ni and Al nanoparticles were also recently used to construct heterostructure photocatalysts for methylene blue degradation and CO_2 reduction, respectively [93, 94].

10.4.3 Design of Metal/Semiconductor Interfaces

The quality and chemical nature of metal/semiconductor interfaces significantly influence the efficiency of interfacial electron transfer. Moreover, interfaces in those heterostructures also provide active sites for certain reactions. Both

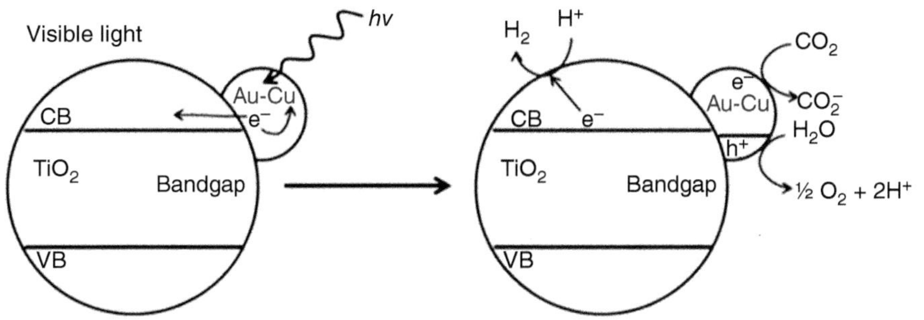

Figure 10.8 AuCu/TiO_2 heterostructures for the CO_2 reduction reaction. Hot electrons generated in AuCu alloy nanoparticles were not only injected to TiO_2 to drive H_2 evolution, but also transferred to Cu sites to retain Cu in a Cu^0 state, thus facilitating the reduction of CO_2 on Cu. *Source*: Figure reprinted with permission from ref. [84]. Copyright © 2014, American Chemical Society.

aspects necessitate the optimization of metal/semiconductor interfaces for better photocatalytic performances.

10.4.3.1 Optimization of the Interfacial Electron Transfer

Strategies have been developed to manipulate the electronic structure of metal/semiconductor heterostructures to optimize the electron transfer efficiency and the photocatalytic performance [52, 53, 95, 96]. One study demonstrated that fabrication procedures significantly affected the interfacial electronic structure of heterostructures [53]. In this study, Au/TiO_2 was prepared by loading Au nanoparticles on TiO_2 support via either the photodeposition method (Au/TiO_2-PD) or the deposition-precipitation method (Au/TiO_2-DP) [52]. Au/TiO_2-PD had a large amount of Ti^{3+} defects at the Au/TiO_2 interface. Under the plasmonic excitation of Au, those Ti^{3+} defects acted as electron-trapping states that inhibited the interfacial electron transfer, leading to a lower rate of H_2 evolution on Au/TiO_2–PD [52].

The control of interfacial electronic structures of plasmonic heterostructures can also be achieved by doping the surface of semiconductors. In one study, the TiO_2 surface in an Au/TiO_2 heterostructure was doped with a low amount of Nb^{5+} (0.1 wt.%) [95]. Such pentavalent dopants increased the density of positive charges near the interface, narrowing the width of the depletion layer inside the TiO_2, and increasing the possibility for the electron transfer via quantum tunneling that was otherwise insignificant at an interface without doping.

More recently, it was discovered that the orbital hybridization between Au and CdSe at the Au/CdSe interface enabled a new pathway for the electron transfer, namely plasmon-induced metal-to-semiconductor interfacial charge transfer transition (PICTT), or direct transfer in short. In this pathway, the decay of SPR in Au directly generated hot electrons in the CB of CdSe and hot holes in Au, circumventing the Schottky barrier and leading to a high transfer efficiency of ca. 24% that did not obey the wavelength dependence predicted based on conventional over-barrier transfer [96]. A similar direct transfer was also observed in an Ag/$CsPbBr_3$ heterostructure, which exhibited an electron transfer efficiency of ca. 50% [97]. Interestingly, such a direct electron transfer was not observed in Au/CdS heterostructures despite the chemical and physical similarity between CdS and CdSe [98], indicating that an appropriate interfacial metal-semiconductor orbital hybridization is crucial for the direct electron transfer [97, 99]. Designing plasmonic metal/semiconductor heterostructures with the direct electron transfer at the interface is expected to produce photocatalysts with high activity due to the astonishingly high electron transfer efficiency of direct electron transfer. However, a systematic prediction of possible metal and semiconductor combinations that would enable the direct electron transfer has yet to be developed.

10.4.3.2 Optimization of Interfacial Active Sites

The interface of a plasmonic heterostructure could also provide important active sites for reactions [11, 12]. One study showed that on an Au/BiOCl heterostructure [12], plasmon-driven oxidation of benzyl alcohol preferentially occurred at the interface between Au and BiOCl surface with oxygen vacancies (V_O). The reaction required the cooperation between a hot-hole-assisted H-abstraction process on Au surface and a hot-electron-driven O_2 activation on a BiOCl-V_O near the Au nanoparticle (Figure 10.9a), which can only occur at the Au/BiOCl interface. In another study, active sites for hot-hole-driven water oxidation on Au/TiO_2 heterostructures were also positioned at the Au/TiO_2 interface using the Kelvin probe force microscopy [48]. These results motivated the development of strategies to improve the photocatalytic activity of plasmonic heterostructures by optimizing the chemical structure of metal/semiconductor interfaces. Calcination of plasmonic heterostructures under the controlled temperature is one strategy. For instance, by calcinating Au/TiO_2(P25) heterostructures fabricated from the deposition-precipitation method at temperatures higher than 300 °C [11]. Au nanoparticles were aggregated and selectively migrated to boundaries between anatase and rutile phases [11]. Such Au/anatase/rutile heterostructures exhibited a significantly higher activity toward the plasmon-driven aerobic oxidation of phenylethanol. This was attributed to the synergy between the easier electron transfer from Au to rutile and the higher O_2 affinity on anatase that led to a more efficient O_2 activation (Figure 10.9b).

10.5 Photocatalytic Reactions Mediated by Plasmonic Heterostructures

To date, plasmon-driven reactions are mostly catalyzed by heterostructures consisting of plasmonic metal nanoparticles and *n*-type semiconductors. A wide collection of reactions has been achieved using plasmonic heterostructures as photocatalysts, including water splitting reaction and various organic transformations (Figure 10.10) [100].

10.5.1 Water Splitting

Autonomous water splitting driven by plasmonic metal/semiconductor heterostructures was first reported on Au/TiO_2 heterostructures with TiO_2 deposited on tips

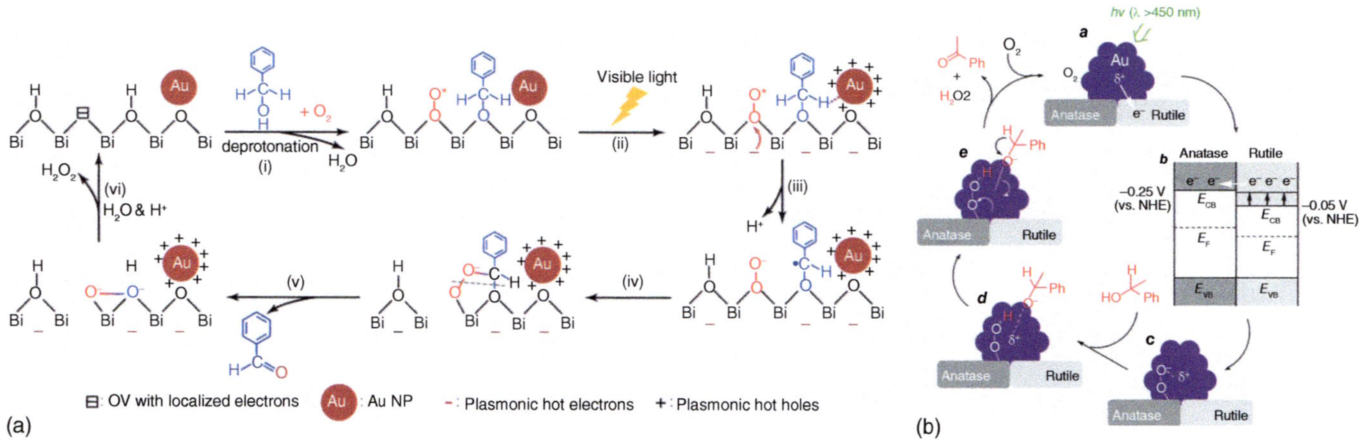

Figure 10.9 Active sites at metal/semiconductor interfaces. (a) The oxidation of benzyl alcohol facilitated by a cooperation between hot-hole-driven H abstraction and hot-electron-driven O_2 activation. (b) Efficient O_2 activation at the Au/anatase/rutile interface due to the synergy between rutile-to-anatase electron transport and O_2 adsorption on anatase. *Source*: Panel A reprinted with permission from ref. [12]. Copyright © 2017, American Chemical Society. Panel B reprinted with permission from ref. [11]. Copyright © 2012, American Chemical Society.

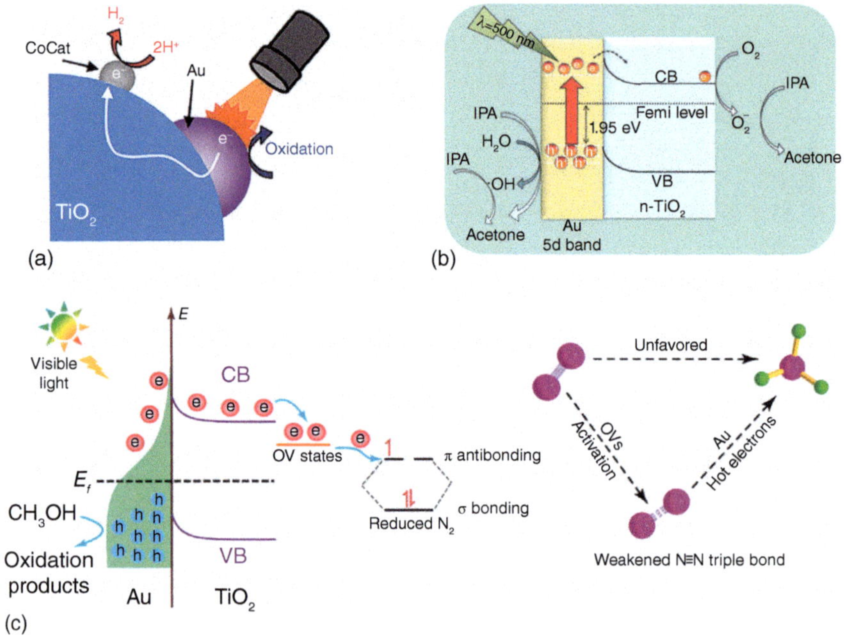

Figure 10.10 Photochemical reactions catalyzed by plasmonic metal/semiconductor heterostructures. (a) H_2 evolution on an Au/TiO_2 heterostructure decorated with Pt co-catalysts. (b) Alcohol oxidation on an Au/TiO_2 heterostructure. The oxidation can be driven by either hot electrons through a $O_2^{\bullet-}$-mediated pathway on TiO_2 (right) or by hot holes through a $^{\bullet}OH$-mediated pathway on Au (left), and the oxidation can also be directly driven by hot holes on Au (left). (c) N_2 fixation on a Au/TiO_2 heterostructure. N_2 was activated by being adsorbed on an oxygen vacancy (OV) on TiO_2, followed by the reduction driven by plasmon-generated electrons accumulated on TiO_2. *Source*: Panel A reprinted with permission from ref. [112]. Copyright © 2013, American Chemical Society. Panel B adapted with permission from ref. [49]. Copyright © 2019, American Chemical Society. Panel C adapted with permission from ref. [117]. Copyright © 2018, American Chemical Society.

of Au nanorods [101]. TiO_2 and Au were further decorated with Pt nanoparticles and cobalt-based oxygen evolution catalysts (Co-OEC), respectively. Under the plasmonic excitation of Au nanorods, hot electrons were first transferred to TiO_2 and then to Pt to drive water reduction (i.e. H_2 evolution), while hot holes left on Au were collected by Co-OEC to drive water oxidation (i.e. O_2 evolution). Recently, the plasmon-driven water splitting has also been achieved on Ag/N-doped TiO_2 and Au/carbon nitride [102, 103].

Autonomous water splitting simply driven by the plasmonic excitation of heterostructures, however, usually suffered from the low conversion efficiency of incident photon energy, mostly due to the slow kinetics of four-hole-driven water oxidation that limits the overall reaction rate [104, 105]. Therefore, most studies focused on half-reactions of water splitting separately, namely H_2 evolution and O_2 evolution. In the plasmon-driven H_2 evolution, plasmon-generated hot electrons are accumulated on the semiconductor surface to reduce H^+ or H_2O to form H_2, while hot holes left on metals are scavenged by electron donors such as methanol (but not water). Conventional Au/TiO_2 has been widely used to drive the H_2 evolution [52, 53, 106, 107]. Recently, the H_2 evolution has also been achieved using other combinations of metals and semiconductors, such as Au/BiOOCl, Au/g-C_3N_4, and Au/MoS_2 [65, 108, 109]. However, the rate of H_2 evolution achieved with plasmonic heterostructures still remained low, likely due to the lack of optimal active sites. Therefore, in some cases, co-catalysts for H_2 evolution, such as Pt nanoparticles, were adopted to further boost the rate of H_2 evolution (Figure 10.10a) [39, 74, 77, 110–112]. In one study, depositing Pt nanoparticles (0.5 wt.%) on TiO_2 in Au/TiO_2 heterostructures led to a six-fold enhancement in the rate of H_2 evolution [39].

Only few studies have reported the plasmon-driven O_2 evolution. In one study, Au/$SrTiO_3$ was successfully used to drive water oxidation under the plasmonic excitation with Ag^+ functioning as scavengers for hot electrons [45]. It was shown that in this system, only energetic hot holes generated from interband transitions of Au nanoparticles significantly contributed to the O_2 evolution. A more recent study on Au/$SrTiO_3$ demonstrated that interfacial states with lower potential energy were formed when $SrTiO_3$ had (100) surface orientation, which allowed the capture of more energetic plasmon-generated holes from Au and further enhanced the activity of water oxidation [113]. Recently, the O_2 evolution was achieved on Au/TiO_2, and its interface was identified as active sites for O_2 evolution [48]. Similar to the case of H_2 evolution, modifying the surface of Au with hole-transport materials such as catechol also enhanced the charge separation and boosted the rate of O_2 evolution [114].

10.5.2 Organic Transformation

Plasmon-driven organic transformations on plasmonic heterostructures can occur through either hot-electron-driven or hot-hole-driven pathways [11, 12, 37, 46, 49, 54, 73, 78, 115–117]. For hot-electron-driven organic transformations, plasmon-generated hot electrons transferred to the CB of semiconductors are subsequently injected into targeting molecules. Plasmon-driven deuteration of aryl halides on Au/CdS was found to follow this pathway [78]. In this process, with SO_3^{2-} used as a hole scavenger, hot electrons accumulated in CdS were injected to aryl substrates, forming aryl halide anions that spontaneously underwent the dehalogenation and deuterium addition. Studies have also shown that the injection of hot electrons into O_2 molecules played a critical role in mediating the oxidation of various organic compounds. In this process, O_2 molecules accepted one electron to form reactive $O_2^{\bullet-}$ radicals that subsequently initiated the oxidation of target reactants. Plasmon-driven oxidation of methylene blue on Au/TiO_2, methyl orange on Au/g-C_3N_4, and different alcohols on Au/BiOCl or Au/TiO_2 were found to occur through this $O_2^{\bullet-}$-mediated pathway (Figure 10.10b) [11, 12, 37, 49, 54].

Hot-hole-driven reactions usually occurred on the surface of metal nanoparticles. For instance, Ag/TiO_2 was reported to catalyze the oxidative dimerization of *p*-aminothiophenol (*p*-ATP), where hot holes remained on Ag nanoparticles oxidized amino groups of *p*-ATP to NH_2^+, and the coupling between two oxidized *p*-ATP led to the formation of *p,p*′-dimercaptoazobenzene [115]. It has also been demonstrated that plasmon-generated hot holes in Au/TiO_2 heterostructures directly drove the oxidation of isopropyl alcohols to form acetone on the surface of Au nanoparticles (Figure 10.10b) [49]. Similarly, hot-hole-driven conversions of cinnamyl alcohol to cinnamaldehyde and benzylamine to benzaldehyde on Au/TiO_2 heterostructures were also observed [46]. Meanwhile, plasmon-generated hot holes have also been reported to initiate organic conversions by first being injected into H_2O molecules to form reactive •OH radicals, which subsequently reacted with target reactants [49, 73]. Plasmon-driven degradation of tetracycline hydrochloride on Au/g-C_3N_4/Pt ternary heterostructures and oxidation of alcohols on Au/TiO_2 were found to follow this •OH-mediated pathway [49, 73]. It should be noted that the concurrence of multiple pathways on a single plasmonic heterostructure is possible [49].

10.5.3 Other Reactions

Plasmon-driven fixation of N_2 to NH_3 has been recently reported on $Au/(BiO)_2CO_3$, $Au/KNbO_3$, $Au/SrTiO_3$, Au/TiO_2, and Au/g-C_3N_4 heterostructures [13, 56, 118–121]. In these systems, hot electrons accumulated in the CB of semiconductors were transferred to antibonding π^* orbitals of N_2, leading to the cleavage of N≡N triple bonds. Noting the high chemical stability of N≡N triple bond, the activation of N_2 molecules is usually necessary for the reaction to proceed. For Au/TiO_2, N_2 was activated when being adsorbed on V_O on TiO_2 near the Au/TiO_2 interface (Figure 10.10c) [118]. Similarly, for Au/g-C_3N_4, nitrogen vacancies on g-C_3N_4 promoted the chemisorption and activation of N_2 [121].

Plasmon-driven CO_2 reduction has been achieved on Au/TiO_2 with CH_4 as the main product, as well as on Ag/TiO_2 with CH_3OH as the main product [51, 110, 122]. Similar to the N_2 fixation, the chemical stability CO_2 necessitates the activation of CO_2 upon adsorption [51]. The CO_2 reduction is well known for its broad spectrum of products ranging from CO to more complicated multicarbon species, and the selectivity of products closely depends on the adsorption geometry and binding strength of CO_2 and related reaction intermediates [123]. While the orbital hybridization between metals and semiconductors at their interfaces was thought to govern the subtle selectivity of CO_2 reductions, a more systematic understanding of the relationship between the selectivity and the property of plasmonic heterostructures has yet to be developed.

Plasmonic heterostructures have also been applied to drive other reactions including O_2 reduction, Pb^{2+} oxidation, and Au oxidation [50, 124, 125], all of which relied on the separation of plasmon-generated electrons and holes at the metal/semiconductor interface and shared similar hot-carrier-mediated pathways with reactions described in this section.

10.6 Outlook

In summary, plasmonic metal/semiconductor heterostructures enable the charge separation by the Schottky barrier formed at their interfaces and have been widely used as catalysts for solar-driven photochemical reactions. Careful designs of semiconductors, metals, and metal/semiconductor interfaces are necessary to optimize the efficiency of interfacial charge transfer and the overall catalytic activity of plasmonic heterostructures.

To date, most plasmonic heterostructures used as photocatalysts consist of metal nanoparticles and *n*-type semiconductors, in which the transfer of hot electrons from metals to semiconductors occurs at the interface, and the surface of semiconductors serves as active sites for electron-driven reactions. In view of the higher inherent catalytic activity and diverse modification methods of semiconductors, developing plasmonic heterostructures consisting of metal nanoparticles and *p*-type semiconductors that allow the collection of hot holes in the VB of semiconductors would further broaden their applications in the overall solar-driven chemical processes.

Meanwhile, transition metal oxides (e.g. TiO_2, ZnO) are still most-used semiconductors for constructing plasmonic metal/semiconductor heterostructures. Taking advantages of new semiconductor materials with supreme electronic and chemical properties may lead to the production of plasmonic heterostructures with improved photocatalytic performance. The use of MOFs as mentioned in this chapter is one example, while the further development of other new materials such as perovskites may offer more opportunities.

Acknowledgments

We acknowledge support from the National Science Foundation under Grants DMR-1352328 and CHE-1308644. W.G. especially appreciates the support of a graduate school fellowship from the University of Florida and the Department of Energy (DOE) Science Graduate Student Research (SCGSR) award.

References

1 Christopher, P. and Moskovits, M. (2017). Hot charge carrier transmission from plasmonic nanostructures. *Annual Review of Physical Chemistry* 68: 379–398.

2 Kim, M., Lin, M., Son, J. et al. (2017). Hot-electron-mediated photochemical reactions: principles, recent advances, and challenges. *Advanced Optical Materials* 5: 1700004.

3 Linic, S., Aslam, U., Boerigter, C., and Morabito, M. (2015). Photochemical transformations on plasmonic metal nanoparticles. *Nature Materials* 14: 567–576.

4 Linic, S., Christopher, P., and Ingram, D.B. (2011). Plasmonic-metal nanostructures for efficient conversion of solar to chemical energy. *Nature Materials* 10: 911–921.

5 Brongersma, M.L., Halas, N.J., and Nordlander, P. (2015). Plasmon-induced hot carrier science and technology. *Nature Nanotechnology* 10: 25–34.

6 Huang, J., Guo, W., Hu, Y., and Wei, W.D. (2020). Plasmonic metal–semiconductor heterostructures for hot-electron-driven photochemistry. *MRS Bulletin* 45: 37–42.

7 Zhang, Y., He, S., Guo, W. et al. (2018). Surface-plasmon-driven hot electron photochemistry. *Chemical Reviews* 118: 2927–2954.

8 Zhang, Z. and Yates, J.T. Jr., (2012). Band bending in semiconductors: chemical and physical consequences at surfaces and interfaces. *Chemical Reviews* 112: 5520–5551.

9 Tung, R.T. (2014). The physics and chemistry of the Schottky barrier height. *Applied Physics Reviews* 1: 011304.

10 DuChene, J.S., Sweeny, B.C., Johnston-Peck, A.C. et al. (2014). Prolonged hot electron dynamics in plasmonic-metal/semiconductor heterostructures with implications for solar photocatalysis. *Angewandte Chemie, International Edition* 53: 7887–7891.

11 Tsukamoto, D., Shiraishi, Y., Sugano, Y. et al. (2012). Gold nanoparticles located at the interface of anatase/rutile TiO_2 particles as active plasmonic photocatalysts for aerobic oxidation. *Journal of the American Chemical Society* 134: 6309–6315.

12 Li, H., Qin, F., Yang, Z. et al. (2017). New reaction pathway induced by plasmon for selective benzyl alcohol oxidation on BiOCl possessing oxygen vacancies. *Journal of the American Chemical Society* 139: 3513–3521.

13 Li, C., Wang, T., Zhao, Z.J. et al. (2018). Promoted fixation of molecular nitrogen with surface oxygen vacancies on plasmon-enhanced TiO2 photoelectrodes. *Angewandte Chemie, International Edition* 57: 5278–5282.

14 Zheng, B.Y., Zhao, H., Manjavacas, A. et al. (2015). Distinguishing between plasmon-induced and photoexcited carriers in a device geometry. *Nature Communications* 6: 7797.

15 Goddeti, K.C., Lee, C., Lee, Y.K., and Park, J.Y. (2018). Three-dimensional hot electron photovoltaic device with vertically aligned TiO_2 nanotubes. *Scientific Reports* 8: 7330.

16 Tian, Y. and Tatsuma, T. (2005). Mechanisms and applications of plasmon-induced charge separation at TiO_2 films loaded with gold nanoparticles. *Journal of the American Chemical Society* 127: 7632–7637.

17 Furube, A., Du, L., Hara, K. et al. (2007). Ultrafast plasmon-induced electron transfer from gold nanodots into TiO_2 nanoparticles. *Journal of the American Chemical Society* 129: 14852–14853.

18 Fowler, R.H. (1931). The analysis of photoelectric sensitivity curves for clean metals at various temperatures. *Physics Review* 38: 45–56.

19 Yu, Y., Sun, Y., Hu, Z. et al. (2019). Fast photoelectric conversion in the near-infrared enabled by plasmon-induced hot-electron transfer. *Advanced Materials* 31: e1903829.

20 Tagliabue, G., Jermyn, A.S., Sundararaman, R. et al. (2018). Quantifying the role of surface plasmon excitation and hot carrier transport in plasmonic devices. *Nature Communications* 9 (1): 3394.

21 Govorov, A.O., Zhang, H., Demir, H.V., and Gun'ko, Y.K. (2014). Photogeneration of hot plasmonic electrons with metal nanocrystals: quantum description and potential applications. *Nano Today* 9: 85–101.

22 Kim, Y., Smith, J.G., and Jain, P.K. (2018). Harvesting multiple electron-hole pairs generated through plasmonic excitation of Au nanoparticles. *Nature Chemistry* 10: 763–769.

23 Schlather, A.E., Manjavacas, A., Lauchner, A. et al. (2017). Hot hole photoelectrochemistry on Au@SiO_2@Au nanoparticles. *Journal of Physical Chemistry Letters* 8: 2060–2067.

24 Barman, T., Hussain, A.A., Sharma, B., and Pal, A.R. (2015). Plasmonic hot hole generation by interband transition in gold-polyaniline. *Scientific Reports* 5: 18276.

25 Bernardi, M., Mustafa, J., Neaton, J.B., and Louie, S.G. (2015). Theory and computation of hot carriers generated by surface plasmon polaritons in noble metals. *Nature Communications* 6: 7044.

26 Brown, A.M., Sundararaman, R., Narang, P. et al. (2016). Nonradiative plasmon decay and hot carrier dynamics: effects of phonons, surfaces, and geometry. *ACS Nano* 10: 957–966.

27 Sundararaman, R., Narang, P., Jermyn, A.S. et al. (2014). Theoretical predictions for hot-carrier generation from surface plasmon decay. *Nature Communications* 5: 5788.

28 Yu, Y., Wijesekara, K.D., Xi, X., and Willets, K.A. (2019). Quantifying wavelength-dependent plasmonic hot carrier energy distributions at metal/semiconductor interfaces. *ACS Nano* 13: 3629–3637.

29 Ratchford, D.C., Dunkelberger, A.D., Vurgaftman, I. et al. (2017). Quantification of efficient plasmonic hot-electron injection in gold nanoparticle-TiO_2 films. *Nano Letters* 17: 6047–6055.

30 Moskovits, M. (2011). Hot electrons cross boundaries. *Science* 332: 676.

31 Mubeen, S., Hernandez-Sosa, G., Moses, D. et al. (2011). Plasmonic photosensitization of a wide band gap semiconductor: converting plasmons to charge carriers. *Nano Letters* 11: 5548–5552.

32 Shiraishi, Y., Yasumoto, N., Imai, J. et al. (2017). Quantum tunneling injection of hot electrons in Au/TiO_2 plasmonic photocatalysts. *Nanoscale* 9: 8349–8361.

33 Govorov, A.O., Zhang, H., and Gun'ko, Y.K. (2013). Theory of photoinjection of hot plasmonic carriers from metal nanostructures into semiconductors and surface molecules. *Journal of Physical Chemistry C* 117: 16616–16631.

34 Sakai, N., Fujiwara, Y., Takahashi, Y., and Tatsuma, T. (2009). Plasmon-resonance-based generation of cathodic photocurrent at electrodeposited gold nanoparticles coated with TiO_2 films. *ChemPhysChem* 10: 766–769.

35 Schüth, F. (2013). Colloidal deposition as method to study the influence of the support on the activity of gold catalysts in CO-oxidation. *Physica Status Solidi B* 250: 1142–1151.

36 Tsubota, S., Nakamura, T., Tanaka, K., and Haruta, M. (1998). Effect of calcination temperature on the catalytic activity of Au colloids mechanically mixed with TiO_2 powder for CO oxidation. *Catalysis Letters* 56: 131–135.

37 Mukhopadhyay, S., Maiti, D., Chatterjee, S. et al. (2016). Design and application of Au decorated ZnO/TiO_2 as a stable photocatalyst for wide spectral coverage. *Physical Chemistry Chemical Physics* 18: 31622–31633.
38 Bian, J., Qu, Y., Zhang, X. et al. (2018). Dimension-matched plasmonic $Au/TiO_2/BiVO_4$ nanocomposites as efficient wide-visible-light photocatalysts to convert CO_2 and mechanistic insights. *Journal of Materials Chemistry A* 6: 11838–11845.
39 Bian, Z., Tachikawa, T., Zhang, P. et al. (2014). Au/TiO_2 superstructure-based plasmonic photocatalysts exhibiting efficient charge separation and unprecedented activity. *Journal of the American Chemical Society* 136: 458–465.
40 Huang, J., He, S., Goodsell, J.L. et al. (2020). Manipulating atomic structures at the Au/TiO_2 interface for O_2 activation. *Journal of the American Chemical Society* 142: 6456–6460.
41 Wang, Y., Fang, H.B., Zheng, Y.Z. et al. (2015). Controllable assembly of well-defined monodisperse Au nanoparticles on hierarchical ZnO microspheres for enhanced visible-light-driven photocatalytic and antibacterial activity. *Nanoscale* 7: 19118–19128.
42 Zhang, P., Fujitsuka, M., and Majima, T. (2017). Hot electron-driven hydrogen evolution using anisotropic gold nanostructure assembled monolayer MoS_2. *Nanoscale* 9: 1520–1526.
43 Jiang, K.Y., Weng, Y.L., Guo, S.Y. et al. (2017). Self-assembly of metal/semiconductor heterostructures via ligand engineering: unravelling the synergistic dual roles of metal nanocrystals toward plasmonic photoredox catalysis. *Nanoscale* 9: 16922–16936.
44 Zhang, H., Wang, C., Sun, H.L. et al. (2017). In situ dynamic tracking of heterogeneous nanocatalytic processes by shell-isolated nanoparticle-enhanced Raman spectroscopy. *Nature Communications* 8: 15447.
45 Liu, L., Li, P., Adisak, B. et al. (2014). Gold photosensitized $SrTiO_3$ for visible-light water oxidation induced by Au interband transitions. *Journal of Materials Chemistry A* 2: 9875–9882.
46 Takahiro, K., Naya, S.-I., and Tada, H. (2014). Highly active supported plasmonic photocatalyst consisting of gold nanoparticle-loaded mesoporous titanium(IV) oxide overlayer and conducting substrate. *Journal of Physical Chemistry C* 118 (46): 26887–26893.
47 Wen, Y., Liu, B., Zeng, W., and Wang, Y. (2013). Plasmonic photocatalysis properties of Au nanoparticles precipitated anatase/rutile mixed TiO_2 nanotubes. *Nanoscale* 5: 9739–9746.
48 Wang, S., Gao, Y., Miao, S. et al. (2017). Positioning the water oxidation reaction sites in plasmonic photocatalysts. *Journal of the American Chemical Society* 139: 11771–11778.
49 Fu, A., Chen, X., Tong, L. et al. (2019). Remarkable visible-light photocatalytic activity enhancement over Au/p-type TiO_2 promoted by efficient interfacial charge transfer. *ACS Applied Materials & Interfaces* 11: 24154–24163.
50 Awa, K., Naya, S.-I., Fujishima, M., and Tada, H. (2020). A three-component plasmonic photocatalyst consisting of gold nanoparticle and TiO_2–SnO_2

nanohybrid with heteroepitaxial junction: hydrogen peroxide synthesis. *Journal of Physical Chemistry C* 124: 7797–7802.
51 Collado, L., Reynal, A., Fresno, F. et al. (2018). Unravelling the effect of charge dynamics at the plasmonic metal/semiconductor interface for CO_2 photoreduction. *Nature Communications* 9: 4986.
52 Priebe, J.B., Radnik, J., Lennox, A.J.J. et al. (2015). Solar hydrogen production by plasmonic Au–TiO_2 catalysts: impact of synthesis protocol and TiO_2 phase on charge transfer efficiency and H_2 evolution rates. *ACS Catalysis* 5: 2137–2148.
53 Priebe, J.B., Karnahl, M., Junge, H. et al. (2013). Water reduction with visible light: synergy between optical transitions and electron transfer in Au-TiO_2 catalysts visualized by in situ EPR spectroscopy. *Angewandte Chemie, International Edition* 52: 11420–11424.
54 Cheng, N., Tian, J., Liu, Q. et al. (2013). Au-nanoparticle-loaded graphitic carbon nitride nanosheets: green photocatalytic synthesis and application toward the degradation of organic pollutants. *ACS Applied Materials & Interfaces* 5: 6815–6819.
55 Han, S., Yu, L., Zhang, H. et al. (2019). Gold plasmon-enhanced solar hydrogen production over $SrTiO_3/TiO_2$ heterostructures. *ChemCatChem* 11: 6203–6207.
56 Xing, P., Wu, S., Chen, Y. et al. (2019). New application and excellent performance of Ag/$KNbO_3$ nanocomposite in photocatalytic NH_3 synthesis. *ACS Sustainable Chemistry & Engineering* 7: 12408–12418.
57 Ma, J., Zhou, C., Long, J. et al. (2018). Reducing the barrier effect of graphene sheets on a Ag cocatalyst to further improve the photocatalytic performance of TiO_2. *RSC Advances* 8: 14056–14063.
58 Lide, D.R. (2008). *CRC Handbook of Chemistry and Physics*, 89e. Boca Raton, FL: Taylor & Francis.
59 Pan, X., Yang, M.Q., Fu, X. et al. (2013). Defective TiO_2 with oxygen vacancies: synthesis, properties and photocatalytic applications. *Nanoscale* 5: 3601–3614.
60 Diebold, U. (2003). The surface science of titanium dioxide. *Surface Science Reports* 48: 53–229.
61 Kudo, A. and Miseki, Y. (2009). Heterogeneous photocatalyst materials for water splitting. *Chemical Society Reviews* 38: 253–278.
62 Grätzel, M. (2001). Photoelectrochemical cells. *Nature* 414: 338–344.
63 Tee, S.Y., Win, K.Y., Teo, W.S. et al. (2017). Recent progress in energy-driven water splitting. *Advancement of Science* 4: 1600337.
64 Ola, O. and Maroto-Valer, M.M. (2015). Review of material design and reactor engineering on TiO_2 photocatalysis for CO2 reduction. *Journal of Photochemistry and Photobiology, C: Photochemistry Reviews* 24: 16–42.
65 Liu, G., Wang, T., Zhou, W. et al. (2015). Crystal-facet-dependent hot-electron transfer in plasmonic-Au/semiconductor heterostructures for efficient solar photocatalysis. *Journal of Materials Chemistry C* 3: 7538–7542.
66 Cai, X., Zhu, M., Elbanna, O.A. et al. (2017). Au nanorod photosensitized $La_2Ti_2O_7$ nanosteps: successive surface heterojunctions boosting visible to near-infrared photocatalytic H_2 evolution. *ACS Catalysis* 8: 122–131.

67 Sun, S., Gao, P., Yang, Y. et al. (2016). N-doped TiO_2 nanobelts with coexposed (001) and (101) facets and their highly efficient visible-light-driven photocatalytic hydrogen production. *ACS Applied Materials & Interfaces* 8: 18126–18131.
68 Shi, H., Zhang, S., Zhu, X. et al. (2017). Uniform gold-nanoparticle-decorated {001}-faceted anatase TiO_2 nanosheets for enhanced solar-light photocatalytic reactions. *ACS Applied Materials & Interfaces* 9: 36907–36916.
69 Lin, Y.-J. and Yang, S.-H. (2013). Carrier transport and photoresponse for heterojunction diodes based on the reduced graphene oxide-based TiO_2 composite and p-type Si. *Applied Physics A: Materials Science & Processing* 116: 91–95.
70 Liu, B., Xu, B., Li, S. et al. (2019). Heptazine-based porous graphitic carbon nitride: a visible-light driven photocatalyst for water splitting. *Journal of Materials Chemistry A* 7: 20799–20805.
71 Bak, T., Nowotny, M.K., Sheppard, L.R., and Nowotny, J. (2008). Mobility of electronic charge carriers in titanium dioxide. *Journal of Physical Chemistry C* 112: 12981–12987.
72 He, Y., Zhang, M., Shi, J.-J. et al. (2019). Two-dimensional g-C_3N_4/InSe heterostructure as a novel visible-light photocatalyst for overall water splitting: a first-principles study. *Journal of Physics D: Applied Physics* 52: 015304.
73 Xue, J., Ma, S., Zhou, Y. et al. (2015). Facile photochemical synthesis of Au/Pt/g-C_3N_4 with plasmon-enhanced photocatalytic activity for antibiotic degradation. *ACS Applied Materials & Interfaces* 7: 9630–9637.
74 Lou, Z., Fujitsuka, M., and Majima, T. (2017). Two-dimensional Au-nanoprism/reduced graphene oxide/pt-nanoframe as plasmonic photocatalysts with multiplasmon modes boosting hot electron transfer for hydrogen generation. *Journal of Physical Chemistry Letters* 8: 844–849.
75 Shi, Y., Wang, J., Wang, C. et al. (2015). Hot electron of Au nanorods activates the electrocatalysis of hydrogen evolution on MoS_2 nanosheets. *Journal of the American Chemical Society* 137: 7365–7370.
76 Wang, S.S., Jiao, L., Qian, Y. et al. (2019). Boosting electrocatalytic hydrogen evolution over metal-organic frameworks by plasmon-induced hot-electron injection. *Angewandte Chemie (International Ed. in English)* 58: 10713–10717.
77 Xiao, J.D., Han, L., Luo, J. et al. (2018). Integration of plasmonic effects and schottky junctions into metal-organic framework composites: steering charge flow for enhanced visible-light photocatalysis. *Angewandte Chemie (International Ed. in English)* 57: 1103–1107.
78 Dong, Y., Su, Y., Du, L. et al. (2019). Plasmon-enhanced deuteration under visible-light irradiation. *ACS Nano* 13: 10754–10760.
79 DuChene, J.S., Niu, W., Abendroth, J.M. et al. (2012). Halide anions as shape-directing agents for obtaining high-quality anisotropic gold nanostructures. *Chemistry of Materials* 25: 1392–1399.
80 Jiang, W., Bai, S., Wang, L. et al. (2016). Integration of multiple plasmonic and co-catalyst nanostructures on TiO_2 nanosheets for visible-near-infrared photocatalytic hydrogen evolution. *Small* 12: 1640–1648.

81 Wu, B., Liu, D., Mubeen, S. et al. (2016). Anisotropic growth of TiO_2 onto gold nanorods for plasmon-enhanced hydrogen production from water reduction. *Journal of the American Chemical Society* 138: 1114–1117.

82 Hong, J.W., Wi, D.H., Lee, S.U., and Han, S.W. (2016). Metal-semiconductor heteronanocrystals with desired configurations for plasmonic photocatalysis. *Journal of the American Chemical Society* 138: 15766–15773.

83 Gawande, M.B., Goswami, A., Felpin, F.X. et al. (2016). Cu and Cu-based nanoparticles: synthesis and applications in catalysis. *Chemical Reviews* 116: 3722–3811.

84 Neatu, S., Macia-Agullo, J.A., Concepcion, P., and Garcia, H. (2014). Gold-copper nanoalloys supported on TiO_2 as photocatalysts for CO_2 reduction by water. *Journal of the American Chemical Society* 136: 15969–15976.

85 Shown, I., Hsu, H.C., Chang, Y.C. et al. (2014). Highly efficient visible light photocatalytic reduction of CO_2 to hydrocarbon fuels by Cu-nanoparticle decorated graphene oxide. *Nano Letters* 14: 6097–6103.

86 Tan, J.Z.Y., Fernández, Y., Liu, D. et al. (2012). Photoreduction of CO_2 using copper-decorated TiO_2 nanorod films with localized surface plasmon behavior. *Chemical Physics Letters* 531: 149–154.

87 Zhang, S., Peng, B., Yang, S. et al. (2015). Non-noble metal copper nanoparticles-decorated TiO_2 nanotube arrays with plasmon-enhanced photocatalytic hydrogen evolution under visible light. *International Journal of Hydrogen Energy* 40: 303–310.

88 DeSario, P.A., Pietron, J.J., Brintlinger, T.H. et al. (2017). Oxidation-stable plasmonic copper nanoparticles in photocatalytic TiO_2 nanoarchitectures. *Nanoscale* 9: 11720–11729.

89 Fan, M., Song, C., Chen, T. et al. (2016). Visible-light-drived high photocatalytic activities of Cu/g-C_3N_4 photocatalysts for hydrogen production. *RSC Advances* 6: 34633–34640.

90 Pacioni, N.L., Filippenko, V., Presseau, N., and Scaiano, J.C. (2013). Oxidation of copper nanoparticles in water: mechanistic insights revealed by oxygen uptake and spectroscopic methods. *Dalton Transactions* 42: 5832–5838.

91 Chatterjee, H. and Ghosh, S.K. (2019). Solvent-specific controlled oxidation of Cu nanospheres to Cu@Cu_2O@Cu nanomatryushkas. *Journal of Physical Chemistry C* 123: 6888–6897.

92 Tan, C.F., Su Su Zin, A.K., Chen, Z. et al. (2018). Inverse stellation of CuAu-ZnO multimetallic-semiconductor nanostartube for plasmon-enhanced photocatalysis. *ACS Nano* 12: 4512–4520.

93 Robatjazi, H., Zhao, H., Swearer, D.F. et al. (2017). Plasmon-induced selective carbon dioxide conversion on earth-abundant aluminum-cuprous oxide antenna-reactor nanoparticles. *Nature Communications* 8: 27.

94 He, S., Huang, J., Goodsell, J.L. et al. (2019). Plasmonic Nickel-TiO_2 Heterostructures for visible-light-driven photochemical reactions. *Angewandte Chemie, International Edition* 58: 6038–6041.

95 Shiraishi, Y., Imai, J., Yasumoto, N. et al. (2019). Doping of Nb^{5+} species at the Au-TiO_2 interface for plasmonic photocatalysis enhancement. *Langmuir* 35: 5455–5462.

96 Wu, K., Chen, J., McBride, J.R., and Lian, T. (2015). Efficient hot-electron transfer by a plasmon-induced interfacial charge-transfer transition. *Science* 349: 632.

97 Huang, X., Li, H., Zhang, C. et al. (2019). Efficient plasmon-hot electron conversion in Ag-$CsPbBr_3$ hybrid nanocrystals. *Nature Communications* 10: 1163.

98 Wu, K., Rodriguez-Cordoba, W.E., Yang, Y., and Lian, T. (2013). Plasmon-induced hot electron transfer from the Au tip to CdS rod in CdS-Au nanoheterostructures. *Nano Letters* 13: 5255–5263.

99 Long, R. and Prezhdo, O.V. (2014). Instantaneous generation of charge-separated state on TiO_2 surface sensitized with plasmonic nanoparticles. *Journal of the American Chemical Society* 136: 4343–4354.

100 Wang, C. and Astruc, D. (2014). Nanogold plasmonic photocatalysis for organic synthesis and clean energy conversion. *Chemical Society Reviews* 43: 7188–7216.

101 Mubeen, S., Lee, J., Singh, N. et al. (2013). An autonomous photosynthetic device in which all charge carriers derive from surface plasmons. *Nature Nanotechnology* 8: 247–251.

102 Bai, J., Lu, B., Han, Q. et al. (2018). (111) Facets-oriented Au-decorated carbon nitride nanoplatelets for visible-light-driven overall water splitting. *ACS Applied Materials & Interfaces* 10: 38066–38072.

103 Ingram, D.B. and Linic, S. (2011). Water splitting on composite plasmonic-metal/semiconductor photoelectrodes: evidence for selective plasmon-induced formation of charge carriers near the semiconductor surface. *Journal of the American Chemical Society* 133: 5202–5205.

104 Mesa, C.A., Francas, L., Yang, K.R. et al. (2020). Multihole water oxidation catalysis on haematite photoanodes revealed by operando spectroelectrochemistry and DFT. *Nature Chemistry* 12: 82–89.

105 Zhang, Y., Zhang, H., Liu, A. et al. (2018). Rate-limiting O-O bond formation pathways for water oxidation on hematite photoanode. *Journal of the American Chemical Society* 140: 3264–3269.

106 Gomes Silva, C., Juárez, R., Marino, T. et al. (2011). Influence of excitation wavelength (UV or visible light) on the photocatalytic activity of titania containing gold nanoparticles for the generation of hydrogen or oxygen from water. *Journal of the American Chemical Society* 133: 595–602.

107 Qian, K., Sweeny, B.C., Johnston-Peck, A.C. et al. (2014). Surface plasmon-driven water reduction: gold nanoparticle size matters. *Journal of the American Chemical Society* 136: 9842–9845.

108 Caux, M., Menard, H., AlSalik, Y.M. et al. (2019). Photo-catalytic hydrogen production over Au/g-C_3N_4: effect of gold particle dispersion and morphology. *Physical Chemistry Chemical Physics* 21: 15974–15987.

109 Chava, R.K., Do, J., and Kang, M. (2019). Strategy for improving the visible photocatalytic H_2 evolution activity of 2D graphitic carbon nitride nanosheets through the modification with metal and metal oxide nanocomponents. *Applied Catalysis. B, Environmental* 248: 538–551.

110 Zhang, Z., Wang, Z., Cao, S.-W., and Xue, C. (2013). Au/Pt nanoparticle-decorated TiO_2 nanofibers with plasmon-enhanced photocatalytic activities for solar-to-fuel conversion. *Journal of Physical Chemistry C* 117: 25939–25947.

111 Choi, S. and Nam, Y.S. (2018). Gold–titanium dioxide half-dome heterostructures for plasmonic hydrogen evolution. *ACS Applied Energy Materials* 1: 5169–5175.

112 Tanaka, A., Sakaguchi, S., Hashimoto, K., and Kominami, H. (2012). Preparation of Au/TiO_2 with metal cocatalysts exhibiting strong surface plasmon resonance effective for photoinduced hydrogen formation under irradiation of visible light. *ACS Catalysis* 3: 79–85.

113 Shi, X., Li, X., Toda, T. et al. (2020). Interfacial structure-modulated plasmon-induced water oxidation on strontium titanate. *ACS Applied Energy Materials* 3: 5675–5683.

114 Zhang, Y., Zhang, Y., Guo, W. et al. (2020). Modulating multi-hole reaction pathways for photoelectrochemical water oxidation on gold nanocatalysts. *Energy & Environmental Science* 13: 1501–1508.

115 Chu, J., Miao, P., Han, X. et al. (2016). Ultrafast surface-plasmon-induced photodimerization of *p*-aminothiophenol on Ag/TiO_2 nanoarrays. *ChemCatChem* 8: 1819–1824.

116 Hao, C.H., Guo, X.N., Pan, Y.T. et al. (2016). Visible-light-driven selective photocatalytic hydrogenation of cinnamaldehyde over au/sic catalysts. *Journal of the American Chemical Society* 138 (30): 9361–9364.

117 Naya, S., Niwa, T., Kume, T., and Tada, H. (2014). Visible-light-induced electron transport from small to large nanoparticles in bimodal gold nanoparticle-loaded titanium(IV) oxide. *Angewandte Chemie, International Edition* 53: 7305–7309.

118 Yang, J., Guo, Y., Jiang, R. et al. (2018). High-efficiency "working-in-tandem" nitrogen photofixation achieved by assembling plasmonic gold nanocrystals on ultrathin titania nanosheets. *Journal of the American Chemical Society* 140: 8497–8508.

119 Xiao, C., Hu, H., Zhang, X., and MacFarlane, D.R. (2017). Nanostructured gold/bismutite hybrid heterocatalysts for plasmon-enhanced photosynthesis of ammonia. *ACS Sustainable Chemistry & Engineering* 5: 10858–10863.

120 Oshikiri, T., Ueno, K., and Misawa, H. (2016). Selective dinitrogen conversion to ammonia using water and visible light through plasmon-induced charge separation. *Angewandte Chemie, International Edition* 55: 3942–3946.

121 Wu, S., Chen, Z., Liu, K. et al. (2020). Chemisorption-induced and plasmon-promoted photofixation of nitrogen on gold-loaded carbon nitride nanosheets. *ChemSusChem* 13: 3455–3461.

122 Hou, W., Hung, W.H., Pavaskar, P. et al. (2011). Photocatalytic conversion of CO_2 to hydrocarbon fuels via plasmon-enhanced absorption and metallic interband transitions. *ACS Catalysis* 1: 929–936.

123 Kumar, B., Llorente, M., Froehlich, J. et al. (2012). Photochemical and photoelectrochemical reduction of CO_2. *Annual Review of Physical Chemistry* 63: 541–569.

124 Teranishi, M., Hoshino, R., Naya, S., and Tada, H. (2016). Gold-nanoparticle-loaded carbonate-modified titanium(IV) oxide surface: visible-light-driven formation of hydrogen peroxide from oxygen. *Angewandte Chemie, International Edition* 55: 12773–12777.

125 Nishi, H., Sakamoto, M., and Tatsuma, T. (2018). Local trapping of energetic holes at gold nanoparticles on TiO_2. *Chemical Communications* 54: 11741–11744.

Epilogue

Suljo Linic

Department of Chemical Engineering and Michigan Catalysis Science and Technology Institute (MiCSTI), University of Michigan, Ann Arbor, Michigan 48109, United States

The book provides us with 10 chapters that cover a range of topics related to plasmonic catalysis. It is the first book dedicated to this emerging field, which has over last 10 years seen a tremendous interest from academic as well as commercial stakeholders. The topics include discussions of the fundamentals of the response of plasmonic nanostructures to resonant light illumination, electronic excitations accompanying these light-matter interactions, synthesis and characterizations of plasmonic nanostructures and the performance of these nanomaterials in photo-driven chemical transformations. In many ways, the book represents a time snapshot of the field and motivates us to think ahead.

The initial interest in plasmonic catalysis on metal nanoparticles was mainly motivated by a desire to design reactive chemical systems that can provide us with alternative mechanisms for activating chemical bonds on metal surfaces compared to the traditional heat-induced chemical reactions [1–3]. I note that metal nanoparticles are in general an important class of heterogeneous thermal catalysts, used for several industrial chemical transformations including dehydrogenations, partial oxidations, reduction reactions, ammonia synthesis, and hydrocarbon reforming, among others. These processes are typically performed at relatively high temperatures to provide sufficient energy for activating chemical bonds on the surfaces of the nanoparticles. An unavoidable side effect of this approach is that this heat energy is deposited into every available reaction coordinate. This can result in the simultaneous activation of unselective reaction pathways leading to the undesirable formation of by-products and chemical waste, which are the main problems associated with thermal catalysis [4].

Unlike these thermally driven processes, plasmonic nanoparticles offer an opportunity to drive chemical transformations by the excitation of energetic charge carriers (typically via photoexcitation) into the reactants (adsorbates) [1, 2, 5]. Under this mechanism, it is in principle possible to have an improved control over the outcome of chemical reactions by specifically targeting electronic excitations into specific adsorbate orbitals that result in the preferential activation of desired

Plasmonic Catalysis: From Fundamentals to Applications, First Edition.
Edited by Pedro H.C. Camargo and Emiliano Cortés.

chemical pathways. This offers an opportunity to potentially discover new, more selective reaction pathways that cannot be accessed in temperature-driven catalysis.

At this point, as described in the book, we have clear experimental demonstrations of visible-light-driven chemical transformations on plasmonic metal nanostructures. It is also clear, as discussed in multiple chapters in the book, that these chemical transformations can proceed by direct or indirect excitation of energetic charge carriers into reacting surface adsorbates. Early on, the focus was on single electron (hole) excitations and reactions. For example, the work of my groups has focused on single electron-driven O_2 dissociation on Ag nanoparticles while Hallas has demonstrated single electron-driven H_2 dissociation on Au [1–3, 5]. The mechanisms associated with these single-electron reactions rely on vibronic excitation of the reactant by energetic charge carriers. In principle, these mechanisms are similar to the processes underpinning the resonant surface-enhanced Raman scattering on metal surfaces [6].

Since these early contributions, the filed has boldly moved forward. For example, there have been a number of recent reports of multi-electron (or multi-hole) reactions, such as CO_2 or N_2 reduction, on plasmonic nanoparticles [7, 8]. These are extremely intriguing and potentially promising developments as they imply that it is possible to generate renewable H_2 from water or renewable fuels and chemicals from CO_2 on plasmonic nanoparticles using sunlight energy. To take full advantage of these directions, however, we need to understand the mechanisms behind these multi-electron reactions on metals. It is well known that to execute these reactions, multiple hot electrons (or holes) need to transfer to reactant (and reaction intermediates) quite rapidly. This rapid transfer is required as the reaction intermediates are unstable, and the reverse reactions are thermodynamically preferred. As discussed in the book, in plasmonic metallic nanostructures the lifetime of photo-generated high-energy charge carriers is short (approximately tens-femtoseconds), and the time between excitation events is relatively long at low light intensities (for example, one Sun). Given these physical constraints, it is difficult to argue that multiple electronic excitations, required to execute multi-electron reactions, can take place faster than the reverse reactions. The central question are: (i) how are these multiple high-energy charge carriers supplied to the high-energy intermediates in short times under moderate light intensities (sunlight) and (ii) what are the upper limits of the quantum efficiency of these processes?

Another critical direction is to try to respond to the initial questions raised during the early days of this field, which is whether it is possible to use plasmonic nanostructures to perform surface chemistry with higher control over the product selectivity than in purely temperature-driven chemical reactions. The experimental observations and models discussed in the book imply that it is in principle possible to selectively excite specific electronic excitations at the molecule/nanoparticle interface [9, 10]. Whether these excitations yield a preferential heating of specific vibrational modes and whether this specific vibrational heating will lead to preferential reactions associated with this reaction's coordinate at the expense of the other reaction coordinates is ultimately related to the distribution of this electronic energy among different vibrational modes, which is in turn governed by the shape of the excited potential energy surface. Clearly, designing selective plasmonic chemistry will require engineering of not only optical properties of the plasmonic metal, but also the electronic structure of the reactants adsorbed on the active centers as well as the ground and excited potential energy surfaces. In principle, this will require a complete, almost atomistic control over the entire multifunctional plasmonic catalyst. While the field is moving in this direction [11], we have not gotten very far on this journey and significant advancements in predictive modelling of these reaction as well as in the chemical synthesis of targeted multicomponent structures are required.

It is also critical to appreciate that the discoveries in the field of plasmonic catalysis have had several initially unintended but nevertheless useful consequences. For example, the fact that the rates of various chemical transformations can be increased by illuminating plasmonic nanoparticles can be exploited to drive chemical transformation in the environments where it is easier to introduce light as opposed to heat stimulus. In addition, the fundamental work performed in the field of plasmonic photocatalysis has stimulated questions about a broader use of plasmonic nanostructures in a range of hot electron (hole) applications, including photovoltaics, photodetection, and sensing. Similar to photo-catalysis, these applications require coupling of a plasmonic component, which amplifies the interaction of light with the material, to an attached non-plasmonic component that extracts this energy in the form of electronic excitations to perform a function. For example, one can envision a multicomponent hybrid material, where a plasmonic component amplifies and concentrates the light energy within the material, and an attached non-plasmonic component extracts this energy in the form of electronic excitations [energetic electron–hole (e–h) pairs] to perform a function. Examples of these hybrid materials include plasmonic–metal/metal, plasmonic–metal/semiconductor and plasmonic–metal/molecule systems. At the core of these applications is a flow of energy across plasmonic/non-plasmonic interfaces.

To take full advantage of these opportunities, we need to understand the mechanisms of energy dissipation in these hybrid materials, including the initial location of generated e–h pairs, their energy distribution, and their flow through the hybrid nanomaterial [11–13]. It is also critical to accurately describe how these energetic charge carriers couple to the phonon modes of the hybrid material.

In conclusion, the last few years have seen an emergence of a completely new field of chemical conversion labelled plasmonic catalysis. The book gives a comprehensive snapshot of this field. I believe that it will serve not only to explain critical phenomena associated with the field but also to guide and inform future interest in this and related fields.

Ann Arbor, MI, USA *Suljo Linic*
20 October 2020

References

1 Christopher, P., Xin, H., and Linic, S. (2011). Visible-light-enhanced catalytic oxidation reactions on plasmonic silver nanostructures. *Nat. Chem.* 3 (6): 467–472. https://doi.org/10.1038/nchem.1032.

2 Christopher, P., Xin, H., Marimuthu, A., and Linic, S. (2012). Singular characteristics and unique chemical bond activation mechanisms of photocatalytic reactions on plasmonic nanostructures. *Nat. Mater.* 11 (12): 1044–1050. https://doi.org/10.1038/nmat3454.

3 Linic, S., Christopher, P., and Ingram, D.B. (2011). Plasmonic-metal nanostructures for efficient conversion of solar to chemical energy. *Nat. Mater.* 10 (12): 911–921. https://doi.org/10.1038/nmat3151.

4 Ertl, G., Knozinger, H., and Weitkamp, J. (2008). Frontmatter. In: *Handbook of Heterogeneous Catalysis* (eds. G. Ertl, H. Knözinger, and J. Weitkamp), I–XIX. Wiley-VCH Verlag GmbH.

5 Mukherjee, S., Libisch, F., Large, N. et al. (2013). Hot electrons do the impossible: plasmon-induced dissociation of H_2 on Au. *Nano Lett.* 13 (1): 240–247. https://doi.org/10.1021/nl303940z.

6 Jiang, Bosnick, K., Maillard, M., and Brus, L. (2003). Single molecule Raman spectroscopy at the junctions of large Ag nanocrystals. *J. Phys. Chem. B* 107 (37): 9964–9972. https://doi.org/10.1021/jp034632u.

7 Hu, C., Chen, X., Jin, J. et al. (2019). Surface plasmon enabling nitrogen fixation in pure water through a dissociative mechanism under mild conditions. *J. Am. Chem. Soc.* https://doi.org/10.1021/jacs.9b01375.

8 Creel, E.B., Corson, E.R., Eichhorn, J. et al. (2019). Directing selectivity of electrochemical carbon dioxide reduction using plasmonics. *ACS Energy Lett.* 1098–1105. https://doi.org/10.1021/acsenergylett.9b00515.

9 Boerigter, C., Aslam, U., Linic, S. (2016). Mechanism of charge transfer from plasmonic nanostructures to chemically attached materials. *ACS Nano* 10 (6): 6108–6115. https://doi.org/10.1021/acsnano.6b01846.

10 Boerigter, C., Campana, R., Morabito, M., and Linic, S. (2016). Evidence and implications of direct charge excitation as the dominant mechanism in plasmon-mediated photocatalysis. *Nat. Commun.* 7, 10545. https://doi.org/10.1038/ncomms10545.

11 Aslam, U., Chavez, S., and Linic, S. (2017). Controlling energy flow in multimetallic nanostructures for plasmonic catalysis. *Nat. Nanotechnol.* 12 (10): 1000–1005. https://doi.org/10.1038/nnano.2017.131.

12 Chavez, S.A., Aslam, U., and Linic, S. (2018). Design principles for directing energy and energetic charge flow in multicomponent plasmonic nanostructures. *ACS Energy Lett.*

13 Chavez, S., Rao, V.G., and Linic, S. (2019). Unearthing the factors governing site specific rates of electronic excitations in multicomponent plasmonic systems and catalysts. *Faraday Discuss.* 214 (0): 441–453. https://doi.org/10.1039/C8FD00143J.

Index

Plasmonic Catalysis: From Fundamentals to Applications, First Edition.
Edited by Pedro H.C. Camargo and Emiliano Cortés.